INTRODUCTION TO THE ARITHMETIC THEORY OF AUTOMORPHIC FUNCTIONS

PUBLICATIONS OF THE MATHEMATICAL SOCIETY OF JAPAN

1. The Construction and Study of Certain Important Algebras. By Claude Chevalley.
2. Lie Groups and Differential Geometry. By Katsumi Nomizu.
3. Lectures on Ergodic Theory. By Paul R. Halmos.
4. Introduction to the Problem of Minimal Models in the Theory of Algebraic Surfaces. By Oscar Zariski.
5. Zur Reduktionstheorie Quadratischer Formen. Von Carl Ludwig Siegel.
6. Complex Multiplication of Abelian Varieties and its Applications to Number Theory. By Goro Shimura and Yutaka Taniyama.
7. Equations Différentielles Ordinaires du Premier Ordre dans le Champ Complexe. Par Masuo Hukuhara, Tosihusa Kimura et Mme Tizuko Matuda.
8. Theory of Q-varieties. By Teruhisa Matsusaka.
9. Stability Theory by Liapunov's Second Method. By Taro Yoshizawa.
10. Fonctions Entières et Transformées de Fourier. Applications. Par Szolem Mandelbrojt.
11. Introduction to the Arithmetic Theory of Automorphic Functions. By Goro Shimura. (Kanô Memorial Lectures 1)

PUBLICATIONS OF THE MATHEMATICAL SOCIETY OF JAPAN
11

INTRODUCTION TO THE ARITHMETIC THEORY OF AUTOMORPHIC FUNCTIONS

BY
GORO SHIMURA

KANÔ MEMORIAL LECTURES 1

Iwanami Shoten, Publishers
and
Princeton University Press
1971

Kanô Memorial Lectures

In 1969, the Mathematical Society of Japan received an anonymous donation to encourage the publication of lectures in mathematics of distinguished quality in commemoration of the late Kôkichi Kanô (1865–1942).

K. Kanô was a remarkable scholar who lived through an era when Western mathematics and philosophy were first introduced to Japan. He began his career as a scholar by studying mathematics and remained a rationalist for his entire life, but enormously enlarged the domain of his interest to include philosophy and history.

In appreciating the sincere intentions of the donor, our Society has decided to publish a series of " Kanô Memorial Lectures " as a part of our Publications. This is the first volume in the series.

Publications of the Mathematical Society of Japan, volumes 1 through 10, should be ordered directly from the Mathematical Society of Japan. Volume 11 and subsequent volumes should be ordered from Princeton University Press, except in Japan, where they should be ordered from Iwanami Shoten, Publishers.

Co-published for
the Mathematical Society of Japan
by
Iwanami Shoten, Publishers
and
Princeton University Press

Printed in the United States of America

PREFACE

There are two major topics treated in this volume:

I. Complex multiplication of elliptic or elliptic modular functions.

II. Applications of the theory of Hecke operators to the zeta-functions of algebraic curves and abelian varieties.

Although these will form the "raison d'être" of the book, I have also attempted, in the first few chapters, to present an introductory account of the theory of automorphic functions of one complex variable, along with the fundamentals of Hecke operators. Our discussion is mainly concerned with elliptic modular functions of arbitrary level and the geometric objects directly related to them, except that we consider automorphic functions of a more general type in the first two and the last two chapters, and abelian varieties of higher dimension with complex multiplication in a few places.

As to the first topic, we shall give two formulations, both in terms of adeles. One is concerned with the behavior of an elliptic curve and its points of finite order under automorphisms of the number field in question. The other is closely connected with the structure of the field $\mathfrak{F}$ of all modular functions of all levels whose Fourier coefficients belong to cyclotomic fields. It will be shown that the group of all automorphisms of $\mathfrak{F}$ is isomorphic to the adelization of $GL_2(\boldsymbol{Q})$ modulo rational scalar matrices and the archimedean part. Then the reciprocity-law in the maximal abelian extension of an imaginary quadratic field is given as a certain commutativity of the action of the adeles with the specialization of the functions of $\mathfrak{F}$.

The second topic is a development of the result of Eichler in his paper appeared in the Archiv der Mathematik vol. 5, 1954. The conjecture of Hasse and Weil will be verified for the algebraic curves uniformized by modular functions. Further we shall show that if a cusp form of weight 2 is a common eigen-function of the Hecke operators, then the product of several Dirichlet series associated with it coincides, up to finitely many Euler factors, with the zeta-function of a certain abelian variety which is specifically given.

As an application of this result, it will be shown that the arithmetic of a real quadratic field——its units, abelian extensions, etc.——is closely connected with the modular forms of "Neben"-type in Hecke's sense. My excuse for including this rather immature subject is that I think it gives a positive, if not complete, answer to the question "Can one construct abelian extensions of a real quadratic field by an analytic means?", which arises naturally after the detailed discussion of the corresponding problem for an imaginary quadratic field in Chapters 5 and 6.

The present book has grown out of my lectures at Princeton University and the University of Tokyo on various occasions during 1963–69. The notes taken by Larry Goldstein (Fall Term 1965) and by Alain Robert (Spring Term 1969) were most helpful in preparing the first draft. Here I gratefully acknowledge my indebtedness to them. I wish to express my hearty thanks to K. Doi, H. Naganuma, and H. Trotter who made the table of eigen-values of Hecke operators in § 7.7; and to W. Casselman, S. Lang, T. Miyake, A. Robert, and A. Weil, who read the manuscript as a whole or in part. Many of their suggestions have been incorporated in the present volume. My thanks are also due to S. Iyanaga and Y. Kawada, who took an interest in this work, and invited me to publish it in Publications of the Mathematical Society of Japan. Finally I would like to extend thanks to the audience of my lectures, whose enthusiasm was very encouraging.

Princeton, May 1970 Goro Shimura

CONTENTS

NOTATION AND TERMINOLOGY

0.1. The symbols $\boldsymbol{Z}, \boldsymbol{Q}, \boldsymbol{R}, \boldsymbol{C}$, and $\boldsymbol{H}$ denote respectively the ring of rational integers, the rational number field, the real number field, the complex number field, and the division ring of Hamilton quaternions. For a rational prime p, $\boldsymbol{Z}_p$ and $\boldsymbol{Q}_p$ denote the ring of p-adic integers and the field of p-adic numbers, respectively. For $z \in \boldsymbol{C}$, we denote by $\bar{z}$, $\mathrm{Re}(z)$, and $\mathrm{Im}(z)$ the complex conjugate, the real part, and the imaginary part of z, respectively. The symbol $\mathfrak{H}$ denotes the upper half complex plane:

$$\mathfrak{H} = \{z \in \boldsymbol{C} \mid \mathrm{Im}(z) > 0\}\,.$$

If we discuss a Fuchsian group of the first kind Γ on $\mathfrak{H}$, then $\mathfrak{H}^*$ denotes the union of $\mathfrak{H}$ and the cusps of Γ, see §§ 1.2, 1.3. (Therefore $\mathfrak{H}^*$ depends on Γ.)

0.2. For an associative ring T with an identity element, we denote by $T^\times$ the group of all invertible elements of T, and by $M_n(T)$ the ring of all square matrices of size n with coefficients in T. Then we put $GL_n(T) = M_n(T)^\times$. The identity element of $M_n(T)$ is denoted by 1_n, and often simply by 1. The transpose of $X \in M_n(T)$ is denoted by ${}^tX$. If T is commutative, we denote by $\det(X)$ and $\mathrm{tr}(X)$ the determinant and trace of $X \in M_n(T)$, and put

$$SL_n(T) = \{X \in GL_n(T) \mid \det(X) = 1\}\,.$$

If there is no risk of confusion, we write T^n for the product of n copies of T, and often consider the elements of T as row-vectors or column-vectors with components in T. This applies especially to the cases $T = \boldsymbol{Z}, \boldsymbol{Q}, \boldsymbol{R}, \boldsymbol{C}$, or $\boldsymbol{H}$. If V is a T-module, $\mathrm{End}(V, T)$ denotes the ring of all T-linear endomorphisms of V.

0.3. For an arbitrary field K, we denote by $\mathrm{Aut}(K)$ the group of all automorphisms of K. If F is a subfield of K, $\mathrm{Aut}(K/F)$ denotes the subgroup of $\mathrm{Aut}(K)$ consisting of the elements which are trivial on F. When K is a finite or an infinite Galois extension of F, we put $\mathrm{Aut}(K/F) = \mathrm{Gal}(K/F)$. If $x_1, \cdots, x_n$ are elements of K, $F(x_1, \cdots, x_n)$ stands for the subfield of K generated over F by $x_1, \cdots, x_n$. (See also Appendix 1.) For subfields $F_1, \cdots, F_m$ of K, we denote by $F_1 \cdots F_m$ the composite of $F_1, \cdots, F_m$, i.e., the smallest subfield of K containing $F_1, \cdots, F_m$. If σ is an isomorphism of K to another field, we denote by x^σ the image of $x \in K$ under σ, so that $(x^\sigma)^\tau = x^{\sigma\tau}$.

0.4. The symbol $\bar{\boldsymbol{Q}}$ denotes the algebraic closure of $\boldsymbol{Q}$ in $\boldsymbol{C}$. By an *algebraic number field*, we understand a subfield of $\bar{\boldsymbol{Q}}$. A *prime divisor*, or simply a *prime*, of an algebraic number field F means an equivalence class of non-trivial valuations of F. The *maximal order* of F is the ring of all algebraic integers in F. If F is of finite degree over $\boldsymbol{Q}$, a non-archimedean prime divisor of F corresponds uniquely to a prime ideal of the maximal order of F, which we simply call a *prime ideal in* F. If $\mathfrak{x}$ is a fractional ideal in F, $N(\mathfrak{x})$ denotes its absolute norm, i. e., the positive rational number which generates the fractional ideal $N_{F/\boldsymbol{Q}}(\mathfrak{x})$ in $\boldsymbol{Q}$. Occasionally, the complex conjugate of an element x of $\bar{\boldsymbol{Q}}$ is denoted by x^ρ.

0.5. If a and b are rational integers, we denote by (a, b) the positive integer d such that $d\boldsymbol{Z}=a\boldsymbol{Z}+b\boldsymbol{Z}$ (unless $a=b=0$). Especially $(a, b)=1$ if and only if a and b have no common divisors other than ± 1.

0.6. The notation $[X:Y]$ means the index of a subgroup Y of a group X, or the dimension of a vector space X over a field Y, especially the degree of an algebraic extension X of a field Y. The distinction will be clear from the context. If f is a homomorphism of a group to a group, the kernel of f is denoted by $\mathrm{Ker}(f)$. Occasionally, an *isomorphism* means an *injective homomorphism*. For example, we speak of an *isomorphism* of a quadratic extension of $\boldsymbol{Q}$ *into* $M_2(\boldsymbol{Q})$, instead of an *isomorphism of K onto a subfield* of $M_2(\boldsymbol{Q})$.

0.7. The symbol id. stands for the identity map for which the set in question is clear from the context. If a map f defined on a set X is the identity map on a subset Y of X, we write $f=\text{id.}$ on Y.

0.8. As for the terminology and notation concerning algebraic geometry, see Appendix at the end of the book.

LIST OF SYMBOLS

(in alphabetical order, except for a few at the end)

SUGGESTIONS TO THE READER

This book is not homogeneously written; it is intended for readers with various mathematical backgrounds.

The reader who is familiar with elementary properties of topological groups and Riemann surfaces will have no difficulty in Chapters 1, 2, 3. In § 2.3, the Riemann-Roch theorem for a compact Riemann surface is needed. Also, in the proof of Prop. 2.15, one needs the divisibility property of the jacobian variety. Further, in § 3.5, a theorem of Wedderburn about an algebra with radical is employed. If the reader is not acquainted with any of these theorems, he is advised simply to accept the statements, since the rest of the chapters does not require them again.

After the first three chapters, the reader may go directly to Chapter 8, which demands only a very elementary knowledge of homology and cohomology of groups and simplicial complexes.

Chapters 4, 5, 6 presuppose the knowledge of elliptic curves and class field theory. The reader is advised to go through the Appendix before reading these chapters, to make sure of the terminology of algebraic geometry, even if he is an expert on the subject.

The last section of Chapter 5 and a large part of Chapters 7, 9 are intended for the most advanced reader. The style is therefore somewhat different from the rest of the book, although the author believes that the degree of sophistication is still tolerable for inexperienced readers.

There are a few exercises at the end of each section. Some of them are routine applications of the material of the text. But they are often statements of secondary importance which could be given as theorems or examples with detailed proofs in a more extensive book. At any rate, there should be no great difficulty in working them out by the methods developed in the text.

Theorems, propositions, lemmas, remarks, and exercises are numbered in *one sequence* throughout each chapter. Displayed formulas, statements, and assumptions are cross-referred to in parentheses such as (3.5.7), which means the seventh of those in Section 3.5.

CHAPTER 1
FUCHSIAN GROUPS OF THE FIRST KIND

1.1. Transformation groups and quotient spaces

In this section we shall discuss some elementary properties of a group of transformations acting on a topological space. All topological groups are assumed Hausdorff.

Let G be a topological group, and S a topological space. We say that G *acts continuously on* S, or G is a *transformation group* on S, if a continuous map $G \times S \ni (g, s) \mapsto gs \in S$ is given and satisfies the following conditions: (i) $(ab)s = a(bs)$ for $a \in G$, $b \in G$, $s \in S$; (ii) $es = s$ for all $s \in S$, where e denotes the identity element of G. We see that, for every $g \in G$, the map $s \mapsto gs$ is a homeomorphism of S onto itself. We shall write also $g(s)$ for gs. For every $s \in S$, we put $Gs = \{gs \mid g \in G\}$, and call it the *orbit of* s *under* G, or simply the *G-orbit of* s. Two points with the same G-orbit are often called *G-equivalent*, or *equivalent under G*. We say that G acts *transitively* on S if there is only one G-orbit, S itself.

Let us denote by $G\backslash S$ the set of all G-orbits of points on S. Let $\pi : S \to G\backslash S$ denote the natural projection defined by $\pi(s) = Gs$. Call a subset X of $G\backslash S$ *open* if $\pi^{-1}(X)$ is open in S. It can easily be verified that this defines a topology on $G\backslash S$, which we call the *quotient topology*. Then π is clearly continuous. Moreover, π is open, since if Y is an open subset of S, then $\pi^{-1}(\pi(Y)) = \bigcup_{g \in G} g(Y)$, and this is obviously open. It should be noted that $G\backslash S$ is not necessarily Hausdorff, even if S is Hausdorff.

Let K be a closed subgroup of G. Consider the action of K on G by right multiplication. Then the K-orbit of an element g of G is just a left coset gK. Introduce the quotient topology in G/K as above. The closedness of K implies that G/K is Hausdorff. To show this, let $aK \neq bK$. Define a continuous map $f : G \times G \to G$ by $f(x, y) = x^{-1}y$. Then $(a, b) \notin f^{-1}(K)$. Since $f^{-1}(K)$ is closed, there exist open sets U resp. V containing a resp. b, such that $(U \times V) \cap f^{-1}(K) = \emptyset$. If $h : G \to G/K$ is the natural projection, this means $h(U) \cap h(V) = \emptyset$, q.e.d.

Now let G act on G/K as usual by the rule $g \cdot (xK) = gxK$ for $g \in G$, $x \in G$. The map $(g, xK) \mapsto gxK$ of $G \times (G/K)$ to G/K is obviously continuous. Furthermore, this action is transitive.

Let S be an arbitrary Hausdorff space on which G acts continuously and transitively. Fix any point t of S, and put $K = \{g \in G \mid gt = t\}$. Then K is a closed subgroup of G, and called *the isotropy subgroup of G at t*, or *the stability group of t.* There is a natural one-to-one map $\lambda: G/K \to S$ defined by $\lambda(gK) = gt$. For any subset X of S, one has $\lambda^{-1}(X) = h(\{g \in G \mid gt \in X\})$, where h is the projection map: $G \to G/K$. This equality shows that $\lambda^{-1}(X)$ is open if X is open. Hence λ is *continuous*. But λ is not necessarily a homeomorphism. One can at least prove the following criterion:

THEOREM 1.1. *The map $\lambda: G/K \to S$ is a homeomorphism if both G and S are locally compact, and G has a countable base of open sets.*

PROOF. Let U be an open set in G, and let $g \in U$. It is sufficient to show that gt is an interior point of Ut. Take a compact neighborhood V of the identity element of G so that $V = V^{-1}$ and $gV^2 \subset U$. If Vt contains an interior point vt with $v \in V$, then $gt = gv^{-1}vt$ is obviously an interior point of Ut. By our assumption, G is a union $\bigcup_n g_n V$ with countably many $\{g_n\} \subset G$. Then $S = \bigcup_n g_n Vt$, and Vt must contain an interior point, on account of the following Lemma, so that our theorem is proved.

LEMMA 1.2. *Let S be a (non-empty) locally compact Hausdorff space, and $V_1, \cdots, V_n, \cdots$ be countably many closed subsets of S such that $S = \bigcup_{n=1}^{\infty} V_n$. Then at least one of the V_n has an interior point.*

PROOF. Assuming that no V_n has interior points, let us derive a contradiction. Take a non-empty open subset W_1 of S whose closure $\overline{W}_1$ is compact. Define $W_2, W_3, \cdots$ successively so that W_n is non-empty and open, and $\overline{W}_{n+1} \subset W_n - V_n$. Then the $\overline{W}_n$ form a decreasing sequence of non-empty compact sets, hence $\bigcap_n \overline{W}_n \neq \emptyset$. But this is a contradiction, since the intersection is disjoint with any V_n, q. e. d.

PROPOSITION 1.3. *Let G be a topological group acting continuously on a locally compact Hausdorff space S. Then $G\backslash S$ is compact if and only if there exists a compact subset C of S such that $GC = S$.*

PROOF. Let π denote the natural map of S to $G\backslash S$. If $GC = S$, we have $\pi(C) = G\backslash S$, so that the 'if'-part is obvious. Conversely, cover S by open sets with compact closures, and map them by π. If $G\backslash S$ is compact, we have $G\backslash S = \bigcup_i \pi(U_i)$ with finitely many open sets U_i whose closures $\bar{U}_i$ are compact. Then $S = G \cdot (\bigcup_i \bar{U}_i)$, q. e. d.

Let G be a topological group. In general, a subset M of G may have limit points in G even if the induced topology of M is discrete. But, for a subgroup of G, we have:

PROPOSITION 1.4. *Let Γ be a subgroup of G. Suppose that the induced topology of Γ is locally compact. Then Γ is closed in G. Especially, if Γ is discrete, then Γ is closed, and has no limit point in G.*

We call Γ a *discrete subgroup* of G, if the induced topology of Γ is discrete.

PROOF. Suppose that Γ has a compact neighborhood C of the identity element e. Take an open neighborhood U of e in G so that $U\cap\Gamma\subset C$. Let x be an element of the closure of Γ. We can find a neighborhood V of x so that $V^{-1}V\subset U$. Then $(V\cap\Gamma)^{-1}(V\cap\Gamma)\subset C$. Note that $V\cap\Gamma\neq\emptyset$, and take an element y of $V\cap\Gamma$. Then $V\cap\Gamma\subset yC$. Now for every neighborhood W of x, we have $W\cap V\cap\Gamma\neq\emptyset$, hence x belongs to the closure of $V\cap\Gamma$. Since yC is compact, $x\in yC\subset\Gamma$, hence Γ is closed. The last assertion is obvious.

PROPOSITION 1.5. *Let G be a locally compact group, and K a compact subgroup of G. Put $S=G/K$, and let $h: G\to S$ be the natural map. If A is a compact subset of S, $h^{-1}(A)$ is compact.*

PROOF. Take an open covering of G whose members have compact closures, and consider their images on S by h. Then we see that $A\subset\bigcup_i h(V_i)$ with finitely many open sets V_i whose closures $\bar{V}_i$ are compact. Hence $h^{-1}(A)\subset\bigcup_i \bar{V}_i K$. Observe that $\bar{V}_i K$ is compact. Therefore, $h^{-1}(A)$, being a closed subset of a compact set, is compact.

PROPOSITION 1.6. *Let G, K, S, and h be as in Prop. 1.5, and Γ a subgroup of G. Then the following two statements are equivalent:*

(1) *Γ is a discrete subgroup of G.*

(2) *For any two compact subsets A and B of S, $\{g\in\Gamma \mid g(A)\cap B\neq\emptyset\}$ is a finite set.*

PROOF. Let A and B be compact subsets of S, and let $C=h^{-1}(A)$, $D=h^{-1}(B)$, $g\in\Gamma$. If $g(A)\cap B\neq\emptyset$, one has $gC\cap D\neq\emptyset$, hence $g\in\Gamma\cap(DC^{-1})$. By Prop. 1.5, C and D are compact, hence DC^{-1} is compact. If Γ is discrete, $\Gamma\cap(DC^{-1})$ is both compact and discrete, hence must be finite. This shows (1)$\Rightarrow$(2). To prove the converse, let V be a compact neighborhood of e in G, and let $t=h(e)$. Then $\Gamma\cap V\subset\{g\in\Gamma \mid gt\in h(V)\}$. Viewing t and $h(V)$ as A and B of (2), we find that $\Gamma\cap V$ is a finite set. Therefore Γ is discrete.

Hereafter till the end of this section, G, K, S, h will be the same as in Prop. 1.5, and Γ a discrete subgroup of G. By (2) of Prop. 1.6, $\{g\in\Gamma \mid g(z)=z\}$ is a finite set for every $z\in S$.

PROPOSITION 1.7. *For every $z\in S$, there exists a neighborhood U of z such that $\{g\in\Gamma \mid g(U)\cap U\neq\emptyset\}=\{g\in\Gamma \mid g(z)=z\}$.*

PROOF. Let V be a compact neighborhood of z. By Prop. 1.6, $\{g\in\Gamma \mid g(V)\cap V\neq\emptyset\}$ is a finite set, say $\{g_1,\cdots,g_r\}$. Suppose that $g_i(z)=z$ or $\neq z$

according as $1 \leqq i \leqq s$ or $s < i \leqq r$. For every $i > s$, take a neighborhood V_i of z and a neighborhood W_i of $g_i(z)$ so that $V_i \cap W_i = \emptyset$, and put $U = V \cap \{\cap_{i>s}(V_i \cap g_i^{-1}(W_i))\}$. Then U has the required property.

PROPOSITION 1.8. *If two points z and w of S are not Γ-equivalent, then there exist neighborhoods U of z and V of w such that $g(U) \cap V = \emptyset$ for every $g \in \Gamma$.*

PROOF. Let X and Y be compact neighborhoods of z and w respectively. By Prop. 1.6, $\{g \in \Gamma \mid g(X) \cap Y \neq \emptyset\}$ is a finite set, say $\{g_1, \cdots, g_r\}$. Since z and w are not Γ-equivalent, we have $g_i(z) \neq w$ for every i. Therefore we find neighborhoods U_i of $g_i(z)$ and V_i of w such that $U_i \cap V_i = \emptyset$. Put $U = X \cap g_1^{-1}(U_1) \cap \cdots \cap g_r^{-1}(U_r)$, $V = Y \cap V_1 \cap \cdots \cap V_r$. Then U and V have the desired property.

Let $\Gamma \backslash S$ denote the set of all Γ-orbits of the points of S. Prop. 1.8 implies that $\Gamma \backslash S$, with the quotient topology, *is a Hausdorff space.* Now we have an obvious commutative diagram:

$$\begin{array}{ccc} G & \longrightarrow & S = G/K \\ \downarrow & & \downarrow \\ \Gamma \backslash G & \longrightarrow & \Gamma \backslash S = \Gamma \backslash G/K . \end{array}$$

We see easily that all maps in this diagram are open and continuous.

PROPOSITION 1.9. *$\Gamma \backslash G$ is compact if and only if $\Gamma \backslash S$ is compact.*

PROOF. By Prop. 1.3, if $\Gamma \backslash S$ is compact, we have $S = \Gamma C$ with a compact subset C of S, so that $G = \Gamma \cdot h^{-1}(C)$. By Prop. 1.5, $h^{-1}(C)$ is compact, hence, by Prop. 1.3, $\Gamma \backslash G$ is compact. The converse part is obvious.

PROPOSITION 1.10. *Let G_1 and G_2 be locally compact groups, Γ a closed subgroup of $G_1 \times G_2$, and Γ_1 the projection of Γ to G_1. Suppose that G_2 is compact. Then the following assertions hold:*

(1) *Γ_1 is closed in G_1.*

(2) *$\Gamma \backslash (G_1 \times G_2)$ is compact if and only if $\Gamma_1 \backslash G_1$ is compact.*

(3) *If Γ is discrete in $G_1 \times G_2$, then Γ_1 is discrete in G_1.*

PROOF. Let V be a compact neighborhood of the identity in G_1. Then $(V \times G_2) \cap \Gamma$ is compact, and $V \cap \Gamma_1$ is its image by the projection map of $G_1 \times G_2$ to G_1. Therefore $V \cap \Gamma_1$ is compact. By Prop. 1.4, Γ_1 must be closed in G_1. If further Γ is discrete, then $(V \times G_2) \cap \Gamma$ is finite, so that $V \cap \Gamma_1$ is finite, hence (3). The assertion (2) follows easily from Prop. 1.3.

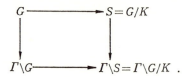

In general, two subgroups Γ and Γ' of a group G are said to be *commensurable* if $\Gamma \cap \Gamma'$ is of finite index in Γ and in Γ'. The following proposition can easily be verified, and may therefore be left to the reader as an exercise.

PROPOSITION 1.11. (1) *If Γ_1 is commensurable with Γ_2, and Γ_2 is commensurable with Γ_3, then Γ_1 is commensurable with Γ_3.*

(2) *Let Γ and Γ' be commensurable subgroups of a topological group G. If Γ is discrete, then Γ' is discrete.*

(3) *Let Γ and Γ' be commensurable closed subgroups of a locally compact group G. If $\Gamma\backslash G$ is compact, then $\Gamma'\backslash G$ is compact.*

1.2. Classification of linear fractional transformations

Although our main interest is in the transformations on the upper half plane, let us first consider more generally a linear fractional transformation on $\boldsymbol{C} \cup \{\infty\}$. For $\sigma = \begin{bmatrix} a & b \\ c & d \end{bmatrix} \in GL_2(\boldsymbol{C})$ and $z \in \boldsymbol{C} \cup \{\infty\}$, put $\sigma(z) = (az+b)/(cz+d)$. Suppose that this is not the identity transformation, i. e., σ is not a scalar matrix. From the theory of the Jordan canonical form, we see that the matrix σ is conjugate to one of the following two forms:

$$\text{(i)} \quad \begin{bmatrix} \lambda & 1 \\ 0 & \lambda \end{bmatrix}; \qquad \text{(ii)} \quad \begin{bmatrix} \lambda & 0 \\ 0 & \mu \end{bmatrix}, \quad \lambda \neq \mu .$$

Therefore, our transformation is essentially of the following types:

$$\text{(i)} \quad z \mapsto z + \lambda^{-1}; \qquad \text{(ii)} \quad z \mapsto cz, \quad c \neq 1 .$$

In the first case, we call σ *parabolic*. In the second case, we call σ *elliptic* if $|c|=1$, *hyperbolic* if c is real and positive, and *loxodromic* otherwise. This definition applies to both matrices and transformations. The identity transformation is excluded from this classification. We see that the number of fixed points of σ is one or two, according as σ is parabolic or not. If we impose the condition $\det(\sigma)=1$, then the classification can be done by means of $\operatorname{tr}(\sigma)$:

PROPOSITION 1.12. *Let $\sigma \in SL_2(\boldsymbol{C})$, $\sigma \neq \pm 1_2$. Then*

$$\begin{aligned} \sigma \textit{ is parabolic} \quad &\Leftrightarrow \operatorname{tr}(\sigma) = \pm 2, \\ \textit{elliptic} \quad &\Leftrightarrow \operatorname{tr}(\sigma) \textit{ is real and } |\operatorname{tr}(\sigma)| < 2, \\ \textit{hyperbolic} \quad &\Leftrightarrow \operatorname{tr}(\sigma) \textit{ is real and } |\operatorname{tr}(\sigma)| > 2, \\ \textit{loxodromic} \quad &\Leftrightarrow \operatorname{tr}(\sigma) \textit{ is not real.} \end{aligned}$$

PROOF. Since $\det(\sigma)=1$, the Jordan canonical form for σ is either

$\begin{bmatrix} \pm 1 & 1 \\ 0 & \pm 1 \end{bmatrix}$ or $\begin{bmatrix} \lambda & 0 \\ 0 & \lambda^{-1} \end{bmatrix}$, $\lambda \neq \pm 1$. Therefore the first three $\Rightarrow$ and the first $\Leftarrow$ can easily be checked. Now suppose that $\sigma = \begin{bmatrix} \lambda & 0 \\ 0 & \lambda^{-1} \end{bmatrix}$ and $\mathrm{tr}(\sigma) = \lambda + \lambda^{-1}$ is real. If λ is real, σ must be hyperbolic. If λ is imaginary, λ and $\bar{\lambda}$ are the roots of the equation $x^2 - \mathrm{tr}(\sigma)x + 1 = 0$, hence $\lambda\bar{\lambda} = 1$. Therefore σ is elliptic. Thus σ cannot be loxodromic if $\mathrm{tr}(\sigma)$ is real. This proves the last $\Rightarrow$. Since the conditions on the right hand sides are mutually exclusive, this completes the proof.

Let us now restrict ourselves to the transformations with real matrices. For $z \in \boldsymbol{C}$ and $\alpha = \begin{bmatrix} p & q \\ r & s \end{bmatrix} \in GL_2(\boldsymbol{R})$, put

$$j(\alpha, z) = rz + s. \tag{1.2.1}$$

If $w = \alpha(z)$, we have

$$\alpha \begin{bmatrix} z \\ 1 \end{bmatrix} = \begin{bmatrix} az+b \\ cz+d \end{bmatrix} = \begin{bmatrix} w \\ 1 \end{bmatrix} \cdot j(\alpha, z).$$

Further if $w' = \alpha(z')$,

$$\alpha \cdot \begin{bmatrix} z & z' \\ 1 & 1 \end{bmatrix} = \begin{bmatrix} w & w' \\ 1 & 1 \end{bmatrix} \begin{bmatrix} j(\alpha, z) & 0 \\ 0 & j(\alpha, z') \end{bmatrix}. \tag{1.2.2}$$

Substituting $\bar{z}$ and $\bar{w}$ for z' and w', and taking the determinant, we obtain

$$\det(\alpha) \cdot \mathrm{Im}(z) = \mathrm{Im}(\alpha(z)) \cdot |j(\alpha, z)|^2. \tag{1.2.3}$$

Let $\mathfrak{H}$ denote the complex upper half plane, i.e.,

$$\mathfrak{H} = \{z \in \boldsymbol{C} \mid \mathrm{Im}(z) > 0\}.$$

Further, put

$$GL_2^+(\boldsymbol{R}) = \{\alpha \in GL_2(\boldsymbol{R}) \mid \det(\alpha) > 0\}.$$

If $\alpha \in GL_2^+(\boldsymbol{R})$, α maps $\mathfrak{H}$ onto itself. It is also well known that every holomorphic automorphism of $\mathfrak{H}$ is obtained from an element of $GL_2^+(\boldsymbol{R})$. Obviously α induces the identity map if and only if it is a scalar matrix. Therefore the group of all holomorphic automorphisms of $\mathfrak{H}$ is isomorphic to $GL_2^+(\boldsymbol{R})/[\boldsymbol{R}^\times \cdot 1_2]$, and to $SL_2(\boldsymbol{R})/\{\pm 1_2\}$.

From (1.2.2) we obtain easily

$$j(\alpha\beta, z) = j(\alpha, \beta(z)) j(\beta, z). \tag{1.2.4}$$

Furthermore, substituting $z + dz$ (formally) for z' in (1.2.2), and taking the determinant, we obtain

$$\frac{d}{dz}\alpha(z) = \det(\alpha) \cdot j(\alpha, z)^{-2}. \tag{1.2.5}$$

If $\alpha=\begin{bmatrix} p & q \\ r & s \end{bmatrix}\in SL_2(\boldsymbol{R})$ and $i=\sqrt{-1}$, we have $\alpha(i)=i$ if and only if $p=s$, $q=-r$, $p^2+q^2=1$. Therefore, the special orthogonal group

$$SO(2)=\{\alpha\in SL_2(\boldsymbol{R}) \mid {}^t\alpha\alpha=1_2\}$$

is the isotropy subgroup of $SL_2(\boldsymbol{R})$ at i. The action of $SL_2(\boldsymbol{R})$ on $\mathfrak{H}$ is transitive, since, for $a>0$, $a^{-1/2}\cdot\begin{bmatrix} a & b \\ 0 & 1 \end{bmatrix}$ sends i to $ai+b$. Therefore, by Th. 1.1, $\mathfrak{H}$ is homeomorphic to $SL_2(\boldsymbol{R})/SO(2)$, through the map $\alpha\mapsto\alpha(i)$.

We shall now study more closely the transformations obtained from the elements of $SL_2(\boldsymbol{R})$. By Prop. 1.12, $SL_2(\boldsymbol{R})$ contains no loxodromic transformations. For every $z\in\mathfrak{H}$, we can find an element τ of $SL_2(\boldsymbol{R})$ so that $\tau(i)=z$. Then

$$\tau\cdot SO(2)\cdot\tau^{-1}=\{\alpha\in SL_2(\boldsymbol{R}) \mid \alpha(z)=z\}\,.$$

Since every element of $SO(2)$ has characteristic roots of absolute value 1, this shows that *an element of $SL_2(\boldsymbol{R})$ with at least one fixed point in $\mathfrak{H}$ must be either $\pm 1_2$ or elliptic.*

For every $s\in\boldsymbol{R}\cup\{\infty\}$, put

$$F(s)=\{\alpha\in SL_2(\boldsymbol{R}) \mid \alpha(s)=s\}\,,$$

$$P(s)=\{\alpha\in F(s) \mid \alpha \text{ parabolic or} = \pm 1_2\}\,.$$

Since $SL_2(\boldsymbol{R})$ acts transitively on $\boldsymbol{R}\cup\{\infty\}$, we can find an element σ of $SL_2(\boldsymbol{R})$ so that $\sigma(\infty)=s$. Then $F(s)=\sigma F(\infty)\sigma^{-1}$, $P(s)=\sigma P(\infty)\sigma^{-1}$. Now we see easily that

$$F(\infty)=\left\{\begin{bmatrix} a & b \\ 0 & a^{-1} \end{bmatrix} \,\middle|\, a\in\boldsymbol{R}^\times,\ b\in\boldsymbol{R}\right\},$$

$$P(\infty)=\left\{\pm\begin{bmatrix} 1 & h \\ 0 & 1 \end{bmatrix} \,\middle|\, h\in\boldsymbol{R}\right\}\cong\boldsymbol{R}\times\{\pm 1\}\,.$$

This shows that *if an element σ of $SL_2(\boldsymbol{R})$, $\neq\pm 1_2$, has at least one fixed point on $\boldsymbol{R}\cup\{\infty\}$, then σ is either parabolic or hyperbolic.* From these considerations, we obtain

PROPOSITION 1.13. *Let $\sigma\in SL_2(\boldsymbol{R})$, $\sigma\neq\pm 1_2$. Then*

σ is parabolic $\Leftrightarrow$ σ has only one fixed point on $\boldsymbol{R}\cup\{\infty\}$,

elliptic $\Leftrightarrow$ σ has one fixed point z in $\mathfrak{H}$, and the other fixed point $\bar{z}$,

hyperbolic $\Leftrightarrow$ σ has two fixed points on $\boldsymbol{R}\cup\{\infty\}$.

PROPOSITION 1.14. *Let $\sigma\in SL_2(\boldsymbol{R})$, $\sigma\neq\pm 1_2$, and let $m\in\boldsymbol{Z}$, $\sigma^m\neq\pm 1_2$. Then σ is parabolic (resp. elliptic, hyperbolic) if and only if σ^m is parabolic (resp. elliptic, hyperbolic).*

PROOF. The 'only if'-part follows immediately from Prop. 1.13 or the Jordan form of σ. Then the 'if'-part is obvious.

EXERCISE 1.15. Let α and β be elements of $SL_2(\boldsymbol{R})$, $\neq \pm 1_2$, such that $\alpha\beta = \beta\alpha$. Prove:

(1) If α is parabolic (resp. elliptic, hyperbolic), then β is parabolic (resp. elliptic, hyperbolic).

(2) If $\alpha(z) = z$ for some $z \in \boldsymbol{C} \cup \{\infty\}$, then $\beta(z) = z$.

Let us now fix a discrete subgroup Γ of $SL_2(\boldsymbol{R})$. A point z of $\mathfrak{H}$ is called an *elliptic point* of Γ if there exists an elliptic element σ of Γ such that $\sigma(z) = z$. Similarly, a point s of $\boldsymbol{R} \cup \{\infty\}$ is called a *cusp* of Γ if there exists a parabolic element τ of Γ such that $\tau(s) = s$. If w is a cusp (resp. an elliptic point) of Γ and $\gamma \in \Gamma$, then we see easily that $\gamma(w)$ is also a cusp (resp. an elliptic point) of Γ.

PROPOSITION 1.16. *If z is an elliptic point of Γ, then $\{\sigma \in \Gamma \mid \sigma(z) = z\}$ is a finite cyclic group.*

PROOF. If $\tau \in SL_2(\boldsymbol{R})$ and $\tau(i) = z$, we have $\{\sigma \in \Gamma \mid \sigma(z) = z\} = \tau SO(2)\tau^{-1} \cap \Gamma$. Since Γ is discrete and $SO(2)$ is compact, this intersection must be a finite group. Now $SO(2)$ is isomorphic to $\boldsymbol{R}/\boldsymbol{Z}$, and its finite subgroups are all cyclic, q. e. d.

PROPOSITION 1.17. *Let s be a cusp of Γ, and $\Gamma_s = \{\sigma \in \Gamma \mid \sigma(s) = s\}$. Then $\Gamma_s/(\Gamma \cap \{\pm 1_2\})$ is isomorphic to $\boldsymbol{Z}$. Moreover, an element of Γ_s is either $\pm 1_2$ or parabolic, i. e., $\Gamma_s = \Gamma \cap P(s)$.*

PROOF. We have seen that $P(s)$ is isomorphic to $\boldsymbol{R} \times \{\pm 1\}$. Therefore, $(P(s) \cap \Gamma)/(\Gamma \cap \{\pm 1\})$ is isomorphic to a non-trivial discrete subgroup of $\boldsymbol{R}$, hence isomorphic to $\boldsymbol{Z}$. Now, without losing generality, we may assume that $s = \infty$. Take a generator $\sigma = \begin{bmatrix} \pm 1 & h \\ 0 & \pm 1 \end{bmatrix}$ (modulo ± 1) of this group. Assume that Γ_s contains a hyperbolic element $\tau = \begin{bmatrix} a & b \\ 0 & a^{-1} \end{bmatrix}$, $|a| \neq 1$. Taking τ^{-1} instead of τ, if necessary, we may assume that $|a| < 1$. Then $\tau\sigma\tau^{-1} = \begin{bmatrix} \pm 1 & a^2 h \\ 0 & \pm 1 \end{bmatrix} \in P(s) \cap \Gamma$. But this is a contradiction, since $|a^2 h| < |h|$. Therefore $\Gamma_s = P(s) \cap \Gamma$.

PROPOSITION 1.18. *The elements of Γ of finite order consist of the elliptic elements together with $\pm 1_2$.*

PROOF. If an element σ of $SL_2(\boldsymbol{R})$ is of finite order, σ is conjugate in $SL_2(\boldsymbol{C})$ to a matrix $\begin{bmatrix} \zeta & 0 \\ 0 & \bar{\zeta} \end{bmatrix}$ with a root of unity ζ. By our definition, such a σ is elliptic if $\zeta \neq \pm 1$. The converse part follows immediately from Prop. 1.16.

PROPOSITION 1.19. *The set of all elliptic points of Γ has no limit point in $\mathfrak{H}$.*

PROOF. Assume that there is a sequence of distinct elliptic points $\{z_n\}$ of Γ converging to $w \in \mathfrak{H}$. By Prop. 1.7, we can find a neighborhood U of w such that, for $\gamma \in \Gamma$, $\gamma(U) \cap U \neq \emptyset$ if and only if $\gamma(w) = w$. For sufficiently large n, we have $z_n \in U$ and $z_n \neq w$. One has $\gamma(z_n) = z_n$ for some elliptic element γ of Γ. Then $\gamma(U) \cap U \neq \emptyset$, hence $\gamma(w) = w$. Thus γ has *two* fixed points on $\mathfrak{H}$, a contradiction.

Each *matrix* of $SL_2(\boldsymbol{R})$ (or of $GL_2^+(\boldsymbol{R})$) should not be confused with the *transformation* on $\mathfrak{H}$ represented by it. Especially one should be careful about the order of an elliptic element:

PROPOSITION 1.20. *Let σ be an elliptic element of Γ. If σ, as a matrix, is of an even order $2h$, then Γ contains -1_2, and the transformation $z \mapsto \sigma(z)$ is of order h.*

PROOF. One can find an element τ of $GL_2(\boldsymbol{C})$ so that $\tau\sigma\tau^{-1} = \begin{bmatrix} \zeta & 0 \\ 0 & \bar{\zeta} \end{bmatrix}$ with a primitive $(2h)$-th root of unity ζ. Then $\zeta^h = -1$, hence $\sigma^h = -1_2$, q. e. d.

COROLLARY 1.21. *If Γ does not contain -1_2, every elliptic element of Γ is of an odd order.*

This is an immediate consequence of Prop. 1.20.

To distinguish the transformation group from the matrix group, we shall denote by $\bar{\Gamma}$ the image of Γ by the natural map

$$SL_2(\boldsymbol{R}) \longrightarrow SL_2(\boldsymbol{R})/\{\pm 1_2\}.$$

For an elliptic point z of Γ, the order of the group

$$\{\sigma \in \bar{\Gamma} \mid \sigma(z) = z\}$$

is called the *order* of the elliptic point z (relative to Γ).

PROPOSITION 1.22. *Neither elliptic nor parabolic element α of $SL_2(\boldsymbol{R})$ is conjugate in $SL_2(\boldsymbol{R})$ to α^{-1}.*

PROOF. Assume that $\gamma\alpha\gamma^{-1} = \alpha^{-1}$ for some $\gamma \in SL_2(\boldsymbol{R})$. If α is elliptic, as is observed above, there exists an element τ of $SL_2(\boldsymbol{R})$ such that $\tau\alpha\tau^{-1} \in SO(2)$. Put $\tau\alpha\tau^{-1} = \begin{bmatrix} p & q \\ -q & p \end{bmatrix}$ and $\tau\gamma\tau^{-1} = \begin{bmatrix} a & b \\ c & d \end{bmatrix}$. Then we have $q \neq 0$ since α is elliptic, and

$$\begin{bmatrix} a & b \\ c & d \end{bmatrix}\begin{bmatrix} p & q \\ -q & p \end{bmatrix}=\begin{bmatrix} p & -q \\ q & p \end{bmatrix}\begin{bmatrix} a & b \\ c & d \end{bmatrix},$$

so that $a=-d$, $b=c$. Then $1=\det(\gamma)=-(a^2+b^2)$, which is impossible since a and b are real. If α is parabolic, we can take τ so that $\tau\alpha\tau^{-1}=\pm\begin{bmatrix} 1 & h \\ 0 & 1 \end{bmatrix}$. Then $\begin{bmatrix} a & b \\ c & d \end{bmatrix}\begin{bmatrix} 1 & h \\ 0 & 1 \end{bmatrix}=\begin{bmatrix} 1 & -h \\ 0 & 1 \end{bmatrix}\begin{bmatrix} a & b \\ c & d \end{bmatrix}$, so that $c=0$, $a=-d$, hence $1=\det(\gamma)$ $=-a^2$, which is again impossible.

Note that a hyperbolic element α is conjugate in $SL_2(\boldsymbol{R})$ to α^{-1}.

1.3. The topological space $\Gamma\backslash\mathfrak{H}^*$

Hereafter till the end of this section, we denote by Γ any discrete subgroup of $SL_2(\boldsymbol{R})$, and by $\mathfrak{H}^*$ the union of $\mathfrak{H}$ and the cusps of Γ. The set $\mathfrak{H}^*$ depends on Γ; of course $\mathfrak{H}^*=\mathfrak{H}$ if Γ has no cusps. We observe that Γ acts on $\mathfrak{H}^*$, hence the quotient space $\Gamma\backslash\mathfrak{H}^*$ is meaningful. We shall consider a structure of Riemann surface on $\Gamma\backslash\mathfrak{H}^*$ in the next section. For that purpose we first define a topology of $\mathfrak{H}^*$. For every $z\in\mathfrak{H}$, as a fundamental system of open neighborhoods of z, we take the usual one. For a fundamental system of open neighborhoods of a cusp $s\neq\infty$, we take all sets of the form:

$\{s\}\cup\{$the interior of a circle in $\mathfrak{H}$ tangent to the real axis at $s\}$.

If ∞ is a cusp, we take the sets

$$\{\infty\}\cup\{z\in\mathfrak{H}\mid \operatorname{Im}(z)>c\}, \tag{1.3.0}$$

for all positive numbers c, as a fundamental system of open neighborhoods of ∞. We shall write (1.3.0) also as $\{z\in\mathfrak{H}^*\mid \operatorname{Im}(z)>c\}$. It can easily be seen that this defines a Hausdorff topology on $\mathfrak{H}^*$, and every element of Γ acts on $\mathfrak{H}^*$ as a homeomorphism. However, $\mathfrak{H}^*$ is *not* locally compact, unless $\mathfrak{H}^*=\mathfrak{H}$.

For a cusp $s\leqq\infty$ of Γ, put

$$P(s)=\{\alpha\in SL_2(\boldsymbol{R})\mid \alpha(s)=s,\ \alpha \text{ parabolic or } =\pm 1_2\},$$

$$\Gamma_s=P(s)\cap\Gamma=\{\gamma\in\Gamma\mid\gamma(s)=s\} \qquad \text{(see Prop. 1.17)}.$$

The neighborhoods of s of the above type are obviously stable under $P(s)$.

To study the structure of $\Gamma\backslash\mathfrak{H}^*$, let us assume that ∞ is a cusp of Γ. We need the formula

$$\operatorname{Im}(\alpha(z))=\det(\alpha)\cdot\operatorname{Im}(z)/|cz+d|^2 \qquad \text{for}\quad \alpha=\begin{bmatrix} a & b \\ c & d \end{bmatrix}\in GL_2(\boldsymbol{R}), \tag{1.3.1}$$

which was proved in § 1.2. For every $\sigma\in\Gamma$, we let c_σ denote the lower left entry of the matrix σ. Then $\Gamma_\infty=\{\sigma\in\Gamma\mid c_\sigma=0\}$. By Prop. 1.17, we can find a generator $\pm\begin{bmatrix} 1 & h \\ 0 & 1 \end{bmatrix}$ of Γ_∞ modulo $\pm 1_2$.

LEMMA 1.23. *$|c_\sigma|$ depends only on the double coset $\Gamma_\infty\sigma\Gamma_\infty$.*

This can be verified by a simple matrix computation.

LEMMA 1.24. *Given $M>0$, there are only finitely many double cosets $\Gamma_\infty\sigma\Gamma_\infty$ such that $\sigma\in\Gamma$ and $|c_\sigma|\leqq M$.*

PROOF. Since $\Gamma_\infty=\{\sigma\in\Gamma\mid c_\sigma=0\}$, it is sufficient to consider only those σ for which $c_\sigma\neq 0$. Take a generator $\tau=\pm\begin{bmatrix}1 & h\\ 0 & 1\end{bmatrix}$ of Γ_∞ modulo $\pm 1_2$. Let $\sigma=\begin{bmatrix}a & b\\ c & d\end{bmatrix}\in\Gamma$, $0\neq|c|\leqq M$. We are going to find an element σ'' in $\Gamma_\infty\sigma\Gamma_\infty$ such that $\sigma''(i)$ is contained in a compact set K which depends only on M and h. First we can find an integer n so that $1\leqq d+nhc\leqq 1+|hc|$. Put $\sigma'=\sigma\tau^n=\begin{bmatrix}a' & b'\\ c' & d'\end{bmatrix}$. Then $|c'|=|c|$, $|d'|=d+nhc$. By (1.3.1), $\mathrm{Im}(\sigma'(i))=1/(c'^2+d'^2)$. We have $1\leqq|d'|\leqq 1+|hc|$, and $|c|\leqq M$, hence $1\leqq c'^2+d'^2<M^2+(1+|h|M)^2$. Therefore $\sigma'(i)$ belongs to the domain

$$1\geqq \mathrm{Im}(z)\geqq 1/[M^2+(1+|h|M)^2]. \tag{1.3.2}$$

Now the transformation $z\mapsto\tau^m(z)=z+mh$ does not change $\mathrm{Im}(z)$. We can take m so that $\tau^m\sigma'(i)$ satisfies (1.3.2) and

$$0\leqq \mathrm{Re}(z)\leqq |h|. \tag{1.3.3}$$

The conditions (1.3.2) and (1.3.3) define a compact set K in $\mathfrak{H}$. We have thus found an element $\sigma''=\tau^m\sigma\tau^n$ such that $\sigma''(i)\in K$. By Prop. 1.6, there are only finitely many such σ'' in Γ. This proves the lemma.

LEMMA 1.25. *There exists a positive number r, depending only on Γ, such that $|c_\sigma|\geqq r$ for all $\sigma\in\Gamma-\Gamma_\infty$. Moreover, for such an r, one has $\mathrm{Im}(z)\cdot\mathrm{Im}(\sigma(z))\leqq 1/r^2$ for all $z\in\mathfrak{H}$ and all $\sigma\in\Gamma-\Gamma_\infty$.*

PROOF. The existence of r follows immediately from Lemma 1.24. If $\sigma=\begin{bmatrix}a & b\\ c & d\end{bmatrix}\in\Gamma$ and $c\neq 0$, we have

$$\mathrm{Im}(\sigma(z))=\mathrm{Im}(z)\cdot|cz+d|^{-2}\leqq\mathrm{Im}(z)\cdot(c\cdot\mathrm{Im}(z))^{-2}\leqq r^{-2}\,\mathrm{Im}(z)^{-1},\qquad \text{q. e. d.}$$

LEMMA 1.26. *For every cusp s of Γ, there exists a neighborhood U of s in $\mathfrak{H}^*$ such that $\Gamma_s=\{\sigma\in\Gamma\mid\sigma(U)\cap U\neq\emptyset\}$.*

PROOF. We may assume that $s=\infty$. Let $U=\{z\in\mathfrak{H}^*\mid\mathrm{Im}(z)>1/r\}$, with a number r of Lemma 1.25. If $\sigma\in\Gamma-\Gamma_\infty$ and $z\in U$, we have, by Lemma 1.25, $\mathrm{Im}(\sigma(z))<1/r$. Thus U has the required property.

Observe that two points of the set U are equivalent under Γ only if they are so under Γ_s, and hence $\Gamma_s\backslash U$ may be identified with a subset of $\Gamma\backslash\mathfrak{H}^*$; moreover U contains no elliptic point of Γ.

LEMMA 1.27. *For every cusp s of Γ and for every compact subset K of $\mathfrak{H}$, there exists a neighborhood U of s such that $U \cap \gamma(K) = \emptyset$ for every $\gamma \in \Gamma$.*

PROOF. Assume again $s = \infty$. We can find two positive numbers A and B so that $A < \mathrm{Im}(z) < B$ for all $z \in K$. Take a number r as in Lemma 1.25, and put

$$U = \{z \in \mathfrak{H}^* \mid \mathrm{Im}(z) > \mathrm{Max}(B, 1/Ar^2)\}.$$

Let $z \in K$. By Lemma 1.25, if $\sigma \in \Gamma - \Gamma_\infty$, $\mathrm{Im}(\sigma(z)) < 1/Ar^2$. If $\sigma \in \Gamma_\infty$, $\mathrm{Im}(\sigma(z)) = \mathrm{Im}(z) < B$. Thus U has the required property.

Let us now consider the quotient topology of $\Gamma \backslash \mathfrak{H}^*$ as defined in § 1.1. Namely we take

$$\{X \subset \Gamma \backslash \mathfrak{H}^* \mid \pi^{-1}(X) \text{ is open in } \mathfrak{H}^*\}$$

to be the class of all open sets in $\Gamma \backslash \mathfrak{H}^*$, where π is the natural projection of $\mathfrak{H}^*$ to $\Gamma \backslash \mathfrak{H}^*$. If U is as in Lemma 1.26 (and its proof), then $\pi(U)$ can be identified with $\Gamma_s \backslash U$, and is a neighborhood of $\pi(s)$.

THEOREM 1.28. *The quotient space $\Gamma \backslash \mathfrak{H}^*$, with the above topology, is a Hausdorff space.*

PROOF. By Prop. 1.8, $\Gamma \backslash \mathfrak{H}$ is a Hausdorff space. Since $\Gamma \backslash \mathfrak{H}^*$ is the union of $\Gamma \backslash \mathfrak{H}$ and the equivalence classes of cusps, it remains to show that an equivalence class of cusps can be separated from an equivalence class of points in $\mathfrak{H}$, and also from another equivalence class of cusps. Lemma 1.27 takes care of the former case. Therefore let us consider two cusps s and t which are not Γ-equivalent. Without losing generality, we may assume $t = \infty$. Let Γ_∞ and $\pm\begin{bmatrix} 1 & h \\ 0 & 1 \end{bmatrix}$ be as before. Define three sets L, K, and V as follows:

$$L = \{z \in \boldsymbol{C} \mid \mathrm{Im}(z) = u\},$$
$$K = \{z \in L \mid 0 \leqq \mathrm{Re}(z) \leqq |h|\},$$
$$V = \{z \in \mathfrak{H}^* \mid \mathrm{Im}(z) > u\},$$

where u is a positive number. Since K is compact, we can find, by Lemma 1.27, a neighborhood U of s so that $K \cap \Gamma U = \emptyset$. We may assume that the boundary of U is a circle tangent to the real line $\boldsymbol{R}$. Let us show that $V \cap \Gamma U = \emptyset$. Assume, on the contrary, that $\gamma(U) \cap V \neq \emptyset$ for some $\gamma \in \Gamma$. Since $\gamma(s) \neq \infty$, the boundary of $\gamma(U)$ is a circle tangent to $\boldsymbol{R}$. Therefore, if $\gamma(U) \cap V \neq \emptyset$, then $\gamma(U) \cap L \neq \emptyset$, hence $\gamma(U)$ intersects some translation of K by an element of Γ_∞, i. e., there exists an element δ of Γ_∞ such that $\gamma(U) \cap \delta(K) \neq \emptyset$. Then $\delta^{-1}\gamma(U) \cap K \neq \emptyset$, a contradiction. This completes the proof.

PROPOSITION 1.29. *The quotient space $\Gamma \backslash \mathfrak{H}^*$ is locally compact.*

PROOF. Our task is to show that if s is a cusp of Γ and if π denotes the natural map of $\mathfrak{H}^*$ to $\Gamma\backslash\mathfrak{H}^*$, then $\pi(s)$ has a compact neighborhood. We may assume that $s=\infty$. By Lemma 1.26 and the remark after it, there is a neighborhood $V=\{z\in\mathfrak{H}^*\mid \mathrm{Im}(z)\geqq c\}$ with a positive constant c such that V/Γ_∞ is identified with $\pi(V)$. If $\begin{bmatrix}1 & h\\ 0 & 1\end{bmatrix}$ is a generator of Γ_∞ (modulo ± 1), we see that $\pi(V)$ coincides with the image of $\{z\in V\mid z=\infty \text{ or } 0\leqq \mathrm{Re}(z)\leqq |h|\}$ by π. The latter set is obviously compact, hence $\pi(V)$ is compact, q. e. d. (See also § 1.5, where we shall show that $\Gamma\backslash\mathfrak{H}^*$ has a structure of a Riemann surface.)

PROPOSITION 1.30. *Let Γ and Γ' be mutually commensurable discrete subgroups of $SL_2(\boldsymbol{R})$ (see* p. 5). *Then Γ and Γ' have the same set of cusps.*

PROOF. It suffices to consider the case in which $\Gamma'\subset\Gamma$ and $[\Gamma:\Gamma']<\infty$. If s is a cusp of Γ', then clearly s is a cusp of Γ. If s is a cusp of Γ, then $\sigma(s)=s$ for some parabolic element σ of Γ. We have $\sigma^e\in\Gamma'$ for some positive integer e. Then σ^e is parabolic, and $\sigma^e(s)=s$. Therefore s is a cusp of Γ', q. e. d.

PROPOSITION 1.31. *Let Γ and Γ' be as in Prop.* 1.30. *Then $\Gamma\backslash\mathfrak{H}^*$ is compact if and only if $\Gamma'\backslash\mathfrak{H}^*$ is compact.*

PROOF. Again we may assume that $\Gamma'\subset\Gamma$, $[\Gamma:\Gamma']<\infty$. If $\Gamma'\backslash\mathfrak{H}^*$ is compact, then, since the natural projection $\Gamma'\backslash\mathfrak{H}^*\to\Gamma\backslash\mathfrak{H}^*$ is continuous, $\Gamma\backslash\mathfrak{H}^*$ must be compact. Conversely, if $\Gamma\backslash\mathfrak{H}^*$ is compact, consider the projection map π (resp. π') of $\mathfrak{H}^*$ to $\Gamma\backslash\mathfrak{H}^*$ (resp. $\Gamma'\backslash\mathfrak{H}^*$). The proof of Prop. 1.29 shows that every point of $\Gamma\backslash\mathfrak{H}^*$ has a neighborhood which is the image of a compact subset of $\mathfrak{H}^*$ under π. Since $\Gamma\backslash\mathfrak{H}^*$ is compact, we can find finitely many compact subsets U_k of $\mathfrak{H}^*$ such that $\Gamma\backslash\mathfrak{H}^*=\bigcup_k\pi(U_k)$. Now we can find finitely many elements α_j of Γ such that $\Gamma=\bigcup_j\Gamma'\alpha_j$. Then $\Gamma'\backslash\mathfrak{H}^*=\bigcup_{j,k}\pi'(\alpha_jU_k)$. Therefore $\Gamma'\backslash\mathfrak{H}^*$ is compact.

PROPOSITION 1.32. *If $\Gamma\backslash\mathfrak{H}^*$ is compact, then the number of Γ-inequivalent cusps (resp. elliptic points) is finite.*

PROOF. Let C (resp. E) denote the set of all cusps (resp. all elliptic points) of Γ. For each $z\in\mathfrak{H}$, take a neighborhood U_z of z in $\mathfrak{H}$ so that $U_z\cap E$ is either empty, or possibly $\{z\}$. This is possible in view of Prop. 1.7. By Lemma 1.26, for each $s\in C$, we can find a neighborhood U_s of s containing no elliptic points. Let π denote the projection map of $\mathfrak{H}^*$ to $\Gamma\backslash\mathfrak{H}^*$. If $\Gamma\backslash\mathfrak{H}^*$ is compact, we can select a finite number of sets of the form $\pi(U_z)$ or $\pi(U_s)$ which cover $\Gamma\backslash\mathfrak{H}^*$. Then the number of points in $\pi(C)$ (resp. $\pi(E)$) is at most the number of $\pi(U_s)$ (resp. $\pi(U_z)$), which are necessary to cover $\Gamma\backslash\mathfrak{H}^*$, q. e. d.

PROPOSITION 1.33. *If $\Gamma\backslash\mathfrak{H}$ is compact, then Γ has no parabolic element.*

PROOF. Let π denote the projection map of $\mathfrak{H}$ to $\Gamma\backslash\mathfrak{H}$. Suppose that ∞ is a cusp of Γ. Take an infinite sequence $\{z_n\}$ of points of $\mathfrak{H}$ such that $\mathrm{Im}(z_n)\to\infty$. By Lemma 1.26, there exists a neighborhood

$$U=\{z\in\mathfrak{H}^* \mid \mathrm{Im}(z)>c\}$$

of ∞ such that $\Gamma_\infty=\{\gamma\in\Gamma \mid \gamma(U)\cap U\neq\emptyset\}$. Then $z_n\in U$ for sufficiently large n. Since no element of Γ_∞ changes $\mathrm{Im}(z)$, if two points of $\{z_n\}$ have distinct and sufficiently large imaginary parts, then they are not Γ-equivalent. Therefore $\{\pi(z_n)\}$ contains a sequence of infinitely many distinct points of $\Gamma\backslash\mathfrak{H}$. If $\Gamma\backslash\mathfrak{H}$ is compact, there exists a point w of $\mathfrak{H}$ such that $\pi(w)$ is a limit point of $\{\pi(z_n)\}$. Let K be a compact neighborhood of w. By Lemma 1.27, there exists a neighborhood V of ∞ such that $K\cap\Gamma V=\emptyset$. This is a contradiction, since $\pi(z_n)\in\pi(K)\cap\pi(V)$ for sufficiently large n.

1.4. The modular group $SL_2(\boldsymbol{Z})$

In this section we shall illustrate the preceding discussion by studying the modular group $SL_2(\boldsymbol{Z})$. It is clear that $SL_2(\boldsymbol{Z})$ is a discrete subgroup of $SL_2(\boldsymbol{R})$. Let us determine its cusps and elliptic points.

First let us show that the cusps of $\Gamma=SL_2(\boldsymbol{Z})$ are exactly the points in $\boldsymbol{Q}\cup\{\infty\}$. It is clear that ∞ is a fixed point under the parabolic element $\begin{bmatrix}1&1\\0&1\end{bmatrix}$ of Γ. If $\begin{bmatrix}a&b\\c&d\end{bmatrix}$ is a parabolic element of Γ, it has only one fixed point s. If s is finite, it satisfies

$$cs^2+(d-a)s-b=0, \qquad c\neq 0.$$

Since the discriminant of this equation vanishes, s must be contained in $\boldsymbol{Q}$. Conversely, for $p/q\in\boldsymbol{Q}$ with $p\in\boldsymbol{Z}$, $q\in\boldsymbol{Z}$, $(p, q)=1$, take integers t and u so that $pt-qu=1$. Then $\sigma=\begin{bmatrix}p&u\\q&t\end{bmatrix}\in\Gamma$, and $\sigma(\infty)=p/q$. Since the image of a cusp under any element of Γ is a cusp, this shows that all points of $\boldsymbol{Q}\cup\{\infty\}$ are cusps of Γ. Moreover we have shown that all cusps are equivalent to the cusp at ∞. Thus $\Gamma\backslash\mathfrak{H}^*=(\Gamma\backslash\mathfrak{H})\cup\{\infty\}$.

Next let us determine the elliptic points of $SL_2(\boldsymbol{Z})$. If σ is an elliptic element of $SL_2(\boldsymbol{Z})$, $|\mathrm{tr}(\sigma)|$ is an integer and <2 by Prop. 1.12. Therefore the characteristic polynomial of σ is either x^2+1 or $x^2\pm x+1$, so that $\sigma^4=1$ or $\sigma^6=1$, and $\sigma^2\neq 1$[1]. If $\sigma^6=1$, we have $\sigma^3=\pm 1$. In the case $\sigma^3=-1$, we have

1) One can reason also as follows: If ζ is a characteristic root of σ, ζ satisfies a quadratic equation with rational coefficients, so that $[\boldsymbol{Q}(\zeta):\boldsymbol{Q}]\leqq 2$. Therefore $\sigma^m=1$ with $m=2, 4, 3$, or 6. This reasoning is applicable to the case with an algebraic number field of higher degree in place of $\boldsymbol{Q}$.

$(-\sigma)^3=1$. Thus, for the determination of elliptic elements (or points), it is sufficient to consider the cases $\sigma^4=1$ and $\sigma^3=1$.

CASE 1: $\sigma^4=1$. Let $\boldsymbol{Z}^2$ denote the module of all column vectors $\begin{bmatrix} a \\ b \end{bmatrix}$ with a and b in $\boldsymbol{Z}$. Let the elements $\boldsymbol{Z}[\sigma]$ act on $\boldsymbol{Z}^2$ by left multiplication. Since $\boldsymbol{Z}[\sigma]$ is isomorphic to $\boldsymbol{Z}[i]$, $\boldsymbol{Z}[\sigma]$ is a principal ideal domain. The module $\boldsymbol{Z}^2$ over $\boldsymbol{Z}[\sigma]$ is torsion-free, since $(a+b\sigma)x=0$ implies $(a^2+b^2)x=0$, hence $x=0$, if $a+b\sigma\neq 0$. Therefore $\boldsymbol{Z}^2$ must be a free $\boldsymbol{Z}[\sigma]$-module of rank 1, i.e., $\boldsymbol{Z}^2=\boldsymbol{Z}[\sigma]u$ for some $u\in\boldsymbol{Z}^2$. Put $v=\sigma u$. Then u and v form a basis of $\boldsymbol{Z}^2$ over $\boldsymbol{Z}$. We have

$$\sigma\cdot[u\ v]=[u\ v]\begin{bmatrix} 0 & -1 \\ 1 & 0 \end{bmatrix}, \qquad \det[u\ v]=\pm 1.$$

If $\det[u\ v]=1$, this shows that σ is conjugate to $\begin{bmatrix} 0 & -1 \\ 1 & 0 \end{bmatrix}$ in $SL_2(\boldsymbol{Z})$. If $\det[u\ v]=-1$, then $\sigma=\tau\begin{bmatrix} 0 & 1 \\ -1 & 0 \end{bmatrix}\tau^{-1}$ with $\tau=[v\ u]$. Thus every elliptic element σ in $SL_2(\boldsymbol{Z})$ of order 4 is conjugate to $\pm\begin{bmatrix} 0 & -1 \\ 1 & 0 \end{bmatrix}$ in $SL_2(\boldsymbol{Z})$. Therefore every elliptic point of order 2 is equivalent to the fixed point of $\begin{bmatrix} 0 & -1 \\ 1 & 0 \end{bmatrix}$, that is i. $\left(\begin{bmatrix} 0 & -1 \\ 1 & 0 \end{bmatrix}\right.$ is not conjugate to $\begin{bmatrix} 0 & 1 \\ -1 & 0 \end{bmatrix}$ on account of Prop. 1.22.$\Big)$

CASE 2: $\sigma^3=1$. We see that $\boldsymbol{Z}[\sigma]$ is isomorphic to $\boldsymbol{Z}[e^{2\pi i/3}]$, which is a principal ideal domain. Therefore we have again $\boldsymbol{Z}^2=\boldsymbol{Z}[\sigma]u$ for some u. Put $v=\sigma u$. Then

$$\sigma\cdot[u\ v]=[u\ v]\begin{bmatrix} 0 & -1 \\ 1 & -1 \end{bmatrix}, \qquad \det[u\ v]=\pm 1.$$

Therefore σ is conjugate to either $\tau=\begin{bmatrix} 0 & -1 \\ 1 & -1 \end{bmatrix}$ or $\tau^2=\begin{bmatrix} -1 & 1 \\ -1 & 0 \end{bmatrix}$ in $SL_2(\boldsymbol{Z})$. Thus every elliptic point of order 3 is equivalent to the point $e^{2\pi i/3}$. (τ is not conjugate to τ^2 in $SL_2(\boldsymbol{R})$, see Prop. 1.22.)

For any discrete subgroup Γ of $SL_2(\boldsymbol{R})$, we call F a *fundamental domain for* $\Gamma\backslash\mathfrak{H}$ (or simply *for* Γ), if (i) F is a connected open subset of $\mathfrak{H}$; (ii) no two points of F are equivalent under Γ; (iii) every point of $\mathfrak{H}$ is equivalent to some point of the closure of F under Γ. It can be shown that every Γ has a fundamental domain. An explicit construction of a fundamental domain for a given Γ and its exact shape have been the object of much research. Here we shall not go into the details of this topic, but just find the standard fundamental domain for $\Gamma=SL_2(\boldsymbol{Z})$.

Let $z \in \mathfrak{H}$, and $\sigma = \begin{bmatrix} a & b \\ c & d \end{bmatrix} \in SL_2(\boldsymbol{Z})$. Then $\mathrm{Im}\,(\sigma(z)) = \mathrm{Im}\,(z)/|cz+d|^2$. Now $\{cz+d \mid c \in \boldsymbol{Z},\ d \in \boldsymbol{Z}\}$ is a lattice in $\boldsymbol{C}$. Therefore $\mathrm{Min}\,|cz+d|$, for $(c, d) \neq (0, 0)$ with $c \in \boldsymbol{Z}$, $d \in \boldsymbol{Z}$, exists. Thus, for a given z, $\mathrm{Max}_{\sigma \in \Gamma}\,\mathrm{Im}\,(\sigma(z))$ exists. If σ is such that $\mathrm{Im}\,(\sigma(z))$ is maximum, and $w = \sigma(z) = x+iy$, $\gamma = \begin{bmatrix} 0 & 1 \\ -1 & 0 \end{bmatrix}$, then

$$\mathrm{Im}\,(\gamma\sigma(z)) = \mathrm{Im}\,(-1/w) = y/|w|^2 \leqq y\,,$$

hence $|w| \geqq 1$. If $\tau = \begin{bmatrix} 1 & 1 \\ 0 & 1 \end{bmatrix}$, we have $\mathrm{Im}\,(\tau^h\sigma(z)) = \mathrm{Im}\,(\sigma(z))$ for every $h \in \boldsymbol{Z}$, hence $|\tau^h\sigma(z)| \geqq 1$. Choosing a suitable h, we see that z is equivalent to a point of the region

$$\{w \in \boldsymbol{C} \mid -1/2 \leqq \mathrm{Re}\,(w) \leqq 1/2,\ |w| \geqq 1\}\,.$$

Let us show that the interior F of this set is a fundamental domain for $SL_2(\boldsymbol{Z})$. Let z and z' be distinct points of F. Assume that $z' = \sigma(z)$ with $\sigma = \begin{bmatrix} a & b \\ c & d \end{bmatrix} \in \Gamma$. We may assume that $\mathrm{Im}\,(z) \leqq \mathrm{Im}\,(z') = \mathrm{Im}\,(z)/|cz+d|^2$. Then

$$(*) \qquad |c| \cdot \mathrm{Im}\,(z) \leqq |cz+d| \leqq 1\,.$$

If $c=0$, then $a=d=\pm 1$, hence $z' = z \pm b$, which is impossible. Therefore $c \neq 0$. Looking at the shape of F, we observe that $\mathrm{Im}\,(z) > \sqrt{3}/2$, hence by $(*)$, $|c|=1$. Then from $(*)$ we obtain $|z \pm d| \leqq 1$. But if $z \in F$ and $|d| \geqq 1$, we have $|z+d| > 1$. Therefore we must have $d=0$, so that $|z| \leqq 1$. This contradicts that $z \in F$. Thus we have proved that F is a fundamental domain for Γ.

It can easily be verified that the set

$$F' = F \cup \{z \in \boldsymbol{C} \mid |z| \geqq 1,\ \mathrm{Re}\,(z) = -1/2\} \cup \{z \in \boldsymbol{C} \mid |z| = 1,\ -1/2 \leqq \mathrm{Re}\,(z) \leqq 0\}$$

is a set of representatives for $\mathfrak{H}$ modulo Γ. It follows that $\Gamma\backslash\mathfrak{H}^* = (\Gamma\backslash\mathfrak{H}) \cup \{\infty\}$ is compact. By Prop. 1.31, *$\Gamma'\backslash\mathfrak{H}^*$ is compact if Γ' is a discrete subgroup of $SL_2(\boldsymbol{R})$ commensurable with $SL_2(\boldsymbol{Z})$.*

EXERCISE 1.34. Give another proof for the results about the elliptic points of $SL_2(\boldsymbol{Z})$ by determining such points belonging to F'.

The modular group $SL_2(\boldsymbol{Z})$ is generated by two elements $\sigma = \begin{bmatrix} 1 & 1 \\ 0 & 1 \end{bmatrix}$ and $\tau = \begin{bmatrix} 0 & -1 \\ 1 & 0 \end{bmatrix}$. To show this, let T be the subgroup of $SL_2(\boldsymbol{Z})$ generated by σ and τ. Then $-1 = \tau^2 \in T$. Observe that every element of $SL_2(\boldsymbol{Z})$ of the form $\begin{bmatrix} * & * \\ 0 & * \end{bmatrix}$ is contained in T, and if $\begin{bmatrix} a & b \\ c & d \end{bmatrix} \in T$, then $\begin{bmatrix} -c & -d \\ a & b \end{bmatrix} = \tau \cdot \begin{bmatrix} a & b \\ c & d \end{bmatrix} \in T$. Suppose $T \neq SL_2(\boldsymbol{Z})$, and take an element $\begin{bmatrix} a & b \\ c & d \end{bmatrix}$ of $SL_2(\boldsymbol{Z}) - T$ so that $\mathrm{Min}\,(|a|, |c|)$ is the smallest among such elements. We may assume

$|a|\geqq|c|>0$. Take integers q and r so that $a=cq+r$ and $0\leqq r<|c|$. Then $\sigma^{-q}\begin{bmatrix} a & b \\ c & d \end{bmatrix}=\begin{bmatrix} r & * \\ c & * \end{bmatrix}\in T$, and $r=\mathrm{Min}(r, |c|)<|c|=\mathrm{Min}(|a|, |c|)$, which is a contradiction.

EXERCISE 1.35. Let $\bar{F}$ denote the closure of F, and A the subgroup of $SL_2(\boldsymbol{Z})$ generated by the elements α such that $\alpha(\bar{F})\cap\bar{F}\neq\emptyset$. Let U be the union of $\gamma(\bar{F})$ for all $\gamma\in A$. Using the connectedness of $\mathfrak{H}$, show that $U=\mathfrak{H}$, $A=SL_2(\boldsymbol{Z})$, and $SL_2(\boldsymbol{Z})$ is generated by σ and τ. Observe that this method is applicable to any Γ for which a fundamental domain is (explicitly) given.

1.5. The quotient $\Gamma\backslash\mathfrak{H}^*$ as a Riemann surface

Throughout this section, Γ will denote a discrete subgroup of $SL_2(\boldsymbol{R})$, and $\mathfrak{H}^*$ the union of $\mathfrak{H}$ and the cusps of Γ. Recall the main result of § 1.3 which asserts that $\Gamma\backslash\mathfrak{H}^*$ is a Hausdorff space.

By a *Riemann surface*, we shall mean, as usual, a one-dimensional connected complex analytic manifold. More specifically, a Riemann surface is a connected Hausdorff space $\mathfrak{W}$ on which there is defined a "complex structure" S with the following properties:

(1) *S is a collection of pairs (U_α, p_α) with α in a set A of indices, where $\{U_\alpha\}_{\alpha\in A}$ is an open covering of $\mathfrak{W}$, and p_α is a homeomorphism of U_α onto an open subset of $\boldsymbol{C}$.*

(2) *If $U_\alpha\cap U_\beta\neq\emptyset$, the map*

$$p_\beta\circ p_\alpha^{-1}:\ p_\alpha(U_\alpha\cap U_\beta)\to p_\beta(U_\alpha\cap U_\beta)$$

is holomorphic.

(3) *S is maximal under the conditions* (1) *and* (2).

The map p_α is often called a *local parameter* at a point contained in U_α. Requirement (3) is not essential, since given any S satisfying (1) and (2), there exists a unique complex structure S' containing S. In fact, S' is given as the set of all pairs (V, q) formed by an open subset V of $\mathfrak{W}$ and a homeomorphism q of V onto an open subset of $\boldsymbol{C}$ such that $p_\alpha\circ q^{-1}$ and $q\circ p_\alpha^{-1}$ are holomorphic whenever $V\cap U_\alpha\neq\emptyset$.

Let us now define a complex structure on $\Gamma\backslash\mathfrak{H}^*$. Denote by φ the natural projection map of $\mathfrak{H}^*$ to $\Gamma\backslash\mathfrak{H}^*$. For each $v\in\mathfrak{H}^*$, put

$$\Gamma_v=\{\gamma\in\Gamma\mid\gamma(v)=v\}.$$

By Prop. 1.7 and Lemma 1.26, there exists an open neighborhood U of v such that

$$\Gamma_v=\{\gamma\in\Gamma\mid\gamma(U)\cap U\neq\emptyset\}.$$

Then we have a natural injection $\Gamma_v\backslash U \to \Gamma\backslash\mathfrak{H}^*$, and $\Gamma_v\backslash U$ is an open neighborhood of $\varphi(v)$ in $\Gamma\backslash\mathfrak{H}^*$. If v is neither an elliptic point nor a cusp, Γ_v contains only 1 and possibly -1, so that the map $\varphi: U \to \Gamma_v\backslash U$ is a homeomorphism. We take $(\Gamma_v\backslash U, \varphi^{-1})$ as a member of the complex structure of $\Gamma\backslash\mathfrak{H}^*$.

Next assume that v is an elliptic point, and denote by $\bar{\Gamma}_v$ the transformation group $(\Gamma_v \cdot \{\pm 1\})/\{\pm 1\}$. Let λ be a holomorphic isomorphism of $\mathfrak{H}$ onto the unit disc D such that $\lambda(v) = 0$. If $\bar{\Gamma}_v$ is of order n, then $\lambda\bar{\Gamma}_v\lambda^{-1}$ consists of the transformations

$$w \mapsto \zeta^k w, \qquad k = 0, 1, \cdots, n-1, \qquad \zeta = e^{2\pi i/n}.$$

Then we can define a map $p: \Gamma_v\backslash U \to \boldsymbol{C}$ by $p(\varphi(z)) = \lambda(z)^n$. We see that p is a homeomorphism onto an open subset of $\boldsymbol{C}$. Thus we include $(\Gamma_v\backslash U, p)$ in our complex structure.

Let s be a cusp of Γ, and let ρ be an element of $SL_2(\boldsymbol{R})$ such that $\rho(s) = \infty$. Then

$$\rho\Gamma_s\rho^{-1} \cdot \{\pm 1\} = \left\{ \pm \begin{bmatrix} 1 & h \\ 0 & 1 \end{bmatrix}^m \,\middle|\, m \in \boldsymbol{Z} \right\}$$

with a *positive* number h. Then we can define a homeomorphism p of $\Gamma_s\backslash U$ into an open subset of $\boldsymbol{C}$ by $p(\varphi(z)) = \exp[2\pi i \rho(z)/h]$, and include $(\Gamma_s\backslash U, p)$ in our complex structure.

It is now easy to check the condition (2) for our complex structure. Thus we have been able to make $\Gamma\backslash\mathfrak{H}^*$ a Riemann surface. By abuse of language, we sometimes call a point of $\Gamma\backslash\mathfrak{H}^*$ an *elliptic point* or a *cusp*, if it corresponds to an elliptic point or a cusp on $\mathfrak{H}^*$ with respect to Γ.

EXERCISE 1.36. Let Γ' be a subgroup of Γ of finite index. Prove that the natural map of $\Gamma'\backslash\mathfrak{H}^*$ to $\Gamma\backslash\mathfrak{H}^*$ is holomorphic.

Let us now recall some elementary properties of the homology groups of a compact Riemann surface $\mathfrak{W}$. If $H_i(\mathfrak{W}, \boldsymbol{Z})$ denotes the i-dimensional homology group of $\mathfrak{W}$ with coefficients in $\boldsymbol{Z}$, we have:

$$H_0(\mathfrak{W}, \boldsymbol{Z}) \cong \boldsymbol{Z},$$
$$H_1(\mathfrak{W}, \boldsymbol{Z}) \cong \boldsymbol{Z}^{2g},$$
$$H_2(\mathfrak{W}, \boldsymbol{Z}) \cong \boldsymbol{Z},$$
$$H_p(\mathfrak{W}, \boldsymbol{Z}) = 0 \qquad \text{for} \quad p > 2.$$

The non-negative integer g is called the *genus* of $\mathfrak{W}$. The *Euler characteristic* χ of $\mathfrak{W}$ is defined by

$$\chi = \textstyle\sum_{p=0}^{2} (-1)^p \dim H_p(\mathfrak{W}, \boldsymbol{Z}) = 2 - 2g.$$

If we take a triangulation of $\mathfrak{W}$ and let c_p denote the number of p-simplexes, then $\chi=c_0-c_1+c_2$.

Let $\mathfrak{W}$ and $\mathfrak{W}'$ be two compact Riemann surfaces, and $f:\mathfrak{W}'\to\mathfrak{W}$ a holomorphic mapping. Then f is either constant or surjective. Suppose that f is surjective. Then $(\mathfrak{W}', f)$ is called a *covering* of $\mathfrak{W}$. If $z_0\in\mathfrak{W}'$, $w_0=f(z_0)$, and if u and t are local parameters at z_0 and w_0, respectively, which map z_0 and w_0 to the origin, then we can express f in the form

$$t(f(z))=a_e u(z)^e+a_{e+1}u(z)^{e+1}+\cdots, \qquad a_e\neq 0$$

in a neighborhood of z_0, with a positive integer e. The integer e is independent of the choice of u and t, and called the *ramification index* of the covering $(\mathfrak{W}', f)$ at z_0. There are only finitely many, say h, inverse images of w_0 by f. If $e_1, \cdots, e_h$ are their respective ramification indices, the number

$$n=e_1+\cdots+e_h$$

depends only on $\mathfrak{W}, \mathfrak{W}', f$, and is independent of w_0. We call n the *degree* of the covering. It is known that the number of ramified points (i. e., those z_0 for which $e>1$) is finite. If g and g' are the genera of $\mathfrak{W}$ and $\mathfrak{W}'$, respectively, then these integers are connected by the *Hurwitz formula*

$$2g'-2=n(2g-2)+\textstyle\sum_{z\in\mathfrak{W}'}(e_z-1)\,, \tag{1.5.1}$$

where e_z is the ramification index at z. This can be proved as follows. Triangulate $\mathfrak{W}$ so that among the 0-simplexes are included all points any of whose inverse images under f is ramified, and so that each 1-simplex lies within a single parametric disc. Taking the inverse image of this triangulation under f, we get a triangulation of $\mathfrak{W}'$. If c_0, c_1, c_2 and c_0', c_1', c_2' denote the number of 0-, 1-, 2-simplexes in these triangulations, then one has

$$2-2g=c_0-c_1+c_2\,, \qquad 2-2g'=c_0'-c_1'+c_2'\,.$$

Observe that $c_2'=nc_2$, $c_1'=nc_1$, $c_0'=nc_0-\sum_{z\in\mathfrak{W}'}(e_z-1)$. The formula now follows immediately.

By a *Fuchsian group of the first kind*, we shall mean a discrete subgroup Γ of $SL_2(\boldsymbol{R})$ (or of $SL_2(\boldsymbol{R})/\{\pm 1\}$) such that $\Gamma\backslash\mathfrak{H}^*$ is compact. Endowed with the complex structure defined above, $\Gamma\backslash\mathfrak{H}^*$ becomes a compact Riemann surface. If Γ' is a subgroup of Γ of finite index, the natural map $\Gamma'\backslash\mathfrak{H}^*\to\Gamma\backslash\mathfrak{H}$ defines a covering in the above sense. Let $\bar{\Gamma}$ and $\bar{\Gamma}'$ denote the images of Γ and Γ' by the natural map

$$SL_2(\boldsymbol{R})\to SL_2(\boldsymbol{R})/\{\pm 1\}\,.$$

Then the degree of the covering is exactly $[\bar{\Gamma}:\bar{\Gamma}']$.

For every $z\in\mathfrak{H}^*$, put

$$\bar{\Gamma}_z = \{\gamma \in \bar{\Gamma} \mid \gamma(z) = z\}, \qquad \bar{\Gamma}'_z = \bar{\Gamma}_z \cap \bar{\Gamma}'.$$

Consider a commutative diagram

$$\begin{array}{ccc} \mathfrak{H}^* & \xrightarrow{\text{identity}} & \mathfrak{H}^* \\ \varphi' \downarrow & & \downarrow \varphi \\ \Gamma' \backslash \mathfrak{H}^* & \xrightarrow{f} & \Gamma \backslash \mathfrak{H}^* \end{array}$$

where each map is a natural projection. Let $z \in \mathfrak{H}^*$, $p = \varphi(z)$, and $f^{-1}(p) = \{q_1, \cdots, q_h\}$. Choose points w_k of $\mathfrak{H}^*$ so that $q_k = \varphi'(w_k)$.

Proposition 1.37. *The ramification index e_k of f at q_k is $[\bar{\Gamma}_{w_k} : \bar{\Gamma}'_{w_k}]$. Moreover, if $w_k = \sigma_k(z)$ with $\sigma_k \in \bar{\Gamma}$, then $e_k = [\bar{\Gamma}_z : \sigma_k^{-1}\bar{\Gamma}'\sigma_k \cap \bar{\Gamma}_z]$, and $\bar{\Gamma} = \bigcup_{k=1}^{h} \bar{\Gamma}'\sigma_k\bar{\Gamma}_z$ (disjoint). Especially if $\bar{\Gamma}'$ is a normal subgroup of $\bar{\Gamma}$, then $e_1 = \cdots = e_h$, and $[\bar{\Gamma} : \bar{\Gamma}'] = e_1 h$.*

Proof. The first assertion follows immediately from the definition of ramification index. Since $\bar{\Gamma}_{w_k} = \sigma_k \bar{\Gamma}_z \sigma_k^{-1}$ and $\bar{\Gamma}'_{w_k} = \bar{\Gamma}' \cap \sigma_k \bar{\Gamma}_z \sigma_k^{-1}$, we obtain the second assertion. Let $\gamma \in \bar{\Gamma}$. Since $f(\varphi'(\gamma(z))) = \varphi(\gamma(z)) = \varphi(z) = p$, we have $\varphi'(\gamma(z)) = q_k$ for some k, hence $\varphi'(\gamma(z)) = \varphi'(\sigma_k(z))$. Therefore, $\gamma(z) = \delta\sigma_k(z)$ for some $\delta \in \bar{\Gamma}'$. Then we have $\gamma^{-1}\delta\sigma_k \in \bar{\Gamma}_z$, so that $\gamma \in \bar{\Gamma}'\sigma_k\bar{\Gamma}_z$. This shows that $\bar{\Gamma} = \bigcup_{k=1}^{h} \bar{\Gamma}'\sigma_k\bar{\Gamma}_z$. If $\varepsilon \in \bar{\Gamma}'\sigma_k\bar{\Gamma}_z$, we have $\varphi'(\varepsilon(z)) = \varphi'(w_k) = q_k$. Therefore the union is disjoint. The remaining part of our proposition is obvious.

1.6. Congruence subgroups of $SL_2(Z)$

The shape of the fundamental domain for $SL_2(\boldsymbol{Z})\backslash\mathfrak{H}^*$, given in § 1.4, tells us that the Riemann surface $SL_2(\boldsymbol{Z})\backslash\mathfrak{H}^*$ is isomorphic to the Riemann sphere. Let us now study $\Gamma\backslash\mathfrak{H}^*$ for certain subgroups Γ of $SL_2(\boldsymbol{Z})$. In this section $\mathfrak{H}^*$ means $\mathfrak{H} \cup \boldsymbol{Q} \cup \{\infty\}$.

For every positive integer N, put

$$(1.6.1) \quad \Gamma_N = \Gamma(N) = \{\gamma \in SL_2(\boldsymbol{Z}) \mid \gamma \equiv 1_2 \mod (N)\}$$

$$= \left\{ \begin{bmatrix} a & b \\ c & d \end{bmatrix} \in SL_2(\boldsymbol{Z}) \,\middle|\, a \equiv d \equiv 1,\ b \equiv c \equiv 0 \mod N\boldsymbol{Z} \right\}.$$

Then $\Gamma(N)$ is a normal subgroup of $SL_2(\boldsymbol{Z})$, and called *the principal congruence subgroup* (of $SL_2(\boldsymbol{Z})$) *of level* N. In general, a subgroup of $SL_2(\boldsymbol{Z})$ is called a *congruence subgroup* of $SL_2(\boldsymbol{Z})$ if it contains $\Gamma(N)$ for some N.

Lemma 1.38. *If $f : SL_2(\boldsymbol{Z}) \to SL_2(\boldsymbol{Z}/N\boldsymbol{Z})$ is defined by $f(\alpha) = \alpha \mod (N)$, then the sequence*

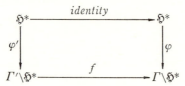

$$1 \longrightarrow \Gamma(N) \longrightarrow SL_2(\mathbf{Z}) \xrightarrow{f} SL_2(\mathbf{Z}/N\mathbf{Z}) \longrightarrow 1$$

is exact.

PROOF. The only non-trivial point is the surjectivity of f. We shall prove more generally that the map $SL_m(\mathbf{Z}) \to SL_m(\mathbf{Z}/N\mathbf{Z})$ is surjective for any positive integer m, i. e., if $A \in M_m(\mathbf{Z})$ and $\det(A) \equiv 1 \bmod (N)$, then $A \equiv B \bmod (N)$ for some $B \in SL_m(\mathbf{Z})$. If $m=1$, this is obvious. Therefore assume the assertion to be true for $m-1$, and $m>1$. Now for such an A, by elementary divisor theory, we can find two elements U and V of $SL_m(\mathbf{Z})$ such that UAV is a diagonal matrix. Let $a_1, \cdots, a_m$ be the diagonal elements of UAV, and $b = a_2 \cdots a_m$. Put

$$W = \begin{bmatrix} b & 1 & & & \\ b-1 & 1 & & & \\ & & 1 & & \\ & & & \ddots & \\ & & & & 1 \end{bmatrix}, \quad X = \begin{bmatrix} 1 & -a_2 & & & \\ 0 & 1 & & & \\ & & 1 & & \\ & & & \ddots & \\ & & & & 1 \end{bmatrix}, \quad A' = \begin{bmatrix} 1 & 0 & & & \\ 1-a_1 & a_1a_2 & & & \\ & & a_3 & & \\ & & & \ddots & \\ & & & & a_m \end{bmatrix}.$$

Since $a_1 b = \det(A) \equiv 1 \bmod (N)$, we see that $WUAVX \equiv A' \bmod (N)$. By the induction assumption, there exists an element C of $SL_{m-1}(\mathbf{Z})$ such that

$$C \equiv \begin{bmatrix} a_1a_2 & & & \\ & a_3 & & \\ & & \ddots & \\ & & & a_m \end{bmatrix} \quad \bmod (N).$$

Put

$$B = U^{-1}W^{-1} \left[\begin{array}{c|c} 1 & 0 \\ \hline 1-a_1 & \\ & C \\ 0 & \end{array}\right] X^{-1}V^{-1}.$$

Then B has the required property.

If $N = \prod_p p^e$ is the decomposition of N into the product of powers of distinct primes p, we see that

$$\mathbf{Z}/N\mathbf{Z} \cong \textstyle\prod_p (\mathbf{Z}/p^e\mathbf{Z}),$$

$$GL_2(\mathbf{Z}/N\mathbf{Z}) \cong \textstyle\prod_p GL_2(\mathbf{Z}/p^e\mathbf{Z}),$$

$$SL_2(\mathbf{Z}/N\mathbf{Z}) \cong \textstyle\prod_p SL_2(\mathbf{Z}/p^e\mathbf{Z}).$$

Now consider an exact sequence

$$1 \longrightarrow X \longrightarrow GL_2(\mathbf{Z}/p^e\mathbf{Z}) \longrightarrow GL_2(\mathbf{Z}/p\mathbf{Z}) \longrightarrow 1.$$

Since X consists of the elements of $M_2(\boldsymbol{Z}/p^e\boldsymbol{Z})$ which are congruent to 1_2 modulo (p), the order of X is $p^{4(e-1)}$. It is well known that the order of $GL_2(\boldsymbol{Z}/p\boldsymbol{Z})$ is $(p^2-1)(p^2-p)$. Therefore,

$$\begin{aligned}\text{the order of } GL_2(\boldsymbol{Z}/p^e\boldsymbol{Z}) &= p^{4(e-1)}(p^2-p)(p^2-1)\\ &= p^{4e}(1-p^{-1})(1-p^{-2}),\\ \text{the order of } SL_2(\boldsymbol{Z}/p^e\boldsymbol{Z}) &= p^{3e}(1-p^{-2}).\end{aligned}$$

By Lemma 1.38, we obtain

$$[\Gamma(1):\Gamma(N)] = N^3\cdot\textstyle\prod_{p|N}(1-p^{-2}).$$

Since $-1_2\in\Gamma(2)$ and $-1_2\notin\Gamma(N)$ for $N>2$, we find

$$(1.6.2)\qquad [\bar{\Gamma}(1):\bar{\Gamma}(N)] = \begin{cases}(N^3/2)\cdot\prod_{p|N}(1-p^{-2}) & \text{if } N>2,\\ 6 & \text{if } N=2.\end{cases}$$

PROPOSITION 1.39. *If $N>1$, $\Gamma(N)$ has no elliptic element.*

PROOF. In § 1.4, we have seen that every elliptic element of $\Gamma(1)$ is conjugate to one of the following elements:

$$\pm\begin{bmatrix}0 & -1\\ 1 & 0\end{bmatrix},\quad \pm\begin{bmatrix}0 & -1\\ 1 & -1\end{bmatrix},\quad \pm\begin{bmatrix}-1 & 1\\ -1 & 0\end{bmatrix}.$$

None of these is congruent to 1_2 modulo (N) if $N>1$. Since $\Gamma(N)$ is a normal subgroup of $\Gamma(1)$, we obtain our proposition.

Let us now find the ramification indices of the covering

$$\Gamma(N)\backslash\mathfrak{H}^*\longrightarrow\Gamma(1)\backslash\mathfrak{H}^*.$$

Let φ_N denote the projection map of $\mathfrak{H}^*$ to $\Gamma(N)\backslash\mathfrak{H}^*$. By Prop. 1.38, the ramification index at $\varphi_N(z)$, for $z\in\mathfrak{H}^*$, is $[\bar{\Gamma}(1)_z:\bar{\Gamma}(N)_z]$. If z is an elliptic point of $\Gamma(1)$, $\bar{\Gamma}(1)_z$ is of order 2 or 3. By the above proposition, $\bar{\Gamma}(N)_z=\{1\}$ if $N>1$. Therefore the ramification index at $\varphi_N(z)$ is 2 or 3 accordingly. Furthermore, putting

$$(1.6.3)\qquad \mu_N=[\bar{\Gamma}(1):\bar{\Gamma}(N)],$$

we see that the number of points on $\Gamma(N)\backslash\mathfrak{H}^*$ lying above $\varphi_1(z)$ is $\mu_N/2$ or $\mu_N/3$ accordingly (if $N>1$).

If s is a cusp, s is $\Gamma(1)$-equivalent to ∞. Now we have

$$\bar{\Gamma}(1)_\infty=\left\{\begin{bmatrix}1 & 1\\ 0 & 1\end{bmatrix}^m \,\middle|\, m\in\boldsymbol{Z}\right\},$$

$$\bar{\Gamma}(N)_\infty=\bar{\Gamma}(N)\cap\bar{\Gamma}(1)_\infty=\left\{\begin{bmatrix}1 & N\\ 0 & 1\end{bmatrix}^m \,\middle|\, m\in\boldsymbol{Z}\right\},$$

so that $[\bar{\Gamma}(1)_\infty : \bar{\Gamma}(N)_\infty] = N$. Therefore $\Gamma(N)$ has exactly μ_N/N inequivalent cusps.

PROPOSITION 1.40. *Let $\bar{\Gamma}'$ be a subgroup of $\bar{\Gamma}(1)$ of index μ, and ν_2, ν_3 the numbers of $\bar{\Gamma}'$-inequivalent elliptic points of order 2, 3, respectively. Further let ν_∞ be the number of $\bar{\Gamma}'$-inequivalent cusps. Then the genus of $\bar{\Gamma}'\backslash\mathfrak{H}^*$ is given by*

$$g = 1 + \frac{\mu}{12} - \frac{\nu_2}{4} - \frac{\nu_3}{3} - \frac{\nu_\infty}{2}.$$

PROOF. Consider the covering $\bar{\Gamma}'\backslash\mathfrak{H}^* \to \bar{\Gamma}(1)\backslash\mathfrak{H}^*$. Let $e_1, \cdots, e_t$ be the ramification indices at the points of $\bar{\Gamma}'\backslash\mathfrak{H}^*$ lying above $\varphi_1(e^{2\pi i/3})$. Then $\mu = e_1 + \cdots + e_t$, and e_i is 1 or 3. The number of i for which $e_i = 1$ is ν_3. If $t = \nu_3 + \nu_3'$, we have $\mu = \nu_3 + 3\nu_3'$, so that $\sum_{i=1}^t (e_i - 1) = \mu - t = 2\nu_3' = 2(\mu - \nu_3)/3$. Similarly, if e_P is the ramification index at a point P of $\bar{\Gamma}'\backslash\mathfrak{H}^*$, we have

$$\sum (e_P - 1) = (\mu - \nu_2)/2 \qquad (P \text{ lying above } \varphi_1(i)),$$

$$\sum (e_P - 1) = \mu - \nu_\infty \qquad (P \text{ lying above } \varphi_1(\infty)).$$

Now we have seen that $\bar{\Gamma}(1)\backslash\mathfrak{H}^*$ is of genus 0. Therefore we obtain our assertion from the Hurwitz formula (1.5.1).

In the case $\bar{\Gamma}' = \bar{\Gamma}(N)$, we have $\nu_2 = \nu_3 = 0$ if $N > 1$, and $\nu_\infty = \mu_N/N$. Thus we obtain the formula for the genus g_N of $\Gamma(N)\backslash\mathfrak{H}^*$:

(1.6.4) $$g_N = 1 + \mu_N \cdot (N-6)/12N \qquad (N > 1).$$

Let us now determine an explicit set of representatives for the cusps modulo $\Gamma(N)$-equivalence.

LEMMA 1.41. *Let a, b, c, d be integers such that $(a, b) = 1$, $(c, d) = 1$, and $\begin{bmatrix} a \\ b \end{bmatrix} \equiv \begin{bmatrix} c \\ d \end{bmatrix} \bmod (N)$. Then there exists an element σ of $\Gamma(N)$ such that $\begin{bmatrix} a \\ b \end{bmatrix} = \sigma \begin{bmatrix} c \\ d \end{bmatrix}$.*

PROOF. (I) Assume $\begin{bmatrix} c \\ d \end{bmatrix} = \begin{bmatrix} 1 \\ 0 \end{bmatrix}$. Then $a \equiv 1 \bmod (N)$. Take integers p and q so that $ap - bq = (1-a)/N$, and put $\sigma = \begin{bmatrix} a & Nq \\ b & 1+Np \end{bmatrix}$. Then σ has the required property.

(II) In the general case, take integers r and s so that $cr + ds = 1$, and put $\tau = \begin{bmatrix} c & -s \\ d & r \end{bmatrix}$. Then $\tau \begin{bmatrix} 1 \\ 0 \end{bmatrix} = \begin{bmatrix} c \\ d \end{bmatrix} \equiv \begin{bmatrix} a \\ b \end{bmatrix} \bmod (N)$, hence $\tau^{-1} \begin{bmatrix} a \\ b \end{bmatrix} \equiv \begin{bmatrix} 1 \\ 0 \end{bmatrix} \bmod (N)$. By the result of (I), we can find an element σ of $\Gamma(N)$ so that $\sigma \begin{bmatrix} 1 \\ 0 \end{bmatrix} = \tau^{-1} \begin{bmatrix} a \\ b \end{bmatrix}$. Then $\tau\sigma\tau^{-1}$ has the required property.

LEMMA 1.42. *Let $s = a/b$ and $s' = c/d$ be cusps of $\Gamma(N)$, with integers*

a, b, c, d such that $(a, b)=1$, $(c, d)=1$. (We understand that $\pm 1/0=\infty$.) Then s and s' are equivalent under $\Gamma(N)$ if and only if $\pm\begin{bmatrix} a \\ b \end{bmatrix} \equiv \begin{bmatrix} c \\ d \end{bmatrix} \bmod (N)$.

PROOF. If $\pm\begin{bmatrix} a \\ b \end{bmatrix} \equiv \begin{bmatrix} c \\ d \end{bmatrix} \bmod (N)$, there exists, by Lemma 1.41, an element $\sigma=\begin{bmatrix} p & q \\ r & s \end{bmatrix}$ of $\Gamma(N)$ such that $\begin{bmatrix} p & q \\ r & s \end{bmatrix}\begin{bmatrix} c \\ d \end{bmatrix}=\pm\begin{bmatrix} a \\ b \end{bmatrix}$. If $bd\neq 0$, we have obviously $\sigma(s')=s$. This is true even if $bd=0$, as can be shown by a simple verification. Conversely, if $\sigma(s')=s$ with $\sigma=\begin{bmatrix} p & q \\ r & s \end{bmatrix}\in\Gamma(N)$, then $a/b=(pc+qd)/(rc+sd)$ again under the assumption $bd\neq 0$. Hence there exists a rational number λ such that $\lambda\begin{bmatrix} a \\ b \end{bmatrix}=\begin{bmatrix} p & q \\ r & s \end{bmatrix}\begin{bmatrix} c \\ d \end{bmatrix}$. Put $\lambda=m/n$ with integers m and n, which are relatively prime. Then $m\begin{bmatrix} a \\ b \end{bmatrix}=n\begin{bmatrix} p & q \\ r & s \end{bmatrix}\begin{bmatrix} c \\ d \end{bmatrix}$. Since $(a, b)=1$ and $(c, d)=1$, we have $m=\pm 1$ and $n=\pm 1$, hence $\lambda=\pm 1$. The verification of the case $bd=0$ is also easy and may therefore be left to the reader.

Thus the $\Gamma(N)$-equivalence classes of cusps are completely determined by Lemma 1.42. For example, if $N=2$, there are three inequivalent cusps, represented by 0, 1, ∞.

Let us now study a family of congruence subgroups of $SL_2(\boldsymbol{Z})$, which are not normal subgroups of $SL_2(\boldsymbol{Z})$. Put, for a positive integer N,

$$A_N=\left\{\begin{bmatrix} a & b \\ c & d \end{bmatrix}\in M_2(\boldsymbol{Z}) \,\middle|\, c\equiv 0 \bmod (N)\right\},$$

$$\Gamma_0(N)=A_N\cap SL_2(\boldsymbol{Z})=\left\{\begin{bmatrix} a & b \\ c & d \end{bmatrix}\in SL_2(\boldsymbol{Z}) \,\middle|\, c\equiv 0 \bmod (N)\right\}. \tag{1.6.5}$$

Then A_N is a subring of $M_2(\boldsymbol{Z})$, and $\Gamma_0(N)$ is a subgroup of $\Gamma(1)$ containing $\Gamma(N)$. We see easily that if $\alpha=\begin{bmatrix} N & 0 \\ 0 & 1 \end{bmatrix}$,

$$\Gamma_0(N)=\alpha^{-1}\Gamma(1)\alpha\cap\Gamma(1). \tag{1.6.6}$$

Note that $-1\in\Gamma_0(N)$. By the map f of Lemma 1.39, $\Gamma_0(N)/\Gamma(N)$ is mapped to the group of all matrices of the form $\begin{bmatrix} a & b \\ 0 & a^{-1} \end{bmatrix}$ in $SL_2(\boldsymbol{Z}/N\boldsymbol{Z})$. This group is clearly of order $N\cdot\varphi(N)$, where φ is Euler's function. Therefore

$$[\Gamma(1):\Gamma_0(N)]=[\bar{\Gamma}(1):\bar{\Gamma}_0(N)]=N\cdot\textstyle\prod_{p\mid N}(1+p^{-1}).$$

This proves the first assertion of

PROPOSITION 1.43. *Let the notation be as in Prop. 1.40. If $\bar{\Gamma}'=\bar{\Gamma}_0(N)$, one has:*

(1) $\mu = N\cdot\prod_{p|N}(1+p^{-1})$.

(2) $\nu_2 = \begin{cases} 0 & \textit{if } N \textit{ is divisible by } 4, \\ \prod_{p|N}\left(1+\left(\frac{-1}{p}\right)\right) & \textit{otherwise}. \end{cases}$

(3) $\nu_3 = \begin{cases} 0 & \textit{if } N \textit{ is divisible by } 9, \\ \prod_{p|N}\left(1+\left(\frac{-3}{p}\right)\right) & \textit{otherwise}. \end{cases}$

(4) $\nu_\infty = \sum_{d|N, d>0}\varphi((d, N/d))$, *where φ is Euler's function.*

Here we understand that $\varphi(1)=1$; $\left(\frac{\quad}{p}\right)$ is the quadratic residue symbol (in the extended sense), so that

$$\left(\frac{-1}{p}\right) = \begin{cases} 0 & \textit{if} \quad p=2, \\ 1 & \textit{if} \quad p\equiv 1 \bmod (4), \\ -1 & \textit{if} \quad p\equiv 3 \bmod (4), \end{cases}$$

$$\left(\frac{-3}{p}\right) = \begin{cases} 0 & \textit{if} \quad p=3, \\ 1 & \textit{if} \quad p\equiv 1 \bmod (3), \\ -1 & \textit{if} \quad p\equiv 2 \bmod (3). \end{cases}$$

PROOF. First we consider all couples $\{c, d\}$ of positive integers satisfying

(*) $\quad (c, d)=1, \quad d|N, \quad 0<c\leqq N/d \quad$ (or c in any set of representatives for $\boldsymbol{Z}$ modulo (N/d)).

For each couple $\{c, d\}$ we take a and b so that $ad-bc=1$, and fix them. Then the elements $\begin{bmatrix} a & b \\ c & d \end{bmatrix}$ for all couples satisfying (*) form a set of representatives for $\Gamma_0(N)\backslash\Gamma(1)$. In fact, it can easily be verified that they are not equivalent under left multiplication by the elements of $\Gamma_0(N)$, and the number of such couples is exactly μ given in (1). Now by Prop. 1.37, ν_∞ is the number of double cosets in $\Gamma_0(N)\backslash\Gamma(1)/\Gamma_s$ for any fixed cusp s. Take s to be 0. Then we see that ν_∞ is the number of couples $\{c, d\}$ satisfying (*) modulo the equivalence $\sim$ defined by

$$\{c, d\}\sim\{c', d'\} \quad \text{if} \quad \begin{bmatrix} * & * \\ c' & d' \end{bmatrix} = \begin{bmatrix} * & * \\ c & d \end{bmatrix}\begin{bmatrix} 1 & 0 \\ m & 1 \end{bmatrix} \quad \text{for some } m\in\boldsymbol{Z}.$$

If we have the last equality, we have $d=d'$, $c'=c+dm$. Therefore, for a fixed d, there are exactly $\varphi((d, N/d))$ inequivalent couples, and hence we obtain (4).

To determine ν_3, denote by S_1 (resp. S_2) the set of all the elliptic elements of $\Gamma(1)$ of order 3 conjugate to $\tau=\begin{bmatrix} 0 & -1 \\ 1 & -1 \end{bmatrix}$ (resp. τ^2) under $\Gamma(1)$ (see § 1.4

and Prop. 1.22). Put $\zeta=e^{2\pi i/3}$, $A=\boldsymbol{Z}[\zeta]$, and

$$L=\boldsymbol{Z}^2=\left\{\begin{bmatrix} x \\ y \end{bmatrix}\,\middle|\, x, y\in \boldsymbol{Z}\right\}, \qquad L_N=\left\{\begin{bmatrix} x \\ Ny \end{bmatrix}\in L\,\middle|\, x, y\in \boldsymbol{Z}\right\}.$$

For every $\sigma\in S_1\cup S_2$, consider L as a $\boldsymbol{Z}[\sigma]$-module. Since $\boldsymbol{Z}[\sigma]$ is isomorphic to A, and A is a principal ideal domain, there exists a $\boldsymbol{Z}$-linear isomorphism f of A to L such that $f(\zeta x)=\sigma f(x)$ for all $x\in A$. Now let T be the set of all $\boldsymbol{Z}$-linear isomorphisms of A to L. Then T is a disjoint union of the subsets

$$T_i=\{f\in T \mid f(\zeta x)=\sigma f(x) \text{ with } \sigma\in S_i\} \qquad (i=1, 2).$$

If $\alpha\in M_2(\boldsymbol{Z})$ and $\det(\alpha)=-1$, then $f\in T_1 \Leftrightarrow \alpha f\in T_2$. For any $f\in T_1$, put $J=f^{-1}(L_N)$. Since

$$\Gamma_0(N)=\{\gamma\in\Gamma(1) \mid \gamma L_N=L_N\},$$

we see that the element σ satisfying $f(\zeta x)=\sigma f(x)$ belongs to $\Gamma_0(N)$ if and only if J is an ideal of A. Moreover, since A/J is isomorphic to $\boldsymbol{Z}/N\boldsymbol{Z}$, we see that

(i) $N_{K/\boldsymbol{Q}}(J)=N$, where $K=\boldsymbol{Q}(\zeta)$,

(ii) J is not divisible by any positive integer other than 1.

Conversely, if J is such an ideal of A, we can find an element f of T such that $f(J)=L_N$. We may assume that $f\in T_1$ by changing f for εf with $\varepsilon=\begin{bmatrix} 1 & 0 \\ 0 & -1 \end{bmatrix}$ if necessary. Then we obtain an element σ of $S_1\cap\Gamma_0(N)$ by $f(\zeta x)=\sigma f(x)$. Let us now show that the correspondence between J and the conjugacy class of σ in $\Gamma_0(N)$ is one-to-one. Let $f\in T_1$, $f'\in T_1$, $f(\zeta x)=\sigma f(x)$, $f'(\zeta x)=\sigma' f'(x)$, and $f(J)=f'(J)=L_N$ with the same ideal J. We can find an element γ of $\Gamma(1)$ so that $f'=\gamma f$. Then $\sigma=\gamma^{-1}\sigma'\gamma$, and $\gamma L_N=L_N$, so that σ is conjugate to σ' in $\Gamma_0(N)$. Conversely, let $f(J)=L_N$, $f'(J')=L_N$, $f(\zeta x)=\sigma f(x)$, $f'(\zeta x)=\gamma^{-1}\sigma\gamma f'(x)$ with $f\in T_1$, $f'\in T_1$, and $\gamma\in\Gamma_0(N)$. Put $h=f^{-1}\gamma f'$. Then h is a $\boldsymbol{Z}$-linear automorphism of the module A, and $h(\zeta x)=\zeta h(x)$. Put $\lambda=h(1)$. Then $h(a+b\zeta)=(a+b\zeta)\lambda$ for $a, b\in\boldsymbol{Z}$. It follows that $\lambda\in A^{\times}$. Therefore $J=f^{-1}(\gamma L_N)=f^{-1}(\gamma f'(J'))=\lambda J'=J'$. Thus we have proved that ν_3 is the number of all ideals J of $\boldsymbol{Z}[\zeta]$ satisfying the above (i) and (ii). Considering the prime ideal decomposition of J, we see that ν_3 is the number given in (3). We obtain ν_2 by applying the same argument to $(-1)^{1/2}$ and $\begin{bmatrix} 0 & -1 \\ 1 & 0 \end{bmatrix}$ instead of ζ and $\begin{bmatrix} 0 & -1 \\ 1 & -1 \end{bmatrix}$.

As a special case, if N is a prime, 0 and ∞ represent the inequivalent cusps of $\Gamma_0(N)$; the degree of the covering

$$\Gamma_0(N)\backslash\mathfrak{H}^* \longrightarrow \Gamma(1)\backslash\mathfrak{H}^*$$

is $[\Gamma(1):\Gamma_0(N)]=N+1$; the ramification indices at 0 and ∞ are N and 1, respectively.

Define an element τ of $SL_2(\boldsymbol{R})$ by

$$\tau=\begin{bmatrix} 0 & -\sqrt{N}^{-1} \\ \sqrt{N} & 0 \end{bmatrix}=\sqrt{N}^{-1}\begin{bmatrix} 0 & -1 \\ N & 0 \end{bmatrix}.$$

Then $\tau^2=-1$, and $\tau^{-1}\Gamma_0(N)\tau=\Gamma_0(N)$. Therefore we can form a group $\Gamma_0^*(N)$ by

$$\Gamma_0^*(N)=\Gamma_0(N)\cup\Gamma_0(N)\tau\,.$$

Then $\Gamma_0^*(N)$ is a discrete subgroup of $SL_2(\boldsymbol{R})$, which is commensurable with $SL_2(\boldsymbol{Z})$, but not necessarily conjugate to a subgroup of $SL_2(\boldsymbol{Z})$.

So far, all the examples of Γ are commensurable with $SL_2(\boldsymbol{Z})$, so that $\Gamma\backslash\mathfrak{H}$ is not compact. There are of course many Γ for which $\Gamma\backslash\mathfrak{H}$ is compact, since it is a classical fact that every compact Riemann surface of genus >1 is holomorphically isomorphic to $\Gamma\backslash\mathfrak{H}$ with a Fuchsian group Γ with neither parabolic nor elliptic elements. We shall discuss in § 9.2 some interesting (and actually important) Fuchsian groups Γ with compact $\Gamma\backslash\mathfrak{H}$, which are defined in a certain arithmetical way.

EXERCISE 1.44. Let Γ' be a discrete subgroup of $SL_2(\boldsymbol{R})$ such that $\Gamma'\backslash\mathfrak{H}^*$ is compact, and Γ a subgroup of Γ' of index m. Suppose that ∞ is the only cusp of Γ modulo Γ-equivalence, and Γ_∞ is generated by $\begin{bmatrix} 1 & 1 \\ 0 & 1 \end{bmatrix}$. Prove that Γ'_∞ is generated by $\begin{bmatrix} 1 & 1/m \\ 0 & 1 \end{bmatrix}$.

EXERCISE 1.45. Use Ex. 1.44 to prove that no discrete subgroup of $SL_2(\boldsymbol{R})$ contains properly $\Gamma_0^*(N)$, if N is a prime or $=1$. (Observe that, if Γ and Γ' are as above, Γ' is generated by Γ and $\begin{bmatrix} 1 & 1/m \\ 0 & 1 \end{bmatrix}$.)

EXERCISE 1.46. Prove that no conjugate of $\Gamma_0^*(N)$ in $SL_2(\boldsymbol{R})$ is contained in $SL_2(\boldsymbol{Z})$, if N is a prime.

CHAPTER 2
AUTOMORPHIC FORMS AND FUNCTIONS

2.1. Definition of automorphic forms and functions

Hereafter, till the end of § 2.6, Γ will always mean a Fuchsian group of the first kind. As we have seen, $\Gamma\backslash\mathfrak{H}^*$ is a compact Riemann surface. It is well known that the set of all meromorphic functions on a compact Riemann surface form a field of algebraic functions of one variable, with the constant field $\boldsymbol{C}$. Now an *automorphic function on* $\mathfrak{H}$ *with respect to* Γ (or simply, a Γ*-automorphic function*) is a function f on $\mathfrak{H}$ of the form $f=g\circ\varphi$, with a meromorphic function g on $\Gamma\backslash\mathfrak{H}^*$, where φ is the natural map of $\mathfrak{H}^*$ to $\Gamma\backslash\mathfrak{H}^*$. A more general notion, an *automorphic form*, can be defined as follows.

For every $\sigma=\begin{bmatrix} a & b \\ c & d \end{bmatrix}\in GL_2(\boldsymbol{R})$, and $z\in\boldsymbol{C}$, we put

$$j(\sigma, z)=cz+d\,.$$

Then, as was shown in § 1.2,

$$j(\sigma\tau, z)=j(\sigma, \tau(z))\cdot j(\tau, z)\,,$$

$$\frac{d}{dz}\sigma(z)=\det(\sigma)\cdot j(\sigma, z)^{-2}\,.$$

For every integer k, $\sigma\in GL_2^+(\boldsymbol{R})$, and a function f on $\mathfrak{H}$, we write

$$f\,|\,[\sigma]_k=\det(\sigma)^{k/2}\cdot f(\sigma(z))\cdot j(\sigma, z)^{-k}\,.$$

Then it is easily verified that

$$f\,|\,[\sigma\tau]_k=(f\,|\,[\sigma]_k)\,|\,[\tau]_k\,.$$

L :t us insert here one word of caution: Two matrices σ and $-\sigma$ induce the same transformation on $\mathfrak{H}$. However, if k is odd,

$$j(-\sigma, z)^k=-j(\sigma, z)^k\,,$$

hence $f\,|\,[-\sigma]_k=-f\,|\,[\sigma]_k$. If k is even, the action of $[-\sigma]_k$ is the same as $[\sigma]_k$.

DEFINITION 2.1. Let k be an integer. A $\boldsymbol{C}$-valued function f on $\mathfrak{H}$ is called an *automorphic form of weight* k *with respect to* Γ (or simply, a Γ*-automorphic form of weight* k), if f satisfies the following three conditions:

(i) *f is meromorphic on* $\mathfrak{H}$;

(ii) $f\,|\,[\gamma]_k=f$ *for all* $\gamma\in\Gamma$;

(iii) *f is meromorphic at every cusp of Γ.*

The precise meaning of the last condition is as follows. First, the condition must be disregarded if Γ has no cusp. Suppose Γ has a cusp s. Take an element ρ of $SL_2(\boldsymbol{R})$ so that $\rho(s)=\infty$. Putting $\Gamma_s=\{\gamma\in\Gamma \mid \gamma(s)=s\}$, we have

$$\rho\Gamma_s\rho^{-1}\cdot\{\pm 1\}=\left\{\pm\begin{bmatrix}1 & h\\ 0 & 1\end{bmatrix}^m \,\middle|\, m\in\boldsymbol{Z}\right\}$$

with a *positive* real number h. In view of (ii), $f\mid[\rho^{-1}]_k$ is invariant under $[\sigma]_k$ for every $\sigma\in\rho\Gamma_s\rho^{-1}$.

CASE I: k even. Since $f\mid[\rho^{-1}]_k$ is invariant under $z\mapsto z+h$, there exists a meromorphic function $\Phi(q)$ in the domain $0<|q|<r$, with a positive real number r, such that

$$f\mid[\rho^{-1}]_k=\Phi(e^{2\pi iz/h}).$$

Then the condition (iii) means that *Φ is meromorphic at $q=0$.*

CASE II: k odd. If Γ contains -1, the condition (ii) implies $f=-f$, so that there is no automorphic form of weight k other than 0. Therefore, we assume that $-1\notin\Gamma$. Then $\rho\Gamma_s\rho^{-1}$ is generated either by $\begin{bmatrix}1 & h\\ 0 & 1\end{bmatrix}$, or by $-\begin{bmatrix}1 & h\\ 0 & 1\end{bmatrix}$. We say that s is *regular* or *irregular*, accordingly. If s is regular, the condition (iii) should be understood in the same way as in Case I. If s is irregular, $g(z)=f\mid[\rho^{-1}]_k$ satisfies $g(z+h)=-g$, hence $g(z+2h)=g(z)$. The condition (iii) means that there is a function Ψ meromorphic in the neighborhood of 0 such that

$$f\mid[\rho^{-1}]_k=\Psi(e^{\pi iz/h}).$$

The function Ψ must be an odd function.

REMARK 2.2. The above condition for f at s does not depend on the choice of ρ. If it is satisfied for some ρ, then it is so for every ρ such that $\rho(s)=\infty$. The classification of s into regular and irregular ones is also independent of the choice of ρ.

REMARK 2.3. If the above condition is satisfied at a cusp s, then it is satisfied at any cusp equivalent to s under Γ. The verification of these facts is straightforward, and may therefore be left to the reader.

The expression of $f\mid[\rho^{-1}]_k$ as a power series in $e^{2\pi iz/h}$ or in $e^{\pi iz/h}$ is often called *the Fourier expansion* of f at s; it has the following form:

$$f\mid[\rho^{-1}]_k=\textstyle\sum_{n\geqq n_0}c_n e^{2\pi inz/h}.$$

The coefficients c_n are naturally called *the Fourier coefficients.*

If $k=0$, we see, in view of our definition of complex structure of $\Gamma\backslash\mathfrak{H}^*$, that f satisfies the above conditions if and only if f is essentially a meromorphic function on $\Gamma\backslash\mathfrak{H}^*$. Thus an *automorphic function* with respect to Γ is an automorphic form of weight 0 with respect to Γ, and vice versa.

Let us denote by $A_k(\Gamma)$ the set of all automorphic forms of weight k with respect to Γ, especially by $A_0(\Gamma)$ the field of all automorphic functions with respect to Γ. Further we denote by $G_k(\Gamma)$ the set of all $f\in A_k(\Gamma)$ such that f is holomorphic on $\mathfrak{H}$ and the function Φ or Ψ in the above definition at each cusp is holomorphic at the origin; the latter condition means that the Fourier coefficient $c_n=0$ for $n<0$. We also denote by $S_k(\Gamma)$ the set of all $f\in G_k(\Gamma)$ such that the function Φ or Ψ at each cusp vanishes at the origin, i.e., the Fourier coefficient $c_n=0$ for $n\leqq 0$. An element of $G_k(\Gamma)$ (resp. $S_k(\Gamma)$) is called an *integral form* (resp. a *cusp form*) of weight k with respect to Γ. If Γ has no cusp, we have $G_k(\Gamma)=S_k(\Gamma)$. (In this case the terminology "cusp form" is unnecessary, and perhaps slightly confusing, but often convenient.)

If Γ is the principal congruence subgroup of $SL_2(\boldsymbol{Z})$ of level N, an automorphic function (resp. form) is usually called a *modular function* (resp. *form*) *of level N.*

Coming back to the general case, we see easily that

(2.1.1 a) $$f\in A_k(\Gamma),\ g\in A_m(\Gamma) \quad\Rightarrow\quad f\cdot g\in A_{k+m}(\Gamma),$$

(2.1.1 b) $$f\in G_k(\Gamma),\ g\in G_m(\Gamma) \quad\Rightarrow\quad f\cdot g\in G_{k+m}(\Gamma),$$

(2.1.1 c) $$f\in G_k(\Gamma),\ g\in S_m(\Gamma) \quad\Rightarrow\quad f\cdot g\in S_{k+m}(\Gamma).$$

PROPOSITION 2.4. *Let Γ' be a subgroup of $SL_2(\boldsymbol{R})$, and α an element of $GL_2^+(\boldsymbol{R})$ such that $\alpha\Gamma\alpha^{-1}$ is a subgroup of Γ' of finite index. Then $f\mapsto f\,|\,[\alpha]_k$ gives a $\boldsymbol{C}$-linear injection of $A_k(\Gamma')$ (resp. $G_k(\Gamma')$, $S_k(\Gamma')$) into $A_k(\Gamma)$ (resp. $G_k(\Gamma)$, $S_k(\Gamma)$), which is surjective if $\Gamma'=\alpha\Gamma\alpha^{-1}$.*

PROOF. Let $\mathfrak{C}$ (resp. $\mathfrak{C}'$) denote the set of cusps of Γ (resp. Γ'). Then $\alpha(\mathfrak{C})=\mathfrak{C}'$, and our assertion follows immediately from the definition.

Put $\mathfrak{H}^{*\prime}=\mathfrak{H}\cup\mathfrak{C}'$. Then we obtain a commutative diagram

$$\begin{array}{ccc} \mathfrak{H}^* & \xrightarrow{\quad\alpha\quad} & \mathfrak{H}^{*\prime} \\ \downarrow & & \downarrow \\ \Gamma\backslash\mathfrak{H}^* & \xrightarrow{\quad T_\alpha\quad} & \Gamma'\backslash\mathfrak{H}^{*\prime} \end{array}$$

with a holomorphic map T_α. In particular, if $\alpha\Gamma\alpha^{-1}=\Gamma$, T_α is a biregular automorphism of $\Gamma\backslash\mathfrak{H}^*$, which corresponds to the automorphism $f\mapsto f\circ\alpha$ of the field $A_0(\Gamma)$.

Let Δ be a subgroup of Γ of finite index. Identify $A_0(\Gamma)$ (resp. $A_0(\Delta)$) with the field of all meromorphic functions on $\Gamma\backslash\mathfrak{H}^*$ (resp. $\Delta\backslash\mathfrak{H}^*$). As is observed in § 1.5, $\Delta\backslash\mathfrak{H}^*$ is a covering of $\Gamma\backslash\mathfrak{H}^*$ of degree $[\bar{\Gamma}:\bar{\Delta}]$, so that $A_0(\Delta)$ is an algebraic extension of $A_0(\Gamma)$ of degree $[\bar{\Gamma}:\bar{\Delta}]$. Suppose now that $\bar{\Delta}$ is a normal subgroup of $\bar{\Gamma}$, and consider the automorphisms of $\Delta\backslash\mathfrak{H}^*$, or of $A_0(\Delta)$, obtained from the elements of Γ as above (taking Δ in place of Γ). Then we see that $A_0(\Delta)$ is a Galois extension of $A_0(\Gamma)$, and $\mathrm{Gal}(A_0(\Delta)/A_0(\Gamma))$ is isomorphic to $\bar{\Gamma}/\bar{\Delta}$.

PROPOSITION 2.5. *Let Γ' be a subgroup of Γ of finite index, and $\mathfrak{F}$ a subfield of $A_0(\Gamma')$, containing $A_0(\Gamma)$, with the following property:*

(P) If $\alpha\in\bar{\Gamma}$ and $f\circ\alpha=f$ for all $f\in\mathfrak{F}$, then $\alpha\in\bar{\Gamma}'$.

Then $\mathfrak{F}=A_0(\Gamma')$.

PROOF. Put $\Delta=\bigcap_{\alpha\in\Gamma}\alpha\Gamma'\alpha^{-1}$. Then Δ is a normal subgroup of Γ of finite index, contained in Γ'. Identify $\mathrm{Gal}(A_0(\Delta)/A_0(\Gamma))$ with $\bar{\Gamma}/\bar{\Delta}$ as above. The property (P) implies that $\bar{\Gamma}'/\bar{\Delta}$ contains the subgroup of $\mathrm{Gal}(A_0(\Delta)/A_0(\Gamma))$ corresponding to $\mathfrak{F}$. Since every element of $A_0(\Gamma')$ is invariant under Γ', we obtain $A_0(\Gamma')\subset\mathfrak{F}$ by Galois theory. But we have assumed that $\mathfrak{F}\subset A_0(\Gamma')$, hence $\mathfrak{F}=A_0(\Gamma')$.

PROPOSITION 2.6. *Let Γ' be a subgroup of Γ of finite index. Then $A_k(\Gamma)$ (resp. $G_k(\Gamma)$, $S_k(\Gamma)$) is the set of all f in $A_k(\Gamma')$ (resp. $G_k(\Gamma')$, $S_k(\Gamma')$) which are invariant under $[\gamma]_k$ for all $\gamma\in\Gamma$.*

The only non-trivial point is about the condition at cusps. But this can be verified in a straightforward way.

PROPOSITION 2.7. *Every Γ-invariant meromorphic function on $\mathfrak{H}$, algebraic over the field $A_0(\Gamma)$, is actually an automorphic function with respect to Γ.*

PROOF. Let g be such a function and let $g^n+\sum_{\lambda=0}^{n-1}f_\lambda g^\lambda=0$, with $f_\lambda\in A_0(\Gamma)$, be an equation for g over $A_0(\Gamma)$. For a cusp s of Γ, take ρ and $q=e^{2\pi iz/h}$ as in our definition of automorphic function. Then $f_\lambda(\rho^{-1}(z))=\Phi_\lambda(q)$ and $g(\rho^{-1}(z))=\Psi(q)$ with meromorphic functions Φ_λ and Ψ in the domain $0<|q|<r$, where r is a positive real number. Since the functions Φ_λ are meromorphic at $q=0$, we can find a positive integer m so that

$$\lim_{q\to 0} q^m\cdot\Phi_\lambda(q)=0 \qquad (\lambda=0,1,\cdots,n-1). \tag{1}$$

Put $V(q)=q^m\Psi(q)$. Then

$$1+\sum_{\lambda=0}^{n-1}q^{m(n-\lambda)}\Phi_\lambda(q)V(q)^{\lambda-n}=0\,. \tag{2}$$

Assume that $\lim_{k\to\infty}V(q_k)=\infty$ for a sequence of points $\{q_k\}$ which tends to 0.

Then, from (1) and (2), we obtain $1=0$, a contradiction. Therefore $V(q)$ must be bounded in a neighborhood of 0, so that Ψ is meromorphic at $q=0$, q. e. d.

EXERCISE 2.8. Let $f \in A_k(\Gamma)$, and $g=(k+1)\cdot(df/dz)^2-k\cdot f\cdot(d^2f/dz^2)$. Show that (i) $g \in A_{2k+4}(\Gamma)$; (ii) $g \in S_{2k+4}(\Gamma)$ if $f \in G_k(\Gamma)$.

2.2. Examples of modular forms and functions

Let us now present some examples of modular forms and functions. Let L be a lattice in $\boldsymbol{C}$, i. e., a free $\boldsymbol{Z}$-submodule of $\boldsymbol{C}$ of rank 2 which is discrete. Take a basis $\{\omega_1, \omega_2\}$ of L over $\boldsymbol{Z}$ so that $\omega_1/\omega_2 \in \mathfrak{H}$. For an *even* integer k, put

$$E_k(L)=E_k(\omega_1, \omega_2)=\sum_{\omega \in L-\{0\}} \omega^{-k}.$$

This is absolutely convergent for $k \geqq 4$. To show this, let P_m denote the parallelogram on the complex plane whose vertices are $\pm m\omega_1 \pm m\omega_2$. Let $r=\mathrm{Min}\{|z|\,;\, z \in P_1\}$. Then $|z| \geqq mr$ for $z \in P_m$. Since $P_m \cap L$ has exactly $8m$ points, we have

$$\sum_{\omega \in P_m \cap L} |\omega|^{-k} \leqq 8m\cdot(mr)^{-k}.$$

Since $L-\{0\}$ is the union of the $P_m \cap L$ for $m=1, 2, \cdots$, and since $\sum_{m=1}^{\infty} m^{-k+1}$ is convergent for $k>2$, we obtain the absolute convergence of $E_k(L)$.

We see easily that $\lambda^k E_k(\lambda\omega_1, \lambda\omega_2)=E_k(\omega_1, \omega_2)$, and

$$E_k(a\omega_1+b\omega_2, c\omega_1+d\omega_2)=E_k(\omega_1, \omega_2) \qquad \text{for} \quad \begin{bmatrix} a & b \\ c & d \end{bmatrix} \in SL_2(\boldsymbol{Z}),$$

hence

$$E_k((a\omega_1+b\omega_2)/(c\omega_1+d\omega_2), 1)(c\cdot(\omega_1/\omega_2)+d)^{-k}=E_k(\omega_1/\omega_2, 1).$$

This means that, if we put $E_k^*(z)=E_k(z, 1)$, E_k^* is invariant under $[\gamma]_k$ for every $\gamma \in SL_2(\boldsymbol{Z})$. Let us now show that E_k^* is an element of $G_k(SL_2(\boldsymbol{Z}))$, by establishing its Fourier expansion at ∞:

(2.2.1) $$E_k^*(z)=2\cdot\zeta(k)+2\cdot\frac{(2\pi i)^k}{(k-1)!}\sum_{n=1}^{\infty}\sigma_{k-1}(n)q^n, \qquad q=e^{2\pi iz},$$

where ζ is Riemann's zeta function, and $\sigma_s(n)$ denotes the sum of d^s for all positive divisors d of n.

To see this, we start with a well known formula

(2.2.2) $$\pi\cdot\cot(\pi z)=z^{-1}+\sum_{m=1}^{\infty}[(z+m)^{-1}+(z-m)^{-1}],$$

which can be found in conventional textbooks on complex analysis. On the other hand, putting $q=e^{2\pi iz}$, we have

$$\pi\cdot\cot(\pi z)=(\pi\cdot\cos(\pi z))/(\sin(\pi z))=\pi i(e^{\pi iz}+e^{-\pi iz})/(e^{\pi iz}-e^{-\pi iz})$$
$$=\pi i(q+1)/(q-1)=\pi i(1-2\sum_{n=0}^{\infty} q^n).$$

Equating (2.2.2) with the last sum, and differentiating successively with respect to z, we obtain

$$\text{(2.2.3)}\qquad \sum_{m=-\infty}^{\infty}(z+m)^{-2}=(2\pi i)^2\cdot\sum_{n=1}^{\infty}nq^n\,,$$
$$-2\cdot\sum_{m=-\infty}^{\infty}(z+m)^{-3}=(2\pi i)^3\cdot\sum_{n=1}^{\infty}n^2q^n\,,$$
$$\cdots\cdots\cdots\cdots\cdots\cdots\cdots\cdots\cdots$$
$$(-1)^k(k-1)!\sum_{m=-\infty}^{\infty}(z+m)^{-k}=(2\pi i)^k\cdot\sum_{n=1}^{\infty}n^{k-1}q^n\qquad(k\geqq 2)\,.$$

Therefore, if k is even and $\geqq 4$,

$$E_k^*(z)=\sum_{\substack{(m,n)\neq(0,0)\\ m\in\boldsymbol{Z},\ n\in\boldsymbol{Z}}}(mz+n)^{-k}$$
$$=2\cdot\sum_{n=1}^{\infty}n^{-k}+2\cdot\sum_{m=1}^{\infty}\sum_{n=-\infty}^{\infty}(mz+n)^{-k}$$
$$=2\cdot\zeta(k)+[2\cdot(2\pi i)^k/(k-1)!]\cdot\sum_{m=1}^{\infty}\sum_{n=1}^{\infty}n^{k-1}q^{mn}\,,$$

hence the formula (2.2.1).

THEOREM 2.9. *Put*

$$g_2(z)=60\cdot E_4^*(z)\,,\qquad g_3(z)=140\cdot E_6^*(z)\,,$$
$$\Delta(z)=g_2(z)^3-27g_3(z)^2\,,\qquad J(z)=12^3\cdot g_2(z)^3/\Delta(z)\,.$$

Then $\Delta(z)$ is a cusp form of weight 12 with respect to $SL_2(\boldsymbol{Z})$, and $J(z)$ is a modular function of level 1 whose Fourier expansion at ∞ is

$$J(z)=q^{-1}\cdot(1+\sum_{n=1}^{\infty}c_nq^n)$$

with integral coefficients c_n. Moreover, the field of all modular functions of level 1 is the rational function field $\boldsymbol{C}(J)$.

PROOF. Let $\Gamma=SL_2(\boldsymbol{Z})$. Since $g_2\in G_4(\Gamma)$ and $g_3\in G_6(\Gamma)$, we see that $\Delta\in G_{12}(\Gamma)$ and $J\in A_0(\Gamma)$. Now if B_r denotes the r-th Bernoulli number, we have

$$\zeta(2r)=\sum_{n=1}^{\infty}n^{-2r}=2^{2r-1}B_r\pi^{2r}/(2r)!\,,$$

so that $120\cdot\zeta(4)=(2\pi)^4/12$, $280\cdot\zeta(6)=(2\pi)^6/216$. Put

$$X=\sum_{n=1}^{\infty}\sigma_3(n)q^n\,,\qquad Y=\sum_{n=1}^{\infty}\sigma_5(n)q^n\,.$$

Then

$$g_2(z)=(2\pi)^4[1/12+20X]\,,\qquad g_3(z)=(2\pi)^6[1/216-7Y/3]\,,$$
$$(2\pi)^{-12}\Delta(z)=(5X+7Y)/12+100X^2+20^3X^3-3\cdot 7^2Y^2$$
$$=\sum_{n=1}^{\infty}\sum_{\substack{d|n\\ d>0}}\frac{5d^3+7d^5}{12}\cdot q^n+\sum_{n>1}a_nq^n$$

with integers a_n. Now $d^5\equiv d^3 \bmod (12)$ for every integer d. Therefore $(2\pi)^{-12}\Delta=\sum_{n=1}^{\infty}b_nq^n$ with integral coefficients, and $b_1=1$. It follows that $\Delta\in S_{12}(\Gamma)$, and J has the Fourier expansion described as above. To prove the last assertion, we need the following fact:

(2.2.4) $\Delta(z) \neq 0$ *for every* $z \in \mathfrak{H}$.

We shall prove this in § 4.2. Assuming this, we observe that $J(z)$ is holomorphic on $\mathfrak{H}$. Therefore, the function J, viewed as a function on $\Gamma\backslash\mathfrak{H}^*$, has a pole only at the point corresponding to ∞, as the Fourier expansion shows. Since this is a simple pole, and $\Gamma\backslash\mathfrak{H}^*$ is of genus 0, we see that $C(J)$ must be the whole field of meromorphic functions on $\Gamma\backslash\mathfrak{H}^*$, on account of (3) of Prop. 2.11 below.

PROPOSITION 2.10. *Let* $\Gamma = SL_2(\mathbf{Z})$ *and* $\alpha \in GL_2(\mathbf{Q})$, $\det(\alpha) > 0$. *Then* $\mathbf{C}(J, J\circ\alpha)$ *is the field of all modular functions with respect to* $\Gamma \cap \alpha^{-1}\Gamma\alpha$. *In particular,* $\mathbf{C}(J(z), J(Nz))$ *(resp.* $\mathbf{C}(J(z), J(z/N))$*) is the field of all modular functions with respect to* $\Gamma_0(N)$ *(resp.* $\Gamma_0'(N)$*), where* $\Gamma_0(N)$ *is as in* (1.6.5), *and*

$$\Gamma_0'(N) = \left\{ \begin{bmatrix} a & b \\ c & d \end{bmatrix} \in SL_2(\mathbf{Z}) \,\middle|\, b \equiv 0 \bmod (N) \right\}.$$

PROOF. Put $\Gamma' = \Gamma \cap \alpha^{-1}\Gamma\alpha$. By Lemma 3.9 below, Γ' is a subgroup of Γ of finite index. It is obvious that $\mathbf{C}(J, J\circ\alpha) \subset A_0(\Gamma')$. Applying Prop. 2.5 to the present situation, we obtain the first part. The last part is just a special case $\alpha = \begin{bmatrix} N & 0 \\ 0 & 1 \end{bmatrix}$ $\left(\text{resp. } \alpha = \begin{bmatrix} 1 & 0 \\ 0 & N \end{bmatrix}\right)$.

In Chapter 6, we shall discuss some generators of $A_0(\Gamma(N))$, which can be explicitly written as division values of elliptic functions.

2.3. The Riemann-Roch theorem

The purpose of the next few sections is to compute the dimensions of the vector spaces $G_k(\Gamma)$ and $S_k(\Gamma)$ over $\mathbf{C}$, by means of the Riemann-Roch theorem. Therefore let us first recall some elementary facts on the divisors of a compact Riemann surface.[2] For details, see, for instance, Weyl [102], Chevalley [7], Iwasawa [35], Springer [85].

Let $\mathfrak{W}$ be a compact Riemann surface, and K the field of all meromorphic functions on $\mathfrak{W}$. We identify $\mathbf{C}$ with the subfield of K consisting of all constant functions. Then K is an algebraic function field of dimension one

2) All the following definitions, propositions, and theorems are applicable to a complete non-singular algebraic curve V over an algebraically closed field (or rather a universal domain) Ω of any characteristic. In fact, it is enough to replace $\mathfrak{W}$, K, and $\mathbf{C}$ by V, $\Omega(V)$, and Ω, where $\Omega(V)$ denotes the field of all (meromorphic) functions on V. The genus of V is defined, for instance, to be $l(\mathrm{div}\,(\omega))$ with any differential form ω on V, or an integer g with which the Riemann-Roch theorem holds. For this, see also Appendix No 9.

over $\boldsymbol{C}$, i.e., if $f \in K$, $\notin \boldsymbol{C}$, K is a finite algebraic extension of a rational function field $\boldsymbol{C}(f)$. Let D be the free $\boldsymbol{Z}$-module generated by the points of $\mathfrak{W}$, i.e., the module of all formal finite sums $\sum_\nu c_\nu P_\nu$ with $c_\nu \in \boldsymbol{Z}$ and $P_\nu \in \mathfrak{W}$. An element of D is called a *divisor* of $\mathfrak{W}$, or of K. For a divisor $A = \sum c_P P$ we put $c_P = \nu_P(A)$, and $\deg(A) = \sum_P c_P$. We write $A \geqq 0$ if $\nu_P(A) \geqq 0$ for all $P \in \mathfrak{W}$, and $A \geqq B$ if $A - B \geqq 0$. For each $P \in \mathfrak{W}$, the set $\{f \in K \mid f(P) \neq \infty\}$ is a discrete valuation ring with K as its quotient field. Let ν_P denote the normalized discrete order function $K \to \boldsymbol{Z} \cup \{\infty\}$ associated with this valuation ring. If t is a local parameter at P, ν_P is defined as follows: let

$$f(Q) = \textstyle\sum_{\nu \geqq \nu_0} a_\nu t(Q)^\nu \qquad a_{\nu_0} \neq 0\,,$$

where Q is a variable point in a small neighborhood of P. Then $\nu_P(f) = \nu_0$.

Associate with each $f \in K^\times$ a divisor $\operatorname{div}(f)$ by

$$\operatorname{div}(f) = \textstyle\sum_{P \in \mathfrak{W}} \nu_P(f) P\,.$$

Then $f \mapsto \operatorname{div}(f)$ is a homomorphism of $K^\times$ to D, i.e., $\operatorname{div}(hf) = \operatorname{div}(h) + \operatorname{div}(f)$, $\operatorname{div}(f^{-1}) = -\operatorname{div}(f)$. We put

$$(f)_0 = \textstyle\sum_{\nu_P(f) > 0} \nu_P(f) P\,,$$

$$(f)_\infty = -\textstyle\sum_{\nu_P(f) < 0} \nu_P(f) P\,.$$

Then $\operatorname{div}(f) = (f)_0 - (f)_\infty$.

PROPOSITION 2.11. *For every $f \in K^\times$, one has:*

(1) $\deg(\operatorname{div}(f)) = 0$.

(2) $\operatorname{div}(f) = 0 \quad \Leftrightarrow \quad f \in \boldsymbol{C}^\times$.

(3) $[K : \boldsymbol{C}(f)] = \deg(f)_0 = \deg(f)_\infty$ *provided that* $f \notin \boldsymbol{C}^\times$.

For a divisor A, we put

$$\begin{aligned} L(A) &= \{f \in K \mid f = 0 \text{ or } \operatorname{div}(f) \geqq -A\} \\ &= \{f \in K \mid \nu_P(f) \geqq -\nu_P(A) \text{ for all } P \in \mathfrak{W}\}\,. \end{aligned}$$

Clearly $L(A)$ is a vector space over $\boldsymbol{C}$. It can be shown that $L(A)$ is finite dimensional over $\boldsymbol{C}$. Denote its dimension by $l(A)$. Put

$$D_l = \{\operatorname{div}(f) \mid f \in K^\times\}\,.$$

Then D_l is a submodule of D. A coset of D modulo D_l is called a *divisor class*. We say that two divisors A and B are *linearly equivalent*, and write $A \sim B$ if they belong to the same divisor class. If $A \sim B$, we have $\deg(A) = \deg(B)$, and $l(A) = l(B)$.

We can construct a one-dimensional vector space $\operatorname{Dif}(\mathfrak{W})$ over K together with an additive map

$$d : K \to \operatorname{Dif}(\mathfrak{W})$$

satisfying the following conditions:

$$d(hf)=h\cdot df+f\cdot dh\,, \qquad df=0 \iff f\in \boldsymbol{C}^{\times}. \tag{2.3.1}$$

An element of Dif ($\mathfrak{W}$) is called a (meromorphic) *differential form* (of degree one) on $\mathfrak{W}$. If $f\in K-\boldsymbol{C}$, we have Dif $(\mathfrak{W})=K\cdot df$, so that every differential form ω on $\mathfrak{W}$ can be written in the form $\omega=h\cdot df$ with $h\in K$. Then we write $h=\omega/df$. In particular, dk/df is a meaningful element of K for every $k\in K$. For each $P\in\mathfrak{W}$, we take an element t of K such that $\nu_P(t)=1$, and put $\nu_P(\omega)=\nu_P(\omega/dt)$. This is independent of the choice of t. Define div (ω) by

$$\operatorname{div}(\omega)=\textstyle\sum_{P\in\mathfrak{W}}\nu_P(\omega)P\,.$$

Then $\operatorname{div}(f\omega)=\operatorname{div}(f)+\operatorname{div}(\omega)$ for every $f\in K$. Thus the div (ω) for all $\omega\in$ Dif $(\mathfrak{W})$, $\neq 0$, form a divisor class, which is called *the canonical class of* $\mathfrak{W}$ (or of K). We say that ω is *holomorphic*, or of *the first kind* if $\operatorname{div}(\omega)\geqq 0$ or $\omega=0$.

THEOREM 2.12 (The Riemann-Roch Theorem). *Let g be the genus of $\mathfrak{W}$ (see* § 1.5), *and ω a non-zero differential form on $\mathfrak{W}$. Then, for any divisor A of $\mathfrak{W}$, one has*

$$l(A)=\deg(A)-g+1+l(\operatorname{div}(\omega)-A)\,.$$

PROPOSITION 2.13. *For every non-zero differential form ω on $\mathfrak{W}$,*

$$\deg(\operatorname{div}(\omega))=2g-2\,.$$

We see easily that $L(0)=\boldsymbol{C}$, so that $l(0)=1$. Therefore, from Th. 2.12 and Prop. 2.13, we obtain

$$l(\operatorname{div}(\omega))=g\,. \tag{2.3.2}$$

Fix any non-zero differential form ω_0. Then

$$\begin{aligned} L(\operatorname{div}(\omega_0)) &= \{f\in K \mid \operatorname{div}(f)\geqq -\operatorname{div}(\omega_0)\}\\ &= \{f\in K \mid \operatorname{div}(f\omega_0)\geqq 0\}\\ &\cong \{\omega\in \mathrm{Dif}(\mathfrak{W}) \mid \operatorname{div}(\omega)\geqq 0\}\,. \end{aligned}$$

Thus, by (2.3.2), we see that the set of all holomorphic differential forms on $\mathfrak{W}$ is a vector space of dimension g.

PROPOSITION 2.14. *Let A be a divisor of $\mathfrak{W}$. Then:*

(1) $\deg(A)<0 \Rightarrow l(A)=0$;

(2) $\deg(A)>2g-2 \Rightarrow l(A)=\deg(A)-g+1$.

PROOF. If $l(A)>0$, $L(A)$ contains at least one $f\neq 0$ such that $\operatorname{div}(f)\geqq -A$.

Then $\deg(A) \geqq \deg(\mathrm{div}(f)) = 0$. Therefore we obtain (1). If $\deg(A) > 2g-2$, we have $\deg(\mathrm{div}(\omega) - A) < 0$ with any non-zero differential form ω on $\mathfrak{W}$, so that $l(\mathrm{div}(\omega) - A) = 0$ by (1). Then the Riemann-Roch theorem implies that $l(A) = \deg(A) - g + 1$.

2.4. The divisor of an automorphic form

We shall now consider the case where $\mathfrak{W} = \Gamma \backslash \mathfrak{H}^*$ with a Fuchsian group Γ of the first kind. Our main interest is in the spaces $G_k(\Gamma)$ and $S_k(\Gamma)$. We suppose $-1 \notin \Gamma$ whenever we speak of $A_k(\Gamma)$ with *odd* k, since $A_k(\Gamma) = \{0\}$ if k is odd and $-1 \in \Gamma$. If $f(t)$ is a meromorphic function in a complex variable t defined in a neighborhood of 0, we denote by $\nu_t(f)$ the order of f at $t = 0$, i. e.,

$$\nu_t(f) = m \quad \text{if} \quad f(t) = \textstyle\sum_{n \geqq m} c_n t^n, \qquad c_m \neq 0.$$

Further we put $K = A_0(\Gamma)$, and identify K with the field of all meromorphic functions on $\mathfrak{W}$.

PROPOSITION 2.15. *$A_k(\Gamma) \neq \{0\}$ for every integer k.*

On account of (2.1.1), this implies that $A_k(\Gamma)$ is a one-dimensional vector space over K.

PROOF. Take any $\phi \in K - \boldsymbol{C}$. Then $\phi(\gamma(z)) = \phi(z)$ for all $\gamma \in \Gamma$. Take the derivative $\phi' = d\phi/dz$. Then we find $\phi'(\gamma(z)) j(\gamma, z)^{-2} = \phi'(z)$ since $d\gamma(z)/dz = j(\gamma, z)^{-2}$. At a cusp s, we have $\phi(\rho^{-1}(z)) = \Phi(q)$ with a meromorphic function Φ at $q = 0$, where ρ and q are as in § 2.1. Then

$$\phi' \mid [\rho]_2 = \phi'(\rho^{-1}(z)) j(\rho^{-1}, z)^2 = \Phi'(q) \cdot (2\pi i/h) \cdot q,$$

hence $\phi' \in A_2(\Gamma)$. Therefore, by (2.1.1), $0 \neq \phi'^n \in A_{2n}(\Gamma)$ for any integer n, which proves our assertion for even k. The case of odd k will be discussed a little later.

For any $F \in A_2(\Gamma)$, we can view $F(z)dz$ as a differential form on $\mathfrak{W}$. In fact, take $\phi \in K - \boldsymbol{C}$ as above. Since $\phi' = d\phi/dz \in A_2(\Gamma)$, $F/\phi' \in A_0(\Gamma) = K$. Then we put $F(z)dz = (F/\phi')d\phi$. This does not depend on the choice of ϕ. Conversely, if $\omega \in \mathrm{Dif}(\mathfrak{W})$, then $f = \omega/d\phi \in K$, $f\phi' \in A_2(\Gamma)$, and $\omega = (f\phi')dz$. Therefore the map $F \mapsto F \cdot dz$ gives an isomorphism of $A_2(\Gamma)$ onto $\mathrm{Dif}(\Gamma \backslash \mathfrak{H}^*)$.

Now, we can construct an associative (graded) algebra

$$\mathfrak{D} = \textstyle\sum_{n=-\infty}^{\infty} \mathrm{Dif}^n(\mathfrak{W}) \qquad \text{(direct sum)}$$

over K under the following conditions:

(a) $\mathrm{Dif}^0(\mathfrak{W}) = K$, $\mathrm{Dif}^1(\mathfrak{W}) = \mathrm{Dif}(\mathfrak{W})$;

(b) $\mathrm{Dif}^n(\mathfrak{W})$, for each n, is a one-dimensional vector space over K;

(c) For $\alpha \in \mathrm{Dif}^m(\mathfrak{W})$ and $\beta \in \mathrm{Dif}^n(\mathfrak{W})$, the product $\alpha\beta$ is defined as an element of $\mathrm{Dif}^{m+n}(\mathfrak{W})$, and $\alpha\beta = \beta\alpha \neq 0$ if $\alpha \neq 0$, $\beta \neq 0$.

The algebra $\mathfrak{D}$ is uniquely determined by these conditions. If $0 \neq \omega \in \mathrm{Dif}(\mathfrak{W})$, ω^n is meaningful and belongs to $\mathrm{Dif}^n(\mathfrak{W})$. Then $\mathrm{Dif}^n(\mathfrak{W}) = K\omega^n$, so that every element ξ of $\mathrm{Dif}^n(\mathfrak{W})$ is of the form $\xi = f\omega^n$ with $f \in K$. If $\xi \neq 0$, we define $\mathrm{div}(\xi)$ by

$$\mathrm{div}(\xi) = \mathrm{div}(f) + n \cdot \mathrm{div}(\omega).$$

We see easily that this does not depend on the choice of ω, and

$$\mathrm{div}(\xi\eta) = \mathrm{div}(\xi) + \mathrm{div}(\eta).$$

By Prop. 2.13, denoting by g the genus of $\mathfrak{W} = \Gamma\backslash\mathfrak{H}^*$, we have

$$\text{(2.4.1)} \qquad \deg(\mathrm{div}(\xi)) = n(2g-2) \qquad \text{for} \quad 0 \neq \xi \in \mathrm{Dif}^n(\mathfrak{W}).$$

Take ψ as above. For $F \in A_{2n}(\Gamma)$, we observe that $F/\psi'^n \in A_0(\Gamma) = K$, and put $F(z)(dz)^n = (F/\psi'^n)(d\psi)^n$. Then $F(z)(dz)^n$ is a well-defined element of $\mathrm{Dif}^n(\mathfrak{W})$, independent of the choice of ψ. We see easily that $F \mapsto F(z)(dz)^n$ is an isomorphism of $A_{2n}(\Gamma)$ onto $\mathrm{Dif}^n(\mathfrak{W})$.

Let $F \in A_k(\Gamma)$, and $P \in \mathfrak{W}$. We define $\nu_P(F)$ as follows. If P corresponds to a point z_0 of $\mathfrak{H}$, take a holomorphic isomorphism λ of $\mathfrak{H}$ onto the unit disc such that $\lambda(z_0) = 0$. If $\{\gamma \in \bar{\Gamma} \mid \gamma(z_0) = z_0\}$ is of order e, the function $t = \lambda(z)^e$ is the standard local parameter at P (see § 1.5). Then put $\nu_P(F) = \nu_{(z-z_0)}(F)/e$. Next, if P corresponds to a cusp s, take ρ and $q = \exp(2\pi iz/h)$ as in § 2.1. Then, as in our definition of automorphic forms, we have

$$F \mid [\rho^{-1}]_k = \begin{cases} \Psi(q^{1/2}) & \text{if } k \text{ is odd and } s \text{ is irregular}, \\ \Phi(q) & \text{otherwise}, \end{cases}$$

where Ψ and Φ are meromorphic functions around 0. Then we put

$$\nu_P(F) = \begin{cases} \nu_t(\Psi)/2 & (t = q^{1/2} = e^{\pi iz/h}), \\ \nu_q(\Phi), \end{cases}$$

accordingly. Note that $\nu_t(\Psi)$ is odd.

Let us put $D_{\mathbf{Q}} = D \otimes_{\mathbf{Z}} \mathbf{Q}$. Then we can associate, with each $F \in A_k(\Gamma)$, an element $\mathrm{div}(F)$ of $D_{\mathbf{Q}}$ by

$$\mathrm{div}(F) = \textstyle\sum_{P \in \mathfrak{W}} \nu_P(F) P.$$

It is clear that

$$\mathrm{div}(F_1F_2) = \mathrm{div}(F_1) + \mathrm{div}(F_2) \qquad (F_1 \in A_{k_1}(\Gamma),\ F_2 \in A_{k_2}(\Gamma)),$$

$$\text{(2.4.2)} \qquad G_k(\Gamma) = \{F \in A_k(\Gamma) \mid \mathrm{div}(F) \geqq 0\},$$

$$S_k(\Gamma)=\begin{cases}\{F\in A_k(\Gamma)\mid \operatorname{div}(F)\geqq \sum_{j=1}^{u}Q_j+\sum_{j=1}^{u'}Q'_j\} & (k:\text{ even}),\\ \{F\in A_k(\Gamma)\mid \operatorname{div}(F)\geqq \sum_{j=1}^{u}Q_j+(1/2)\sum_{j=1}^{u'}Q'_j\} & (k:\text{ odd}),\end{cases} \tag{2.4.3}$$

where $Q_1, \cdots, Q_u$ (resp. $Q'_1, \cdots, Q'_{u'}$) are the points of $\mathfrak{W}$ corresponding to the regular (resp. irregular) cusps of Γ. Note that the relation $\geqq$ and deg( ) can be extended to D_Q in a natural way.

PROPOSITION 2.16. *Let $P_1, \cdots, P_r$ be the points of $\Gamma\backslash\mathfrak{H}^*$ corresponding to all the elliptic points of Γ, of order $e_1, \cdots, e_r$, respectively, and $Q_1, \cdots, Q_u$, $Q'_1, \cdots, Q'_{u'}$ be as above. Let $0\neq F\in A_k(\Gamma)$, and if k is even, let $\eta=F(z)(dz)^{k/2}$. Then*

$$\operatorname{div}(F)=\operatorname{div}(\eta)+(k/2)\cdot\{\textstyle\sum_{i=1}^{r}(1-e_i^{-1})P_i+\sum_{j=1}^{u}Q_j+\sum_{j=1}^{u'}Q'_j\}\quad (k:\ \textit{even}),$$

$$\deg(\operatorname{div}(F))=(k/2)\cdot\{(2g-2)+\textstyle\sum_{i=1}^{r}(1-e_i^{-1})+u+u'\}\quad (k:\ \textit{even or odd}).$$

PROOF. Let P be a point of $\mathfrak{W}=\Gamma\backslash\mathfrak{H}^*$. If P corresponds to a point z_0 of $\mathfrak{H}$, take $t=\lambda(z)^e$ as above. Then $dt/dz=e\cdot\lambda(z)^{e-1}(d\lambda/dz)$, and $\nu_t(dt/dz)=1-e^{-1}$. Therefore, assuming k to be even, we have

$$\nu_P(\eta)=\nu_P(F\cdot(dz/dt)^{k/2})=\nu_P(F)+(k/2)\cdot(e^{-1}-1). \tag{2.4.4}$$

If P corresponds to a cusp s, take ρ and $q=e^{2\pi iz/h}$ as in p. 29. Then putting $z=\rho(w)$, we have

$$F(w)(dw)^{k/2}=(F\mid[\rho^{-1}]_k)(dz)^{k/2}=\Phi(q)(dz/dq)^{k/2}(dq)^{k/2}=\Phi(q)(2\pi iq/h)^{-k/2}(dq)^{k/2},$$

hence

$$\nu_P(\eta)=\nu_q(\Phi)-k/2=\nu_P(F)-k/2. \tag{2.4.5}$$

Our first formula now follows immediately from (2.4.4) and (2.4.5); the second one for even k from (2.4.1) and the first one. If k is odd, we have $\operatorname{div}(F)=(1/2)\cdot\operatorname{div}(F^2)$, so that we can derive the desired formula of $\deg(\operatorname{div}(F))$ for odd k from that for even k.

The above proposition means that we can make calculation of divisors of automorphic forms by putting formally

$$\operatorname{div}(dz)=-\{\textstyle\sum_{i=1}^{r}(1-e_i^{-1})P_i+\sum_{j=1}^{u}Q_j+\sum_{j=1}^{u'}Q'_j\}.$$

The number $(2g-2)+\sum_{i=1}^{r}(1-e_i^{-1})+u+u'$ occurring in the second formula has an important geometric meaning, which will be studied in the next section.

COROLLARY 2.17. *$S_2(\Gamma)$ is isomorphic to the vector space of all holomorphic differential forms on $\mathfrak{W}=\Gamma\backslash\mathfrak{H}^*$, through the map $F\mapsto F\cdot dz$. It follows especially that $S_2(\Gamma)$ is of dimension g.*

PROOF. If $F\in A_2(\Gamma)$ and $\omega=F\cdot dz$, we see, from Prop. 2.16, that $\operatorname{div}(\omega)\geqq 0$

if and only if $\operatorname{div}(F) \geqq \sum_{j=1}^{u} Q_j + \sum_{j=1}^{u'} Q'_j$. Therefore our assertion follows from (2.4.3).

PROOF of Prop. 2.15 for odd k. Take any non-zero differential form ω on $\mathfrak{W}$ and any point R_0 of $\mathfrak{W}$. Then $\deg[\operatorname{div}(\omega) - 2(g-1)R_0] = 0$. Now it is a classical fact that all the divisor classes of $\mathfrak{W}$ of degree 0 form an abelian group isomorphic to a complex torus of complex dimension g (which is called the jacobian variety of $\mathfrak{W}$). Therefore we can find a divisor B on $\mathfrak{W}$ such that $\operatorname{div}(\omega) - 2(g-1)R_0 \sim 2B$, i.e.,

$$2B - \operatorname{div}(\omega) + 2(g-1)R_0 = \operatorname{div}(f)$$

for some $f \in K^\times$. Put $B' = B + (g-1)R_0$. We can define an element F of $A_2(\Gamma)$ by $F(z)dz = f\omega$. By Prop. 2.16, we have

$$\operatorname{div}(F) = 2B' + \sum_{i=1}^{r} (1 - e_i^{-1}) P_i + \sum_{j=1}^{u} Q_j + \sum_{j=1}^{u'} Q'_j .$$

By Cor. 1.21, the e_i are all odd, if $-1 \notin \Gamma$. Therefore we see that the function F has an even order at every point of $\mathfrak{H}$. Hence we can define a meromorphic function G on $\mathfrak{H}$ so that $G^2 = F$. Since $F \in A_2(\Gamma)$, we have $G \mid [\gamma]_1 = \chi(\gamma) G$ for every $\gamma \in \Gamma$ with $\chi(\gamma) = \pm 1$. Put $\Gamma' = \{\gamma \in \Gamma \mid \chi(\gamma) = 1\}$. Then Γ' is a subgroup of Γ of index $\leqq 2$. Since $F \in A_2(\Gamma)$, we see that G is meromorphic at every cusp of Γ', so that $G \in A_1(\Gamma')$. If $\Gamma = \Gamma'$, this settles the question, since $0 \neq G^k \in A_k(\Gamma)$ for any integer k. Suppose $[\Gamma : \Gamma'] = 2$, and let $\Gamma = \Gamma' \cup \Gamma'\varepsilon$. Since $A_0(\Gamma')$ is a quadratic extension of $A_0(\Gamma)$, and $\operatorname{Gal}(A_0(\Gamma')/A_0(\Gamma))$ is isomorphic to Γ/Γ' as is seen in p. 31 (after Prop. 2.4), there exists a non-zero element h of $A_0(\Gamma')$ such that $h(\varepsilon(z)) = -h(z)$. Then $h \cdot G$ belongs to $A_1(\Gamma')$ and is invariant under $[\varepsilon]_1$, so that $h \cdot G \in A_1(\Gamma)$ by Prop. 2.6. Therefore $0 \neq (h \cdot G)^k \in A_k(\Gamma)$ for any integer k, q. e. d.

2.5. The measure of $\Gamma \backslash \mathfrak{H}$

For every differential form ω on $\mathfrak{H}$ and $\sigma \in SL_2(\boldsymbol{R})$, let us denote by $\omega \circ \sigma$ the transform of ω by σ in an obvious sense: if ω is of degree 0, and hence a function, $\omega \circ \sigma$ is of course meaningful; in general, $d(\omega \circ \sigma) = (d\omega) \circ \sigma$, $(\omega \wedge \eta) \circ \sigma = (\omega \circ \sigma) \wedge (\eta \circ \sigma)$.

PROPOSITION 2.18. *Let η be a differential form on $\mathfrak{H}$ defined by $\eta = y^{-1} dz$, $z = x + iy$. Then:*

(1) $\eta \circ \sigma - \eta = -2i \cdot d \log [j(\sigma, z)]$ *for every* $\sigma \in SL_2(\boldsymbol{R})$.

(2) $d\eta = y^{-2} dx \wedge dy = (i/2y^2) \cdot dz \wedge d\bar{z}$.

(3) $y^{-2} dx \wedge dy$ *is invariant under* $SL_2(\boldsymbol{R})$.

PROOF. If $\sigma = \begin{bmatrix} p & q \\ r & s \end{bmatrix} \in SL_2(\boldsymbol{R})$, we have $dz \circ \sigma = j(\sigma, z)^{-2} dz$ and by (1.2.3),

$y\circ\sigma=|j(\sigma,z)|^{-2}y$, so that $\eta\circ\sigma=[(r\bar{z}+s)/(rz+s)]\cdot\eta$. Therefore

$$\eta\circ\sigma-\eta=[(r\bar{z}+s)/(rz+s)-1]\eta=-[2ir/(rz+s)]\cdot dz=-2i\cdot d\log(rz+s)\,.$$

The formula (2) is obtained in a straightforward way. Taking the exterior differentiation of (1), we obtain (3).

Now let us define a measure m on $\mathfrak{H}$ by

$$m(A)=\int_A y^{-2}dx\,dy$$

for a subset A of $\mathfrak{H}$. By (3) of Prop. 2.18, m is an invariant measure, i. e., $m(A)=m(\sigma(A))$ for every $\sigma\in SL_2(\boldsymbol{R})$ and for a (measurable) set A.

We can use this measure to introduce a measure μ on $\Gamma\backslash\mathfrak{H}^*$. Let $\varphi:\mathfrak{H}^*\to\Gamma\backslash\mathfrak{H}^*$ be the projection map, and for $v\in\mathfrak{H}^*$, put

$$\Gamma_v=\{\gamma\in\Gamma\mid\gamma(v)=v\}\,.$$

For each v, we can find an open neighborhood U of v such that

$$\Gamma_v=\{\gamma\in\Gamma\mid\gamma(U)\cap U\neq\emptyset\}\,,$$

and $\gamma(U)=U$ for all $\gamma\in\Gamma_v$. Then $\Gamma_v\backslash U$ may be identified with an open neighborhood of $\varphi(v)$ in $\Gamma\backslash\mathfrak{H}^*$. If v is not a cusp and if $\bar{\Gamma}_v$ is of order e (v is elliptic if $e>1$), we can divide U into e angular sectors $U_1,\cdots,U_e$ such that $\gamma(U_i)=U_{i+1}$ for $1\leqq i<e$ and $\gamma(U_e)=U_1$, where γ is a generator of $\bar{\Gamma}_v$. Then, for $A'\subset\Gamma_v\backslash U$, we can find a set A of representatives for A' in U_1, and define $\mu(A')=m(A)$. Similarly if v is a cusp, say, at ∞, $\bar{\Gamma}_v$ is generated by an element of the form $\begin{bmatrix}1&h\\0&1\end{bmatrix}$, and U may have the form $U=\{\infty\}\cup\{z\mid \mathrm{Im}(z)>c\}$. Then, for $A'\subset\Gamma_v\backslash U$, we can find a set A of representatives for A' in the region

$$\{z=x+iy\mid y>c,\ 0\leqq x<h\}\,,\qquad\text{(excluding }\infty)\,,$$

and define $\mu(A')=m(A)$. Now we can cover $\Gamma\backslash\mathfrak{H}^*$ by open sets of the form $\Gamma_v\backslash U$. If $\{h_\lambda\}_{\lambda=1}^{\infty}$ is a C^∞-partition of unity subordinate to this covering $\{W_\lambda\}_{\lambda=1}^{\infty}$, then for a continuous function f on $\Gamma\backslash\mathfrak{H}^*$, we can put

$$(*)\qquad\int_{\Gamma\backslash\mathfrak{H}^*}f\cdot d\mu=\textstyle\sum_{i=1}^{\infty}\displaystyle\int_{W_\lambda}fh_\lambda\cdot d\mu\,.$$

It can easily be seen that this measure on $\Gamma\backslash\mathfrak{H}^*$ does not depend on the choice of $\{W_\lambda\}$ and $\{h_\lambda\}$. We shall also write $\int_{\Gamma\backslash\mathfrak{H}}(f\circ\varphi)y^{-2}dx\,dy$ for the integral $(*)$.

PROPOSITION 2.19. *If $\Gamma\backslash\mathfrak{H}^*$ is compact, then $\mu(\Gamma\backslash\mathfrak{H}^*)<\infty$. (In other words, $(*)$ defines a Radon measure.)*

PROOF. If $\Gamma\backslash\mathfrak{H}^*$ is compact, the covering of the above type can be made by finitely many sets of the form $\Gamma_v\backslash U$ for which the closure of U is compact. If v is not a cusp, then clearly $\mu(\Gamma_v\backslash U)<\infty$. If v is a cusp, the finiteness follows from

$$\iint_{\substack{0<x<h\\c<y}} y^{-2}dx\,dy<\infty\,.$$

THEOREM 2.20. *Let g be the genus of the compact Riemann surface $\Gamma\backslash\mathfrak{H}^*$, m the number of inequivalent cusps of Γ, and $e_1,\cdots,e_r$ the orders of the inequivalent elliptic points of Γ. Then*

$$\frac{1}{2\pi}\int_{\Gamma\backslash\mathfrak{H}} y^{-2}dx\,dy=2g-2+m+\textstyle\sum_{\nu=1}^{r}(1-1/e_\nu)\,.$$

PROOF. It is well-known that $\Gamma\backslash\mathfrak{H}^*$, being a compact Riemann surface, can be triangulated using piecewise analytic curves, which, without loss of generality, can be chosen so as not to contain any cusps or elliptic points. Using this triangulation, $\Gamma\backslash\mathfrak{H}^*$ can be represented as a normal form $a_1b_1a_1^{-1}b_1^{-1}\cdots a_gb_ga_g^{-1}b_g^{-1}$ consisting of a $4g$-sided polygon all of whose vertices are identified, and whose boundary consists of $2g$ curves a_i, b_i traced once in each direction in the above order, and such that the elliptic points and cusps are in the interior of the polygon. Next, draw non-intersecting piecewise analytic paths connecting one of the vertices of the polygon to the respective elliptic points and cusps. Also draw a small circle around each elliptic point or cusp, whose radius will be made to tend to zero. Cutting the polygon along these paths and small circles, we now get a "polygon" with $4g+2m+2r$ sides, neglecting the small circles. Now take a small open disc in this polygon, and map it into $\mathfrak{H}$ by the inverse of the projection map $\mathfrak{H}^*\to\Gamma\backslash\mathfrak{H}^*$. This is a holomorphic map, and can be continued holomorphically to the whole inside of the polygon. Therefore, the polygon can be mapped onto a polygon on $\mathfrak{H}$, which we call Π. From our construction, we see that the boundary $\partial\Pi$ of Π can be written in the form

(1) $$\partial\Pi=\textstyle\sum_{\lambda=1}^{n}(S_\lambda-\gamma_\lambda(S_\lambda))+\sum_{\nu=1}^{m+r}T_\nu \qquad (n=2g+m+r)\,.$$

Here the T_ν are the curves corresponding to the small circles, the S_λ correspond to the "sides", and γ_λ is a certain element of Γ for each λ. The interior of Π, when each T_ν is shrunk to a point, is certainly a fundamental domain for Γ. But the S_λ are not necessarily "straight lines" in the sense of non-Euclidean geometry. One can actually construct a fundamental domain for Γ which is a polygon whose sides are straight lines in that sense. For our present need, however, the polygon Π, in the loosest sense, is completely adequate.

Now we consider the differential form η defined in Prop. 2.18. Since $d\eta = y^{-2}dxdy$, we have

$$\mu(\Gamma\backslash\mathfrak{H}^*) = \lim \int_{\Pi} y^{-2}dxdy = \lim \int_{\partial\Pi} \eta \tag{2}$$

by Stokes' theorem. The limit procedure is taken by shrinking the circles. By (1),

$$\int_{\partial\Pi} \eta = \sum_{\lambda=1}^{n} \int_{S_\lambda} (\eta - \eta\circ\gamma_\lambda) + \sum_{\nu=1}^{m+r} \int_{T_\nu} \eta .$$

Let F be a non-zero element of $A_2(\Gamma)$. Define a differential form ξ on $\mathfrak{H}$ by $\xi = d(\log F) = F^{-1}F'dz$. Taking the logarithmic derivative of

$$F(\sigma(z)) = F(z)j(\sigma, z)^2 \qquad (\sigma \in \Gamma),$$

we obtain

$$\xi\circ\sigma - \xi = 2 \cdot d(\log j(\sigma, z)) \qquad (\sigma \in \Gamma).$$

By (1) of Prop. 2.18, we have

$$\eta\circ\sigma - \eta = -i(\xi\circ\sigma - \xi) \qquad (\sigma \in \Gamma).$$

Therefore

$$\int_{\partial\Pi} \eta = -i\int_{\partial\Pi} \xi + \sum_{\nu=1}^{m+r} \int_{T_\nu} \eta + i\sum_{\nu=1}^{m+r} \int_{T_\nu} \xi . \tag{3}$$

If T_ν corresponds to an elliptic point v of order e, then clearly $\int_{T_\nu} \eta$ tends to 0. As for $\int_{T_\nu} \xi$, take a holomorphic map τ of $\mathfrak{H}$ onto the unit disc such that $\tau(v) = 0$, and put $t(z) = \tau(z)^e$. Then t is a local parameter, and we may assume that T_ν is the image of a small circle C_ν in the t-plane with origin as its center. This circle should be taken in the negative direction, since the exterior of the circle corresponds to the interior of Π. Therefore putting $\omega = F(z)dz$ and $\psi(t) = F(z)(dz/dt)$, we have

$$\begin{aligned}\int_{T_\nu} \xi &= -\int_{C_\nu} [d(\log\psi) + d(\log(dt/dz))] = -2\pi i[\nu_t(\psi) + \nu_t(dt/dz)] \\ &= -2\pi i[\nu_P(\omega) + 1 - e^{-1}] = -2\pi i \cdot \nu_P(F)\end{aligned}$$

by (2.4.4), where P is the point of $\Gamma\backslash\mathfrak{H}^*$ corresponding to the elliptic point in question.

Next, assume that T_ν corresponds to a cusp s. Let ρ be an element of $SL_2(\boldsymbol{R})$ such that $\rho(s) = \infty$ and let $q = e^{2\pi i\rho(z)/h}$. Then we may assume that T_ν is the image of a small circle C_ν in the q-plane with origin as its center. Putting $w = \rho(z)$ and $F(\rho^{-1}(w))j(\rho^{-1}, w)^2 = \Phi(q)$, we have $F(z)dz = \Phi(q)dw$. We can take $\rho = \begin{bmatrix} 0 & 1 \\ -1 & s \end{bmatrix}$ if $s \neq \infty$. Then $dw/dz = w^2$, so that $F(z) = \Phi(q)w^2$, and

$$\begin{aligned}\int_{T_\nu} d(\log F) &= -\int_{C_\nu} [d(\log \Phi(q)) + 2 \cdot d(\log w)] \\ &= -2\pi i \cdot \nu_q(\Phi) - \int_{w_0}^{w_0+h} 2 \cdot d(\log w) \\ &\to \quad -2\pi i \cdot \nu_q(\Phi) = -2\pi i \cdot \nu_P(F) \qquad (w_0 \to \infty).\end{aligned}$$

Here P is the point of $\Gamma\backslash\mathfrak{H}^*$ corresponding to s. If $s=\infty$, we can take ρ to be the identity matrix, and obtain the same result. As for η, we have

$$\int_{T_\nu} \eta = \int_{\rho(T_\nu)} \eta \circ \rho^{-1} = \int_{\rho(T_\nu)} \{\eta - 2i \cdot d \log (j(\rho^{-1}, w))\}$$

by (1) of Prop. 2.18. We understand, as above, that $\rho(T_\nu)$ is the segment from w_0 to w_0+h. Then

$$\int_{T_\nu} \eta = \int_{w_0}^{w_0+h} [dz/y - 2i \cdot d(\log w)] \to 0 \qquad (w_0 \to \infty).$$

Thus combining all these computations, we obtain, from (2) and (3),

$$\mu(\Gamma\backslash\mathfrak{H}^*) = -i \int_{\partial\Pi} d(\log F) + 2\pi \textstyle\sum_{i=1}^r \nu_{P_i}(F) + 2\pi \sum_{j=1}^m \nu_{Q_j}(F).$$

Now $(2\pi i)^{-1} \int_{\partial\Pi} d(\log F)$ is the sum of $\nu_P(F)$ for all P in the interior of the polygon Π. Therefore we obtain $\mu(\Gamma\backslash\mathfrak{H}^*) = 2\pi \cdot \deg(\mathrm{div}(F))$, which, together with Prop. 2.16, proves our theorem.

From this theorem, we see that

(2.5.1) $$2g-2+m+\textstyle\sum_{i=1}^r (1-e_i^{-1}) > 0.$$

If $g>1$, this inequality is trivial. If $g=1$, one has $m+r \geqq 1$. If $g=0$, one has $m+\sum_{i=1}^r (1-e_i^{-1}) > 2$, hence $m+r \geqq 3$. One can show, without difficulties, that the case $g=0$, $m=0$, $(e_1, e_2, e_3)=(2, 3, 7)$ gives rise to a Γ with the smallest $\mu(\Gamma\backslash\mathfrak{H}^*)$. Thus

$$(2\pi)^{-1} \int_{\Gamma\backslash\mathfrak{H}} dx\,dy/y^2 \geqq 1/42$$

for every Fuchsian group Γ of the first kind. Using this fact, it can be shown that the group of all automorphisms of any compact Riemann surface of genus $g>1$ has order $\leqq 84(g-1)$. For this topic, we refer the reader to Hurwitz, Werke I, pp. 391-430, Fricke-Klein [22, 606-621], and [77, 3.18].

It was also shown by Siegel [83] that the converse of Prop. 2.19 is true, i. e., if $\mu(\Gamma\backslash\mathfrak{H}) < \infty$ for a discrete subgroup Γ of $SL_2(\mathbf{R})$, then $\Gamma\backslash\mathfrak{H}^*$ is compact.

2.6. The dimension of the space of cusp forms

Let F_0 be a non-zero element of $A_k(\Gamma)$, and $B=\operatorname{div}(F_0)$. Every element F of $A_k(\Gamma)$ can be written in the form $F=f\cdot F_0$ with $f\in K$. Then $\operatorname{div}(F)\geqq 0$ if and only if $\operatorname{div}(f)\geqq -B$. Therefore, by (2.4.2), we have

$$\dim G_k(\Gamma)=\dim\{f\in K\mid \operatorname{div}(f)\geqq -B\}\,, \tag{2.6.1}$$

and similarly, by (2.4.3),

$$\dim S_k(\Gamma)=\dim\{f\in K\mid \operatorname{div}(f)\geqq -B+\textstyle\sum_{j=1}^{u}Q_j+\mu\sum_{j=1}^{u'}Q'_j\}\,, \tag{2.6.2}$$

where $\mu=1$ or $1/2$ according as k is even or odd. To compute these dimensions, we are going to apply the Riemann-Roch theorem to the divisors

$$-B\,,\qquad -B+\textstyle\sum_{j=1}^{u}Q_j+\sum_{j=1}^{u'}Q'_j\,.$$

However, these divisors are elements of $D_{\boldsymbol{Q}}$, but do not necessarily belong to D. To dispose of this difficulty, we consider the "integral part" of an element of $D_{\boldsymbol{Q}}$. For $x\in\boldsymbol{R}$, let $[x]$ denote the largest integer $\leqq x$; if $p=[x]$, one has $p\leqq x<p+1$. Then, for $A=\sum c_P P\in D_{\boldsymbol{Q}}$ with $c_P\in\boldsymbol{Q}$, we put

$$[A]=\textstyle\sum_P [c_P]P.$$

LEMMA 2.21. *For $f\in K^\times$ and $A\in D_{\boldsymbol{Q}}$,*

$$\operatorname{div}(f)\geqq -A \quad\Leftrightarrow\quad \operatorname{div}(f)\geqq -[A]\,.$$

PROOF. Put $A=\sum_P c_P P$. Then

$$\begin{aligned}\operatorname{div}(f)\geqq -A &\Leftrightarrow \nu_P(f)\geqq -c_P \Leftrightarrow -\nu_P(f)\leqq c_P \Leftrightarrow -\nu_P(f)\leqq [c_P]\\ &\Leftrightarrow \nu_P(f)\geqq -[c_P] \Leftrightarrow \operatorname{div}(f)\geqq -[A]\,,\qquad \text{q. e. d.}\end{aligned}$$

Let us first suppose that k is even and put $n=k/2$, $F_0=\xi/(dz)^n$ with $\xi\in\operatorname{Dif}^n(\mathfrak{W})$. By Prop. 2.16,

$$B=\operatorname{div}(F_0)=\operatorname{div}(\xi)+n\cdot\{\textstyle\sum_{i=1}^{r}(1-e_i^{-1})P_i+\sum_{j=1}^{u}Q_j+\sum_{j=1}^{u'}Q'_j\}\,.$$

By (2.6.1) and Lemma 2.21, we have

$$\dim G_k(\Gamma)=l([B])\,.$$

Putting $u+u'=m$, we see that

$$\deg([B])=n(2g-2+m)+\textstyle\sum_{i=1}^{r}[n(e_i-1)/e_i]\,. \tag{2.6.3}$$

LEMMA 2.22. *Let $k\in\boldsymbol{Z}$, $e\in\boldsymbol{Z}$, $e>0$. If $k(e-1)$ is even,*

$$[k(e-1)/2e]\geqq (k-2)(e-1)/2e\,.$$

PROOF. Put $p=[k(e-1)/2e]$. Then $k(e-1)/2e<p+1$, so that $k(e-1)<2ep+2e$. Since both sides of this inequality are even, we have $k(e-1)\leqq 2ep+2e-2$, so that $(k-2)(e-1) \leqq 2ep$, and hence $(k-2)(e-1)/2e \leqq p$, q. e. d. (Note that the inequality is false if $k(e-1)$ is odd.)

By virtue of Lemma 2.22 and (2.5.1), we obtain, if $n > 1$,

$$\deg([B])-(2g-2) \geqq (n-1)\{(2g-2)+\textstyle\sum_{i=1}^{r}(1-e_i^{-1})+m\}+m \\ > m . \tag{2.6.4}$$

Therefore, by (2) of Prop. 2.14,

$$l([B]) = \deg([B])-g+1 .$$

If $n=1$ and $m>0$, we obtain the same result. If $n=1$ and $m=0$, we have $G_k(\Gamma)=S_k(\Gamma)$, and Cor. 2.17 answers the question. If $n=0$, then $B=0$, so that $l([B])=1$. If $n<0$, we have

$$\deg([B]) \leqq \deg(B) = n\{(2g-2)+\textstyle\sum_{i=1}^{r}(1-e_i^{-1})+m\} < 0$$

by (2.5.1), hence, by (1) of Prop. 2.14, $l([B])=0$. Thus we have proved

THEOREM 2.23. *Let g be the genus of $\Gamma\backslash\mathfrak{H}^*$, m the number of inequivalent cusps of Γ, and $e_1, \cdots, e_r$ the order of the inequivalent elliptic elements of $\bar{\Gamma}$. Then the dimension of the vector space $G_k(\Gamma)$, for an even integer k, is given by*

$$\dim G_k(\Gamma) = \begin{cases} (k-1)(g-1)+(k/2)\cdot m+\sum_{i=1}^{r}[k(e_i-1)/2e_i] & (k>2), \\ g+m-1 & (k=2,\ m>0), \\ g & (k=2,\ m=0), \\ 1 & (k=0), \\ 0 & (k<0). \end{cases}$$

Applying the same reasoning to the divisor $B' = B-\sum_{j=1}^{u} Q_j - \sum_{j=1}^{u'} Q'_j$, and observing that $\deg([B']) > 2g-2$ if $n \geqq 2$ in view of (2.6.4), we obtain

THEOREM 2.24. *The dimension of the vector space $S_k(\Gamma)$, for an even integer k, is given by*

$$\dim S_k(\Gamma) = \begin{cases} (k-1)(g-1)+\left(\frac{k}{2}-1\right)m+\sum_{i=1}^{r}[k(e_i-1)/2e_i] & (k>2), \\ g & (k=2), \\ 1 & (k=0,\ m=0), \\ 0 & (k=0,\ m>0), \\ 0 & (k<0). \end{cases}$$

Let us now suppose that k is odd. The notation F_0 and B being as above, put $\eta = F_0^2/(dz)^k$. Then $\eta \in \mathrm{Dif}^k(\mathfrak{W})$, and by Prop. 2.16,

(2.6.5) $\quad \operatorname{div}(F_0)=(1/2)\operatorname{div}(\eta)+(k/2)\cdot\{\sum_{i=1}^{r}(1-e_i^{-1})P_i+\sum_{j=1}^{u}Q_j+\sum_{j=1}^{u'}Q'_j\}$.

From our definition of $\operatorname{div}(F_0)$, we see that

$$\nu_P(F_0)\equiv\begin{cases}1/2 & (P=Q'_j),\\ (\text{integer})/e_i \quad \bmod \boldsymbol{Z} & (P=P_i),\\ 0 & \text{otherwise}.\end{cases}$$

Therefore, from (2.6.5), we obtain

$$(1/2)\cdot\nu_P(\eta)\equiv\begin{cases}1/2 & (P=Q_j),\\ \quad\bmod \boldsymbol{Z} & \\ 0 & \text{otherwise}.\end{cases}$$

This is obvious if $P\neq P_i$. For $P=P_i$, put $\nu_{P_i}(\eta)=c_i$. Then $c_i\in\boldsymbol{Z}$, and

$$\nu_{P_i}(F_0)=c_i/2+k(e_i-1)/2e_i=(e_ic_i+k(e_i-1))/2e_i.$$

Since e_i is odd and $e_i\cdot\nu_{P_i}(F_0)\in\boldsymbol{Z}$, c_i must be even.

Thus we obtain

$$[B]=(1/2)\operatorname{div}(\eta)+\sum_{i=1}^{r}[k(e_i-1)/2e_i]P_i+(k/2)\sum_{j=1}^{u}Q_j+((k-1)/2)\sum_{j=1}^{u'}Q'_j,$$

$$\begin{aligned}[B-\textstyle\sum_{j=1}^{u}Q_j-(1/2)\sum_{j=1}^{u'}Q'_j]&=(1/2)\operatorname{div}(\eta)+\textstyle\sum_{i=1}^{r}[k(e_i-1)/2e_i]\cdot P_i\\&\quad+((k-2)/2)\textstyle\sum_{j=1}^{u}Q_j+((k-1)/2)\sum_{j=1}^{u'}Q'_j.\end{aligned}$$

If $k<0$,

$$\deg([B])\leqq\deg(B)=(k/2)\cdot\{2g-2+\textstyle\sum_{i=1}^{r}(1-e_i^{-1})+m\}<0,$$

hence $G_k(\Gamma)=S_k(\Gamma)=\{0\}$. Suppose that $k\geqq 3$. Then, by Lemma 2.22,

$$\begin{aligned}&\deg([B-\textstyle\sum_{j=1}^{u}Q_j-(1/2)\sum_{j=1}^{u'}Q'_j])-(2g-2)\\&\quad=(k-2)(2g-2)/2+u(k-2)/2+u'(k-1)/2+\textstyle\sum_{i=1}^{r}[k(e_i-1)/2e_i]\\&\quad\geqq\frac{k-2}{2}\{2g-2+u+u'+\textstyle\sum_{i=1}^{r}(1-e_i^{-1})\}\\&\quad>0.\end{aligned}$$

(The e_i are all odd, since we are assuming that $-1\notin\Gamma$, see Cor. 1.21.) Therefore, by (2) of Prop. 2.14, we obtain

THEOREM 2.25. *The notation being as in Th. 2.23, suppose that* $-1\notin\Gamma$. *Let* u *(resp.* u'*) be the number of inequivalent regular (resp. irregular) cusps of* Γ. *Then, for an odd integer* k, *one has*

$$\dim G_k(\Gamma)=\begin{cases}(k-1)(g-1)+uk/2+u'(k-1)/2+\sum_{i=1}^{r}[k(e_i-1)/2e_i] & (k\geqq 3),\\ 0 & (k<0),\end{cases}$$

$$\dim S_k(\Gamma)=\begin{cases}(k-1)(g-1)+u(k-2)/2+u'(k-1)/2+\sum_{i=1}^{r}[k(e_i-1)/2e_i] & (k\geqq 3),\\ 0 & (k<0).\end{cases}$$

We observe that the number u must be even.

For an obvious reason, our method is not effective in the case $k=1$. If $k=1$, we have $\deg([B])=g-1+u/2$, so that

(2.6.6) $$\dim G_1(\Gamma) \geqq u/2,$$

(2.6.7) $$\dim G_1(\Gamma)=u/2 \quad \text{if} \quad u>2g-2.$$

Further, we have

$$\deg[B-\textstyle\sum_{j=1}^{u} Q_j-(1/2)\sum_{j=1}^{u'} Q'_j]=g-1-u/2.$$

Therefore, by (1) of Prop. 2.14, we obtain

(2.6.8) $$S_1(\Gamma)=\{0\} \quad \text{if} \quad u>2g-2.$$

For example, consider the group Γ_N of (1.6.1) for $N>2$. Clearly $-1 \notin \Gamma_N$. Since every parabolic element of Γ_N is conjugate to a power of $\begin{bmatrix} 1 & N \\ 0 & 1 \end{bmatrix}$ under Γ_1, we see that every cusp of Γ_N is regular. If $\mu_N=[\bar{\Gamma}_1 : \bar{\Gamma}_N]$, we have $u=\mu_N/N$, and $g=1+\mu_N/12-u/2$ as is shown in § 1.6, so that $u/2-g+1 = u(1-N/12)$. Therefore,

(2.6.9) $$\dim G_1(\Gamma_N)=\mu_N/2N \quad \text{and} \quad S_1(\Gamma_N)=\{0\} \quad \text{for} \quad 3 \leqq N \leqq 11.$$

It is an open problem to determine $\dim G_1(\Gamma)$ and $\dim S_1(\Gamma)$ in a more effective way.

Coming back to even k, if $\Gamma=SL_2(\boldsymbol{Z})$, we have $g=0$, $m=1$, and $\{e_1, e_2\} = \{2, 3\}$, so that, by an easy calculation, we obtain

PROPOSITION 2.26. *If $\Gamma=SL_2(\boldsymbol{Z})$, for even $k \geqq 2$,*

$$\dim G_k(\Gamma)=\begin{cases} [k/12] & (k \equiv 2 \bmod (12)), \\ [k/12]+1 & (k \not\equiv 2 \bmod (12)), \end{cases}$$

$$\dim S_k(\Gamma)=\begin{cases} 0 & (k=2), \\ [k/12]-1 & (k>2,\ k \equiv 2 \bmod (12)), \\ [k/12] & (k \not\equiv 2 \bmod (12)). \end{cases}$$

For example, we see that $\dim G_k(\Gamma)=1$ and $\dim S_k(\Gamma)=0$ for $k=4, 6, 8, 10$. We have seen in § 2.2 that $G_k(\Gamma)$ contains a non-trivial element E_k^*. Therefore,

$$G_k(\Gamma)=\boldsymbol{C} \cdot E_k^*, \qquad S_k(\Gamma)=\{0\} \qquad (k=4, 6, 8, 10).$$

For $k=12$, $\dim S_{12}(\Gamma)=1$, and $\dim G_{12}(\Gamma)=2$. The form $\Delta(z)$ considered in Th. 2.9 generates $S_{12}(\Gamma)$. As is shown in (2.2.1), E_k^* is not a cusp form. Therefore $G_{12}(\Gamma)$ is spanned by $\Delta(z)$ and E_{12}^*. By a similar reasoning, we can show that

$$S_{14}(\Gamma)=\{0\},$$

$$S_k(\Gamma)=\boldsymbol{C}\cdot\Delta\cdot E^*_{k-12} \qquad (k=16, 18, 20, 22),$$

$$S_{24}(\Gamma)=\boldsymbol{C}\cdot\Delta\cdot E^*_{12}+\boldsymbol{C}\cdot\Delta^2.$$

More generally, we have

PROPOSITION 2.27. *If* $\Gamma=SL_2(\boldsymbol{Z})$, *the space* $G_k(\Gamma)$ *is spanned over* $\boldsymbol{C}$ *by the functions* $g_2^a g_3^b$ *with non-negative integers* a *and* b *such that* $4a+6b=k$, *and* $S_k(\Gamma)=\Delta\cdot G_{k-12}(\Gamma)$, *where* Δ, g_2, *and* g_3 *are as in Th.* 2.9.

PROOF. Put $\boldsymbol{g}_2(\omega_1, \omega_2)=60\cdot E_4(\omega_1, \omega_2)$, $\boldsymbol{g}_3(\omega_1, \omega_2)=140\cdot E_6(\omega_1, \omega_2)$ with E_4 and E_6 of § 2.2. Then $g_2(z)=\boldsymbol{g}_2(z, 1)$ and $g_3(z)=\boldsymbol{g}_3(z, 1)$. We shall later (in § 4.2) show that $\boldsymbol{g}_2(\omega_1, \omega_2)$ and $\boldsymbol{g}_3(\omega_1, \omega_2)$ are algebraically independent over $\boldsymbol{C}$. It follows from this that the monomials $g_2(z)^a g_3(z)^b$, with $4a+6b=k$ for a fixed k, are linearly independent over $\boldsymbol{C}$, since

$$\omega_2^{-k}g_2(z)^a g_3(z)^b=\boldsymbol{g}_2(\omega_1, \omega_2)^a\boldsymbol{g}_3(\omega_1, \omega_2)^b \qquad (z=\omega_1/\omega_2).$$

Now it can easily be verified that the number of non-negative integral solutions (a, b) of $4a+6b=k$ is $[k/12]$ or $[k/12]+1$ according as $k\equiv 2$ or $\not\equiv 2$ mod (12). Therefore we obtain the first assertion in view of Prop. 2.26. Since $\Delta(z)\cdot G_{k-12}(\Gamma)\subset S_k(\Gamma)$ and $\dim S_k(\Gamma)=\dim G_{k-12}(\Gamma)$ by Prop. 2.26, we obtain the second assertion.

As an example of $S_k(\Gamma')$ with a congruence subgroup Γ' of $SL_2(\boldsymbol{Z})$, we have

EXAMPLE 2.28. *Let* N *be one of the integers* 2, 3, 5, *and* 11, *and let* $k=24/(N+1)$. *Then* $S_k(\Gamma_0(N))$ *is one-dimensional, and spanned by* $(\Delta(z)\Delta(Nz))^{1/(N+1)}$.

PROOF. The first assertion follows from Th. 2.24 and Prop. 1.43 by a simple computation. Since $\Delta(z)\neq 0$ everywhere on $\mathfrak{H}$, we can define $\Delta(z)^{1/m}$, for any positive integer m, as a holomorphic function on $\mathfrak{H}$. Put $g(z)=\Delta(z)\Delta(Nz)$. By (1.6.6) and Prop. 2.4, $\Delta(Nz)\in S_{12}(\Gamma_0(N))$, so that $g\in S_{24}(\Gamma_0(N))$. Since $\Delta(z)=q\phi(q)$ with a holomorphic function $\phi(q)$ in $q=e^{2\pi iz}$ such that $\phi(0)\neq 0$, we have $g(z)=q^{N+1}\phi(q)\phi(q^N)$, so that g has a zero of order $N+1$ at the cusp ∞. By Prop. 1.43, 0 and ∞ are the only inequivalent cusps of $\Gamma_0(N)$, since N is a prime. Put $\tau=N^{-1/2}\begin{bmatrix}0 & -1\\ N & 0\end{bmatrix}$. Then τ permutes 0 and ∞, and

$$g\,|\,[\tau]_k=\Delta(-1/Nz)\Delta(-1/z)(Nz)^{-12}z^{-12}=\Delta(Nz)\Delta(z)=g(z).$$

Since $\tau\begin{bmatrix}1 & 1\\ 0 & 1\end{bmatrix}\tau^{-1}=\begin{bmatrix}1 & 0\\ -N & 1\end{bmatrix}$ is a generator of

$$\{\gamma\in\Gamma_0(N)\,|\,\gamma(0)=0\},$$

we see that g has a zero of order $N+1$ also at the cusp 0. Now let f be a non-zero element of $S_k(\Gamma_0(N))$. Then both g and f^{N+1} belong to $S_{24}(\Gamma_0(N))$, so that $f^{N+1}/g \in A_0(\Gamma_0(N))$. Since $g \neq 0$ on $\mathfrak{H}$, we see that f^{N+1}/g is a holomorphic function on $\mathfrak{H}$. Moreover, since f has zero at 0 and ∞, f^{N+1}/g is holomorphic even at cusps. Therefore f^{N+1}/g must be a constant, which completes the proof.

It is a classical fact that $\Delta(z)$ has an expression

$$(2\pi)^{-12}\Delta(z) = q \prod_{n=1}^{\infty} (1-q^n)^{24} \qquad (q = e^{2\pi i z}).$$

Actually if we put $\eta(z) = e^{2\pi i z/24} \prod_{n=1}^{\infty} (1-q^n)$, η satisfies

$$\eta((az+b)/(cz+d)) = \lambda \cdot (cz+d)^{1/2}\eta(z) \qquad \left(\begin{bmatrix} a & b \\ c & d \end{bmatrix} \in SL_2(\boldsymbol{Z})\right)$$

with a certain constant λ depending on a, b, c, d. On this and other related topics, we refer the reader to Dedekind [8], Hermite [31], Hurwitz [32], Weber [89, pp. 112-130], Siegel [84], and Weil [100].

EXERCISE 2.29. Let N be one of the integers 2, 3, 4, 6, 12, and let $k = 12/N$. Prove that $S_k(\Gamma(N)) = \boldsymbol{C} \cdot \Delta(z)^{1/N}$.

REMARK 2.30. We can associate a function φ on $SL_2(\boldsymbol{R})$ with any element f of $G_k(\Gamma)$ by $\varphi(\alpha) = f(\alpha(i))j(\alpha, i)^{-k}$ for $\alpha \in SL_2(\boldsymbol{R})$. Then it can easily be verified that $\varphi(\gamma \cdot \alpha) = \varphi(\alpha)$ for every $\gamma \in \Gamma$, and $\varphi(\alpha \cdot \sigma(\theta)) = e^{ik\theta} \cdot \varphi(\alpha)$ for every $\sigma(\theta) = \begin{bmatrix} \cos\theta & \sin\theta \\ -\sin\theta & \cos\theta \end{bmatrix} \in SO(2)$. It is often convenient and essential to deal with φ instead of f. We shall not, however, pursue this view-point further in this book.

CHAPTER 3
HECKE OPERATORS AND THE ZETA-FUNCTIONS ASSOCIATED WITH MODULAR FORMS

3.1. Definition of the Hecke ring

Let G be a multiplicative group, and Γ, Γ' be subgroups of G. Let us write $\Gamma \sim \Gamma'$ if Γ and Γ' are commensurable, i. e., if $\Gamma \cap \Gamma'$ is of finite index in Γ and in Γ' (see § 1.1, especially Prop. 1.11). Fix a subgroup Γ of G, and put

$$\tilde{\Gamma} = \{\alpha \in G \mid \alpha\Gamma\alpha^{-1} \sim \Gamma\}.$$

By (1) of Prop. 1.11, we see that $\tilde{\Gamma}$ is a subgroup of G containing Γ, and also the center of G. Moreover, if Γ' is a subgroup of G commensurable with Γ, then $\tilde{\Gamma}' = \tilde{\Gamma}$. We call $\tilde{\Gamma}$ *the commensurator of* Γ *in* G.

In the following discussion, we shall fix Γ and a family $\{\Gamma_\lambda\}_{\lambda \in \Lambda}$ of subgroups of G which are commensurable with Γ, where Λ is a set of indices. Note that $\alpha\Gamma_\lambda\alpha^{-1} \sim \Gamma_\mu$ for every $\alpha \in \tilde{\Gamma}$ and every λ, $\mu \in \Lambda$.

PROPOSITION 3.1. *If* $\alpha \in \tilde{\Gamma}$, *one has disjoint coset decompositions*

$$\Gamma_\lambda \alpha \Gamma_\mu = \textstyle\bigcup_{i=1}^{d} \Gamma_\lambda \alpha_i \quad \textit{with} \quad d = [\Gamma_\mu : \Gamma_\mu \cap \alpha^{-1}\Gamma_\lambda \alpha],$$

$$\Gamma_\lambda \alpha \Gamma_\mu = \textstyle\bigcup_{j=1}^{e} \beta_j \Gamma_\mu \quad \textit{with} \quad e = [\Gamma_\lambda : \Gamma_\lambda \cap \alpha\Gamma_\mu\alpha^{-1}].$$

PROOF. Consider a disjoint coset decomposition

$$\Gamma_\mu = \textstyle\bigcup_i (\Gamma_\mu \cap \alpha^{-1}\Gamma_\lambda\alpha)\delta_i .$$

Then $\alpha^{-1}\Gamma_\lambda\alpha\Gamma_\mu = \bigcup_i \alpha^{-1}\Gamma_\lambda\alpha\delta_i$, hence $\Gamma_\lambda\alpha\Gamma_\mu = \bigcup_i \Gamma_\lambda\alpha\delta_i$. If $\Gamma_\lambda\alpha\delta_i = \Gamma_\lambda\alpha\delta_j$, then $\delta_i\delta_j^{-1} \in \Gamma_\mu \cap \alpha^{-1}\Gamma_\lambda\alpha$, and so $i=j$. This proves the first relation. A similar argument applies to the second one.

Now let us consider a $\boldsymbol{Z}$-module $R_{\lambda\mu}$ consisting of all formal finite sums of the form $\sum_k c_k \cdot (\Gamma_\lambda\alpha_k\Gamma_\mu)$ with $c_k \in \boldsymbol{Z}$, $\alpha_k \in \tilde{\Gamma}$. For every $\Gamma_\lambda\alpha\Gamma_\mu$ with $\alpha \in \tilde{\Gamma}$, denote by $\deg(\Gamma_\lambda\alpha\Gamma_\mu)$ the number of cosets $\Gamma_\lambda\varepsilon$ contained in $\Gamma_\lambda\alpha\Gamma_\mu$. Further, for $x = \sum_k c_k \cdot (\Gamma_\lambda\alpha_k\Gamma_\mu) \in R_{\lambda\mu}$, define $\deg(x)$ by $\deg(x) = \sum_k c_k \cdot \deg(\Gamma_\lambda\alpha_k\Gamma_\mu)$, and call it the *degree* of x. (We can actually define another degree by considering cosets $\delta\Gamma_\mu$ contained in $\Gamma_\lambda\alpha\Gamma_\mu$. This may not be equal to the above one.)

We shall now introduce a law of multiplication: $R_{\lambda\mu} \times R_{\mu\nu} \to R_{\lambda\nu}$. First consider disjoint coset decompositions

$$\Gamma_\lambda\alpha\Gamma_\mu = \cup_i \Gamma_\lambda\alpha_i\,, \qquad \Gamma_\mu\beta\Gamma_\nu = \cup_j \Gamma_\mu\beta_j$$

(of course with α and β in $\tilde{\Gamma}$). Then $\Gamma_\lambda\alpha\Gamma_\mu\beta\Gamma_\nu = \cup_j \Gamma_\lambda\alpha\Gamma_\mu\beta_j = \cup_{i,j}\Gamma_\lambda\alpha_i\beta_j$, therefore $\Gamma_\lambda\alpha\Gamma_\mu\beta\Gamma_\nu$ is a finite union of double cosets of the form $\Gamma_\lambda\xi\Gamma_\nu$. With $u=\Gamma_\lambda\alpha\Gamma_\mu$, $v=\Gamma_\mu\beta\Gamma_\nu$, and $w=\Gamma_\lambda\xi\Gamma_\nu$, we define the "product" $u\cdot v$ to be an element of $R_{\lambda\nu}$ given by

$$u\cdot v = \sum m(u\cdot v\,;\, w)w\,,$$

where the sum is extended over all $w=\Gamma_\lambda\xi\Gamma_\nu \subset \Gamma_\lambda\alpha\Gamma_\mu\beta\Gamma_\nu$, and

(3.1.1) $m(u\cdot v\,;\,w) =$ *the number of* (i,j) *such that* $\Gamma_\lambda\alpha_i\beta_j = \Gamma_\lambda\xi$ (*for a fixed* ξ).

To make this definition meaningful, one has to show that the right hand side of (3.1.1) depends only on u, v, and w, and not on the choice of representatives $\{\alpha_i\}$, $\{\beta_j\}$, and ξ. For that purpose, let $\sharp(S)$ denote the number of elements in a finite set S. We see that $\Gamma_\lambda\alpha_i\beta_j=\Gamma_\lambda\xi$ if and only if $\Gamma_\lambda\alpha_i = \Gamma_\lambda\xi\beta_j^{-1}$. Further, for a given j, the last equality holds for exactly one i. Therefore

$$\begin{aligned}&\sharp\{(i,j)\mid \Gamma_\lambda\alpha_i\beta_j=\Gamma_\lambda\xi\} = \sharp\{j\mid \xi\beta_j^{-1}\in\Gamma_\lambda\alpha\Gamma_\mu\}\\ &=\sharp\{j\mid \beta_j\in\Gamma_\mu\alpha^{-1}\Gamma_\lambda\xi\} = \sharp\{j\mid \Gamma_\mu\beta_j\subset\Gamma_\mu\alpha^{-1}\Gamma_\lambda\xi\}\\ &=\text{the number of cosets of the form } \Gamma_\mu\varepsilon \text{ in } \Gamma_\mu\beta\Gamma_\nu\cap\Gamma_\mu\alpha^{-1}\Gamma_\lambda\xi\,.\end{aligned}$$

The last number is obviously independent of the choice of $\{\alpha_i\}$ and $\{\beta_j\}$. Now, if $\Gamma_\lambda\xi\Gamma_\nu=\Gamma_\lambda\eta\Gamma_\nu$, then $\xi=\delta'\eta\delta$ with $\delta'\in\Gamma_\lambda$ and $\delta\in\Gamma_\nu$, hence

$$\Gamma_\mu\beta\Gamma_\nu\cap\Gamma_\mu\alpha^{-1}\Gamma_\lambda\xi = (\Gamma_\mu\beta\Gamma_\nu\cap\Gamma_\mu\alpha^{-1}\Gamma_\lambda\eta)\delta\,.$$

Therefore the number in question is independent of the choice of ξ.

After this verification, we can now define the law of multiplication $R_{\lambda\mu}\times R_{\mu\nu}\to R_{\lambda\nu}$ by extending $\boldsymbol{Z}$-linearly the map $(u,v)\mapsto u\cdot v$ in an obvious way.

PROPOSITION 3.2. *Let* u, v, w, $\{\alpha_i\}$, $\{\beta_j\}$, *and* ξ *be as above. Then*

$$\deg(w)\cdot m(u\cdot v\,;\,w) = \sharp\{(i,j)\mid \Gamma_\lambda\alpha_i\beta_j\Gamma_\nu = \Gamma_\lambda\xi\Gamma_\nu\}\,.$$

PROOF. Let $\Gamma_\lambda\xi\Gamma_\nu = \cup_{k=1}^{f}\Gamma_\lambda\xi_k$ be a disjoint coset decomposition. Then $\Gamma_\lambda\alpha_i\beta_j\Gamma_\nu=\Gamma_\lambda\xi\Gamma_\nu$ if and only if $\Gamma_\lambda\alpha_i\beta_j=\Gamma_\lambda\xi_k$ for some k. Observing that the last equality holds for exactly one k, we have therefore

$$\begin{aligned}\sharp\{(i,j)\mid \Gamma_\lambda\alpha_i\beta_j\Gamma_\nu=\Gamma_\lambda\xi\Gamma_\nu\} &= \textstyle\sum_{k=1}^{f}\sharp\{(i,j)\mid\Gamma_\lambda\alpha_i\beta_j=\Gamma_\lambda\xi_k\}\\ &= f\cdot m(u\cdot v\,;\,w)\,, \qquad \text{q. e. d.}\end{aligned}$$

PROPOSITION 3.3. *For every* $x\in R_{\lambda\mu}$ *and every* $y\in R_{\mu\nu}$, *one has*

$$\deg(x\cdot y) = \deg(x)\cdot\deg(y)\,.$$

PROOF. Let the notation be the same as in Prop. 3.2. Taking the summation over all $w=\Gamma_\lambda\xi\Gamma_\nu\subset\Gamma_\lambda\alpha\Gamma_\mu\beta\Gamma_\nu$, we have

$$\begin{aligned}\deg(u\cdot v)=\textstyle\sum_w \deg(w)\cdot m(u\cdot v\,;w)&=\text{the number of all }(i,j)\\&=\deg(u)\cdot\deg(v)\,.\end{aligned}$$

By linearity we obtain the formula in the general case.

PROPOSITION 3.4. *The above multiplication law is associative in the sense that* $(x\cdot y)\cdot z=x\cdot(y\cdot z)$ *for* $x\in R_{\kappa\lambda}$, $y\in R_{\lambda\mu}$, $z\in R_{\mu\nu}$.

PROOF. Let M_μ denote the $\boldsymbol{Z}$-module of all formal finite sums $\sum_k c_k\cdot\Gamma_\mu\xi_k$ with $c_k\in\boldsymbol{Z}$ and $\xi_k\in\tilde{\Gamma}$. Let $u=\Gamma_\lambda\alpha\Gamma_\mu=\bigcup_i\Gamma_\lambda\alpha_i$ (disjoint). We can assign to u a $\boldsymbol{Z}$-linear map of M_μ into M_λ (which we denote again by u) by means of the action $u\cdot\sum_k c_k\Gamma_\mu\xi_k=\sum_{i,k}c_k\Gamma_\lambda\alpha_i\xi_k$. It can easily be seen that this does not depend on the choice of $\{\alpha_i\}$ and $\{\xi_k\}$. By linearity we obtain a map of $R_{\lambda\mu}$ into $\mathrm{Hom}(M_\mu, M_\lambda)$. This map is injective. In fact, if $\sum_\alpha c_\alpha\cdot(\Gamma_\lambda\alpha\Gamma_\mu)\cdot\Gamma_\mu\xi=0$ is a non-trivial cancellation, we have $\Gamma_\lambda\alpha_1\xi=\Gamma_\lambda\alpha_2\xi$ for some α_1 and α_2. But this implies $\Gamma_\lambda\alpha_1\Gamma_\mu=\Gamma_\lambda\alpha_2\Gamma_\mu$, hence no such cancellation is possible. Thus we get the injectivity. Now consider disjoint coset decompositions $\Gamma_\lambda\alpha\Gamma_\mu=\bigcup_i\Gamma_\lambda\alpha_i$, $\Gamma_\mu\beta\Gamma_\nu=\bigcup_j\Gamma_\mu\beta_j$, and $\Gamma_\lambda\xi\Gamma_\nu=\bigcup_k\Gamma_\lambda\xi_k$ for each $\Gamma_\lambda\xi\Gamma_\nu\subset\Gamma_\lambda\alpha\Gamma_\mu\beta\Gamma_\nu$. Then

$$\begin{aligned}(\Gamma_\lambda\alpha\Gamma_\mu)\cdot\{(\Gamma_\mu\beta\Gamma_\nu)\cdot(\Gamma_\nu\eta)\}&=\textstyle\sum_{i,j}\Gamma_\lambda\alpha_i\beta_j\eta\\&=\textstyle\sum_{\xi,k}m(\Gamma_\lambda\alpha\Gamma_\mu\cdot\Gamma_\mu\beta\Gamma_\nu\,;\Gamma_\lambda\xi\Gamma_\nu)\Gamma_\lambda\xi_k\\&=\{(\Gamma_\lambda\alpha\Gamma_\mu)\cdot(\Gamma_\mu\beta\Gamma_\nu)\}\cdot\Gamma_\nu\eta\,.\end{aligned}$$

This shows that $(y\cdot z)\cdot a=y\cdot(z\cdot a)$ for $y\in R_{\lambda\mu}$, $z\in R_{\mu\nu}$ and $a\in M_\nu$. If further $x\in R_{\kappa\lambda}$, we have $((x\cdot y)\cdot z)\cdot a=(x\cdot y)\cdot(z\cdot a)=x\cdot(y\cdot(z\cdot a))=x\cdot((y\cdot z)\cdot a)=(x\cdot(y\cdot z))\cdot a$. By the injectivity proved above, we obtain $(x\cdot y)\cdot z=x\cdot(y\cdot z)$, q. e. d.

LEMMA 3.5. *Let* $\alpha\in\tilde{\Gamma}$. *Suppose that the number of cosets of the form* $\Gamma_\lambda\xi$ *in* $\Gamma_\lambda\alpha\Gamma_\mu$ *is equal to the number of cosets of the form* $\eta\Gamma_\mu$ *in* $\Gamma_\lambda\alpha\Gamma_\mu$. *Then there exists a common set of representatives* $\{\alpha_i\}$ *such that* $\Gamma_\lambda\alpha\Gamma_\mu=\bigcup_i\Gamma_\lambda\alpha_i=\bigcup_i\alpha_i\Gamma_\mu$.

PROOF. Let $\Gamma_\lambda\xi\subset\Gamma_\lambda\alpha\Gamma_\mu$ and $\eta\Gamma_\mu\subset\Gamma_\lambda\alpha\Gamma_\mu$. Then $\xi\in\Gamma_\lambda\alpha\Gamma_\mu=\Gamma_\lambda\eta\Gamma_\mu$, hence $\xi=\delta\eta\varepsilon$ with $\delta\in\Gamma_\lambda$ and $\varepsilon\in\Gamma_\mu$. Put $\zeta=\delta^{-1}\xi$. Then $\Gamma_\lambda\xi=\Gamma_\lambda\zeta$, $\eta\Gamma_\mu=\zeta\Gamma_\mu$, i. e., ζ is a common representative for $\Gamma_\lambda\xi$ and $\eta\Gamma_\mu$. Our assertion can easily be derived from this fact.

We shall now show that this phenomenon takes place when Γ is a discrete subgroup of $SL_2(\boldsymbol{R})$ with $\mu(\Gamma\backslash\mathfrak{H})<\infty$. As G, we take

$$GL_2^+(\boldsymbol{R})=\{\alpha\in GL_2(\boldsymbol{R})\mid \det(\alpha)>0\}.$$

PROPOSITION 3.6. *Let Γ_λ and Γ_μ be commensurable with Γ, and let $\alpha\in\tilde{\Gamma}$. If $\mu(\Gamma_\lambda\backslash\mathfrak{H})=\mu(\Gamma_\mu\backslash\mathfrak{H})$, the number of cosets of the form $\Gamma_\lambda\xi$ in $\Gamma_\lambda\alpha\Gamma_\mu$ is equal to the number of cosets of the form $\eta\Gamma_\mu$ in $\Gamma_\lambda\alpha\Gamma_\mu$.*

PROOF. Let $d=[\Gamma_\mu:\Gamma_\mu\cap\alpha^{-1}\Gamma_\lambda\alpha]$, $e=[\Gamma_\lambda:\Gamma_\lambda\cap\alpha\Gamma_\mu\alpha^{-1}]$. Then $e=[\alpha^{-1}\Gamma_\lambda\alpha:\alpha^{-1}\Gamma_\lambda\alpha\cap\Gamma_\mu]$, hence $d\cdot\mu(\Gamma_\mu\backslash\mathfrak{H})=\mu(\Gamma_\mu\cap\alpha^{-1}\Gamma_\lambda\alpha\backslash\mathfrak{H})=e\cdot\mu(\alpha^{-1}\Gamma_\lambda\alpha\backslash\mathfrak{H})$ $=e\cdot\mu(\Gamma_\lambda\backslash\mathfrak{H})$. Therefore we have $d=e$, which proves our assertion on account of Prop. 3.1.

Coming back to the general case, we obtain:

PROPOSITION 3.7. *Let $\alpha\in\tilde{\Gamma}$, $\beta\in\tilde{\Gamma}$. Then*
(1) $\Gamma_\lambda\alpha\beta\Gamma_\mu=(\Gamma_\lambda\alpha\Gamma_\lambda)\cdot(\Gamma_\lambda\beta\Gamma_\mu)$ *if* $\Gamma_\lambda\alpha=\alpha\Gamma_\lambda$;
(2) $\Gamma_\lambda\alpha\beta\Gamma_\mu=(\Gamma_\lambda\alpha\Gamma_\mu)\cdot(\Gamma_\mu\beta\Gamma_\mu)$ *if* $\Gamma_\mu\beta=\beta\Gamma_\mu$.

This follows immediately from our definition of the law of multiplication. Let us now fix any semi-group Δ such that $\Gamma\subset\Delta\subset\tilde{\Gamma}$. Let $R(\Gamma,\Delta)$ denote the $\boldsymbol{Z}$-module of all formal finite sums $\sum_k c_k\cdot\Gamma\alpha_k\Gamma$ with $c_k\in\boldsymbol{Z}$ and $\alpha_k\in\Delta$. With respect to the law of multiplication introduced above, $R(\Gamma,\Delta)$ becomes an associative ring, which we call *the Hecke ring with respect to Γ and Δ*. Obviously $\Gamma=\Gamma\cdot1\cdot\Gamma$ is the identity element.

PROPOSITION 3.8. *If G has an anti-automorphism $\alpha\mapsto\alpha^*$ such that $\Gamma^*=\Gamma$ and $(\Gamma\alpha\Gamma)^*=\Gamma\alpha\Gamma$ for every $\alpha\in\Delta$, then $R(\Gamma,\Delta)$ is commutative.* (Here an anti-automorphism of G means a one-to-one map of G onto itself satisfying $(\alpha\beta)^*=\beta^*\alpha^*$.)

PROOF. Applying $*$ to $\Gamma\alpha\Gamma$, we find that the number of right cosets in $\Gamma\alpha\Gamma$ is the same as the number of left cosets. Therefore, by Lemma 3.5, for any $\alpha,\beta\in\Delta$, we can put $\Gamma\alpha\Gamma=\bigcup_i\Gamma\alpha_i=\bigcup_i\alpha_i\Gamma$ and $\Gamma\beta\Gamma=\bigcup_j\Gamma\beta_j$ $=\bigcup_j\beta_j\Gamma$ (all disjoint). Then $\Gamma\alpha\Gamma=\Gamma\alpha^*\Gamma=\bigcup_i\Gamma\alpha_i^*$ and $\Gamma\beta\Gamma=\Gamma\beta^*\Gamma$ $=\bigcup_j\Gamma\beta_j^*$. If $\Gamma\alpha\Gamma\beta\Gamma=\bigcup_\xi\Gamma\xi\Gamma$, then $\Gamma\beta\Gamma\alpha\Gamma=\Gamma\beta^*\Gamma\alpha^*\Gamma=(\Gamma\alpha\Gamma\beta\Gamma)^*$ $=\bigcup_\xi\Gamma\xi\Gamma$. Therefore we have

$$(\Gamma\alpha\Gamma)\cdot(\Gamma\beta\Gamma)=\textstyle\sum_\xi c_\xi(\Gamma\xi\Gamma),$$

$$(\Gamma\beta\Gamma)\cdot(\Gamma\alpha\Gamma)=\textstyle\sum_\xi c'_\xi(\Gamma\xi\Gamma)$$

with the same components $\Gamma\xi\Gamma$. By Prop. 3.2, we have

$$c_\xi\cdot\deg(\Gamma\xi\Gamma)=\#\{(i,j)\mid\Gamma\alpha_i\beta_j\Gamma=\Gamma\xi\Gamma\}\qquad\text{(applying }*)$$

$$=\#\{(i,j)\mid\Gamma\beta_j^*\alpha_i^*\Gamma=\Gamma\xi\Gamma\}=c'_\xi\cdot\deg(\Gamma\xi\Gamma),$$

so that $c_\xi=c'_\xi$. This completes the proof.

So far, no motivation has been given to our discussion. First we take the simplest case as an example. Let F be an algebraic number field of finite degree, J the ring of integers in F, and $E=J^{\times}$ (see 0.2). For simplicity, let us assume that the class number of F is one. Then to every ideal $A=\alpha J$ in F, we can associate a coset $\alpha E=E\alpha E$. Thus in this case we put $E=\Gamma$ and $\Delta=J-\{0\}$ (or $\Delta=F-\{0\}$). Our multiplication is just ideal multiplication. If the class number is greater than one, we can make the same type of consideration by means of the ideles. Let us now take a non-commutative (say simple) algebra X over an algebraic number field. Let S be an order in X, i. e., a finitely generated $\boldsymbol{Z}$-submodule of X of maximal rank, which is a ring with identity. If $\Gamma=S^{\times}$, every left principal ideal $S\alpha$ is determined by $\Gamma\alpha$. Since we do not have commutativity, multiplication of ideals does not go so smoothly. Therefore, instead of $\Gamma\alpha$, we can take the double coset $\Gamma\alpha\Gamma$ which has less variances than $\Gamma\alpha$. This point of view will be clarified more explicitly in the following sections, by taking X to be a matrix algebra $M_n(\boldsymbol{Q})$, especially $M_2(\boldsymbol{Q})$. We shall also see in § 7.1 the connection of $\Gamma\alpha\Gamma$ with algebraic correspondences on algebraic curves.

3.2. A formal Dirichlet series with an Euler product

Let us confine ourselves to the case $G=GL_n(\boldsymbol{Q})$ and $\Gamma=SL_n(\boldsymbol{Z})$. For every integer $N\neq 0$, put

$$\Gamma_N=\{\gamma\in\Gamma \mid \gamma\equiv 1_n \mod (N)\}\,.$$

LEMMA 3.9. *Let* $\beta\in M_n(\boldsymbol{Z})$, $\det(\beta)=b\neq 0$. *Then* $\Gamma_{Nb}\subset\beta^{-1}\Gamma_N\beta\cap\beta\Gamma_N\beta^{-1}$.

PROOF. Put $\beta'=b\beta^{-1}$. Since $\beta'\in M_n(\boldsymbol{Z})$, if $\gamma\equiv 1_n \mod (Nb)$, then we have $\beta'\gamma\beta\equiv\beta'\beta=b\cdot 1_n \mod (Nb)$, hence $\beta^{-1}\gamma\beta\equiv 1_n \mod (N)$. This shows especially that $\beta^{-1}\gamma\beta\in M_n(\boldsymbol{Z})$. If $\gamma\in\Gamma_{Nb}$, we have $\det(\beta^{-1}\gamma\beta)=1$, so that $\beta^{-1}\gamma\beta\in\Gamma_N$, hence $\gamma\in\beta\Gamma_N\beta^{-1}$. Similarly $\gamma\in\beta^{-1}\Gamma_N\beta$.

LEMMA 3.10. $\tilde{\Gamma}=GL_n(\boldsymbol{Q})$.

PROOF. If $\alpha\in GL_n(\boldsymbol{Q})$, then $\alpha=c\beta$ with some $c\in\boldsymbol{Q}$ and $\beta\in M_n(\boldsymbol{Z})$. We have $\alpha\Gamma\alpha^{-1}=\beta\Gamma\beta^{-1}$. By Lemma 3.9, $\Gamma\cap\beta\Gamma\beta^{-1}$ contains Γ_b with $b=\det(\beta)$. Since $[\Gamma:\Gamma_b]<\infty$, we have $[\Gamma:\Gamma\cap\alpha\Gamma\alpha^{-1}]<\infty$. Transforming it by the inner automorphism $\xi\mapsto\alpha^{-1}\xi\alpha$, and then substituting α^{-1} for α, we obtain $[\alpha\Gamma\alpha^{-1}:\alpha\Gamma\alpha^{-1}\cap\Gamma]<\infty$, so that $\alpha\in\tilde{\Gamma}$.

Put $\Delta=\{\alpha\in M_n(\boldsymbol{Z}) \mid \det(\alpha)>0\}$. Obviously Δ is a semi-group, and $\Gamma\subset\Delta\subset\tilde{\Gamma}$. We shall now determine the structure of $R(\Gamma,\Delta)$. For n integers $a_1,\cdots,a_n$, let $\mathrm{diag}[a_1,\cdots,a_n]$ denote the diagonal matrix with diagonal elements $a_1,\cdots,a_n$. By virtue of the theory of elementary divisors (see

Lemma 3.11 below), we know that the representatives for $\Gamma\backslash\Delta/\Gamma$ are given by the diag $[a_1, \cdots, a_n]$ with positive integers $a_1, \cdots, a_n$ such that a_i divides a_{i+1}. Then we see that the transposition $\xi\mapsto {}^t\xi$ is an anti-automorphism of G, and ${}^t(\Gamma\alpha\Gamma)=\Gamma\alpha\Gamma$ for every double coset $\Gamma\alpha\Gamma$ with $\alpha\in G$, since we may assume α to be diagonal. By Prop. 3.8, this proves that $R(\Gamma, \Delta)$ is commutative.

Our next task is to obtain a sort of multiplication table for the elements of $R(\Gamma, \Delta)$. The main idea is to assign a lattice to each coset $\Gamma\alpha$, and to count the number of lattices instead of counting the number of cosets. For that purpose, put

$$V=\boldsymbol{Q}^n=\text{the vector space of all } n\text{-dimensional row vectors with components in } \boldsymbol{Q},$$

and let $G=GL_n(\boldsymbol{Q})$ act on the right of V. We call a submodule L of V a *lattice* (more specifically a $\boldsymbol{Z}$-*lattice*) in V, if L is finitely generated over $\boldsymbol{Z}$, and V is spanned by L over $\boldsymbol{Q}$. It can easily be seen that L is a lattice in V if and only if L is a free $\boldsymbol{Z}$-module of rank n. If $\alpha\in G$ and L is a lattice in V, then $L\alpha$ is a lattice in V. Note also that if W is a subspace of V and L is a lattice in V, then $L\cap W$ is a lattice in W. Further, if L and M are lattices in V, then (i) $L+M$ and $L\cap M$ are lattices in V; (ii) there exists a positive integer c such that $cL\subset M$.

LEMMA 3.11. *Let L and M be lattices in V. Then there exist n elements $u_1, \cdots, u_n$ of V and n positive rational numbers $b_1, \cdots, b_n$ such that $L=\sum_{i=1}^n \boldsymbol{Z}u_i$, $M=\sum_{i=1}^n \boldsymbol{Z}b_iu_i$, and $b_{i+1}\in b_i\boldsymbol{Z}$.*

This is just (a restatement of) the fundamental theorem of elementary divisors. Obviously $M\subset L$ if and only if all $b_i\in\boldsymbol{Z}$. We call $\{b_1\boldsymbol{Z}, \cdots, b_n\boldsymbol{Z}\}$ *the set of elementary divisors of M relative to L*, and write

$$\{L:M\}=\{b_1, \cdots, b_n\}=\{b_1\boldsymbol{Z}, \cdots, b_n\boldsymbol{Z}\}.$$

If $M\subset L$, one has $[L:M]=b_1\cdots b_n$. Especially if $\alpha=\mathrm{diag}[b_1, \cdots, b_n]$, then $\{L:L\alpha\}=\{b_1, \cdots, b_n\}$.

Hereafter let us denote exclusively by L the standard lattice $\boldsymbol{Z}^n$. Then

$$\Gamma=SL_n(\boldsymbol{Z})=\{\alpha\in G\mid L\alpha=L,\ \det(\alpha)>0\}.$$

For α and β in Δ, we have $\Gamma\alpha=\Gamma\beta$ if and only if $L\alpha=L\beta$.

LEMMA 3.12. *Let M and N be lattices in V. Then $\{L:M\}=\{L:N\}$ if and only if there exists an element α of Γ such that $M\alpha=N$.*

PROOF. The "if"-part is obvious. To prove the "only if"-part, let $\{L:M\}=\{L:N\}=\{a_1, \cdots, a_n\}$. Then there exist $2n$ elements u_i and v_i of V

such that $L=\sum_i \boldsymbol{Z}u_i=\sum_i \boldsymbol{Z}v_i$, $M=\sum_i \boldsymbol{Z}a_iu_i$, $N=\sum_i \boldsymbol{Z}a_iv_i$. Define an element α of G by $u_i\alpha=v_i$ for $i=1, \cdots, n$. Then $L\alpha=L$, $M\alpha=N$, and $\det(\alpha)=\pm 1$. If $\det(\alpha)=-1$, take $-v_1$ in place of v_1.

Let $a_1, \cdots, a_n$ be positive integers such that a_{i+1} is divisible by a_i. Define an element $T(a_1, \cdots, a_n)$ of $R(\Gamma, \Delta)$ by

$$T(a_1, \cdots, a_n)=\Gamma\alpha\Gamma, \qquad \alpha=\mathrm{diag}[a_1, \cdots, a_n].$$

As is remarked above, the ring $R(\Gamma, \Delta)$ is spanned by the $T(a_1, \cdots, a_n)$ over $\boldsymbol{Z}$.

LEMMA 3.13. *Let $\Gamma\alpha\Gamma=T(a_1, \cdots, a_n)$. Then $\Gamma\xi\mapsto L\xi$ gives a one-to-one correspondence between the cosets $\Gamma\xi$ in $\Gamma\alpha\Gamma$ and the lattices M such that $\{L:M\}=\{a_1, \cdots, a_n\}$.*

PROOF. We may assume that $\alpha=\mathrm{diag}[a_1, \cdots, a_n]$. If $\Gamma\xi=\Gamma\alpha\delta$ with $\delta\in\Gamma$, we have $\{L:L\xi\}=\{L:L\alpha\delta\}=\{L:L\alpha\}=\{a_1, \cdots, a_n\}$. Conversely, if $\{L:M\}=\{a_1, \cdots, a_n\}$, then, by Lemma 3.12, there exists an element γ of Γ such that $M=L\alpha\gamma$. Obviously $\Gamma\alpha\gamma\subset\Gamma\alpha\Gamma$. This correspondence $\Gamma\xi\mapsto L\xi$ is one-to-one, since $\Gamma\xi=\Gamma\eta$ if and only if $L\xi=L\eta$.

PROPOSITION 3.14. *The degree of $T(a_1, \cdots, a_n)$ coincides with the number of lattices M such that $\{L:M\}=\{a_1, \cdots, a_n\}$.*

This is an immediate consequence of Lemma 3.13.

PROPOSITION 3.15. *If $(\Gamma\alpha\Gamma)\cdot(\Gamma\beta\Gamma)=\sum_\xi c_\xi\cdot\Gamma\xi\Gamma$ with $c_\xi\in\boldsymbol{Z}$, then c_ξ is the number of lattices M such that $\{L:M\}=\{L:L\beta\}$ and $\{M:L\xi\}=\{L:L\alpha\}$.*

PROOF. Let $\Gamma\alpha\Gamma=\bigcup_i\Gamma\alpha_i$ and $\Gamma\beta\Gamma=\bigcup_j\Gamma\beta_j$ (disjoint). Then

$$c_\xi=\#\{(i,j)\mid\Gamma\alpha_i\beta_j=\Gamma\xi\}=\#\{(i,j)\mid L\alpha_i\beta_j=L\xi\}.$$

Here note that i is uniquely determined by ξ and j. Assume $L\alpha_i\beta_j=L\xi$ and put $M=L\beta_j$. Then $\{L:M\}=\{L:L\beta\}$, and $\{M:L\xi\}=\{L\beta_j:L\alpha_i\beta_j\}=\{L:L\alpha_i\}$ $=\{L:L\alpha\}$. Conversely, let M be a lattice such that $\{L:M\}=\{L:L\beta\}$ and $\{M:L\xi\}=\{L:L\alpha\}$. By Lemma 3.13, $M=L\beta_j$ for one and only one j. Then $\{L:L\xi\beta_j^{-1}\}=\{L\beta_j:L\xi\}=\{L:L\alpha\}$. By Lemma 3.13, $L\xi\beta_j^{-1}=L\alpha_i$ for some i, and $L\xi=L\alpha_i\beta_j$. Thus each M determines a pair (i,j) and conversely. This proves our assertion.

PROPOSITION 3.16. *Let α and β be elements of Δ such that $\det(\alpha)$ is prime to $\det(\beta)$. Then $(\Gamma\alpha\Gamma)\cdot(\Gamma\beta\Gamma)=\Gamma\alpha\beta\Gamma$. In other words,*

$$T(a_1, \cdots, a_n)\cdot T(b_1, \cdots, b_n)=T(a_1b_1, \cdots, a_nb_n) \qquad \textit{if} \quad (a_n, b_n)=1.$$

PROOF. Let $\xi\in\Gamma\alpha\Gamma\beta\Gamma$. Let M and M' be such that $\{L:M\}=\{L:M'\}=$

$=\{L:L\beta\}$ and $\{M:L\xi\}=\{M':L\xi\}=\{L:L\alpha\}$. We have $[M+M':M]=[M':M\cap M']$. The left hand side is a divisor of $[L:M]=\det(\beta)$, and the right hand side is a divisor of $[L:L\alpha]=\det(\alpha)$, since $M+M'\subset L$ and $L\xi\subset M\cap M'$. Since $\det(\alpha)$ is prime to $\det(\beta)$, we have $M+M'=M$ and $M'=M\cap M'$, so that $M=M'$. On account of Prop. 3.15, this implies that the multiplicity of $\Gamma\xi\Gamma$ in $(\Gamma\alpha\Gamma)\cdot(\Gamma\beta\Gamma)$ is one. Now if $\xi\in\Gamma\alpha\Gamma\beta\Gamma$, we can find at least one M as above. Then $L\xi\subset M\subset L$, and $L/L\xi$ is isomorphic to $L/M\oplus M/L\xi$, hence to $L/L\alpha\oplus L/L\beta$, since $\det(\alpha)$ is prime to $\det(\beta)$. Therefore the elementary divisors of $L\xi$ relative to L are completely determined by α and β. This shows that $\Gamma\alpha\Gamma\beta\Gamma$ consists of only one double coset, which is obviously $\Gamma\alpha\beta\Gamma$, q. e. d.

From the above proposition, it follows that every $T(a_1,\cdots,a_n)$ is a product of elements of the form $T(p^{e_1},\cdots,p^{e_n})$ with a prime p and exponents $0\leqq e_1\leqq e_2\leqq\cdots\leqq e_n$, and such an expression is unique (so long as we take at most one factor for each prime). For each prime p, let $R_p^{(n)}$ denote the subring of $R(\Gamma,\Delta)$ generated by the $T(p^{e_1},\cdots,p^{e_n})$. Then our question is reduced to the study of the structure of $R_p^{(n)}$. Before proceeding with this task, we notice a simple fact:

PROPOSITION 3.17. $T(c,\cdots,c)T(b_1,\cdots,b_n)=T(cb_1,\cdots,cb_n)$.

This follows immediately from our definition of the multiplication-law in $R(\Gamma,\Delta)$. In particular, we see that $T(c,\cdots,c)$ is not a zero divisor in the ring $R(\Gamma,\Delta)$. (Actually, we shall see later that $R(\Gamma,\Delta)$ is an integral domain.)

Now we fix a prime p, and will study the structure of $R_p^{(n)}$. Consider $(\mathbf{Z}/p\mathbf{Z})^n=L/pL$ as a vector space of dimension n over the prime field $\mathbf{Z}/p\mathbf{Z}$.

PROPOSITION 3.18. *Let $c_k^{(n)}$ be the number of k-dimensional subspaces of $(\mathbf{Z}/p\mathbf{Z})^n$. Then*

$$c_k^{(n)}=c_{n-k}^{(n)}=\frac{(p^n-1)(p^n-p)\cdots(p^n-p^{k-1})}{(p^k-1)(p^k-p)\cdots(p^k-p^{k-1})}$$

$$=\deg(T(\underbrace{1,\cdots,1}_{n-k},\underbrace{p,\cdots,p}_{k}))\,.$$

PROOF. The equality $c_k^{(n)}=c_{n-k}^{(n)}$ and the expression of $c_k^{(n)}$ as a rational function in p are well-known. To connect this with $\deg(T)$, we use Prop. 3.14. Let M be a lattice in V such that $\{L:M\}=\{1,\cdots,1,p,\cdots,p\}$ with $n-k$ 1's and k p's. Then $pL\subset M\subset L$, and M/pL is an $(n-k)$-dimensional subspace of L/pL. Conversely, for every $(n-k)$-dimensional subspace K of L/pL, we can find such an M uniquely so that $M/pL=K$. This fact together with Prop. 3.14 proves the equality.

Define a $\boldsymbol{Z}$-linear map $\psi : R_p^{(n+1)} \to R_p^{(n)}$ by

$$\psi(T(1, p^{a_1}, \cdots, p^{a_n})) = T(p^{a_1}, \cdots, p^{a_n}) ,$$

$$\psi(T(p^{a_0}, p^{a_1}, \cdots, p^{a_n})) = 0 \qquad \text{if} \quad a_0 > 0 .$$

LEMMA 3.19. *ψ is a surjective homomorphism, and* Ker(ψ) *coincides with* $T(p, \cdots, p) \cdot R_p^{(n+1)}$.

PROOF. The surjectivity is obvious; the assertion about Ker(ψ) follows from Prop. 3.17 and the definition of ψ. Therefore, to complete the proof, it suffices to verify the multiplicativity for the elements $T(1, p^{a_1}, \cdots, p^{a_n})$. Let us put, for simplicity,

$$e' = \{1, p^{a_1}, \cdots, p^{a_n}\} , \qquad e = \{p^{a_1}, \cdots, p^{a_n}\} ,$$

$$f' = \{1, p^{b_1}, \cdots, p^{b_n}\} , \qquad f = \{p^{b_1}, \cdots, p^{b_n}\} ,$$

$$g' = \{1, p^{c_1}, \cdots, p^{c_n}\} , \qquad g = \{p^{c_1}, \cdots, p^{c_n}\} ,$$

$$\mu_g = m(T(e) \cdot T(f) ; T(g)) , \qquad \mu_{g'} = m(T(e') \cdot T(f') ; T(g')) .$$

We are going to show that $\mu_g = \mu_{g'}$. Let $L' = \boldsymbol{Z}^{n+1} = \sum_{i=0}^n \boldsymbol{Z} u_i$, $L = \sum_{i=1}^n \boldsymbol{Z} u_i$, $N' = \boldsymbol{Z} u_0 + \sum_{i=1}^n \boldsymbol{Z} p^{c_i} u_i$, $N = \sum_{i=1}^n \boldsymbol{Z} p^{c_i} u_i$. Then $\{L : N\} = g$, $\{L' : N'\} = g'$, and by Prop. 3.15,

$$\mu_g = \# \{M \mid \{L : M\} = f, \ \{M : N\} = e\} ,$$

$$\mu_{g'} = \# \{M' \mid \{L' : M'\} = f', \ \{M' : N'\} = e'\} .$$

Suppose $\{L' : M'\} = f'$, $\{M' : N'\} = e'$. Then $u_0 \in N' \subset M'$. Put $M = M' \cap L$. Then $M' = \boldsymbol{Z} u_0 + M$, and clearly $\{L : M\} = f$, $\{M : N\} = e$. Conversely, if M is a lattice in $\boldsymbol{Q}^n = \sum_{i=1}^n \boldsymbol{Q} u_i$ such that $\{L : M\} = f$, $\{M : N\} = e$, then put $M' = \boldsymbol{Z} u_0 + M$. It can easily be verified that $M = M' \cap L$, $\{L' : M'\} = f'$, $\{M' : N'\} = e'$. This shows $\mu_g = \mu_{g'}$. Now we have

$$T(e) \cdot T(f) = \sum \mu_g T(g) ,$$

$$T(e') \cdot T(f') = \sum \mu_{g'} T(g') + T(p, \cdots, p) \cdot X$$

with an element X of $R_p^{(n+1)}$. Since $\psi(T(p, \cdots, p)) = 0$, we see that ψ maps $T(e') \cdot T(f')$ to $T(e) \cdot T(f)$. This completes the proof.

THEOREM 3.20. *The ring $R_p^{(n)}$ is the polynomial ring over $\boldsymbol{Z}$ in n elements*

$$T(1, \cdots, 1, p), \ T(1, \cdots, 1, p, p), \cdots, T(p, \cdots, p) ,$$

which are algebraically independent. Especially $R_p^{(n)}$ has no zero-divisors (other than 0).

PROOF. We shall use induction on n. For $n = 1$, our assertion is clear since $T(p^a)=T(p)^a$ by Prop. 3.17. Let us therefore assume that $n>1$, and the assertion is true for $n-1$. For every $\Gamma\alpha\Gamma$ with $\det(\alpha)=p^\nu$, put $w(\Gamma\alpha\Gamma)=\nu$, and for $X=\sum_k c_k\cdot\Gamma\alpha_k\Gamma\in R_p^{(n)}$, define $w(X)$ to be the maximum of $w(\Gamma\alpha_k\Gamma)$ with non-vanishing c_k. Call X *homogeneous* if the $w(\Gamma\alpha_k\Gamma)$ are the same for all $c_k\neq 0$. In particular $T(p^{a_1},\cdots,p^{a_n})$ is homogeneous, and $w(T(p^{a_1},\cdots,p^{a_n}))=a_1+\cdots+a_n$. The product of two homogeneous elements is clearly homogeneous. Put $T_k^{(n)}=T(1,\cdots,1,p,\cdots,p)$ with $n-k$ 1's and k p's. We are going to prove, by induction on w, that every element X of $R_p^{(n)}$ is a polynomial in $T_1^{(n)},\cdots,T_n^{(n)}$. It is sufficient to consider the elements of the form $X=T(p^{a_1},\cdots,p^{a_n})$. If $a_1>0$, we have, by Lemma 3.17,

$$T(p^{a_1},\cdots,p^{a_n})=T(p,\cdots,p)T(p^{a_1-1},\cdots,p^{a_n-1}),$$

so that the question is reduced to an element with smaller w. (Note that $w(X)=0$ if and only if X is a constant, i.e., an element of $\boldsymbol{Z}$.) Therefore assume $a_1=0$. Consider the homomorphism $\psi: R_p^{(n)}\to R_p^{(n-1)}$ obtained in Lemma 3.19. By the assumption of induction, we have

$$\psi(X)=T(p^{a_2},\cdots,p^{a_n})=\sum_k u_k\cdot M_k(T_i^{(n-1)}),$$

where $u_k\in\boldsymbol{Z}$, and the $M_k(T_i^{(n-1)})$ are monomials in $T_1^{(n-1)},\cdots,T_{n-1}^{(n-1)}$. Note that each $M_k(T_i^{(n-1)})$ is homogeneous. Therefore we may assume that $w(M_k(T_i^{(n-1)}))=w(X)$ for all k, since there is no cancellation between homogeneous elements with distinct w's. Substituting $T_i^{(n)}$ for $T_i^{(n-1)}$, put

$$Y=\sum_k u_k\cdot M_k(T_1^{(n)},\cdots,T_{n-1}^{(n)}).$$

We see easily that $w(M_k(T_i^{(n)}))=w(X)$. Since $\psi(X-Y)=0$, there exists an element Z of $R_p^{(n)}$ such that $X-Y=T(p,\cdots,p)\cdot Z$. It is clear that $w(Z)<w(X)$. By induction, Z is a polynomial in $T_i^{(n)}$, hence $X\in\boldsymbol{Z}[T_1^{(n)},\cdots,T_n^{(n)}]$.

To prove the algebraic independence of the $T_i^{(n)}$, let P be a polynomial such that $P(T_1^{(n)},\cdots,T_n^{(n)})=0$, $P\neq 0$. We can express P in the form

$$P(T_1^{(n)},\cdots,T_n^{(n)})=\sum_{i=k}^{l}(T_n^{(n)})^iP_i(T_1^{(n)},\cdots,T_{n-1}^{(n)}),$$

where $0\leq k\leq l$, and $P_k\neq 0$. Since $T_n^{(n)}$ is not a zero-divisor (see Prop. 3.17), we have $0=\sum_{i=k}^{l}(T_n^{(n)})^{i-k}P_i(T_1^{(n)},\cdots,T_{n-1}^{(n)})$. Applying ψ, we obtain $P_k(T_1^{(n-1)},\cdots,T_{n-1}^{(n-1)})=0$. By induction, we have $P_k=0$, a contradiction. This completes the proof.

From Th. 3.20, it follows that the whole ring $R(\Gamma,\Delta)$ is a polynomial ring over $\boldsymbol{Z}$ with infinitely many indeterminates of the form $T(1,\cdots,1,p,\cdots,p)$, p being any prime. In particular, $R(\Gamma,\Delta)$ is an integral domain.

For every positive integer m, let $T(m)$ denote the sum of all $\Gamma\alpha\Gamma$ with $\alpha\in\Delta$ and $\det(\alpha)=m$. Now we consider a formal Dirichlet series (with coefficients in $R(\Gamma,\Delta)$)

$$D(s)=\sum_{m=1}^{\infty} T(m)m^{-s}=\sum_{\Gamma\backslash\Delta/\Gamma}(\Gamma\alpha\Gamma)\cdot\det(\alpha)^{-s},$$

where the last sum is taken over all distinct double cosets $\Gamma\alpha\Gamma$ with α in Δ. From Prop. 3.16, we can easily derive

$$T(mm')=T(m)T(m') \qquad \text{if} \quad (m, m')=1\,. \tag{3.2.1}$$

Therefore, $D(s)$ can be (formally) expressed as an infinite product

$$D(s)=\prod_p \left[\sum_{k=0}^{\infty} T(p^k)p^{-ks}\right],$$

where p runs over all primes. By our definition of $T(m)$, we have

$$\sum_{k=0}^{\infty} T(p^k)X^k=\sum_{0\leqq e_1\leqq\cdots\leqq e_n} T(p^{e_1},\cdots,p^{e_n})X^{e_1+\cdots+e_n}$$

with any indeterminate X. We shall now prove that this formal power series is actually a rational expression in X:

THEOREM 3.21. *Let $T_i^{(n)}=T(1,\cdots,1,p,\cdots,p)$ with $n-i$ 1's and i p's, and let X be an indeterminate. Then*

$$\sum_{k=0}^{\infty} T(p^k)X^k=\left[\sum_{i=0}^{n}(-1)^i p^{i(i-1)/2}T_i^{(n)}X^i\right]^{-1},$$

and therefore

$$\sum_{m=1}^{\infty} T(m)m^{-s}=\prod_p \left[\sum_{i=0}^{n}(-1)^i p^{i(i-1)/2}T_i^{(n)}p^{-is}\right]^{-1},$$

where the product is extended over all primes p.

First we prove two lemmas.

LEMMA 3.22. *Let the integers $c_i^{(k)}$ be as in Prop. 3.18. Then*

$$T_i^{(n)}X^i\cdot\left(\sum_{m=0}^{\infty} T(p^m)X^m\right)$$
$$=\sum_{k=0}^{n} c_i^{(k)}\cdot\left\{\sum_{1\leqq d_1\leqq\cdots\leqq d_k} T(1,\cdots,1,p^{d_1},\cdots,p^{d_k})X^{d_1+\cdots+d_k}\right\}.$$

Here we understand that $c_i^{(k)}=0$ if $i>k$, and $c_0^{(0)}=1$.

PROOF. Fix a set of exponents $\{d_1,\cdots,d_k\}$, and denote by $\mu(d)$ the coefficient of $T(1,\cdots,1,p^{d_1},\cdots,p^{d_k})X^{d_1+\cdots+d_k}$ in the product $T_i^{(n)}X^i\cdot(\sum_{m=0}^{\infty}T(p^m)X^m)$. We observe that such a term can occur in $T_i^{(n)}X^i\cdot T(p^m)X^m$ only if $i+m=d_1+\cdots+d_k$. Fix a lattice N such that $\{L:N\}=\{1,\cdots,1,p^{d_1},\cdots,p^{d_k}\}$. By Prop. 3.15,

$$\mu(d)=\sum_\alpha \#\Big\{M \,\Big|\, \{L:M\}=\{1,\cdots,1,p,\cdots,p\},\ \{M:N\}=\{L:L\alpha\}\Big\},$$

where the sum is extended over all $\Gamma\alpha\Gamma$ such that $\det(\alpha)=p^m$ and $\alpha\in\Delta$. (Here and in the following, the number of repetitions of p is always i.) If $\{L:M\}=\{1,\cdots,1,p,\cdots,p\}$ and $N\subset M$, we can find an element α of Δ such that $\{M:N\}=\{L:L\alpha\}$, and obviously $\det(\alpha)=p^m$. Therefore $\mu(d)$ is the number of lattices M such that

$$(*)\qquad N\subset M,\qquad \{L:M\}=\{1,\cdots,1,p,\cdots,p\}.$$

Take a basis $\{u_i\}$ so that $L=\sum_{\nu=1}^{n}Zu_\nu$, and

$$N=\sum_{\nu=1}^{n-k}Zu_\nu+\sum_{\nu=1}^{k}Zp^{d_\nu}u_{n-k+\nu}.$$

Then $pL+N=\sum_{\nu=1}^{n-k}Zu_\nu+\sum_{\nu=1}^{k}Zpu_{n-k+\nu}$, hence $L/(pL+N)$ is isomorphic to $(Z/pZ)^k$. If M satisfies $(*)$, we have $pL+N\subset M$, and L/M is isomorphic to $(Z/pZ)^i$. Therefore $\mu(d)\neq 0$ only if $i\leqq k$. Assuming $i\leqq k$, we see that $M/(pL+N)$ is a $(k-i)$-dimensional subspace of $L/(pL+N)$. Conversely, any $(k-i)$-dimensional subspace of $L/(pL+N)$ can be written in the form $M/(pL+N)$ with a unique M satisfying $(*)$. We have thus $\mu(d)=c_i^{(k)}$, which completes the proof.

LEMMA 3.23. *$\sum_{i=0}^{k}(-1)^i p^{i(i-1)/2}c_i^{(k)}=0$ if $k>0$.*

PROOF. Put $f(X)=\prod_{i=0}^{k-1}(X-p^i)$. Then we have

$$1=\sum_{i=0}^{k-1}f(X)/[f'(p^i)(X-p^i)],$$

since the right hand side is a polynomial of degree $<k$ which takes the value 1 at k points $p^0, p^1,\cdots,p^{k-1}$. Substitute p^k for X. Then we find

$$1=\sum_{i=0}^{k-1}c_i^{(k)}(-1)^{k-i-1}p^{(k-i)(k-i-1)/2}=\sum_{j=1}^{k}c_j^{(k)}(-1)^{j-1}p^{j(j-1)/2},$$

q. e. d.

PROOF of Th. 3.21. We simply take the product

$$[\sum_{i=0}^{n}(-1)^i p^{i(i-1)/2}T_i^{(n)}X^i]\cdot[\sum_{m=0}^{\infty}T(p^m)X^m].$$

By Lemma 3.22, this equals

$$\sum_{i=0}^{n}(-1)^i p^{i(i-1)/2}\sum_{k=0}^{n}c_i^{(k)}\cdot\{\sum T(1,\cdots,1,p^{d_1},\cdots,p^{d_k})X^{d_1+\cdots+d_k}\}.$$

By Lemma 3.23, only the term with $k=0$ is non-vanishing, and that term is just 1, q. e. d.

It is worth while restating Th. 3.21 in the special cases $n=1, 2$. If $n=1$,

$$\sum_{m=1}^{\infty}T(m)m^{-s}=\prod_p[1-T(p)p^{-s}]^{-1},$$

and if $n=2$,

$$(3.2.2)\qquad \sum_{m=1}^{\infty}T(m)m^{-s}=\prod_p[1-T(1,p)p^{-s}+T(p,p)p^{1-2s}]^{-1}.$$

(Note also that $T(1,p)=T(p)$.)

Th. 3.21, in the case $n=2$, is due to Hecke [29], although he did not discuss the abstract ring $R(\Gamma,\Delta)$, but its representations in the space of modular forms, see below. The abstract ring $R(\Gamma, \Delta)$ was introduced in [71]. The result of Th. 3.21 for arbitrary n is due to Tamagawa [86].

THEOREM 3.24. *If* $n=2$, *and* p *denotes a prime, then the following formulas hold.*

(1) $T(m)=\sum_{ad=m,\, a|d} T(a, d)$.

(2) $T(1, p^k)=T(p^k)-T(p, p)T(p^{k-2})$ $\qquad (k \geqq 2)$.

(3) $T(m)T(n)=\sum_{d|(m,n)} d \cdot T(d, d)T(mn/d^2)$.

(4) $T(p^r)T(p^s)=\sum_{t=0}^{r} p^t T(p^t, p^t)T(p^{r+s-2t})$ $\qquad (r \leqq s)$,
especially $T(p)T(p^k)=T(p^{k+1})+pT(p, p)T(p^{k-1})$ $\qquad (k>0)$.

(5) $T(p)T(1, p^k)=T(1, p^{k+1})+\begin{cases} (p+1)T(p, p) & (k=1), \\ pT(p, p^k) & (k>1). \end{cases}$

(6) $\deg(T(1, p^k))=\deg(T(p^i, p^{i+k}))=p^{k-1}(p+1)$ $\qquad (k>0)$.

(7) $\deg(T(m))=$ *the sum of all positive divisors of* m.

PROOF. The first two relations are obvious. Since $R_p^{(2)}$ is a polynomial ring $\boldsymbol{Z}[T(p), T(p, p)]$, we can embed $R_p^{(2)}$ into a polynomial ring $\boldsymbol{Q}[A, B]$ with two indeterminates A and B so that

$$1-T(p)X+pT(p, p)X^2=(1-AX)(1-BX).$$

Then

$$\sum_{m=0}^{\infty} T(p^m)X^{m+1}=[(1-AX)^{-1}-(1-BX)^{-1}]/(A-B)$$
$$=\sum_{m=0}^{\infty}(A^m-B^m)X^m/(A-B),$$

so that $T(p^m)=(A^{m+1}-B^{m+1})/(A-B)=\sum_{t=0}^{m} A^{m-t}B^t$. Therefore

$$\begin{aligned} T(p^r)T(p^s) &= [A^{s+1}T(p^r)-B^{s+1}T(p^r)]/(A-B) \\ &= (A^{s+1}\textstyle\sum_{t=0}^{r} A^{r-t}B^t-B^{s+1}\sum_{t=0}^{r} A^tB^{r-t})/(A-B) \\ &= \textstyle\sum_{t=0}^{r} A^tB^t(A^{r+s-2t+1}-B^{r+s-2t+1})/(A-B) \\ &= \textstyle\sum_{t=0}^{r} p^tT(p^t, p^t)T(p^{r+s-2t}), \end{aligned}$$

which proves (4). Observe that (4) is a special case of (3). Therefore (3) follows from (4) and (3.2.1). If $k=1$, (5) is a special case of (4). If $k>1$, we obtain, from (2) and (4),

$$\begin{aligned} T(p)T(1, p^k) &= T(p^{k+1})+T(p, p)[pT(p^{k-1})-T(p)T(p^{k-2})] \\ &= T(1, p^{k+1})+T(p, p)[(p+1)T(p^{k-1})-T(p)T(p^{k-2})]. \end{aligned}$$

The last term $T(p)T(p^{k-2})$ is given by (3). Then we obtain (5). By Prop. 3.18, we have $\deg(T(p))=c_1^{(2)}=p+1$, and $\deg(T(p, p))=1$. Applying Prop. 3.3 to (4), we obtain

$$(p+1)\cdot \deg (T(p^k)) = \deg (T(p^{k+1})) + p\cdot \deg (T(p^{k-1}))\,.$$

Then, by induction on k, we can easily verify that

$$(*) \qquad \deg (T(p^k)) = 1+p+\cdots+p^k\,.$$

From this relation, Prop. 3.3, and (3.2.1), we obtain (7). Then (6) follows from (*) and (2).

A few remarks are in order concerning the meaning of the Euler product of Th. 3.21. Since we have been working only with the abstract ring $R(\Gamma, \Delta)$, the Euler product is valid only formally. It is not an analytic statement, but rather an arithmetic statement about the properties of the coefficients of the Dirichlet series. The use of the symbol m^{-s} (so far) involves no analysis; rather m^{-s} is just an indeterminate.

Now let us introduce some analysis. Suppose that we represent the ring $R(\Gamma, \Delta)$ on some vector space over $\boldsymbol{C}$. Then the ring elements $T(m)$ act as matrices with complex coefficients. Through such a representation, the above result concerning the Euler product, if it converges, gives an analytic statement about a certain matrix-valued function of a complex variable s, which has certain multiplicative properties. If we diagonalize the matrices $T(m)$ simultaneously, then the diagonal elements of $D(s)$ are ordinary Dirichlet series, each of which has an Euler product. This will actually be done in §§ 3.4, 3.5.

As an example, consider the simplest representation

$$R(\Gamma, \Delta) \to \boldsymbol{Z}$$
$$\Gamma\alpha\Gamma \mapsto \deg(\Gamma\alpha\Gamma) \qquad \text{(see Prop. 3.3)}.$$

Then we obtain

$$\textstyle\sum_{m=1}^{\infty} \deg (T(m)) m^{-s} = \prod_p \left[\sum_{i=0}^{n} (-1)^i p^{i(i-1)/2} c_i^{(n)} p^{-is}\right]^{-1}.$$

But we have an equality

$$(3.2.3) \qquad \textstyle\sum_{i=0}^{n} (-1)^i p^{i(i-1)/2} c_i^{(n)} X^i = (1-X)(1-pX)\cdots(1-p^{n-1}X)\,,$$

which can easily be proved by induction on n. Therefore

$$\textstyle\sum_{m=1}^{\infty} \deg (T(m)) m^{-s} = \zeta(s)\zeta(s-1)\cdots\zeta(s-n+1)\,,$$

with the Riemann zeta-function ζ.

REMARK 3.25. Let F be a local field, i.e., a finite algebraic extension of the p-adic field $\boldsymbol{Q}_p$, or the field of power series in one variable over a finite field. Let $\mathfrak{r}$ be the maximal compact subring of F, $G = GL_n(F)$, $\Gamma = GL_n(\mathfrak{r})$, and $\Delta = \{\alpha \in M_n(\mathfrak{r}) \mid \det(\alpha) \neq 0\}$. In this case, $R(\Gamma, \Delta)$ is essentially a subalgebra of the group algebra of G. To see this, first note that G is locally compact, and Γ is an open compact subgroup of G. Let R' denote the module of all complex valued continuous functions f with compact support such that $f(axb)=f(x)$ for all $a\in\Gamma$ and $b\in\Gamma$. Fix a Haar measure μ of G so that $\mu(\Gamma)=1$. For f and g in R', define the product $f*g$ by

$$f*g(x)=\int_G f(xy^{-1})g(y)d\mu(y) \qquad (x\in G).$$

It can easily be verified that $f*g\in R'$, and this law of multiplication is associative. Now, to each double coset $\Gamma\alpha\Gamma$, assign its characteristic function. Extending this correspondence $\boldsymbol{C}$-linearly, we obtain a $\boldsymbol{C}$-linear map of $R(\Gamma, G)\otimes_{\boldsymbol{Z}}\boldsymbol{C}$ onto R', which is actually a ring-isomorphism. Furthermore we can develop a theory of formal Dirichlet series (or formal power series) analogous to the above one. We only have to take, instead of p, the number of elements in the residue field of $\mathfrak{r}$ modulo the maximal ideal.

EXERCISE 3.26. (A) Let $\{e_1, \cdots, e_n\}$ be the standard basis of $L=\boldsymbol{Z}^n$, and let $L_\nu=\sum_{i=1}^{\nu}\boldsymbol{Z}e_i$. Prove (by induction on n) that for every $\alpha\in\Delta$, we can find representatives $\{\alpha_j\}$ so that $\Gamma\alpha\Gamma=\bigcup_j\Gamma\alpha_j$ and $L_\nu\alpha_j\subset L_\nu$ for $\nu=1, \cdots, n$.

(B) The notation being as in (A), for every lattice $M\subset L$ such that $[L:M]$ is a power of p, put $[L_\nu:L_\nu\cap M]=p^{a_\nu}$ and $\lambda(M)=\prod_{\nu=1}^{n}X_\nu^{a_\nu-a_{\nu-1}}$. Here $X_1, \cdots, X_n$ are indeterminates, and $a_0=0$. For $\Gamma\alpha\Gamma=\bigcup_i\Gamma\alpha_i$ with $\alpha\in\Delta$ such that $\det(\alpha)$ is a power of p, put $\Phi(\Gamma\alpha\Gamma)=\sum_i\lambda(L\alpha_i)$, and extend $\boldsymbol{Z}$-linearly Φ to a map of $R_p^{(n)}$ into $\boldsymbol{Z}[X_1, \cdots, X_n]$. Prove that Φ is a surjective ring-isomorphism.

EXERCISE 3.27. Let f be a positive integer, and χ a character of $(\boldsymbol{Z}/f\boldsymbol{Z})^\times$. Find an expression for

$$\sum_{m=1}^{\infty}\chi(m)\cdot\deg(T(m))m^{-s}$$

in terms of the L-function with character χ. (Put $\chi(m)=0$ if m is not prime to f.)

EXERCISE 3.27′. Prove that, if $n=2$,

$$T(p)^m=\sum_{0\leq r\leq m/2}\left[\binom{m}{r}-\binom{m}{r-1}\right]\cdot p^rT(p,p)^rT(p^{m-2r}),$$

where $\binom{m}{r}=m!/r!(m-r)!$.

3.3. The Hecke ring for a congruence subgroup

Let Γ, Δ, and Γ_N be as in § 3.2. We shall now study $R(\Gamma', \Delta')$ with a subgroup Γ' of Γ containing Γ_N for some N, and a certain subset Δ' of Δ.

First we prove a simple

LEMMA 3.28. *Let a and b be positive integers, and c the greatest common divisor of a and b. Then $\Gamma_c = \Gamma_a \cdot \Gamma_b$.*

PROOF. If $\alpha \in \Gamma_c$, there exists an element β of $M_n(\boldsymbol{Z})$ such that $\beta \equiv 1 \bmod (a)$ and $\beta \equiv \alpha \bmod (b)$ by the Chinese remainder theorem. Then $\det(\beta) \equiv 1 \bmod (ab/c)$. By Lemma 1.38 (or by its proof), there exists an element γ of Γ such that $\gamma \equiv \beta \bmod (ab/c)$. Then $\gamma \in \Gamma_a$, $\gamma^{-1}\alpha \in \Gamma_b$, and $\alpha = \gamma \cdot \gamma^{-1}\alpha$, so that $\Gamma_c \subset \Gamma_a \Gamma_b$. Since the opposite inclusion is clear, we obtain the equality.

Let us fix a positive integer N, and put

$$\Delta_N = \{\alpha \in M_n(\boldsymbol{Z}) \mid \det(\alpha) > 0,\ (\det(\alpha), N) = 1\},$$

so that $\Delta = \Delta_1$. Denote by λ_N the natural map of $M_n(\boldsymbol{Z})$ to $M_n(\boldsymbol{Z}/N\boldsymbol{Z})$. We fix a subgroup Γ' of Γ containing Γ_N, and put

$$\Phi = \{\alpha \in \Delta_N \mid \lambda_N(\Gamma'\alpha) = \lambda_N(\alpha\Gamma')\}.$$

We see that $\Phi = \Delta_N$ if $\Gamma' = \Gamma_N$.

LEMMA 3.29. *The notation being as above, let $\alpha, \beta \in \Delta_N$. Then the following assertions hold.*

(1) $\Gamma'\alpha\Gamma' = \{\xi \in \Gamma\alpha\Gamma \mid \lambda_N(\xi) \in \lambda_N(\Gamma'\alpha)\}$ *if* $\alpha \in \Phi$.

(2) $\Gamma_N\alpha\Gamma_N = \Gamma_N\beta\Gamma_N$ *if and only if* $\Gamma\alpha\Gamma = \Gamma\beta\Gamma$ *and* $\alpha \equiv \beta \bmod (N)$.

(3) $\Gamma\alpha\Gamma = \Gamma\alpha\Gamma' = \Gamma'\alpha\Gamma$.

(4) $\Gamma'\alpha\Gamma' = \Gamma'\alpha\Gamma_N = \Gamma_N\alpha\Gamma'$ *if* $\alpha \in \Phi$.

(5) *If $\alpha \in \Phi$ and $\Gamma'\alpha\Gamma' = \bigcup_i \Gamma'\alpha_i$ is a disjoint union, then $\Gamma\alpha\Gamma = \bigcup_i \Gamma\alpha_i$ is a disjoint union.*

PROOF. To show (3), put $a = \det(\alpha)$. By Lemma 3.28 and Lemma 3.9, we have $\Gamma = \Gamma_a\Gamma_N \subset \alpha^{-1}\Gamma\alpha\Gamma_N$, so that $\alpha^{-1}\Gamma\alpha\Gamma \subset \alpha^{-1}\Gamma\alpha\Gamma_N$. Hence $\Gamma\alpha\Gamma \subset \Gamma\alpha\Gamma_N \subset \Gamma\alpha\Gamma'$. Since the opposite inclusion is obvious, we obtain (3). Next, to see (1), let $\xi \in \Gamma\alpha\Gamma$, and $\lambda_N(\xi) \in \lambda_N(\Gamma'\alpha)$. Then $\xi \equiv \gamma\alpha \bmod (N)$ with $\gamma \in \Gamma'$. By (3), $\xi \in \Gamma\alpha\Gamma_N$, hence $\xi = \delta\alpha\varepsilon$ with $\delta \in \Gamma$ and $\varepsilon \in \Gamma_N$. Then $\gamma \equiv \delta \bmod (N)$. Since $\Gamma_N \subset \Gamma'$, we see that $\delta \in \Gamma'$, hence $\xi \in \Gamma'\alpha\Gamma_N \subset \Gamma'\alpha\Gamma'$. Conversely if $\xi \in \Gamma'\alpha\Gamma'$, we have clearly $\xi \in \Gamma\alpha\Gamma$, and by the definition of Φ, $\lambda_N(\xi) \in \lambda_N(\Gamma'\alpha)$. This proves (1). At the same time, we have proved that $\Gamma'\alpha\Gamma' \subset \Gamma'\alpha\Gamma_N$. Since the opposite inclusion is obvious, we obtain (4). The assertion (2) is a special case of (1). Finally, let $\alpha \in \Phi$, and $\Gamma'\alpha\Gamma' = \bigcup_i \Gamma'\alpha_i$ (disjoint). Then $\Gamma\alpha\Gamma = \Gamma\alpha\Gamma' = \bigcup_i \Gamma\alpha_i$. Assume $\Gamma\alpha_i = \Gamma\alpha_j$. Then $\alpha_i = \gamma\alpha_j$ with $\gamma \in \Gamma$. By (1), $\alpha_i \equiv \delta\alpha_j \bmod (N)$ with $\delta \in \Gamma'$. Then $\gamma \equiv \delta \bmod (N)$. Since $\Gamma_N \subset \Gamma'$, we have $\gamma \in \Gamma'$, so that $\Gamma'\alpha_i = \Gamma'\alpha_j$. This proves (5).

PROPOSITION 3.30. *Let the notation be as above. Then the correspondence* $\Gamma'\alpha\Gamma' \mapsto \Gamma\alpha\Gamma$, *with* $\alpha \in \Phi$, *defines a homomorphism of* $R(\Gamma', \Phi)$ *into* $R(\Gamma, \Delta)$.

PROOF. Let $\alpha, \beta \in \Phi$, and let $\Gamma'\alpha\Gamma' = \bigcup_i \Gamma'\alpha_i$, $\Gamma'\beta\Gamma' = \bigcup_j \Gamma'\beta_j$ be disjoint unions. By (5) of Lemma 3.29, $\Gamma\alpha\Gamma = \bigcup_i \Gamma\alpha_i$ and $\Gamma\beta\Gamma = \bigcup_j \Gamma\beta_j$ are disjoint unions. Put $(\Gamma'\alpha\Gamma')(\Gamma'\beta\Gamma') = \sum_\xi c'_\xi \cdot (\Gamma'\xi\Gamma')$ with $c'_\xi \in \boldsymbol{Z}$. Then $\Gamma\alpha\Gamma\beta\Gamma = \Gamma\alpha\Gamma\beta\Gamma' = \Gamma\alpha\Gamma'\beta\Gamma' = \bigcup_\xi \Gamma\xi\Gamma'$ with the same ξ's. Moreover, since $\alpha, \beta \in \Phi$, we have $\lambda_N(\Gamma'\xi) = \lambda_N(\Gamma'\alpha\beta)$ for every $\xi \in \Gamma'\alpha\Gamma'\beta\Gamma'$, so that, by (1) of Lemma 3.29,

$$\Gamma'\xi\Gamma' = \{\zeta \in \Gamma\xi\Gamma \mid \lambda_N(\zeta) \in \lambda_N(\Gamma'\alpha\beta)\}.$$

It follows that $\Gamma'\xi\Gamma' \mapsto \Gamma\xi\Gamma$ is one-to-one. Therefore, put $(\Gamma\alpha\Gamma)(\Gamma\beta\Gamma) = \sum_\xi c_\xi \cdot (\Gamma\xi\Gamma)$ with $c_\xi \in \boldsymbol{Z}$. Then

$$c_\xi = \# \{(i, j) \mid \Gamma\alpha_i\beta_j = \Gamma\xi\},$$

$$c'_\xi = \# \{(i, j) \mid \Gamma'\alpha_i\beta_j = \Gamma'\xi\}.$$

Therefore it is sufficient to show that $\Gamma'\alpha_i\beta_j = \Gamma'\xi$ if and only if $\Gamma\alpha_i\beta_j = \Gamma\xi$. Assume $\Gamma\alpha_i\beta_j = \Gamma\xi$. Then $\xi = \gamma\alpha_i\beta_j$ with $\gamma \in \Gamma$. Since $\lambda_N(\xi) \in \lambda_N(\Gamma'\alpha_i\beta_j)$, we have $\xi \equiv \delta\alpha_i\beta_j$ with $\delta \in \Gamma'$. Then $\delta \equiv \gamma \bmod (N)$, hence $\gamma \in \Gamma'$, so that $\Gamma'\alpha_i\beta_j = \Gamma'\xi$. Since the converse is obvious, this completes the proof.

Hereafter we consider only the case $n = 2$. Let t be a positive divisor of N, and $\mathfrak{h}$ a subgroup of $(\boldsymbol{Z}/N\boldsymbol{Z})^\times$. We shall often denote by the same letter $\mathfrak{h}$ the set of all the integers whose residue classes modulo (N) belong to $\mathfrak{h}$. Define semi-groups Δ_N^*, Δ'_N and a group Γ' by

$$\Delta_N^* = \left\{\alpha \in \Delta \,\middle|\, \lambda_N(\alpha) = \begin{bmatrix} 1 & 0 \\ 0 & x \end{bmatrix} \text{ with } x \in (\boldsymbol{Z}/N\boldsymbol{Z})^\times\right\}, \tag{3.3.1}$$

$$\Delta'_N = \left\{\begin{bmatrix} u & v \\ w & z \end{bmatrix} \in \Delta \,\middle|\, u \in \mathfrak{h},\ v \equiv 0\ (t),\ w \equiv 0\ (N),\ (z, N) = 1\right\}, \tag{3.3.1'}$$

$$\Gamma' = \left\{\begin{bmatrix} a & b \\ c & d \end{bmatrix} \in SL_2(\boldsymbol{Z}) \,\middle|\, a \in \mathfrak{h},\ b \equiv 0\ (t),\ c \equiv 0\ (N)\right\}. \tag{3.3.2}$$

For instance, $\Gamma_0(N)$ and Γ_N are of this type. (But there are some groups between Γ and Γ_N which can not be transformed to this type of group by any conjugacy in Γ.) We see easily that $\Delta'_N = \Delta_N^*\Gamma' = \Gamma'\Delta_N^*$ and $\Delta'_N \subset \Phi$.

PROPOSITION 3.31. *The notation being as above, the correspondence* $\Gamma'\alpha\Gamma' \mapsto \Gamma\alpha\Gamma$, *with* $\alpha \in \Delta'_N$, *defines an isomorphism of* $R(\Gamma', \Delta'_N)$ *onto* $R(\Gamma, \Delta_N)$.

PROOF. On account of Prop. 3.30, it is sufficient to prove the injectivity and the surjectivity of the map in question. Let $\eta \in \Delta_N$, and $b = \det(\eta)$. Take an integer c so that $bc \equiv 1 \bmod (N)$, and put $\varphi = \begin{bmatrix} 1 & 0 \\ 0 & c \end{bmatrix}$. Then

$\det(\eta\varphi)\equiv 1 \bmod (N)$. By Lemma 1.38, there exists an element γ of Γ such that $\gamma\equiv\eta\varphi \bmod (N)$. Then $\gamma^{-1}\eta\equiv\begin{bmatrix}1 & 0\\ 0 & b\end{bmatrix} \bmod (N)$, hence $\gamma^{-1}\eta\in\Delta_N^*$, and $\Gamma\gamma^{-1}\eta\Gamma=\Gamma\eta\Gamma$. This proves the surjectivity. To prove the injectivity, let $\alpha, \beta\in\Delta_N^*$ and $\alpha\equiv\begin{bmatrix}1 & 0\\ 0 & c\end{bmatrix}$, $\beta\equiv\begin{bmatrix}1 & 0\\ 0 & d\end{bmatrix} \bmod (N)$. If $\Gamma\alpha\Gamma=\Gamma\beta\Gamma$, we have $c\equiv\det(\alpha)=\det(\beta)\equiv d \bmod (N)$, hence $\alpha\equiv\beta \bmod (N)$. Therefore, by (1) of Lemma 3.29, $\Gamma'\alpha\Gamma'=\Gamma'\beta\Gamma'$. This proves the injectivity, since $R(\Gamma', \Delta_N')$ (resp. $R(\Gamma, \Delta_N)$) is a free $\boldsymbol{Z}$-module generated by the $\Gamma'\alpha\Gamma'$ (resp. $\Gamma\alpha\Gamma$) with $\alpha\in\Delta_N^*$.

Let us now consider a set

$$\Delta'=\left\{\begin{bmatrix}a & b\\ c & d\end{bmatrix}\in\Delta \,\middle|\, a\in\mathfrak{h},\ b\equiv 0\ (t),\ c\equiv 0\ (N)\right\}. \tag{3.3.3}$$

Then Δ' is a semi-group containing Γ' and Δ_N'. We shall now determine the structure of $R(\Gamma', \Delta')$.

For each prime p, put $E_p=GL_2(\boldsymbol{Z}_p)$. Then, for every $\alpha\in\Delta$, the double coset $E_p\alpha E_p$ is completely determined by the p-part of elementary divisors of α, and vice versa. Further, for a positive integer m, we write $m\mid N^\infty$ if all prime factors of m divide N. Then every positive integer can be uniquely written in the form mq with $m\mid N^\infty$ and $(q, N)=1$.

PROPOSITION 3.32. *Let* $\alpha\in\Delta'$, $\det(\alpha)=mq$, $m\mid N^\infty$, $(q, N)=1$. *Then the following assertions hold.*

(1) $\Gamma'\alpha\Gamma'$
$=\{\beta\in\Delta' \mid \det(\beta)=mq,\ E_p\beta E_p=E_p\alpha E_p$ *for all prime factors* p *of* $q\}$.

(2) *There exists an element* ξ *of* Δ' *such that* $\det(\xi)=q$ *and* $E_p\xi E_p=E_p\alpha E_p$ *for all prime factors* p *of* q.

(3) *If* ξ *is as in* (2), *and* $\eta=\begin{bmatrix}1 & 0\\ 0 & m\end{bmatrix}$, *then*

$$\Gamma'\alpha\Gamma'=(\Gamma'\xi\Gamma')\cdot(\Gamma'\eta\Gamma')=(\Gamma'\eta\Gamma')\cdot(\Gamma'\xi\Gamma').$$

(4) *The element* ξ *of* (2) *can be taken from* Δ_N^*.

PROOF. Let $X(\alpha)$ denote the set defined by the right hand side of (1). Clearly $\Gamma'\alpha\Gamma'\subset X(\alpha)$. To prove the opposite inclusion, let $\beta=\begin{bmatrix}a & *\\ * & *\end{bmatrix}\in X(\alpha)$. Since a is prime to mN, $ae\equiv 1 \bmod (mN)$ for some $e\in\boldsymbol{Z}$. By Lemma 1.38, there exists an element γ of $SL_2(\boldsymbol{Z})$ such that $\gamma\equiv\begin{bmatrix}e & 0\\ 0 & a\end{bmatrix} \bmod (mN)$. Since $\beta\in\Delta'$, we see that $\gamma\in\Gamma'$, and $\gamma\beta\equiv\begin{bmatrix}1 & tb\\ fN & *\end{bmatrix} \bmod (mN)$ with integers b and f. Put $\delta=\begin{bmatrix}1 & 0\\ -fN & 1\end{bmatrix}$. Then $\delta\in\Gamma'$, and $\delta\gamma\beta\equiv\begin{bmatrix}1 & tb\\ 0 & g\end{bmatrix} \bmod (mN)$ with $g\in\boldsymbol{Z}$.

Taking the determinant, we have $mq \equiv g \bmod (mN)$, so that $\delta\gamma\beta \equiv \begin{bmatrix} 1 & tb \\ 0 & mq \end{bmatrix}$ mod (mN). Put $\eta = \begin{bmatrix} 1 & 0 \\ 0 & m \end{bmatrix}$, $\varepsilon = \begin{bmatrix} 1 & tb \\ 0 & 1 \end{bmatrix}$, $\xi = \delta\gamma\beta\varepsilon^{-1}\eta^{-1}$. Then $\det(\xi) = q$, $\xi \equiv \begin{bmatrix} 1 & 0 \\ 0 & q \end{bmatrix}$ mod (N), so that $\xi \in \Delta_N^*$. Moreover, we see that $\beta \in \Gamma'\xi\eta\Gamma'$. By our construction, $E_p\xi E_p = E_p\alpha E_p$ for all p dividing q. This proves (2) and (4). The element ξ may depend on β. Let us now show that $\Gamma'\xi\eta\Gamma'$ is determined only by α and independent of the choice of β. To show this, let ξ_1 be an element of Δ_N^* such that $\det(\xi_1) = q$ and $E_p\xi_1 E_p = E_p\alpha E_p$ for all p dividing q. Then ξ and ξ_1 have the same set of elementary divisors, hence $\Gamma\xi\Gamma = \Gamma\xi_1\Gamma$. Since $\xi \equiv \xi_1 \equiv \begin{bmatrix} 1 & 0 \\ 0 & q \end{bmatrix}$ mod (N), we have $\Gamma_N\xi\Gamma_N = \Gamma_N\xi_1\Gamma_N$ by (2) of Lemma 3.29, so that $\xi_1 = \varphi\xi\psi$ with φ and ψ in Γ_N. By the Chinese remainder theorem, we can find an element θ of $M_2(\boldsymbol{Z})$ so that

$$\theta \equiv 1 \bmod (mN),$$

$$\theta \equiv \eta^{-1}\psi^{-1}\eta \bmod q \cdot M_2(\boldsymbol{Z}_p) \qquad \text{for all } p \text{ dividing } q.$$

Then $\det(\theta) \equiv 1 \bmod (mqN)$. By Lemma 1.38, we can assume that $\theta \in SL_2(\boldsymbol{Z})$. Then $\theta \in \Gamma_N$. Put $\omega = \xi\psi\eta\theta(\xi\eta)^{-1}$. Then, $\det(\omega) = 1$, and

$$\omega \equiv 1 \bmod N \cdot M_2(\boldsymbol{Z}_p) \qquad \text{for all } p \text{ dividing } N,$$

$$\omega \equiv 1 \bmod M_2(\boldsymbol{Z}_p) \qquad \text{for all } p \text{ dividing } q.$$

Therefore $\omega \in M_2(\boldsymbol{Z}_p)$ for all p, so that $\omega \in M_2(\boldsymbol{Z})$, hence $\omega \in \Gamma_N$. Since $\xi\psi\eta = \omega\xi\eta\theta^{-1}$, we have $\Gamma'\xi_1\eta\Gamma' = \Gamma'\xi\psi\eta\Gamma' = \Gamma'\xi\eta\Gamma'$. This shows that $\Gamma'\xi\eta\Gamma'$ is determined only by α. Moreover, we have seen that $\Gamma'\alpha\Gamma' \subset X(\alpha) \subset \Gamma'\xi\eta\Gamma'$. Then obviously these three sets must coincide, hence (1). Now, for any ξ as in (2), we see, from our definition of $X(\alpha)$, that both $\Gamma'\xi\Gamma'\eta\Gamma'$ and $\Gamma'\eta\Gamma'\xi\Gamma'$ are contained in $X(\alpha)$. Therefore

$$\Gamma'\alpha\Gamma' = \Gamma'\xi\Gamma'\eta\Gamma' = \Gamma'\eta\Gamma'\xi\Gamma'.$$

To prove that the multiplicity of $\Gamma'\alpha\Gamma'$ in $(\Gamma'\xi\Gamma')\cdot(\Gamma'\eta\Gamma')$ is 1, we first show

(*) *If $\alpha_1 \in \Delta'$, $\alpha_2 \in \Delta'$ and $\Gamma\alpha_1 = \Gamma\alpha_2$, then $\Gamma'\alpha_1 = \Gamma'\alpha_2$.*

In fact, put $\alpha_1 = \gamma\alpha_2$ with $\gamma \in \Gamma$, and $\lambda_N(\alpha_i) = \begin{bmatrix} a_i & tb_i \\ 0 & a_i^{-1} \end{bmatrix}$, $\lambda_N(\gamma) = \begin{bmatrix} u & w \\ v & x \end{bmatrix}$. Then we have $\begin{bmatrix} a_1 & tb_1 \\ 0 & * \end{bmatrix} = \begin{bmatrix} ua_2 & utb_2 + wa_2^{-1} \\ va_2 & * \end{bmatrix}$ so that $v = 0$, $u = a_1a_2^{-1} \in \mathfrak{h}$, and $t \mid w$, hence $\gamma \in \Gamma'$. This proves (*). Now let $\Gamma'\xi\Gamma' = \bigcup_i \Gamma'\xi_i$, $\Gamma'\eta\Gamma' = \bigcup_j \Gamma'\eta_j$ be disjoint unions. By (*), the $\Gamma\xi_i$ are distinct, and the $\Gamma\eta_j$ are distinct. Moreover, by Prop. 3.16,

$$(\Gamma\xi\Gamma)\cdot(\Gamma\eta\Gamma)=\Gamma\xi\eta\Gamma=\Gamma\alpha\Gamma .$$

Therefore the number of (i, j) such that $\Gamma\xi_i\eta_j=\Gamma\alpha$ is at most one. (Note that $\Gamma\eta\Gamma$ may contain cosets other than $\Gamma\eta_j$.) It follows that the number of (i, j) such that $\Gamma'\xi_i\eta_j=\Gamma'\alpha$ is at most one, hence the multiplicity of $\Gamma'\alpha\Gamma'$ in $(\Gamma'\xi\Gamma')\cdot(\Gamma'\eta\Gamma')$ is one. The product $(\Gamma'\eta\Gamma')\cdot(\Gamma'\xi\Gamma')$ can be treated by the same type of argument.

PROPOSITION 3.33. *Let* $\alpha\in\Delta'$, $\det(\alpha)=m$ *with* $m\mid N^\infty$. *Then*

$$\Gamma'\alpha\Gamma'=\{\beta\in\Delta'\mid\det(\beta)=m\}=\cup_{r=0}^{m-1}\Gamma'\begin{bmatrix}1 & tr\\0 & m\end{bmatrix}\qquad(\text{disjoint}).$$

PROOF. The coincidence of the first two sets is a special case of Prop. 3.32. The last union is obviously contained in the second one. Now let $\beta\in\Delta'$, $\det(\beta)=m$. Consider the special case $q=1$ in the proof of Prop. 3.32. Then we see that $\delta\gamma\beta=\xi\begin{bmatrix}1 & tb\\0 & m\end{bmatrix}$ with an element ξ of Γ_N, an integer b, and elements γ and δ of Γ'. If $b=mh+r$ and $0\leqq r<m$, then $\begin{bmatrix}1 & tb\\0 & m\end{bmatrix}=\begin{bmatrix}1 & th\\0 & 1\end{bmatrix}\begin{bmatrix}1 & tr\\0 & m\end{bmatrix}$. Therefore β is contained in the last set. To show the disjointness, assume $0\leqq r<s\leqq m-1$ and $\begin{bmatrix}1 & tr\\0 & m\end{bmatrix}=\gamma\begin{bmatrix}1 & ts\\0 & m\end{bmatrix}$ with $\gamma=\begin{bmatrix}a & tb\\c & d\end{bmatrix}\in\Gamma'$. Then we have $\begin{bmatrix}1 & tr\\0 & m\end{bmatrix}=\begin{bmatrix}a & ats+tbm\\c & cts+dm\end{bmatrix}$, so that γ must be 1. This completes the proof.

For each positive integer n, let $T'(n)$ denote the sum of all $\Gamma'\alpha\Gamma'$ with $\alpha\in\Delta'$ and $\det(\alpha)=n$. By Prop. 3.33, we obtain

$$\deg(T'(m))=m\qquad\text{if}\quad m\mid N^\infty. \tag{3.3.4}$$

Further, for two positive integers a and d such that

$$a\mid d,\qquad(d, N)=1, \tag{3.3.5}$$

let $T'(a, d)$ denote the element of $R(\Gamma', \Delta'_N)$ which is sent to $T(a, d)=\Gamma\begin{bmatrix}a & 0\\0 & d\end{bmatrix}\Gamma$ by the isomorphism of $R(\Gamma', \Delta'_N)$ onto $R(\Gamma, \Delta)$ of Prop. 3.31. Then we obtain

THEOREM 3.34. (1) *$R(\Gamma', \Delta')$ is a polynomial ring over $\mathbf{Z}$ of the elements of the form*

$T'(p)$ for all primes p dividing N,

$T'(1, p)$, $T'(p, p)$ for all primes p not dividing N.

These elements are algebraically independent.

(2) *Every element $\Gamma'\alpha\Gamma'$ with $\alpha\in\Delta'$ is uniquely expressed as a product $T'(m)T'(a, d)=T'(a, d)T'(m)$ with $m\mid N^\infty$, $a\mid d$, $(d, N)=1$.*

(3) *$T'(m)T'(n)=T'(mn)$ if $m \mid N^\infty$, $n \mid N^\infty$.*
(4) *$T'(n_1n_2)=T'(n_1)T'(n_2)$ if $(n_1, n_2)=1$.*
(5) *$R(\Gamma', \Delta')\otimes_{\mathbf{Z}} \mathbf{Q}$ is generated by the $T'(n)$ for all n over $\mathbf{Q}$.*

PROOF. The assertion (2) follows from Prop. 3.32. By Prop. 3.33, if $m|N^\infty$ and $n|N^\infty$, we see that $T'(m)T'(n)=cT'(mn)$ with a positive integer c. By (3.3.4) and Prop. 3.3, we obtain $c=1$, which proves (3). Therefore we see, in view of Prop. 3.31, that $R(\Gamma', \Delta')$ is generated by the elements listed in (1). The proof of the algebraic independence of these elements is straightforward, and may be left to the reader. Finally, if $n=mq$ with $m \mid N^\infty$ and $(q, N)=1$, we have $T'(n)=T'(m)T'(q)=T'(q)T'(m)$ by (2). Therefore, by Prop. 3.16, Prop. 3.31, and (3), we obtain (4). By (1) and (5) of Th. 3.24, and by Prop. 3.31, we have

$$pT'(p, p)=T'(p)^2-T'(p^2)$$

for every prime p not dividing N, which together with (1) proves (5).

Thus the multiplication of the elements $T'(n)$ can be reduced to that of $T'(p^k)$ with a prime p. If p divides N, we have $T'(p^k)=T'(p)^k$. If $(p, N)=1$, the elements $T'(p^k)$ satisfy the same formulas as in Th. 3.24, on account of Prop. 3.31. We can express these facts as

THEOREM 3.35. *$R(\Gamma', \Delta')$ is a homomorphic image of $R(\Gamma, \Delta)$ through the map*

$$T(n)\mapsto T'(n) \quad \textit{for all positive integers } n\,,$$

$$T(p, p)\mapsto T'(p, p) \quad \textit{for all primes } p \textit{ prime to } N\,,$$

$$T(p, p)\mapsto 0 \quad \textit{for all primes } p \textit{ dividing } N\,.$$

Therefore, from (3) of Th. 3.24, we obtain

(3.3.6) $$T'(m)T'(n)=\textstyle\sum_d d\cdot T'(d, d)T'(mn/d^2)$$

(the summation over all positive divisors d of (m, n) prime to N).

Moreover, if we define a formal Dirichlet series $D'(s)$ by

(3.3.7) $$D'(s)=\textstyle\sum_{\Gamma'\backslash\Delta'/\Gamma'}(\Gamma'\alpha\Gamma')\cdot\det(\alpha)^{-s}=\sum_{n=1}^{\infty}T'(n)n^{-s}\,,$$

then, from the above observation, we obtain

(3.3.8) $$D'(s)=\textstyle\prod_{p\mid N}[1-T'(p)p^{-s}]^{-1} \times\prod_{p\nmid N}[1-T'(p)p^{-s}+T'(p, p)p^{1-2s}]^{-1}\,.$$

By our definition, we have

(3.3.9) $$T'(p)=\Gamma'\begin{bmatrix}1 & 0\\ 0 & p\end{bmatrix}\Gamma' \qquad \text{for every prime } p\,.$$

Let us now study $T'(q, q)$ for a positive integer q prime to N. By Lemma 1.39, there exists an element σ_q of $SL_2(\boldsymbol{Z})$ such that

$$\lambda_N(\sigma_q) = \begin{bmatrix} q^{-1} & 0 \\ 0 & q \end{bmatrix}. \tag{3.3.10}$$

Then $\lambda_N(q \cdot \sigma_q) = \begin{bmatrix} 1 & 0 \\ 0 & q^2 \end{bmatrix}$, and $\Gamma q \cdot \sigma_q \Gamma = T(q, q)$. Therefore

$$T'(q, q) = \Gamma' q \cdot \sigma_q \Gamma'. \tag{3.3.11}$$

There is a simple property of $\Gamma' \alpha \Gamma'$ which can be described by means of the "main involution" of the matrix algebra. For $\alpha = \begin{bmatrix} a & b \\ c & d \end{bmatrix} \in M_2(\boldsymbol{C})$, put

$$\alpha^\iota = \begin{bmatrix} d & -b \\ -c & a \end{bmatrix} = \varepsilon \cdot {}^t\alpha \varepsilon^{-1} \qquad \left(\varepsilon = \begin{bmatrix} 0 & 1 \\ -1 & 0 \end{bmatrix}\right).$$

Then it can easily be verified that

$$(\alpha + \beta)^\iota = \alpha^\iota + \beta^\iota, \qquad (\alpha\beta)^\iota = \beta^\iota \alpha^\iota, \qquad (c\alpha)^\iota = c\alpha^\iota \qquad (c \in \boldsymbol{C}),$$

$$\alpha + \alpha^\iota = \mathrm{tr}(\alpha) \cdot 1_2, \qquad \alpha\alpha^\iota = \det(\alpha) \cdot 1_2.$$

The map ι is called *the main involution* of $M_2(\boldsymbol{C})$. Obviously $M_2(\boldsymbol{Q})$ and $M_2(\boldsymbol{R})$ are stable under it.

Let $\alpha \in \Delta_N^*$ and $\det(\alpha) = q$. Then

$$\lambda_N(\alpha) = \begin{bmatrix} 1 & 0 \\ 0 & q \end{bmatrix}, \qquad \lambda_N(\alpha^\iota) = \begin{bmatrix} q & 0 \\ 0 & 1 \end{bmatrix},$$

so that $\alpha \equiv \sigma_q \alpha^\iota \equiv \alpha^\iota \sigma_q \bmod (N)$. Therefore, since α and α^ι have the same set of elementary divisors, by (2) of Lemma 3.29, we have

$$\Gamma' \alpha \Gamma' = \Gamma' \sigma_q \alpha^\iota \Gamma' = \Gamma' \alpha^\iota \sigma_q \Gamma' \qquad \text{if} \quad \alpha \in \Delta_N^*, \quad \det(\alpha) = q. \tag{3.3.12}$$

Moreover, we can easily verify that $\Gamma' \sigma_q = \sigma_q \Gamma'$, hence, by Prop. 3.7,

$$\Gamma' \alpha \Gamma' = (\Gamma' \sigma_q \Gamma') \cdot (\Gamma' \alpha^\iota \Gamma') = (\Gamma' \alpha^\iota \Gamma') \cdot (\Gamma' \sigma_q \Gamma'). \tag{3.3.13}$$

From this we obtain

$$\Gamma' \alpha \Gamma' \text{ commutes with } \Gamma' \alpha^\iota \Gamma' \text{ if } \alpha \in \Delta_N^*. \tag{3.3.14}$$

PROPOSITION 3.36. *For each positive integer a prime to N, fix an element σ_a of $SL_2(\boldsymbol{Z})$ as in* (3.3.10). *Then, for every positive integer n, one has*

$$\{\alpha \in \Delta' \mid \det(\alpha) = n\} = \bigcup_a \bigcup_{b=0}^{d-1} \Gamma' \sigma_a \cdot \begin{bmatrix} a & bt \\ 0 & d \end{bmatrix} \qquad (a > 0,\ ad = n,\ (a, N) = 1),$$

and the right hand side is a disjoint union.

PROOF. The right hand side is clearly contained in the left hand side.

To show the disjointness of the right hand side, suppose $\gamma\sigma_a\cdot\begin{bmatrix} a & bt \\ 0 & d \end{bmatrix}$ $=\sigma_u\cdot\begin{bmatrix} u & vt \\ 0 & w \end{bmatrix}$ with $\gamma\in\Gamma'$. Put $\sigma_u^{-1}\gamma\sigma_a=\begin{bmatrix} e & f \\ g & h \end{bmatrix}$. Then $\begin{bmatrix} e & f \\ g & h \end{bmatrix}\begin{bmatrix} a & bt \\ 0 & d \end{bmatrix}$ $=\begin{bmatrix} u & vt \\ 0 & w \end{bmatrix}$, so that $g=0$. Since $\det(\sigma_u^{-1}\gamma\sigma_a)=1$ and $au>0$, we have $e=h=1$, hence $a=u$, $d=w$, and $vt=bt+fd$. Since $\gamma\in\Gamma'$, we have $f=f't$ with some $f'\in\boldsymbol{Z}$. Then $v=b+f'd$, so that $v=b$. This proves the disjointness. Now let $n=mq$ with $m|N^\infty$ and $(q,N)=1$. Then $\deg(T'(n))=m\cdot\deg(T'(q))$. By (7) of Th. 3.24 and by (5) of Lemma 3.29, $\deg(T'(q))=\deg(T(q))=\sum_{c|q,c>0}c$. Therefore it is easily seen that $\deg(T'(n))$ coincides with the number of the cosets of our disjoint union. This completes the proof.

3.4. Action of double cosets on automorphic forms

So far our discussion of double cosets has been purely algebraic or arithmetic. Let us now come back to the situation of Chapter 2, and consider the representation of double cosets in the space of automorphic forms, as is indicated at the end of § 3.2. First we recall our notation:

$$j(\sigma,z)=cz+d \qquad \left(z\in\mathfrak{H},\ \sigma=\begin{bmatrix} a & b \\ c & d \end{bmatrix}\in GL_2(\boldsymbol{R})\right),$$

$$f\,|\,[\sigma]_k=\det(\sigma)^{k/2}\cdot f(\sigma(z))j(\sigma,z)^{-k}$$

for a function f on $\mathfrak{H}$.

Let Γ_1 and Γ_2 be commensurable Fuchsian groups of the first kind, $\tilde{\Gamma}$ the commensurator of Γ_1 and Γ_2 in $GL_2^+(\boldsymbol{R})$ in the sense of § 3.1, and $\alpha\in\tilde{\Gamma}$. For $f\in A_k(\Gamma_1)$, we put

$$f\,|\,[\Gamma_1\alpha\Gamma_2]_k=\det(\alpha)^{k/2-1}\cdot\textstyle\sum_{\nu=1}^{d}f\,|\,[\alpha_\nu]_k\,, \tag{3.4.1}$$

where

$$\Gamma_1\alpha\Gamma_2=\textstyle\bigcup_{\nu=1}^{d}\Gamma_1\alpha_\nu \qquad \text{(disjoint)}.$$

It is clear that $f\,|\,[\Gamma_1\alpha\Gamma_2]_k$ is independent of the choice of the representatives α_ν.

PROPOSITION 3.37. $[\Gamma_1\alpha\Gamma_2]_k$ *sends* $A_k(\Gamma_1)$, $G_k(\Gamma_1)$, $S_k(\Gamma_1)$ *into* $A_k(\Gamma_2)$, $G_k(\Gamma_2)$, $S_k(\Gamma_2)$, *respectively.*

PROOF. Let $\delta\in\Gamma_2$. Then $\{\Gamma_1\alpha_\nu\delta\}_\nu$ coincides with $\{\Gamma_1\alpha_\nu\}_\nu$ as a whole. Therefore, if $g=f\,|\,[\Gamma_1\alpha\Gamma_2]_k$,

$$g\,|\,[\delta]_k=\det(\alpha)^{k/2-1}\cdot\textstyle\sum_\nu f\,|\,[\alpha_\nu\delta]_k=\det(\alpha)^{k/2-1}\cdot\sum_\nu f\,|\,[\alpha_\nu]_k=g\,.$$

On the other hand, by Prop. 2.4, $f\,|\,[\alpha_\nu]_k\in A_k(\alpha_\nu^{-1}\Gamma_1\alpha_\nu)$. Put

$$\Gamma_3 = \bigcap_\nu \alpha_\nu^{-1}\Gamma_1\alpha_\nu \cap \Gamma_2 .$$

Then Γ_3 is a subgroup of Γ_2 of finite index, and $g \in A_k(\Gamma_3)$. By Prop. 2.6, we see that $g \in A_k(\Gamma_2)$. The same argument applies to $G_k(\Gamma_i)$ and $S_k(\Gamma_i)$.

Consider the module R_{12} generated by $\Gamma_1\alpha\Gamma_2$ with $\alpha \in \tilde{\Gamma}$ (see § 3.1). For every $X = \sum c_\alpha \cdot \Gamma_1\alpha\Gamma_2 \in R_{12}$ with $c_\alpha \in \boldsymbol{Z}$, we define

$$f \mid [X]_k = \sum c_\alpha f \mid [\Gamma_1\alpha\Gamma_2]_k \qquad (f \in A_k(\Gamma_1)) .$$

PROPOSITION 3.38. $[XY]_k = [X]_k[Y]_k$ *for every* $X \in R_{12}$ *and every* $Y \in R_{23}$.

PROOF. It is sufficient to show that

$$(f \mid [\Gamma_1\alpha\Gamma_2]_k) \mid [\Gamma_2\beta\Gamma_3]_k = f \mid [(\Gamma_1\alpha\Gamma_2)\cdot(\Gamma_2\beta\Gamma_3)]_k .$$

Let $(\Gamma_1\alpha\Gamma_2)\cdot(\Gamma_2\beta\Gamma_3) = \sum c_\xi(\Gamma_1\xi\Gamma_3)$ with $c_\xi \in \boldsymbol{Z}$, and let

$$\Gamma_1\alpha\Gamma_2 = \bigcup_i \Gamma_1\alpha_i , \qquad \Gamma_2\beta\Gamma_3 = \bigcup_j \Gamma_2\beta_j , \qquad \Gamma_1\xi\Gamma_3 = \bigcup_h \Gamma_1\xi_h$$

be disjoint unions. By our definition of multiplication, we see that

$$\sum_{i,j} \Gamma_1\alpha_i\beta_j = \sum_{\xi,h} c_\xi \cdot \Gamma_1\xi_h .$$

Therefore

$$\begin{aligned}(f \mid [\Gamma_1\alpha\Gamma_2]_k) &\mid [\Gamma_2\beta\Gamma_3]_k \\ &= \det(\alpha\beta)^{k/2-1}\sum_{i,j} f \mid [\alpha_i\beta_j]_k = \det(\alpha\beta)^{k/2-1}\sum_{\xi,h} c_\xi \cdot f \mid [\xi_h]_k \\ &= f \mid [(\Gamma_1\alpha\Gamma_2)\cdot(\Gamma_2\beta\Gamma_3)]_k , \qquad \text{q. e. d.}\end{aligned}$$

In particular, fix a Fuchsian group Γ of the first kind. Then we see that the action of $R(\Gamma, \tilde{\Gamma})$ on $A_k(\Gamma)$ (resp. $G_k(\Gamma)$, $S_k(\Gamma)$) defines a representation of the ring $R(\Gamma, \tilde{\Gamma})$.

We shall now fix our attention to $S_k(\Gamma)$, and introduce an inner product in the space $S_k(\Gamma)$. For two elements f and g of $S_k(\Gamma)$, we put

$$\text{(3.4.2)} \qquad \langle f, g \rangle = \int_{\Gamma\backslash\mathfrak{H}} f(z)\overline{g(z)}\cdot y^{k-2}dx\,dy \qquad (z = x+iy \in \mathfrak{H}) .$$

Here note that $f(z)\overline{g(z)}y^k$ and $y^{-2}dxdy$ are invariant under Γ, on account of Prop. 2.18, and (1.2.3). Therefore the integral is well-defined if it converges. To prove the convergence, it is sufficient to show that $f(z)\overline{g(z)}y^k$, as a function on $\Gamma\backslash\mathfrak{H}^*$, is continuous at the points corresponding to cusps. Let s be a cusp of Γ, ρ an element of $SL_2(\boldsymbol{R})$ such that $\rho(s) = \infty$, and $\Gamma_s = \{\gamma \in \Gamma \mid \gamma(s) = s\}$. Then

$$\rho\Gamma_s\rho^{-1}\cdot\{\pm 1\} = \left\{\pm\begin{bmatrix}1 & h\\ 0 & 1\end{bmatrix}^m \,\middle|\, m \in \boldsymbol{Z}\right\}$$

with a positive real number h. Then there are holomorphic functions $\Phi(q)$

and $\Psi(q)$ at $q=0$, such that

$$f\,|\,[\rho^{-1}]_k=\Phi(e^{\pi iz/h}), \qquad g\,|\,[\rho^{-1}]_k=\Psi(e^{\pi iz/h}).$$

Then we have

$$\begin{aligned} f(w)\overline{g(w)}\,\mathrm{Im}\,(w)^k &= f(\rho^{-1}(z))\overline{g(\rho^{-1}(z))}\,\mathrm{Im}\,(\rho^{-1}(z))^k \qquad (w=\rho^{-1}(z)) \\ &= \Phi(e^{\pi iz/h})\overline{\Psi(e^{\pi iz/h})}\,\mathrm{Im}\,(z)^k. \end{aligned}$$

Since $\Phi(0)=\Psi(0)=0$, we see that this function is continuous around the point of $\Gamma\backslash\mathfrak{H}^*$ corresponding to s, q. e. d.

The inner product $\langle f, g\rangle$ is of course hermitian and positive definite; it is called *the Petersson inner product* (or *the Petersson metric*) in $S_k(\Gamma)$. We shall now determine the adjoint of $[\Gamma_1\alpha\Gamma_2]_k$ with respect to the inner product.

PROPOSITION 3.39. *Let Γ_1 and Γ_2 be commensurable Fuchsian groups of the first kind, and let $\alpha\in\tilde{\Gamma}_1$. If* $\det(\alpha)=1$, *one has*

$$\langle f\,|\,[\Gamma_1\alpha\Gamma_2]_k, g\rangle_2=\langle f, g\,|\,[\Gamma_2\alpha^{-1}\Gamma_1]_k\rangle_1$$

for every $f\in S_k(\Gamma_1)$ and $g\in S_k(\Gamma_2)$, where $\langle\ ,\ \rangle_i$ denotes the Petersson inner product in $S_k(\Gamma_i)$ for $i=1, 2$.

PROOF. First note that, for any $\alpha\in SL_2(\boldsymbol{R})$ and for any measurable set A on $\mathfrak{H}$, we have

$$\int_{\alpha(A)} f\cdot\bar{g}\cdot y^{k-2}dxdy=\int_A (f\,|\,[\alpha]_k)\cdot\overline{(g\,|\,[\alpha]_k)}\cdot y^{k-2}dxdy. \tag{3.4.3}$$

Now let P be a fundamental domain for $\Gamma_2\backslash\mathfrak{H}$. (For example, one can take P to be the polygon Π on $\mathfrak{H}$ considered in the proof of Th. 2.20.) Let

$$\Gamma_2=\textstyle\bigcup_\nu(\Gamma_2\cap\alpha^{-1}\Gamma_1\alpha)\varepsilon_\nu$$

be a disjoint union. Then $\Gamma_1\alpha\Gamma_2=\bigcup\Gamma_1\alpha\varepsilon_\nu$ is a disjoint union. By (3.4.3), we have

$$\begin{aligned} &\int_P (f\,|\,[\Gamma_1\alpha\Gamma_2]_k)\cdot\bar{g}\cdot y^{k-2}dxdy \\ &\quad=\textstyle\sum_\nu\displaystyle\int_P (f\,|\,[\alpha\varepsilon_\nu]_k)\cdot\bar{g}\cdot y^{k-2}dxdy=\textstyle\sum_\nu\displaystyle\int_{\varepsilon_\nu(P)}(f\,|\,[\alpha]_k)\cdot\bar{g}\cdot y^{k-2}dxdy \\ &\quad=\int_Q (f\,|\,[\alpha]_k)\cdot\bar{g}\cdot y^{k-2}dxdy=\int_{\alpha(Q)} f\cdot\overline{(g\,|\,[\alpha^{-1}]_k)}\cdot y^{k-2}dxdy, \end{aligned}$$

where $Q=\bigcup_\nu\varepsilon_\nu(P)$. It can easily be seen that Q is a fundamental domain for $\Gamma_2\cap\alpha^{-1}\Gamma_1\alpha$, hence $\alpha(Q)$ is a fundamental domain for $\alpha\Gamma_2\alpha^{-1}\cap\Gamma_1$. If $\langle\ ,\ \rangle'$ (resp. $\langle\ ,\ \rangle''$) denotes the Petersson inner product in $S_k(\Gamma_2\cap\alpha^{-1}\Gamma_1\alpha)$ (resp. $S_k(\alpha\Gamma_2\alpha^{-1}\cap\Gamma_1)$), then, we have shown that

$$\langle f\,|\,[\Gamma_1\alpha\Gamma_2]_k, g\rangle_2 = \langle f\,|\,[\alpha]_k, g\rangle' = \langle f, g\,|\,[\alpha^{-1}]_k\rangle''.$$

Interchanging f and g, and taking α^{-1} in place of α, we obtain

$$\langle f, g\,|\,[\Gamma_2\alpha^{-1}\Gamma_1]_k\rangle_1 = \langle f, g\,|\,[\alpha^{-1}]_k\rangle'',$$

which completes the proof.

In view of our definition of $j(\sigma, z)$ and $[\sigma]_k$, we have $f\,|\,[c]_k = f$ for every $c \in \boldsymbol{R}^\times$, so that

$$f\,|\,[\Gamma_1 c\Gamma_1]_k = c^{k-2}f \qquad (c \in \boldsymbol{R}^\times). \tag{3.4.4}$$

Therefore, the above proposition needs a modification by a scalar factor if $\det(\alpha) \neq 1$. However, by means of the main involution ι of $M_2(\boldsymbol{R})$ introduced in § 3.3, we have, for any $\alpha \in \tilde{\Gamma}_1$ (not necessarily satisfying $\det(\alpha) = 1$),

$$\langle f\,|\,[\Gamma_1\alpha\Gamma_2]_k, g\rangle_2 = \langle f, g\,|\,[\Gamma_2\alpha^\iota\Gamma_1]_k\rangle_1. \tag{3.4.5}$$

This can easily be verified, since if $\alpha = c\beta$ with $c \in \boldsymbol{R}^\times$ and $\beta \in SL_2(\boldsymbol{R})$, then $\alpha^\iota = c\beta^{-1}$.

PROPOSITION 3.40. *Let Γ_0 be a normal subgroup of finite index of a Fuchsian group Γ of the first kind. Then the linear transformation $[\Gamma_0\sigma\Gamma_0]_k$ on $S_k(\Gamma_0)$, with any $\sigma \in \Gamma$, is unitary with respect to the Petersson inner product on $S_k(\Gamma_0)$.*

This is an immediate consequence of Prop. 3.7 and Prop. 3.39.

A linear transformation of $S_k(\Gamma_1)$ of the type $[\Gamma_1\alpha\Gamma_1]_k$ is called a *Hecke operator* (in a generalized sense). In the next section we shall discuss in detail the Hecke operators in the form by which Hecke originally defined them.

Let us now briefly mention that the double coset $\Gamma_1\alpha\Gamma_2$ can be interpreted as an "algebraic correspondence", of which a more detailed discussion will be made in Chapter 7. Let $\Gamma' = \Gamma_2 \cap \alpha^{-1}\Gamma_1\alpha$, and let φ_1, φ_2, and φ' denote the projection maps of $\mathfrak{H}^*$ to $\Gamma_1\backslash\mathfrak{H}^*$, $\Gamma_2\backslash\mathfrak{H}^*$, and $\Gamma'\backslash\mathfrak{H}^*$ respectively. We can define two holomorphic maps

$$P_1: \Gamma'\backslash\mathfrak{H}^* \longrightarrow \Gamma_1\backslash\mathfrak{H}^*, \qquad P_2: \Gamma'\backslash\mathfrak{H}^* \longrightarrow \Gamma_2\backslash\mathfrak{H}^*$$

by $P_1 \circ \varphi' = \varphi_1 \circ \alpha$, $P_2 \circ \varphi' = \varphi_2$. Note that P_2 is the natural projection, and P_1 the composed map of the natural projection of $\Gamma'\backslash\mathfrak{H}^*$ to $(\alpha^{-1}\Gamma_1\alpha)\backslash\mathfrak{H}^*$ with the isomorphism of $(\alpha^{-1}\Gamma_1\alpha)\backslash\mathfrak{H}^*$ to $\Gamma_1\backslash\mathfrak{H}^*$ obtained from $z \mapsto \alpha(z)$. Now let $\Gamma_2 = \bigcup_{i=1}^d \Gamma'\varepsilon_i$ be a disjoint coset decomposition. Then $\Gamma_1\alpha\Gamma_2 = \bigcup_{i=1}^d \Gamma_1\alpha\varepsilon_i$ (disjoint, see Proof of Prop. 3.1). Therefore if $\varphi_2(z)$ is a point of $\Gamma_2\backslash\mathfrak{H}^*$ with $z \in \mathfrak{H}^*$, we have

$$P_2^{-1}(\varphi_2(z)) = \{\varphi'(\varepsilon_i(z)) \mid i=1, \cdots, e\},$$

$$P_1[P_2^{-1}(\varphi_2(z))] = \{\varphi_1(\alpha\varepsilon_i(z)) \mid i=1, \cdots, e\}.$$

Since $\varphi_1(\beta(z))$ depends only on $\Gamma_1\beta$, we can assert the following:

If $\Gamma_1\alpha\Gamma_2 = \bigcup_{i=1}^e \Gamma_1\alpha_i$, then, through $P_1 \circ P_2^{-1}$, the point $\varphi_2(z)$ corresponds to the points $\varphi_1(\alpha_i(z))$ for $i=1, \cdots, e$.

This is the most primitive form of what we call an *algebraic correspondence*, especially a *modular correspondence* when Γ's are congruence subgroups of $SL_2(\boldsymbol{Z})$. As to the historical account of this topic, the reader is referred to Hurwitz, Mathematische Werke I.

3.5. Hecke operators and their connection with Fourier coefficients

Let us now consider the case of $\Gamma = SL_2(\boldsymbol{Z})$ and its congruence subgroups. Let N be a fixed positive integer, and let Δ_N^*, Γ', and Δ' be as in (3.3.1-3). From (3.3.14) and (3.4.5), we obtain

THEOREM 3.41. *The linear transformations $[\Gamma'\alpha\Gamma']_k$ on $S_k(\Gamma')$, with $\alpha \in \Delta_N^*$, are mutually commutative, and normal with respect to the Petersson inner product on $S_k(\Gamma')$.*

Here we call a linear transformation *normal* if it commutes with its adjoint with respect to the inner product in question. If $N=1$, we have $\Gamma\alpha\Gamma = \Gamma\alpha^\iota\Gamma$ for every $\alpha \in \Delta$, since α and α^ι have the same elementary divisors. Therefore we obtain, from (3.4.5),

THEOREM 3.42. *The linear transformations $[\Gamma\alpha\Gamma]_k$ on $S_k(\Gamma)$, with $\alpha \in \Delta$, are mutually commutative, and self-adjoint with respect to the Petersson inner product on $S_k(\Gamma)$.*

It is a well-known fact that mutually commutative normal linear transformations are simultaneously diagonalizable, i. e., there exists a basis of the vector space in question whose members are common eigen-vectors of the transformations. Therefore we can find common eigen-functions of the $[\Gamma'\alpha\Gamma']_k$ for all $\alpha \in \Delta_N^*$, which form a basis of $S_k(\Gamma')$. In particular, *if $N=1$, the eigen-values are real,* since the $[\Gamma\alpha\Gamma]_k$ are self-adjoint.

Suppose $N=1$. Since $f \mid T(p, p)_k = p^{k-2}f$, an element of $S_k(\Gamma)$ is a common eigen-function of the $[\Gamma\alpha\Gamma]_k$ for all $\alpha \in \Delta$ if and only if it is a common eigen-function of the $T(p)$ for all primes p. Let f be such an eigen-function, and let $f \mid T(n)_k = \mu_n f$ with $\mu_n \in \boldsymbol{R}$ for each positive integer n. By (3.2.2), we have (formally)

$$\textstyle\sum_{n=1}^\infty \mu_n n^{-s} = \prod_p [1-\mu_p p^{-s} + p^{k-1-2s}]^{-1}.$$

In the next §, we shall show that this Dirichlet series is convergent on some half plane, and can be holomorphically continued to the whole complex s-plane; further it will be shown that it satisfies a functional equation analogous to that of the Riemann zeta-function. We shall also prove similar results for congruence subgroups of Γ.

We shall restrict our discussion to $S_k(\Gamma')$. Actually one can consider the Dirichlet series associated with the elements of $G_k(\Gamma')$. It is known that $G_k(\Gamma')$ is spanned by $S_k(\Gamma')$ and the "Eisenstein series" belonging to Γ', which we studied in § 2.2 in the special case $N=1$. And it can be shown that the Dirichlet series associated with the Eisenstein series of level N are essentially of the form

$$L(s, \chi_1)L(s-k+1, \chi_2),$$

where $L(s, \chi)$ is an L-function defined by

$$L(s, \chi) = \sum_{m=1}^{\infty} \chi(m)m^{-s}$$

with a character χ of $(\boldsymbol{Z}/N\boldsymbol{Z})^\times$. For details, see Hecke [27], [29]. Therefore the nature of the coefficients of such Dirichlet series is rather simple. As compared with this, the arithmetical meaning of the Dirichlet series associated with cusp forms is still quite mysterious.

Let us now consider Γ' and Δ' in a somewhat specialized form. We fix a positive divisor t of N, and consider two extreme cases $\mathfrak{h}=(\boldsymbol{Z}/N\boldsymbol{Z})^\times$ and $\mathfrak{h}=\{1\}$ in (3.3.2, 3). Namely we put

$$(3.5.1)\quad \Gamma_0' = \left\{\gamma \in SL_2(\boldsymbol{Z}) \,\middle|\, \lambda_N(\gamma) = \begin{bmatrix} a & tb \\ 0 & a^{-1} \end{bmatrix} \text{ with } a \in (\boldsymbol{Z}/N\boldsymbol{Z})^\times,\ b \in \boldsymbol{Z}/N\boldsymbol{Z}\right\},$$

$$(3.5.1')\quad \Gamma'' = \left\{\gamma \in \Gamma' \,\middle|\, \lambda_N(\gamma) = \begin{bmatrix} 1 & tb \\ 0 & 1 \end{bmatrix} \text{ with } b \in \boldsymbol{Z}/N\boldsymbol{Z}\right\},$$

$$\Delta_0' = \left\{\alpha \in \Delta \,\middle|\, \lambda_N(\alpha) = \begin{bmatrix} a & tb \\ 0 & d \end{bmatrix} \text{ with } a \in (\boldsymbol{Z}/N\boldsymbol{Z})^\times,\ b \in \boldsymbol{Z}/N\boldsymbol{Z},\ d \in \boldsymbol{Z}/N\boldsymbol{Z}\right\},$$

$$\Delta'' = \left\{\alpha \in \Delta \,\middle|\, \lambda_N(\alpha) = \begin{bmatrix} 1 & tb \\ 0 & d \end{bmatrix} \text{ with } b \in \boldsymbol{Z}/N\boldsymbol{Z},\ d \in \boldsymbol{Z}/N\boldsymbol{Z}\right\},$$

where λ_N is the natural map of $M_2(\boldsymbol{Z})$ to $M_2(\boldsymbol{Z}/N\boldsymbol{Z})$. Clearly Γ'' is a normal subgroup of Γ_0', and Γ_0'/Γ'' is isomorphic to $(\boldsymbol{Z}/N\boldsymbol{Z})^\times$. Let ψ be a character of $(\boldsymbol{Z}/N\boldsymbol{Z})^\times$, i.e., a homomorphism of $(\boldsymbol{Z}/N\boldsymbol{Z})^\times$ into $\{z \in \boldsymbol{C} \mid |z|=1\}$. For convenience, we put, for $a \in \boldsymbol{Z}$,

$$\psi(a) = \begin{cases} 0 & \text{if } (a, N) \neq 1, \\ \psi(a \bmod N\boldsymbol{Z}) & \text{if } (a, N) = 1. \end{cases}$$

Further, for $\xi = \begin{bmatrix} a & b \\ c & d \end{bmatrix} \in \Delta$, we put $a(\xi) = a_\xi = a$. Now we denote by

$S_k(\Gamma_0', \psi)$ the set of all elements f of $S_k(\Gamma'')$ satisfying

(3.5.2) $$f \mid [\gamma]_k = \psi(a_\gamma)^{-1} f \qquad \textit{for all} \quad \gamma \in \Gamma_0'.$$

If σ_q is an element of $SL_2(\boldsymbol{Z})$ as in (3.3.10), this is equivalent to

(3.5.3) $$f \mid [\sigma_q]_k = \psi(q) f \qquad \textit{for every } q \textit{ prime to } N.$$

Since $S_k(\Gamma'')$ may be viewed as a (Γ_0'/Γ'')-module, we see that $S_k(\Gamma'')$ is the direct sum of the spaces $S_k(\Gamma_0', \psi)$ for all characters ψ of $(\boldsymbol{Z}/N\boldsymbol{Z})^\times$. We see also that $S_k(\Gamma_0', \psi) = \{0\}$ unless $\psi(-1) = (-1)^k$. From Prop. 3.40, we obtain immediately

(3.5.4) *The subspaces* $S_k(\Gamma_0', \psi)$ *of* $S_k(\Gamma'')$ *are mutually orthogonal with respect to the Petersson inner product.*

Let Γ', $\mathfrak{h}$, and Δ' be as in (3.3.2) and (3.3.3). We observe that $\Gamma'' \subset \Gamma' \subset \Gamma_0'$, $\Delta'' \subset \Delta' \subset \Delta_0'$, and $S_k(\Gamma')$ is the direct sum of the spaces $S_k(\Gamma_0', \psi)$ for all ψ such that $\psi(\mathfrak{h}) = 1$.

For every $\alpha \in \Delta_0'$, we can define a linear transformation $[\Gamma_0'\alpha\Gamma_0']_{k,\psi}$ on $S_k(\Gamma_0', \psi)$ as follows. Take a disjoint decomposition $\Gamma_0'\alpha\Gamma_0' = \bigcup_\nu \Gamma_0'\alpha_\nu$, and for $f \in S_k(\Gamma_0', \psi)$, put

(3.5.5) $$f \mid [\Gamma_0'\alpha\Gamma_0']_{k,\psi} = \det(\alpha)^{k/2-1} \textstyle\sum_\nu \psi(a(\alpha_\nu)) \cdot f \mid [\alpha_\nu]_k .$$

It is easy to see that the right hand side does not depend on the choice of $\{\alpha_\nu\}$, and satisfies (3.5.2). Now we have

(3.5.6) $[\Gamma_0'\beta\Gamma_0']_{k,\psi}$ *is the restriction of* $[\Gamma'\beta\Gamma']_k$ *to* $S_k(\Gamma_0', \psi)$ *for every* $\beta \in \Delta'$, *if* $\psi(\mathfrak{h}) = 1$.

In fact, by Prop. 3.36, we can find a disjoint decomposition $\Gamma'\beta\Gamma' = \bigcup_\nu \Gamma'\beta_\nu$ with elements β_ν of the form $\sigma_a \cdot \begin{bmatrix} a & tb \\ 0 & d \end{bmatrix}$ as described there. Then, by the same proposition, we obtain a disjoint decomposition $\Gamma_0'\beta\Gamma_0' = \bigcup_\nu \Gamma_0'\beta_\nu$. Since $\psi(a(\beta)) = 1$ for $\beta \in \Delta'$ if $\psi(\mathfrak{h}) = 1$, we obtain (3.5.6). Observe that, for any $\alpha \in \Delta_0'$, there exists an element β of Δ'' such that $\Gamma_0'\alpha\Gamma_0' = \Gamma_0'\beta\Gamma_0'$. Therefore (3.5.6) implies that $f \mid [\Gamma_0'\alpha\Gamma_0']_{k,\psi}$ belongs to $S_k(\Gamma_0', \psi)$.

Now we see that

$$\Gamma_0'\alpha\Gamma_0' \mapsto [\Gamma_0'\alpha\Gamma_0']_{k,\psi}$$

defines a representation of the ring $R(\Gamma_0', \Delta_0')$ on $S_k(\Gamma_0', \psi)$. Let us denote by $T'(a, d)_{k,\psi}$ and $T'(n)_{k,\psi}$ the action of $T'(a, d)$ and $T'(n)$ on $S_k(\Gamma_0', \psi)$ defined by (3.5.5). By Prop. 3.36, we have

(3.5.7) $$f \mid T'(n)_{k,\psi} = n^{k-1} \textstyle\sum_a \sum_{b=0}^{d-1} \psi(a) f((az+tb)/d) d^{-k} \qquad (a > 0,\ ad = n).$$

(Note that, by virtue of our agreement $\psi(a)=0$ for $(a, N)\neq 1$, we can drop the condition $(a, N)=1$.) By (3.3.11) and (3.4.4), we have

(3.5.8) $$f\mid T'(q, q)_{k,\psi}=q^{k-2}\psi(q)\cdot f \quad \text{for } f\in S_k(\Gamma_0', \psi) \text{ and } (q, N)=1.$$

Therefore, from (3.3.8) and (3.3.6), we obtain formally

(3.5.9) $$\sum_{m=1}^{\infty} T'(m)_{k,\psi}\cdot m^{-s}$$
$$=\prod_{p\mid N}[1-T'(p)_{k,\psi}p^{-s}]^{-1}\cdot\prod_{p\nmid N}[1-T'(p)_{k,\psi}p^{-s}+\psi(p)p^{k-1-2s}]^{-1},$$

(3.5.10) $$T'(m)_{k,\psi}T'(n)_{k,\psi}=\sum_{d\mid(m,n)} d^{k-1}\psi(d)T'(mn/d^2)_{k,\psi}.$$

Note that, in the last formula, d runs over all the positive divisors of (m, n), since $\psi(d)=0$ if $(d, N)\neq 1$. The convergence of (3.5.9) will be proved in the next section (Lemma 3.62, see also Remark 3.46).

Observe that the cusp ∞ of Γ'' is regular, and the stability subgroup of ∞ is generated by $\begin{bmatrix}1 & t\\ 0 & 1\end{bmatrix}$. Let $f\in S_k(\Gamma_0', \psi)$, and $g=f\mid T'(m)_{k,\psi}$ with a fixed positive integer m. Consider the Fourier expansion of f and g at ∞:

$$f(z)=\sum_{n=1}^{\infty} c(n)e^{2\pi inz/t}, \qquad g(z)=\sum_{n=1}^{\infty} c'(n)e^{2\pi inz/t}.$$

By (3.5.7), we obtain

(3.5.11) $$g(z)=\sum_{n=1}^{\infty}\sum_a\sum_{b=0}^{d-1}(ad)^{k-1}d^{-k}\psi(a)c(n)e^{2\pi in(az+tb)/dt} \qquad (ad=m).$$

Since $\sum_{b=0}^{d-1} e^{2\pi inb/d}=d$ or 0 according as d divides n or not, we obtain, comparing the coefficients of $e^{2\pi ilz/t}$ of both sides of (3.5.11),

(3.5.12) $$c'(l)=\sum_{a\mid(l,m)}\psi(a)a^{k-1}c(lm/a^2).$$

THEOREM 3.43. *Let* $f(z)=\sum_{n=1}^{\infty} c(n)e^{2\pi inz/t}$ *be a non-zero element of* $S_k(\Gamma_0', \psi)$. *Suppose that* f *is a common eigen-function of the* $T'(n)_{k,\psi}$ *for all* n: $f\mid T'(n)_{k,\psi}=\lambda_n f$ *with* $\lambda_n\in\mathbf{C}$. *Then* $c(1)\neq 0$, $c(n)=\lambda_n c(1)$, *and*

(3.5.13) $$\sum_{n=1}^{\infty}\lambda_n n^{-s}=\prod_p[1-\lambda_p p^{-s}+\psi(p)p^{k-1-2s}]^{-1} \qquad (\textit{formally}).$$

Conversely, if one has formally

(3.5.14) $$\sum_{n=1}^{\infty} c(n)n^{-s}=\prod_p[1-c(p)p^{-s}+\psi(p)p^{k-1-2s}]^{-1},$$

then $f\mid T'(n)_{k,\psi}=c(n)f$ *for all* n.

PROOF. If $f\mid T'(m)_{k,\psi}=\lambda_m f$, then, taking l to be 1 in (3.5.12), we obtain

(3.5.15) $$\lambda_m c(1)=c(m).$$

Therefore $c(1)\neq 0$, since $f\neq 0$, and (3.5.13) follows from (3.5.9). Conversely, if one has (3.5.14), we see, by the same reasoning as in the proof of Th. 3.24, that

$$c(l)c(m)=\sum_{a|(l,m)} a^{k-1}\psi(a)c(lm/a^2).$$

Therefore, by (3.5.12), we have $c'(l)=c(l)c(m)$, so that $f \mid T'(m)_{k,\psi}=g=c(m)f$.

As an immediate consequence, we obtain

COROLLARY 3.44. *If two functions of $S_k(\Gamma_0', \psi)$ are common eigen-functions of $T'(n)_{k,\psi}$ for all n, and belong to the same eigen-values, then they differ only by a constant factor.*

Let us fix any basis $\{f_1, \cdots, f_\kappa\}$ of $S_k(\Gamma_0', \psi)$ over $\boldsymbol{C}$, where $\kappa=\dim(S_k(\Gamma_0', \psi))$. Put

$$\boldsymbol{f}(z)=\begin{bmatrix} f_1(z) \\ \vdots \\ f_\kappa(z) \end{bmatrix}=\sum_{n=1}^{\infty} \boldsymbol{c}(n)e^{2\pi inz/t}$$

with complex column vectors $\boldsymbol{c}(n)$. Then the $\boldsymbol{c}(n)$, for $n=1, 2, \cdots$, span the space $\boldsymbol{C}^\kappa$ of all κ-dimensional complex column vectors. In fact, if they don't, there exists a non-zero $\boldsymbol{C}$-linear map ξ of $\boldsymbol{C}^\kappa$ into $\boldsymbol{C}$ such that $\xi(\boldsymbol{c}(n))=0$ for all n. Then we have $\xi(\boldsymbol{f})=0$, which is a contradiction, since $f_1, \cdots, f_\kappa$ are linearly independent over $\boldsymbol{C}$.

For every positive integer m, define an element $\Lambda(m)$ of $M_\kappa(\boldsymbol{C})$ by

$$\boldsymbol{f} \mid T'(m)_{k,\psi}=\Lambda(m)\boldsymbol{f}.$$

Then, by the same type of computation as in (3.5.11), (3.5.12), and (3.5.15), we obtain

$$\Lambda(m)\boldsymbol{c}(1)=\boldsymbol{c}(m). \tag{3.5.16}$$

THEOREM 3.45. *If $\kappa=\dim(S_k(\Gamma_0', \psi))$, the linear transformations $T'(n)_{k,\psi}$, for all positive integers n, generate a commutative algebra over $\boldsymbol{C}$ of rank κ. Moreover, the identity map of the algebra (or the map $T'(n)_{k,\psi} \mapsto \Lambda(n)$) is equivalent to a regular representation of the algebra.*

PROOF. Let A be the subalgebra of $M_\kappa(\boldsymbol{C})$ generated by the $\Lambda(m)$ for all m. Consider a $\boldsymbol{C}$-linear map L of A into $\boldsymbol{C}^\kappa$ defined by $L(X)=X\cdot\boldsymbol{c}(1)$ for $X\in A$. By (3.5.16), L is surjective. If $L(X)=0$, we have $X\cdot\boldsymbol{c}(n)=X\Lambda(n)\boldsymbol{c}(1)$ $=\Lambda(n)X\boldsymbol{c}(1)=0$ for every n, so that $X=0$, since the $\boldsymbol{c}(n)$ span the whole $\boldsymbol{C}^\kappa$. Therefore L is injective, and hence L gives an A-linear isomorphism of A onto $\boldsymbol{C}^\kappa$, q. e. d.

REMARK 3.46. Put $\Lambda(n)=(\lambda_{ij}(n))$ with $\lambda_{ij}(n)\in\boldsymbol{C}$, and $g_{ij}(z)=\sum_{n=1}^{\infty}\lambda_{ij}(n)e^{2\pi inz/t}$ (formally, for the moment). By the above theorem, the g_{ij} span, over $\boldsymbol{C}$, a vector space of rank κ. From (3.5.16), we obtain $(g_{ij}(z))\boldsymbol{c}(1)=\boldsymbol{f}(z)$. Since the components of $\boldsymbol{f}$ span $S_k(\Gamma_0', \psi)$, we see that the g_{ij} are actually holomorphic functions and span $S_k(\Gamma_0', \psi)$. Therefore, to prove the convergence of (3.5.9) (or, of $\sum_{n=1}^{\infty} \Lambda(n)n^{-s}$), it is sufficient to show the convergence of $\sum_{n=1}^{\infty} a_n n^{-s}$ for every $\sum_{n=1}^{\infty} a_n e^{2\pi i n z/t} \in S_k(\Gamma_0', \psi)$. This will be done in Lemma 3.62.

To obtain further information on eigen-values, we consider a somewhat general situation. Let A be an arbitrary commutative algebra over a field F, of finite rank, and with an identity element, R the radical of A, and P a unitary A-module. Suppose that the simple components of A/R are all separably algebraic extensions of F, which is always the case if F is of characteristic 0. By a theorem of Wedderburn, there exists a semi-simple subalgebra B of A such that $A = B \oplus R$. Note that B has the same identity element as A. (This is true even if A is non-commutative. In fact, if $1 = b + r$ with $b \in B$ and $r \in R$, then $b = b^2 + br$, so that $br = 0$. Therefore $r = br + r^2 = r^2$. Since r is nilpotent, $r = 0$, q. e. d.) Let $B_1, \cdots, B_s$ be the simple components of B, and e_i the identity element of B_i. Put $P_i = e_i P$. Then $P = P_1 \oplus \cdots \oplus P_s$. Since A is commutative, P_i is an A-submodule of P. Since R is nilpotent, we have a finite decreasing sequence of A-submodules

$$P_i \supset RP_i \supset R^2 P_i \supset \cdots \supset R^{m_i - 1} P_i \supsetneqq R^{m_i} P_i = \{0\} \qquad (m_i \geqq 0).$$

We understand that $R^0 = A$, and $m_i = 0$ if $P_i = \{0\}$.

Let us now take A to be the algebra generated by the $T'(n)_{k,\psi}$ for all n over $\boldsymbol{C}$, and P to be $S_k(\Gamma_0', \psi)$. In this case every B_i is isomorphic to $\boldsymbol{C}$. Take a basis $\{f_1, \cdots, f_\kappa\}$ of $P = S_k(\Gamma_0', \psi)$ so as to contain a basis of $R^h P_i$ for every i and every h, and define $\Lambda(n)$ as above with respect to this $\{f_1, \cdots, f_\kappa\}$. Then $\Lambda(n)$ is clearly a triangular matrix for every n. Let $\lambda_1(n), \cdots, \lambda_\kappa(n)$ be the diagonal elements of $\Lambda(n)$. Then we see that, for each ν,

$$T'(n)_{k,\psi} \mapsto \lambda_\nu(n) \tag{3.5.17}$$

defines a homomorphism of A onto $\boldsymbol{C}$. Of course these κ homomorphisms, as a whole, are independent of the choice of $\boldsymbol{f}$. Now we have

PROPOSITION 3.47. *The notation being as above, there exists, for each ν, a non-zero element g_ν of $S_k(\Gamma_0', \psi)$ such that $g_\nu \mid T'(n)_{k,\psi} = \lambda_\nu(n) g_\nu$ for all n; in other words, there exists an element h_ν of $S_k(\Gamma_0', \psi)$ such that $h_\nu(z) = \sum_{n=1}^{\infty} \lambda_\nu(n) e^{2\pi i n z/t}$.*

In fact, we can find i so that $f_\nu \in P_i$. Then, for this i, take any non-zero element g_ν in $R^{m_i - 1} P_i$. In view of our definition of P_i, we see that the homomorphism (3.5.17) is the same as the map

$$\textstyle\sum_{j=1}^{s} a_j e_j + r \mapsto a_i \qquad (a_j \in \boldsymbol{C},\ r \in R).$$

Since $(\sum_{j=1}^{s} a_j e_j + r) g_\nu = a_i g_\nu$, the element g_ν has the required property.

From Th. 3.45, we see easily that

(3.5.18) *Every non-zero $\boldsymbol{C}$-linear homomorphism of A into $\boldsymbol{C}$ coincides with one of the homomorphisms $\lambda_1, \cdots, \lambda_\kappa$ of* (3.5.17).

Let us now consider operators $[\Gamma'\alpha\Gamma']_k$ with a group Γ' of type (3.3.2) with arbitrarily fixed $N, t, \mathfrak{h}$. Let Δ' be as in (3.3.3), and $\varepsilon = \begin{bmatrix} -1 & 0 \\ 0 & 1 \end{bmatrix}$. We see that $\varepsilon\Gamma' = \Gamma'\varepsilon$. For every $X = \sum c_\nu \cdot \Gamma'\alpha_\nu\Gamma' \in R(\Gamma', \Delta) \otimes_{\boldsymbol{Z}} \boldsymbol{C}$ with $c_\nu \in \boldsymbol{C}$, put $X^\varepsilon = \sum c_\nu \cdot \Gamma'\varepsilon\alpha_\nu\varepsilon^{-1}\Gamma'$, and

$$\mathfrak{B} = \{X \in R(\Gamma', \Delta) \mid X^\varepsilon = X\}.$$

We see that $X \mapsto X^\varepsilon$ is an automorphism of the ring $R(\Gamma', \Delta)$, and hence $\mathfrak{B}$ is a subring of $R(\Gamma', \Delta)$. Now let us prove

(3.5.19) $$R(\Gamma', \Delta') \subset \mathfrak{B}.$$

If $\alpha \in \Delta_N^*$, we see that $\Gamma\varepsilon\alpha\varepsilon^{-1}\Gamma = \Gamma\alpha\Gamma$ and $\varepsilon\alpha\varepsilon^{-1} \equiv \alpha \bmod (N)$, so that $\Gamma'\varepsilon\alpha\varepsilon^{-1}\Gamma' = \Gamma'\alpha\Gamma'$ by (2) of Lemma 3.29. If α is diagonal, we have obviously $\Gamma'\varepsilon\alpha\varepsilon^{-1}\Gamma' = \Gamma'\alpha\Gamma'$. Therefore we obtain (3.5.19) on account of Prop. 3.32.

In the following discussion, we shall denote by $\mathrm{End}(V, K)$ the ring of all K-linear endomorphisms of a vector space V over a field K. For each $X \in R(\Gamma', \Delta) \otimes_{\boldsymbol{Z}} \boldsymbol{C}$, let $[X]_k$ denote the element of $\mathrm{End}(S_k(\Gamma'), \boldsymbol{C})$ corresponding to X.

THEOREM 3.48. *The notation being as above, let B (resp. B_0) denote the algebra generated by the $[X]_k$ for all $X \in \mathfrak{B}$ over $\boldsymbol{C}$ (resp. over $\boldsymbol{Q}$). Suppose that $k \geqq 2$. Then the following assertions hold.*

(1) *B_0 is a semi-simple algebra of finite rank.*

(2) *$B = B_0 \otimes_{\boldsymbol{Q}} \boldsymbol{C}$.*

(3) *The characteristic polynomial of $[X]_k$ for every $X \in \mathfrak{B}$ has rational integral coefficients.*

PROOF. For $X = \sum c_\nu \cdot \Gamma'\alpha_\nu\Gamma' \in R(\Gamma', \Delta) \otimes_{\boldsymbol{Z}} \boldsymbol{C}$ with $c_\nu \in \boldsymbol{C}$, put $X^* = \sum \bar{c}_\nu \cdot \Gamma'\alpha_\nu^\iota\Gamma'$. Then we see easily that $(X^\varepsilon)^* = (X^*)^\varepsilon$. By (3.4.5), $[X^*]_k$ is the adjoint of $[X]_k$ with respect to the Petersson inner product of $S_k(\Gamma')$. Therefore $\mathrm{tr}([X^*X]_k) > 0$ unless $[X]_k = 0$. Now $\mathfrak{B}$ is stable under the map $X \mapsto X^*$. Suppose that B_0 has a nilpotent right ideal N. If $[X]_k \in N$, then $[XX^*]_k$ is contained in N, hence nilpotent, so that $\mathrm{tr}([XX^*]_k) = 0$. Therefore $[X]_k = 0$. Thus B_0 has no nilpotent right ideal $\neq \{0\}$, which proves (1). Now, for $f \in S_k(\Gamma')$, put $f^\varepsilon = \overline{f(-\bar{z})}$. We see easily that $f^\varepsilon \in S_k(\Gamma')$, and $f^\varepsilon \mid [X]_k = (f \mid [X^\varepsilon]_k)^\varepsilon$ for every $X \in R(\Gamma', \Delta)$. Put

$$W = \{f \in S_k(\Gamma') \mid f^\varepsilon = f\}.$$

Then W is an $\boldsymbol{R}$-linear subspace of $S_k(\Gamma')$, and $S_k(\Gamma') = W \otimes_{\boldsymbol{R}} \boldsymbol{C}$. (In fact $2f = (f+f^\varepsilon) - i((if)+(if)^\varepsilon)$.) Let B_1 be the $\boldsymbol{R}$-linear span of B_0. Then we see that W is stable under B_1. Since $\mathrm{End}(S_k(\Gamma'), \boldsymbol{C}) = \mathrm{End}(W, \boldsymbol{R}) \otimes_{\boldsymbol{R}} \boldsymbol{C}$, we see that elements of B_1 are linearly independent over $\boldsymbol{R}$ if and only if they are so over $\boldsymbol{C}$. From this we can conclude that $B = B_1 \otimes_{\boldsymbol{R}} \boldsymbol{C}$. Therefore, in order to prove (2), it is sufficient to prove $B_1 = B_0 \otimes_{\boldsymbol{Q}} \boldsymbol{R}$. The proof of this fact and (3) is based on the following statement which we shall prove in § 8.4.

(3.5.20) *There is a discrete $\boldsymbol{Z}$-submodule L of $S_k(\Gamma')$ of maximal rank which is stable under the $[\Gamma'\alpha\Gamma']_k$ for all $\alpha \in \Delta$.*

Assuming this, let V be the $\boldsymbol{Q}$-linear span of L. Then $S_k(\Gamma') = V \otimes_{\boldsymbol{Q}} \boldsymbol{R}$, and hence $\mathrm{End}(S_k(\Gamma'), \boldsymbol{R}) = \mathrm{End}(V, \boldsymbol{Q}) \otimes_{\boldsymbol{Q}} \boldsymbol{R}$. Therefore elements of B_0 are linearly independent over $\boldsymbol{Q}$ if and only if they are so over $\boldsymbol{R}$. This shows that $B_1 = B_0 \otimes_{\boldsymbol{Q}} \boldsymbol{R}$, hence (2). Let r be the dimension of $S_k(\Gamma')$ over $\boldsymbol{C}$. Then we obtain three faithful representations

$$\rho_0 : B_0 \to M_{2r}(\boldsymbol{Q}) \cong \mathrm{End}(V, \boldsymbol{Q}),$$
$$\rho_1 : B_1 \to M_r(\boldsymbol{R}) \cong \mathrm{End}(W, \boldsymbol{R}),$$
$$\rho\ : B \to M_r(\boldsymbol{C}) \cong \mathrm{End}(S_k(\Gamma'), \boldsymbol{C}).$$

Restrict ρ and ρ_1 to B_0. By Lemma 3.49 below, ρ_0 is equivalent to the direct sum of ρ and its complex conjugate. From the above discussion we see that ρ_1 is equivalent to ρ. Since ρ_1 is a real representation, we see that ρ_0 is equivalent to the direct sum of two copies of ρ. Define ρ_0 with respect to a basis of L over $\boldsymbol{Z}$. If $\xi = [X]_k$ with $X \in \mathfrak{B}$, then ξ sends the lattice L into itself, so that $\rho_0(\xi) \in M_{2r}(\boldsymbol{Z})$. Therefore the characteristic polynomial of $\rho(\xi)$ must have integral coefficients, hence (3).

In the above proof we needed the following elementary

LEMMA 3.49. *Let $\boldsymbol{C}^r$ denote the vector space of all r-dimensional complex column vectors, and $\{x_1, \cdots, x_{2r}\}$ a basis of $\boldsymbol{C}^r$ over $\boldsymbol{R}$. For every $U \in M_r(\boldsymbol{C})$, define an element $\lambda(U) = (\lambda_{ij}(U))$ of $M_{2r}(\boldsymbol{R})$ by $Ux_j = \sum_{i=1}^{2r} \lambda_{ij}(U)x_i$. Then there exists an element Y of $GL_{2r}(\boldsymbol{C})$, independent of U, such that $Y^{-1}\begin{bmatrix} U & 0 \\ 0 & \bar{U} \end{bmatrix} Y = \lambda(U)$.*

PROOF. Let X be the $r \times 2r$ matrix whose i-th column is x_i. Put $Y = \begin{bmatrix} X \\ \bar{X} \end{bmatrix}$. Since $UX = X\lambda(U)$, we have $\bar{U}\bar{X} = \bar{X}\lambda(U)$, so that $\begin{bmatrix} U & 0 \\ 0 & \bar{U} \end{bmatrix} Y = Y\lambda(U)$. Assume that $\det(Y) = 0$. Then there exists a set of $2r$ complex numbers $(a_1, \cdots, a_{2r}) \neq (0, \cdots, 0)$ such that $\sum_i a_i x_i = \sum_i a_i \bar{x}_i = 0$. Then $\sum_{i=1}^{2r} (ca_i + \overline{ca_i}) x_i = 0$ for all $c \in \boldsymbol{C}$. Since $\{x_1, \cdots, x_{2r}\}$ is a basis of $\boldsymbol{C}^r$ over $\boldsymbol{R}$, we have $a_1 = \cdots = a_{2r} = 0$, which is a contradiction. Thus Y is invertible, hence our assertion.

REMARK 3.50. If we take the ring $R(\Gamma', \Delta)$ instead of $\mathfrak{B}$, the assertion (1) will remain true, but the assertions (2) and (3) will not. For example, take $\Gamma(6)$ to be Γ', and let $k=2$. Then $S_2(\Gamma(6))$ is of dimension 1 over $\boldsymbol{C}$, and spanned by $\Delta^{1/6}$ (see Ex. 2.29). Let $\alpha=\begin{bmatrix} 1 & 1 \\ 0 & 1 \end{bmatrix}$. We see easily that $\Delta^{1/6} \mid [\Gamma'\alpha\Gamma']_2 = e^{2\pi i/6}\Delta^{1/6}$. Therefore the $\boldsymbol{Q}$-linear span of the $[X]_k$ for all $X \in R(\Gamma', \Delta)$ is two-dimensional over $\boldsymbol{Q}$, but the $\boldsymbol{C}$-linear span is obviously one-dimensional.

THEOREM 3.51. *Let Γ' and Δ' be as in* (3.3.2-3), *r the dimension of $S_k(\Gamma')$ over $\boldsymbol{C}$, and D (resp. D_0) the algebra generated by the $[\Gamma'\alpha\Gamma']_k$ for all $\alpha \in \Delta'$ over $\boldsymbol{C}$ (resp. over $\boldsymbol{Q}$). Suppose that $k \geqq 2$. Then*

(1) $[D_0 : \boldsymbol{Q}] = r$.

(2) $D = D_0 \otimes_{\boldsymbol{Q}} \boldsymbol{C}$.

(3) *The identity injection of D_0 into* $\mathrm{End}(S_k(\Gamma'), \boldsymbol{C})$ *is equivalent to a regular representation of D_0 over $\boldsymbol{Q}$.*

PROOF. Since $R(\Gamma', \Delta') \subset \mathfrak{B}$, the assertion (2) follows from (2) of Th. 3.48. Let $T'(n)_k$ denote the sum of $[\Gamma'\alpha\Gamma']_k$ with $\alpha \in \Delta'$ and $\det(\alpha) = n$ (cf. § 3.3). By (5) of Th. 3.34, D_0 is generated over $\boldsymbol{Q}$ by the $T'(n)_k$ for all n. Now observe that $S_k(\Gamma')$ is the direct sum of the spaces $S_k(\Gamma'_0, \psi)$ for all the characters ψ of $(\boldsymbol{Z}/N\boldsymbol{Z})^\times$ such that $\psi(\mathfrak{h}) = 1$, where $\mathfrak{h}$ is the subgroup of $(\boldsymbol{Z}/N\boldsymbol{Z})^\times$ in the definition (3.3.2) of Γ'. Let $\psi_1, \cdots, \psi_\mu$ be all such characters. For each ψ_ν take a basis $\{f_1, \cdots, f_\kappa\}$ of $S_k(\Gamma'_0, \psi_\nu)$, and put

$$\boldsymbol{f}^{(\nu)} = \begin{bmatrix} f_1 \\ \vdots \\ f_\kappa \end{bmatrix} = \textstyle\sum_{n=1}^{\infty} \boldsymbol{c}^{(\nu)}(n) e^{2\pi i n z/t} \qquad (\nu = 1, \cdots, \mu),$$

$$\boldsymbol{f} = \begin{bmatrix} \boldsymbol{f}^{(1)} \\ \vdots \\ \boldsymbol{f}^{(\mu)} \end{bmatrix} = \textstyle\sum_{n=1}^{\infty} \boldsymbol{a}(n) e^{2\pi i n z/t}$$

with $\boldsymbol{c}^{(\nu)}(n) \in \boldsymbol{C}^\kappa$ and $\boldsymbol{a}(n) \in \boldsymbol{C}^r$. Define an element $\omega(n)$ of $GL_r(\boldsymbol{C})$ by $\boldsymbol{f} \mid T'(n)_k = \omega(n)\boldsymbol{f}$. From (3.5.16), we obtain $\omega(n)\boldsymbol{a}(1) = \boldsymbol{a}(n)$ for every n. By the same type of reasoning as in the proof of Th. 3.45, we can show that the map $T'(n)_k \mapsto \omega(n)$ is equivalent to a regular representation of D over $\boldsymbol{C}$, and hence $[D : \boldsymbol{C}] = r$. This together with (2) proves (1) and (3).

THEOREM 3.52. *If Γ' is as in* (3.3.2), *and $k \geqq 2$, then $S_k(\Gamma')$ has a basis consisting of cusp forms of which the Fourier coefficients at ∞ are rational integers.*

PROOF. Put $\omega(n)=(\omega_{pq}(n))$ and $f_{pq}=\sum_{n=1}^{\infty}\omega_{pq}(n)e^{2\pi inz/t}$. Since the $\boldsymbol{C}$-linear span of the $\omega(n)$ is r-dimensional, we see that the f_{pq} span a vector space over $\boldsymbol{C}$ of dimension at most r. But we have $(f_{pq})\boldsymbol{a}(1)=\boldsymbol{f}$, so that the f_{pq} span $S_k(\Gamma')$ over $\boldsymbol{C}$, since the components of $\boldsymbol{f}$ form a basis of $S_k(\Gamma')$. Let L be as in (3.5.20), and

$$E=\{\xi\in D_0\mid \xi L\subset L\}\,.$$

Then D_0 is spanned by E over $\boldsymbol{Q}$, and E is a free $\boldsymbol{Z}$-module of rank r. Define a regular representation Φ of D_0 over $\boldsymbol{Q}$ with respect to a basis of E over $\boldsymbol{Z}$. Since E is a subring of D_0 containing $[\Gamma'\alpha\Gamma']_k$ for all $\alpha\in\Delta'$, we see that Φ maps $T'(n)_k$ into $M_r(\boldsymbol{Z})$. Put $\Phi_n=\Phi(T'(n)_k)$. By (3) of Th. 3.51, there exists an element U of $GL_r(\boldsymbol{C})$ such that $U\omega(n)U^{-1}=\Phi_n$. Put $(g_{pq}(z))=U(f_{pq})U^{-1}=\sum_{n=1}^{\infty}\Phi_n e^{2\pi inz/t}$. Then $S_k(\Gamma')$ is spanned by the g_{pq} over $\boldsymbol{C}$, and the g_{pq} have integral Fourier coefficients.

PROPOSITION 3.53. *If f is an element of $S_k(\Gamma')$ which is a common eigen-function of the $T'(n)_k$ for all n, then f belongs to $S_k(\Gamma_0', \psi)$ with a unique character ψ of $(\boldsymbol{Z}/N\boldsymbol{Z})^{\times}$ such that $\psi(\mathfrak{h})=1$, and f is a common eigen-function of all $T'(n)_{k,\psi}$.*

PROOF. By (5) of Th. 3.34, f is an eigen-function of $T'(q,q)_k$ for every q prime to N, so that, by (3.3.11), f is an eigen-function of $[\sigma_q]_k$. Therefore we can define a character ψ of $(\boldsymbol{Z}/N\boldsymbol{Z})^{\times}$ by $f\mid[\sigma_q]_k=\psi(q)f$. Then $f\in S_k(\Gamma_0',\psi)$. The last assertion follows from the formula (3.5.6), which implies that $T'(n)_{k,\psi}$ is the restriction of $T'(n)_k$ to $S_k(\Gamma_0',\psi)$.

PROPOSITION 3.54. *Let $\tau=\begin{bmatrix}0 & -t\\ N & 0\end{bmatrix}$. Then, for every $\alpha\in\Delta_N^*$,*

$$(\Gamma'\alpha^{\iota}\Gamma')(\Gamma'\tau\Gamma')=(\Gamma'\tau\Gamma')(\Gamma'\alpha\Gamma')\,.$$

PROOF. First note that

$$\tau\cdot\begin{bmatrix}a & tb\\ c & d\end{bmatrix}=\begin{bmatrix}d & -tc/N\\ -Nb & a\end{bmatrix}\cdot\tau\,,$$

and hence $\tau\Gamma'=\Gamma'\tau$. For a given $\alpha\in\Delta_N^*$, put $q=\det(\alpha)$, and $\beta=\tau\alpha\tau^{-1}$. Then $\beta\equiv\begin{bmatrix}q & tb\\ 0 & 1\end{bmatrix}$ mod (N) with $b\in\boldsymbol{Z}$. Put $\gamma=\begin{bmatrix}1 & -tb\\ 0 & 1\end{bmatrix}$. Then $\gamma\beta\equiv\begin{bmatrix}q & 0\\ 0 & 1\end{bmatrix}$ $\equiv\alpha^{\iota}$ mod (N). Since q is prime to N, we see that β has the same elementary divisors as α. Therefore, by (2) of Lemma 3.29, $\Gamma'\beta\Gamma'=\Gamma'\gamma\beta\Gamma'=\Gamma'\alpha^{\iota}\Gamma'$. On the other hand, by Prop. 3.7,

$$(\Gamma'\beta\Gamma')(\Gamma'\tau\Gamma')=\Gamma'\beta\tau\Gamma'=\Gamma'\tau\alpha\Gamma'=(\Gamma'\tau\Gamma')(\Gamma'\alpha\Gamma')\,,$$

q. e. d.

PROPOSITION 3.55. *Let* $\tau=\begin{bmatrix}0 & -t\\ N & 0\end{bmatrix}$. *Then* $[\tau]_k$ $(=(tN)^{1-k/2}\cdot[\Gamma''\tau\Gamma'']_k)$ *sends* $S_k(\Gamma_0', \psi)$ *onto* $S_k(\Gamma_0', \bar{\psi})$, *and* $[\tau]_k^2=1$. *Moreover, for every n prime to N, one has*

$$T'(n)_{k,\psi}\cdot[\tau]_k=\psi(n)\cdot[\tau]_k\cdot T'(n)_{k,\bar{\psi}}.$$

PROOF. Let $\alpha\in\Delta_N^*$ and $\det(\alpha)=n$. By Prop. 3.54 and (3.3.13), if $f\in S_k(\Gamma_0', \psi)$, we have

$$(*)\qquad \begin{aligned} f\mid[\tau]_k[\Gamma''\alpha\Gamma'']_k&=f\mid[\Gamma''\alpha^\iota\Gamma'']_k[\tau]_k\\ &=f\mid[\Gamma''\sigma_n^{-1}\Gamma'']_k[\Gamma''\alpha\Gamma'']_k[\tau]_k\\ &=\psi(n)^{-1}f\mid[\Gamma''\alpha\Gamma'']_k[\tau]_k. \end{aligned}$$

Take α to be $q\cdot\sigma_q$. Then

$$f\mid[\tau]_k[\sigma_q]_k=\psi(q)^{-2}\psi(q)f\mid[\tau]_k=\bar{\psi}(q)f\mid[\tau]_k,$$

so that $f\mid[\tau]_k\in S_k(\Gamma_0', \bar{\psi})$. Then from (*) and (3.5.6) we obtain the desired formula. The relation $[\tau]_k^2=1$ is obvious.

PROPOSITION 3.56. *If* λ *is an eigen-value of* $[\Gamma_0'\alpha\Gamma_0']_{k,\psi}$ *with* $\alpha\in\Delta_N^*$, *then* $\lambda=\psi(\det(\alpha))\bar{\lambda}$.

PROOF. If $q=\det(\alpha)$, and $f\mid[\Gamma_0'\alpha\Gamma_0']_{k,\psi}=\lambda f$, we have, by (3.3.13) and (3.5.6), $\lambda f=\psi(q)f\mid[\Gamma''\alpha^\iota\Gamma'']_k$. Denoting by $\langle\ ,\ \rangle$ the Petersson inner product on $S_k(\Gamma'')$, we obtain, by (3.4.5),

$$\lambda\cdot\langle f,f\rangle=\langle f\mid[\Gamma''\alpha\Gamma'']_k, f\rangle=\langle f, f\mid[\Gamma''\alpha^\iota\Gamma'']_k\rangle=\psi(q)\bar{\lambda}\cdot\langle f,f\rangle,$$

and hence $\lambda=\psi(q)\bar{\lambda}$.

PROPOSITION 3.57. *Let* $f\in S_k(\Gamma_0', \psi)$, $f\mid T'(n)_{k,\psi}=a_nf$ *with a positive integer n prime to N, and* $g=f\mid[\tau]_k$. *Then* $g\mid T'(n)_{k,\bar{\psi}}=\bar{a}_ng$.

This is an immediate consequence of Prop. 3.55 and Prop. 3.56.

REMARK 3.58. Put $\sigma=\begin{bmatrix}t & 0\\ 0 & 1\end{bmatrix}$. Then $\begin{bmatrix}a & tb\\ c & d\end{bmatrix}=\sigma\begin{bmatrix}a & b\\ tc & d\end{bmatrix}\sigma^{-1}$, especially $\begin{bmatrix}0 & -t\\ N & 0\end{bmatrix}=\sigma\begin{bmatrix}0 & -1\\ tN & 0\end{bmatrix}\sigma^{-1}$. Therefore we see that $\sigma\Gamma_0'\sigma^{-1}=\Gamma_0(tN)$, and the map $f(z)\mapsto f(tz)$ sends $S_k(\Gamma_0', \psi)$ onto $S_k(\Gamma_0(tN), \psi)$. By means of this map, the discussion of $R(\Gamma', \Delta')$ and the operators $T'(n)_{k,\psi}$ with respect to Γ_0' can be reduced to the case $t=1$, by changing the level N for tN. Note that N and tN have the same prime factors, since t divides N. Therefore we could put $t=1$ in our definition of Γ_0' and Δ_0', without losing much generality. (It should of course be noticed that $(\mathbf{Z}/tN\mathbf{Z})^\times$ may have more characters than $(\mathbf{Z}/N\mathbf{Z})^\times$.)

REMARK 3.59. Let p be a prime not dividing N, and f an eigen-function of $T'(p)_{k,\psi}$ in $S_k(\Gamma_0(N), \psi)$, and let $f \mid T'(p)_{k,\psi}=c_p f$. Put $f_m(z)=f(p^m z)$ for $m=0, 1, 2, \cdots$, and denote by $T''(p)_{k,\varphi}$ the operator in $S_k(\Gamma_0(p^l N), \varphi)$, where $\varphi(a)=\psi(a)$ for $(a, pN)=1$. Then we can easily verify

$$f \mid T''(p)_{k,\varphi}=c_p f-p^{k-1}\psi(p)f_1$$

$$f_m \mid T''(p)_{k,\varphi}=f_{m-1} \qquad (m=1, 2, \cdots, l).$$

It follows that $T''(p)_{k,\varphi}$ is not semi-simple if $l \geqq 3$.

Now let λ and μ be the roots of the quadratic equation

$$x^2-c_p x+\psi(p)p^{k-1}=0.$$

Then $f-\lambda f_1$ is an eigen-function of $T''(p)_{k,\varphi}$ with μ as the eigen-value.

Further put $\tau=\begin{bmatrix} 0 & -1 \\ N & 0 \end{bmatrix}$, $\tau'=\begin{bmatrix} 0 & -1 \\ pN & 0 \end{bmatrix}$. If $f \mid [\tau]_k=g$, we obtain easily

$$f \mid [\tau']_k=p^{k/2}g(pz), \qquad f_1 \mid [\tau']_k=p^{-k/2}g.$$

Suppose that ψ is the trivial character, and $f \mid [\tau]_k=\varepsilon f$ with $\varepsilon=\pm 1$. Then $(f-\lambda f_1) \mid [\tau']_k$ is not an eigen-function of $T''(p)_{k,\varphi}$ unless $c_p=p^{k/2}(1+p^{-1})$, which is usually not the case. (At least it contradicts the Ramanujan conjecture, see below.)

REMARK 3.60. Let A (resp. A_ψ) be the ring generated by all the $T'(n)_k$ (resp. $T'(n)_{k,\psi}$) over $\boldsymbol{C}$, and B (resp. B_ψ) the subalgebra of A (resp. A_ψ) generated by the $T'(n)_k$ (resp. $T'(n)_{k,\psi}$) for all n prime to N. Then A (resp. B) can be identified with the direct sum of the algebras A_ψ (resp. B_ψ) for all ψ such that $\psi(\mathfrak{h})=1$. As for A, this follows immediately from Th. 3.45 and Th. 3.51. As for B, take a diagonalization of the $T'(n)_{k,\psi}$ and define a homomorphism of B onto $\boldsymbol{C}$ by assigning a diagonal element of $T'(n)_{k,\psi}$ to $T'(n)$. In view of (3.5.8), one can not obtain the same homomorphism from two distinct ψ. This shows that $[B:\boldsymbol{C}]=\sum_{\psi(\mathfrak{h})=1}[B_\psi:\boldsymbol{C}]$, and hence B must be the direct sum of such B_ψ.

From Th. 3.41 and (3.5.4), we see that B and B_ψ are commutative semi-simple algebras. Moreover, by Prop. 3.54, we have

$$[\tau]_k^{-1}T'(n)_k[\tau]_k=n^{2-k}T'(n, n)_k T'(n)_k$$

if n is prime to N. Therefore $[\tau]_k^{-1}\cdot B\cdot[\tau]_k=B$, and similarly $[\tau]_k^{-1}\cdot B_\psi\cdot[\tau]_k =B_{\bar\psi}$. Thus $[\tau]_k$ sends a common eigen-function of B (resp. B_ψ) to a common eigen-function of B (resp. $B_{\bar\psi}$). These facts are not necessarily true for A and A_ψ, as shown in Remark 3.59. However, Hecke proved the following facts:

Suppose that $t=1$, *i.e.*, $\Gamma_0'=\Gamma_0(N)$. *Then* $A_\psi=B_\psi$ *(at least) in the following two cases.*

(I) $\psi = 1$, *N is a prime, and* $S_k(\Gamma(1)) = 0$. *(By Prop. 2.26, the last condition is satisfied if and only if* $k < 12$ *or* $k = 14$.)

(II) ψ *is a primitive character modulo* (N).

For details, see [30, Satz 22, Satz 24a].

Historically, the connection of a cusp form with an Euler-product was first mentioned by Ramanujan [60]. He considered the Fourier coefficients c_n of the function

$$(2\pi)^{-12}\Delta(z) = q\prod_{n=1}^{\infty}(1-q^n)^{24} = \sum_{n=1}^{\infty} c_n q^n \qquad (q = e^{2\pi i z})$$

and made two conjectures:

$$(X) \qquad \sum_{n=1}^{\infty} c_n n^{-s} = \prod_p (1 - c_p p^{-s} + p^{11-2s})^{-1};$$

$$(Y) \qquad c_n = O(n^{11/2+\varepsilon}) \text{ for any } \varepsilon > 0.$$

The latter is equivalent to the inequality

$$(Z) \qquad |c_p| \leqq 2p^{11/2} \text{ for all primes } p.$$

The first conjecture (X) was proved by Mordell [51]. Since $S_{12}(\Gamma)$ is one-dimensional and spanned by Δ, Δ must be a common eigen-function of all Hecke operators, and hence (X) follows from Th. 3.43.

It was Hecke who made the first systematic investigation of the relation of modular forms with Dirichlet series having Euler-product, in its full scope. In the above, we have presented the easier part of Hecke's theory [29], [30], along with some new results. The idea of diagonalization of the Hecke operators by means of the inner product in the space of cusp forms is due to Petersson [55]. He also generalized the conjecture (Z) in the following form:

(Z') *Every eigen-value* λ_p *of* $T'(p)_{k,\psi}$, *for any prime p not dividing the level N, satisfies* $|\lambda_p| \leqq 2p^{(k-1)/2}$.

If $k = 2$, we shall be able to prove, in § 7.4, that (Z') is true for almost all p. In the general case, it was shown by Rankin [61] that $c_n = O(n^{k/2-1/5})$ for every $\sum_{n=1}^{\infty} c_n e^{2\pi i n z/N} \in S_k(\Gamma(N))$. Various methods for the estimate of c_n are discussed in Selberg [64].

3.6. The functional equations of the zeta-functions associated with modular forms

Let us first prove two fundamental lemmas for an arbitrary Fuchsian group Γ of the first kind.

LEMMA 3.61. *If $f \in S_k(\Gamma)$, one has $|f(x+iy)| \leqq My^{-k/2}$ with a constant M independent of x. Conversely, if an element f of $A_k(\Gamma)$ is holomorphic on $\mathfrak{H}$, and $|f(x+iy)| \leqq My^{-k/2}$ with a constant M independent of x, then $f \in S_k(\Gamma)$.*

PROOF. For any holomorphic element f of $A_k(\Gamma)$, define a real valued function h on $\mathfrak{H}$ by $h(z) = h(x+iy) = |f(z)|y^{k/2}$. Since $\operatorname{Im}(\gamma(z)) = \operatorname{Im}(z)|j(\gamma, z)|^{-2}$ for $\gamma \in SL_2(\boldsymbol{R})$, we see that h is Γ-invariant. If s is a cusp of Γ, take ρ and $q = e^{2\pi iz/h}$ (or $= e^{\pi iz/h}$) as in p. 29. Then $f|[\rho^{-1}]_k = \Phi(q)$ with a holomorphic function Φ in the domain $0 < |q| < r$ with a positive real number r, so that $h(\rho^{-1}(z)) = \Phi(q) \operatorname{Im}(z)^{k/2}$. Note that $|q| = e^{-2\pi y/h}$ (or $= e^{-\pi y/h}$). Suppose that $f \in S_k(\Gamma)$. Then $\Phi(q) \to 0$ as $q \to 0$. Therefore $h(w) \to 0$ as $w \to s$ (with respect to the topology of $\mathfrak{H}^*$). Thus h can be viewed as a continuous function on $\Gamma \backslash \mathfrak{H}^*$. Since $\Gamma \backslash \mathfrak{H}^*$ is compact, $h(z)$ must be bounded. Conversely, if $h(z)$ is bounded, Φ must be holomorphic at $q = 0$, and $\Phi(0) = 0$. This proves our proposition.

LEMMA 3.62. *Suppose that ∞ is a cusp of Γ, and let*

$$\{\gamma \in \Gamma \cdot \{\pm 1\} \mid \gamma(\infty) = \infty\} = \{\pm 1\} \cdot \left\{ \begin{bmatrix} 1 & h \\ 0 & 1 \end{bmatrix}^m \,\middle|\, m \in \boldsymbol{Z} \right\}$$

with a positive real number h. Let $f \in S_k(\Gamma)$, and

$$f(z) = \begin{cases} \sum_{n=1}^{\infty} c_n e^{\pi inz/h} & \text{if } k \text{ is odd and } \infty \text{ is irregular}, \\ \sum_{n=1}^{\infty} c_n e^{2\pi inz/h} & \text{otherwise.} \end{cases}$$

(See §2.1). Then there is a constant B independent of n such that $|c_n| \leqq B \cdot n^{k/2}$ for all n.

PROOF. If k is even, put $q = e^{2\pi iz/h}$, and $F(q) = \sum_{n=1}^{\infty} c_n q^n$. Then

$$c_n = (2\pi i)^{-1} \int F(q) q^{-n-1} dq,$$

where the integral is taken on the circle $|q| = r$ in the positive direction, for a small $r > 0$. If $\operatorname{Im}(z) = y = h/2\pi n$, then $|e^{2\pi iz/h}| = e^{-1/n}$. By Lemma 3.61, $|F(q)| \leqq My^{-k/2}$ with a constant M. Therefore, taking r to be $e^{-1/n}$, we have $|c_n| \leqq Me \cdot (h/2\pi n)^{-k/2}$. The case of odd k can be treated in a similar way.

Our task is to prove a functional equation for the Dirichlet series $\sum_{n=1}^{\infty} a_n n^{-s}$ attached to any $f(z) = \sum_{n=1}^{\infty} a_n e^{2\pi inz/t}$ of $S_k(\Gamma', \psi)$. For the reason explained in Remark 3.58, it is sufficient to consider the case $t = 1$, i.e., $\Gamma' = \Gamma_0(N)$. We shall generalize our question by considering $\sum_{n=1}^{\infty} \chi(n) a_n n^{-s}$ with any character χ of $(\boldsymbol{Z}/r\boldsymbol{Z})^{\times}$, where r is a positive integer prime to N. Therefore let us first recall a few elementary facts on the Gauss sum associated with χ.

Let us fix a positive integer r, and χ a character of $(\boldsymbol{Z}/r\boldsymbol{Z})^\times$, i.e., a homomorphism of $(\boldsymbol{Z}/r\boldsymbol{Z})^\times$ into $\boldsymbol{C}^\times$. We assume that χ is a *primitive character* modulo (r), by which we mean that there is no character ξ of $(\boldsymbol{Z}/s\boldsymbol{Z})^\times$ with a proper divisor s of r satisfying $\xi(x)=\chi(x)$ for $(x, r)=1$. Then we put, for $c\in\boldsymbol{Z}$,

$$\chi(c)=\begin{cases}\chi(c \bmod r\boldsymbol{Z}) & \text{if } (c, r)=1,\\ 0 & \text{if } (c, r)\neq 1,\end{cases}$$

and define the Gauss sum $W(\chi)$ by

$$W(\chi)=\sum_{c=0}^{r-1}\chi(c)\zeta^c, \qquad \zeta=e^{2\pi i/r}.$$

LEMMA 3.63. *The notation being as above, one has:*

(1) $\sum_{c=0}^{r-1}\chi(c)\zeta^{bc}=\bar{\chi}(b)W(\chi)$ *for every* $b\in\boldsymbol{Z}$;

(2) $W(\chi)W(\bar{\chi})=\chi(-1)r$;

(3) $|W(\chi)|^2=r$;

(4) $\overline{W(\chi)}=\chi(-1)W(\bar{\chi})$.

PROOF. If $(b, r)=1$, denoting by b^{-1} the inverse of b mod $r\boldsymbol{Z}$, we have

$$\sum_c \chi(c)\zeta^{bc}=\sum_a \chi(b^{-1}a)\zeta^a=\chi(b^{-1})\sum_a \chi(a)\zeta^a=\chi(b^{-1})W(\chi).$$

Suppose that $s=r/(r, b)<r$, and put

$$H=\{a\in(\boldsymbol{Z}/r\boldsymbol{Z})^\times \mid a\equiv 1 \bmod s\boldsymbol{Z}\},$$

and let $(\boldsymbol{Z}/r\boldsymbol{Z})^\times=\bigcup_{y\in Y}Hy$ be a disjoint decomposition. Since $bs\equiv 0 \bmod (r)$, we have $xb\equiv b \bmod (r)$ for $x\in H$. Further, since χ is a primitive character modulo (r), χ can not be trivial on H. Therefore

$$\sum_c \chi(c)\zeta^{bc}=\sum_{y\in Y}\sum_{x\in H}\chi(yx)\zeta^{yxb}=\sum_{y\in Y}\zeta^{yb}\chi(y)\sum_{x\in H}\chi(x)=0.$$

Now, by (1),

$$W(\chi)W(\bar{\chi})=\sum_c W(\chi)\bar{\chi}(c)\zeta^c=\sum_{b,c}\chi(b)\zeta^{bc}\zeta^c=\sum_b \chi(b)\sum_c \zeta^{c(b+1)}=\chi(-1)r,$$

since $\sum_c \zeta^{ac}=r$ or 0 according as $a\equiv 0$ or $a\not\equiv 0 \bmod (r)$. Note that $\chi(-1)=\pm 1$. Therefore

$$\overline{W(\chi)}=\sum_c \bar{\chi}(c)\zeta^{-c}=\sum_c \bar{\chi}(-c)\zeta^c=\bar{\chi}(-1)W(\bar{\chi})=\chi(-1)W(\bar{\chi}),$$

and

$$W(\chi)\overline{W(\chi)}=W(\chi)W(\bar{\chi})\chi(-1)=r.$$

Let us also recall a definition of the Γ-function:[3)]

$$\Gamma(s)=\int_0^\infty e^{-x}x^{s-1}dx \qquad (s\in\boldsymbol{C}).$$

3) We have two gammas: one for a discrete subgroup of $SL_2(\boldsymbol{R})$, and the other for the gamma function. Since the distinction will be clear from the context, we use the same letter for both objects.

Substituting ax for x, we obtain

(3.6.1) $$a^{-s}\Gamma(s)=\int_0^\infty e^{-ax}x^{s-1}dx \qquad (s\in \boldsymbol{C},\ a\in \boldsymbol{R},\ a>0).$$

PROPOSITION 3.64. *Let N and r be positive integers, s a positive divisor of N, and M the least common multiple of N, r^2, and rs. Let χ (resp. ψ) be a primitive character of $(\boldsymbol{Z}/r\boldsymbol{Z})^\times$ (resp. $(\boldsymbol{Z}/s\boldsymbol{Z})^\times$). Further let $f(z)=\sum_{n=1}^\infty a_n e^{2\pi i n z}$ be an element of $S_k(\Gamma_0(N),\psi)$. Then $h(z)=\sum_{n=1}^\infty \chi(n)a_n e^{2\pi i n z}$ belongs to $S_k(\Gamma_0(M),\psi\chi^2)$.*

PROOF. Put $\zeta=e^{2\pi i/r}$, and $\alpha_u=\begin{bmatrix}1 & u/r\\ 0 & 1\end{bmatrix}$ for $u\in \boldsymbol{Z}$. Then

$$f\,|\,[\alpha_u]_k=\sum_{n=1}^\infty a_n e^{2\pi i n(z+u/r)}=\sum_{n=1}^\infty \zeta^{nu}a_n e^{2\pi i n z},$$

so that, by (1) of Lemma 3.63,

$$W(\bar\chi)h(z)=\sum_{u=1}^r \bar\chi(u)f\,|\,[\alpha_u]_k.$$

By Prop. 2.4 and Lemma 3.9, we see that $h\in S_k(\Gamma(r^2N))$. Therefore, to prove our assertion, it is sufficient to check the behavior of h under an element $\gamma=\begin{bmatrix}a & b\\ Mc & d\end{bmatrix}$ of $\Gamma_0(M)$. Put

$$a'=a+cuM/r,$$

$$b'=b+du(1-ad)/r-cd^2u^2M/r^2,$$

$$d'=d-cd^2uM/r.$$

Then a, b, c, d are integers, $d\equiv d' \bmod (s)$, and

$$\begin{bmatrix}1 & u/r\\ 0 & 1\end{bmatrix}\begin{bmatrix}a & b\\ Mc & d\end{bmatrix}=\begin{bmatrix}a' & b'\\ Mc & d'\end{bmatrix}\begin{bmatrix}1 & d^2u/r\\ 0 & 1\end{bmatrix}.$$

Therefore, putting $v=d^2u$, we have $f\,|\,[\alpha_u\gamma]_k=\psi(d)f\,|\,[\alpha_v]_k$, so that

$$h\,|\,[\gamma]_k=W(\bar\chi)^{-1}\psi(d)\chi(d^2)\sum_v \bar\chi(v)f\,|\,[\alpha_v]_k=\psi(d)\chi(d^2)h, \qquad \text{q. e. d.}$$

PROPOSITION 3.65. *The notation being as in Prop. 3.64, suppose that r is prime to N, and put $\tau=\begin{bmatrix}0 & -1\\ N & 0\end{bmatrix}$, $\tau'=\begin{bmatrix}0 & -1\\ r^2N & 0\end{bmatrix}$, and $f\,|\,[\tau]_k=\sum_{n=1}^\infty b_n e^{2\pi i n z}$. Then*

$$h\,|\,[\tau']_k=\psi(r)\chi(N)W(\chi)^2r^{-1}\sum_{n=1}^\infty \bar\chi(n)b_n e^{2\pi i n z}.$$

PROOF. Let us use the same notation as in the above proof. Suppose that $(u, r)=1$. Then we can find two integers d and w so that $dr-Nuw=1$. Then $\alpha_u\tau'=r\tau\begin{bmatrix}r & -w\\ -Nu & d\end{bmatrix}\alpha_w$. Put $g=f\,|\,[\tau]_k$. Then

$$W(\bar{\chi})h \mid [\tau']_k = \sum_u \bar{\chi}(u)f \mid [\alpha_u \tau']_k = \sum_u \bar{\chi}(u)\psi(r)g \mid [\alpha_w]_k$$
$$= \psi(r) \sum_w \chi(-Nw)g \mid [\alpha_w]_k = \psi(r)\chi(-N)W(\chi) \sum_{n=1}^{\infty} \bar{\chi}(n)b_n e^{2\pi i n z}.$$

This together with (2) of Lemma 3.63 proves our assertion.

THEOREM 3.66. *Let r be a positive integer prime to N, χ a primitive character of $(\boldsymbol{Z}/r\boldsymbol{Z})^{\times}$, and ψ an arbitrary character of $(\boldsymbol{Z}/N\boldsymbol{Z})^{\times}$. For every $f(z) = \sum_{n=1}^{\infty} a_n e^{2\pi i n z}$ of $S_k(\Gamma_0(N), \psi)$, put*

$$L(s, f, \chi) = \sum_{n=1}^{\infty} \chi(n) a_n n^{-s},$$
$$R(s, f, \chi) = (r^2 N)^{s/2} (2\pi)^{-s} \Gamma(s) L(s, f, \chi).$$

Then $L(s, f, \chi)$ is absolutely convergent for $\mathrm{Re}(s) > 1 + (k/2)$, and can be holomorphically continued to the whole s-plane. Moreover, it satisfies a functional equation

$$R(s, f, \chi) = i^k \psi(r) \chi(N) W(\chi)^2 r^{-1} R(k-s, f \mid [\tau]_k, \bar{\chi}),$$

where $\tau = \begin{bmatrix} 0 & -1 \\ N & 0 \end{bmatrix}$.

PROOF. In view of Prop. 3.64 and Prop. 3.65, it is sufficient to treat the case $r = 1$, and $\chi = 1$. The absolute convergence of $L(s, f, 1)$ for $\mathrm{Re}(s) > k/2 + 1$ follows from Lemma 3.62. By (3.6.1), we obtain formally

$$(*) \qquad \int_0^{\infty} f(iy) y^{s-1} dy = \sum_{n=1}^{\infty} a_n \int_0^{\infty} e^{-2\pi n y} y^{s-1} dy = (2\pi)^{-s} \Gamma(s) L(s, f, 1).$$

To see that this formal computation is actually valid, we note that

$$\left| \int_0^{\varepsilon} f(iy) y^{s-1} dy \right| \leqq A \int_0^{\varepsilon} y^{-k/2} y^{k/2} dy \to 0 \qquad (\varepsilon \to 0)$$

if $\mathrm{Re}(s) > k/2 + 1$, by virtue of Lemma 3.61, and

$$\left| \int_E^{\infty} f(iy) y^{s-1} dy \right| \leqq B \int_E^{\infty} e^{-2\pi y} y^{\mathrm{Re}(s)-1} dy \to 0 \qquad (E \to \infty)$$

for any $s \in \boldsymbol{C}$. (A and B are constants.) Now we have

$$\int_{\varepsilon}^{E} f(iy) y^{s-1} dy = \sum_{n=1}^{\infty} a_n \int_{\varepsilon}^{E} e^{-2\pi n y} y^{s-1} dy,$$

since $\sum_n a_n e^{-2\pi n y}$ is uniformly convergent for $y \geqq \varepsilon$. For any small $\eta > 0$. we can take M so large that

$$\left| \sum_{n>M} a_n \int_{\varepsilon}^{E} e^{-2\pi n y} y^{s-1} dy \right| \leqq \sum_{n>M} |a_n| \int_0^{\infty} e^{-2\pi n y} y^{\sigma-1} dy$$
$$= \Gamma(\sigma)(2\pi)^{-\sigma} \sum_{n>M} |a_n| n^{-\sigma} < \eta \qquad (\mathrm{Re}(s) = \sigma).$$

Therefore we see that

$$\left|\int_0^\infty f(iy)y^{s-1}dy-\textstyle\sum_{n=1}^M a_n \displaystyle\int_0^\infty e^{-2\pi ny}y^{s-1}dy\right|$$

$$=\lim_{\varepsilon\to 0,E\to\infty}\left|\int_\varepsilon^E f(iy)y^{s-1}dy-\textstyle\sum_{n=1}^M a_n \displaystyle\int_\varepsilon^E e^{-2\pi ny}y^{s-1}dy\right|\leqq \eta\,.$$

This proves the validity of (*) for $\mathrm{Re}(s)>k/2+1$. For the same reason, if $g=f\,|\,[\tau]_k$, we obtain

$$(**)\qquad \int_0^\infty g(iy)y^{s-1}dy=\Gamma(s)(2\pi)^{-s}L(s,g,1)\,.$$

Put $A=N^{-1/2}$. Then

$$\int_0^\infty f(iy)y^{s-1}dy=\int_0^A f(iy)y^{s-1}dy+\int_A^\infty f(iy)y^{s-1}dy\,.$$

As is seen above, the first term is convergent for $\mathrm{Re}(s)>k/2+1$, and the second term is convergent for any s. Changing y for $1/Ny$, we obtain, since $f(i/Ny)=N^{k/2}(iy)^k g(iy)$,

$$\int_0^A f(iy)y^{s-1}dy=\int_A^\infty f(i/Ny)N^{-s}y^{-s-1}dy=i^kN^{k/2-s}\int_A^\infty g(iy)y^{k-1-s}dy\,.$$

The last integral is convergent for any s. Similarly

$$\int_A^\infty f(iy)y^{s-1}dy=i^kN^{k/2-s}\int_0^A g(iy)y^{k-1-s}dy\qquad(\mathrm{Re}(s)>k/2+1)\,.$$

Therefore if we put $R'(s,f)=\Gamma(s)(2\pi)^{-s}L(s,f,1)$, we see that $R'(s,f)$ can be holomorphically continued to the whole s-plane, and

$$R'(s,f)=i^kN^{k/2-s}R'(k-s,g)\,.$$

Note that $\Gamma(s)^{-1}$ is an entire function. Therefore we obtain our theorem.

In the above discussion we have obtained a Dirichlet series $L(s)=\sum_{n=1}^\infty a_n n^{-s}$ from a function $f(z)=\sum_{n=1}^\infty a_n e^{2\pi inz}$ by means of the "Mellin inverse transformation"

$$\int_0^\infty f(iy)y^{s-1}dy=\Gamma(s)(2\pi)^{-s}L(s)=R(s)\,.$$

One can actually obtain $f(z)$ from $L(s)$ by the "Mellin transformation"

$$f(iy)=(2\pi i)^{-1}\int R(s)x^{-s}ds\,,$$

where the integral is taken on the vertical line $\mathrm{Re}(s)=\sigma$ for some $\sigma>0$. Hecke employed this correspondence between $f(z)$ and $L(s)$ to prove that $R(s)$ satisfies a functional equation of the above type if and only if $f(z)$ is an automorphic form with respect to a certain discrete subgroup Γ of $SL_2(\boldsymbol{R})$. This result is not completely satisfactory, since $\Gamma\backslash\mathfrak{H}^*$ is often non-compact. A more complete result was recently obtained by Weil, who showed that if one assumes the functional equations for $\sum_{n=1}^{\infty}\chi(n)a_n n^{-s}$ for sufficiently many characters χ, then f belongs to $S_k(\Gamma_0(N), \psi)$ for some N and ψ. For details of these results, we refer the reader to [28], [98], [101].

In our treatment, we have defined an automorphic form to be a complex analytic function. More generally, Maass considered real analytic automorphic forms on $\mathfrak{H}$ which are eigen-functions of some invariant differential operators. For such forms, he developed the theory of Hecke operators and generalized the above correspondence between $f(z)$ and $R(s)$. Here we content ourselves with mentioning only [44], [45], [46] among his numerous papers on this subject.

There are also (at least) three important topics which we do not touch in this book. The first one is the connection of modular forms with quadratic forms. If $P(x)=\sum_{1\leq i\leq j\leq 2k} p_{ij}x_ix_j$ is a positive definite quadratic form with p_{ij} in $\boldsymbol{Z}$, then $\sum_{x\in\boldsymbol{Z}^n} e^{2\pi i P(x)z}$, called a *theta-series*, is a modular form of weight k with respect to some congruence subgroup of $SL_2(\boldsymbol{Z})$. Here the Eisenstein series play an essential role. The reader may be referred to Hecke [26], [30], and Schoeneberg [62]. One should also note many of Siegel's works on quadratic forms, and its generalizations, which are now accessible in three volumes of his collected works. A treatise of this topic, in the adele language, is given in Weil [97]. For this see also Shalika and Tanaka [67].

The second is the explicit computation of the trace of Hecke operators, for which we only mention Selberg [63], Eichler [17], [18], [19], [20], and Shimizu [68]. Finally there is an aspect in which the theory of group representations plays an essential role. For this the reader is referred to a recent work of Jacquet and Langlands [37], and also to the earlier works quoted in the volume. Although we mention these topics separately, they are closely connected with each other, and with what we consider in this book.

Our discussion has been restricted to the case of congruence subgroups of $SL_2(\boldsymbol{Z})$. Actually one can construct zeta-functions from automorphic forms with respect to a unit group of a simple algebra over a number field. They have Euler products of the form of Th. 3.21. For details, the reader is referred to Maass [45], Godement [23], Tamagawa [86], Shimura [74], Shimizu [68], Weil [101], and Jacquet and Langlands [37]. Simple division algebras of an arbitrary degree are treated in [23] and [86], while the remaining articles are concerned with quaternion algebras (in the general sense, including matrix algebras of degree 2).

CHAPTER 4
ELLIPTIC CURVES

4.1. Elliptic curves over an arbitrary field

In this section we give a brief review of a few elementary facts about elliptic curves (without detailed proofs)[4]. An elliptic curve is an abelian variety (a projective non-singular variety with a structure of algebraic group, necessarily commutative) of dimension one, or what amounts to the same, a projective non-singular curve of genus one with a specific point, called the *origin* or the *neutral element.* If the curve is defined over a field k, and the origin is rational over k, then the group law is automatically defined over k. Therefore, when we speak of an *elliptic curve defined over* k, we understand that the curve and the origin are rational over k.

Let E and E' be elliptic curves defined over k. By a *homomorphism* of E into E' (defined over k), we mean a rational map (defined over k) of E into E' that is a group homomorphism. The module of all homomorphisms of E into E' is denoted by $\mathrm{Hom}(E, E')$. Any rational map of E into E' transforming the origin of E to the origin of E' is automatically a homomorphism. An element λ of $\mathrm{Hom}(E, E')$ is called an *isogeny,* if it satisfies the following mutually equivalent conditions: (i) $\lambda \neq 0$; (ii) $\mathrm{Ker}(\lambda)$ is finite; (iii) λ is surjective. (Note that we always identify E with the *set of all points* on the curve *rational over the universal domain,* see Appendix.) If there exists an isogeny of E to E', then there exists an isogeny of E' to E, and we say that E and E' are *isogenous.* This is an equivalence relation. Now we define

$$\begin{aligned} \mathrm{End}(E) &= \text{the ring of all endomorphisms of } E \\ &\quad\ \text{(over the universal domain)} \\ &= \mathrm{Hom}(E, E), \\ \mathrm{End}_{\boldsymbol{Q}}(E) &= \mathrm{End}(E) \otimes_{\boldsymbol{Z}} \boldsymbol{Q}. \end{aligned}$$

---

4) As for the terminology and notation concerning algebraic geometry, see Appendix. Although our discussion in this and next chapters is restricted to elliptic curves, the theory cannot be fully understood unless one considers them as special cases of abelian varieties. Therefore the reader is advised (though not required) to have some acquaintance with the definition and elementary properties of abelian varieties, as given in (the easier part of) Weil [92], [95], and Lang [43]. See also Appendix Nos 10–13. We borrowed, for example, the construction of the roots of unity e_N in § 4.3 from [92, pp. 150–153]. For a detailed discussion of abelian varieties with complex multiplications, see [81].

For $k=\boldsymbol{C}$ it will be shown that $\mathrm{End}(E)$ is a free $\boldsymbol{Z}$-module of finite rank, and $\mathrm{End}_{\boldsymbol{Q}}(E)$ is a division ring of finite rank over $\boldsymbol{Q}$. The same is known to be true for all k. All possible types of $\mathrm{End}_{\boldsymbol{Q}}(E)$, and even of $\mathrm{End}(E)$, have been determined by Deuring [10]: $\mathrm{End}_{\boldsymbol{Q}}(E)$ is isomorphic to either $\boldsymbol{Q}$, or an imaginary quadratic field, or a quaternion algebra over $\boldsymbol{Q}$ ramified at a prime p and ∞; the last case can occur only when the characteristic of the universal domain is p. But we shall not discuss this topic in full generality; we shall treat only the case of characteristic 0 in § 4.4.

From now on, we shall assume that the characteristic is not 2 or 3. Then an elliptic curve defined over a field k is always isomorphic, over k, to a projective curve

$$E:\ Y^2Z=4X^3-g_2XZ^2-g_3Z^3 \tag{4.1.1}$$

with g_i in k, and $\Delta=g_2^3-27g_3^2\neq 0$. (The non-vanishing of Δ is equivalent to the non-singularity of the curve defined by the equation.) We can take the point $(X, Y, Z)=(0, 1, 0)$ as the origin. Conversely, every curve of this form with $\Delta\neq 0$ is an elliptic curve. Hereafter, for convenience, we shall write the equation in the affine form

$$E:\ y^2=4x^3-g_2x-g_3, \tag{4.1.2}$$

but always regard it as a complete curve, by adjoining the point $(x, y)=(\infty, \infty)$, which is the origin of E. Then the map $(x, y)\mapsto(x, -y)$ gives the automorphism -1 of E.

Now such a curve is characterized by its *invariant*

$$j(E)=j_E=g_2^3/\Delta$$

(or $J_E=2^63^3j_E$ which has nicer integrality properties) in the sense that two curves E and E' defined respectively by the equations $y^2=4x^3-g_2x-g_3$ and $y^2=4x^3-g_2'x-g_3'$ are isomorphic over the universal domain if and only if

$$j_E=g_2^3/(g_2^3-27g_3^2)=g_2'^3/(g_2'^3-27g_3'^2)=j_{E'}.$$

One can state this fact in a somewhat stronger form:

PROPOSITION 4.1. *Let E and E' be defined by $y^2=4x^3-g_2x-g_3$ and $y^2=4x^3-g_2'x-g_3'$, respectively, and let λ be an isomorphism of E onto E'. Then there exists an element μ such that*

$$g_2'=\mu^4g_2,\qquad g_3'=\mu^6g_3,\qquad \lambda(x, y)=(\mu^2x, \mu^3y).$$

Observe that j_E belongs to any field of definition for E. Let k_0 be the prime field. Then, *for any j in the universal domain, there exists an elliptic curve E defined over $k_0(j)$ with invariant j:*

for $j=0$, take $g_2=0$ and $g_3=1$;
for $j=1$, take $g_2=1$ and $g_3=0$;
for $j\neq 0, 1$, solve $g/(g-27)=j$, and take $g_2=g_3=g$ $(g=27j/(j-1)\in k_0(j))$. (This is just one of many possible choices, and therefore should not be regarded as standard.)

For E as above, and for an automorphism σ of the universal domain, we define an elliptic curve E^σ by

$$E^\sigma: \quad y^2=4x^3-g_2^\sigma x-g_3^\sigma.$$

Then clearly $j(E^\sigma)=j(E)^\sigma$. *Therefore E is isomorphic to E^σ if and only if σ is the identity map on $k_0(j_E)$*. The field $k_0(j_E)$ is characterized by this property if the characteristic is 0, and called *the field of moduli of E*. We have just shown that E has a model defined over its field of moduli. One can define the field of moduli for any "polarized" abelian variety, see § 5.4. However, it is an open question to know whether any polarized abelian variety has a model defined over its field of moduli.

4.2. Elliptic curves over C

Let us now consider the case where the universal domain is the complex number field $\boldsymbol{C}$. Every elliptic curve defined over (a subfield of) $\boldsymbol{C}$, as a complex analytic manifold, is isomorphic to a one-dimensional complex torus $\boldsymbol{C}/L$, where L is a *lattice* in $\boldsymbol{C}$, by which we mean a discrete submodule of $\boldsymbol{C}$ of rank 2 over $\boldsymbol{Z}$. Conversely, let L be an arbitrary lattice in $\boldsymbol{C}$. Then, an *elliptic function* with periods in L is, by definition, a meromorphic function on $\boldsymbol{C}$ invariant under the translation by the elements of L; we can regard such a function as a meromorphic function on $\boldsymbol{C}/L$ and vice versa. Let F_L denote the field of all elliptic functions with periods in L. It is known that F_L is generated by the Weierstrass functions $\wp$ and $\wp'$, defined by

$$\wp(u)=\wp(u\,;L)=u^{-2}+\textstyle\sum_{\omega\in L'}[(u-\omega)^{-2}-\omega^{-2}],$$

$$\wp'(u)=\frac{d}{du}\wp(u)=-2u^{-3}-2\textstyle\sum_{\omega\in L'}(u-\omega)^{-3} \qquad (L'=L-\{0\}).$$

(It is easy to see that $\wp$ and $\wp'$ are contained in F_L, and have a pole only at $u=0$ (modulo L), of degree 2 and 3, respectively. Therefore, by (3) of Prop. 2.11, we have $[F_L:\boldsymbol{C}(\wp)]=2$, and $[F_L:\boldsymbol{C}(\wp')]=3$, hence $F_L=\boldsymbol{C}(\wp,\wp')$ as asserted.)

The Laurent expansions of $\wp$ and $\wp'$ at $u=0$ have the form:

$$\wp(u)=u^{-2}+\textstyle\sum_{n=2}^{\infty}(2n-1)G_{2n}(L)u^{2n-2},$$

$$\wp'(u)=-2u^{-3}+\textstyle\sum_{n=2}^{\infty}(2n-1)(2n-2)G_{2n}(L)u^{2n-3},$$

$$G_{2n}(L)=\textstyle\sum_{\omega\in L'}\omega^{-2n}.$$

Then we have an equality

$$\wp'^2 = 4\wp^3 - g_2(L)\wp - g_3(L) \tag{4.2.1}$$

with

$$g_2(L) = 60 \cdot G_4(L), \qquad g_3(L) = 140 \cdot G_6(L).$$

(The difference $\wp'^2 - (4\wp^3 - g_2(L)\wp - g_3(L))$ is holomorphic on $\boldsymbol{C}/L$ except at 0; but from the expansions given above, we see that it is holomorphic and vanishes at 0; hence it must be identically equal to 0.)

Since $F_L = \boldsymbol{C}(\wp, \wp')$ is a function field of genus 1, we have

$$g_2(L)^3 - 27g_3(L)^2 \neq 0. \tag{4.2.2}$$

In fact, if this is 0, the equation (4.2.1) defines a curve of genus 0.

For a given L, define an elliptic curve E by

$$E: \ y^2 = 4x^3 - g_2(L)x - g_3(L). \tag{4.2.3}$$

Then the map $u \mapsto (\wp(u), \wp'(u))$ gives an isomorphism of $\boldsymbol{C}/L$ onto E.

Let $\mathfrak{H}$ denote the complex upper half plane as before. For two complex numbers ω_1 and ω_2 such that $\omega_1/\omega_2 \in \mathfrak{H}$, we obtain a lattice $L = \boldsymbol{Z}\omega_1 + \boldsymbol{Z}\omega_2$. Conversely, any lattice in $\boldsymbol{C}$ can be given in this form. We then write

$$\wp(u; \omega_1, \omega_2) = \wp(u; L),$$

$$\Delta(\omega_1, \omega_2) = g_2(\omega_1, \omega_2)^3 - 27g_3(\omega_1, \omega_2)^2,$$

$$g_2(\omega_1, \omega_2) = g_2(L), \qquad g_3(\omega_1, \omega_2) = g_3(L) \qquad (L = \boldsymbol{Z}\omega_1 + \boldsymbol{Z}\omega_2).$$

In § 2.2, we defined a modular form $\Delta(z)$ and a modular function $j(z)$ by

$$\Delta(z) = \Delta(z, 1) = g_2(z, 1)^3 - 27g_3(z, 1)^2,$$

$$j(z) = g_2(\omega_1, \omega_2)^3 / (g_2(\omega_1, \omega_2)^3 - 27g_3(\omega_1, \omega_2)^2) \qquad (z = \omega_1/\omega_2)$$

(or rather $J(z) = 2^6 3^3 \cdot j(z)$), and proved some fundamental properties of these functions. We observe that *$j(z)$ is the invariant j_E of the elliptic curve* (4.2.3), *which is isomorphic to $\boldsymbol{C}/(\boldsymbol{Z}\omega_1 + \boldsymbol{Z}\omega_2)$.* We shall later discuss the connection of modular functions of higher level with the points of finite order on E. Now (4.2.2) shows the non-vanishing of $\Delta(z)$ on $\mathfrak{H}$, which was stated but not proved in § 2.2.

Let us now show that, for any $r, s \in \boldsymbol{C}$ such that $r^3 - 27s^2 \neq 0$, there exists a lattice L in $\boldsymbol{C}$ satisfying $g_2(L) = r$ and $g_3(L) = s$. To see this, consider an elliptic curve $E: y^2 = 4x^3 - rx - s$. Then E is isomorphic to a torus $\boldsymbol{C}/L'$ with a suitable lattice L', and hence isomorphic to the curve $y^2 = 4x^3 - g_2(L')x - g_3(L')$. By Prop. 4.1, we have $g_2(L') = \mu^4 r$ and $g_3(L') = \mu^6 s$ for some $\mu \in \boldsymbol{C}$. Then the lattice $L = \mu L'$ has the desired property. This implies especially that $g_2(\omega_1, \omega_2)$ and $g_3(\omega_1, \omega_2)$ are algebraically independent over $\boldsymbol{C}$, which we needed in the proof of Prop. 2.27.

4.3. Points of finite order on an elliptic curve and the roots of unity

Let E be an elliptic curve defined over a field of characteristic p (which may be 0), and N a positive integer. Put

$$\mathfrak{g}(N) = \mathfrak{g}(N, E) = \{t \in E \mid Nt = 0\}.$$

It can be shown that $\mathfrak{g}(N)$ is isomorphic to a subgroup of $(\boldsymbol{Z}/N\boldsymbol{Z})^2$, the product of two copies of $\boldsymbol{Z}/N\boldsymbol{Z}$. Especially if p does not divide N, $\mathfrak{g}(N)$ is isomorphic to $(\boldsymbol{Z}/N\boldsymbol{Z})^2$. (This is obvious if the universal domain is $\boldsymbol{C}$, since E is then isomorphic to a complex torus.) It should also be remembered that

(4.3.1) *If E is defined over k, then the coordinates of every point of E of finite order are algebraic over k.*

This is obvious, since the number of the images of such a point t under isomorphisms over k is $\leqq N^2$, if $t \in \mathfrak{g}(N)$.

Now fix a rational prime l, and put

$$\mathfrak{g}^{(l)} = \bigcup_{n=1}^{\infty} \mathfrak{g}(l^n).$$

If p does not divide l, it can be shown that $\mathfrak{g}^{(l)}$ is isomorphic to $(\boldsymbol{Q}_l/\boldsymbol{Z}_l)^2$, where $\boldsymbol{Q}_l$ denotes the l-adic number field, and $\boldsymbol{Z}_l$ the ring of l-adic integers. Let $\alpha \in \mathrm{End}(E)$. Then α induces an endomorphism of $\mathfrak{g}^{(l)}$. Since every endomorphism of $(\boldsymbol{Q}_l/\boldsymbol{Z}_l)^2$ is represented by an element of $M_2(\boldsymbol{Z}_l)$ in an obvious way, we thus obtain an injective homomorphism of $\mathrm{End}(E)$ into $M_2(\boldsymbol{Z}_l)$, which can be extended to an injective homomorphism R_l of $\mathrm{End}_{\boldsymbol{Q}}(E)$ into $M_2(\boldsymbol{Q}_l)$. We call R_l an l-adic representation of $\mathrm{End}_{\boldsymbol{Q}}(E)$. It can be shown that the characteristic polynomial of $R_l(\alpha)$, for any $\alpha \in \mathrm{End}_{\boldsymbol{Q}}(E)$, has rational coefficients (integral coefficients if $\alpha \in \mathrm{End}(E)$), and is independent of l.

We shall now associate an N-th root of unity $e_N(s, t)$ with two elements s and t of $\mathfrak{g}(N)$. Let D_0 be the module of all divisors of degree 0 on E, and D_H the submodule of D_0 consisting of the divisors of all functions on E, so that D_0/D_H is the module of all divisor classes of E of degree 0. For each $t \in E$, let (t) denote the divisor associated to the point t. It is a well-known fact that the map $t \mapsto (t)-(0) \in D_0$ defines an isomorphism of E onto D_0/D_H. (Actually the group law on E is defined by means of this one-to-one correspondence between E and D_0/D_H.) Therefore we have

(4.3.2) *If $t_1, \cdots, t_m \in E$, $c_1, \cdots, c_m \in \boldsymbol{Z}$, $\sum_{i=1}^m c_i = 0$, and $\sum_{i=1}^m c_i t_i = 0$, then $\sum_{i=1}^m c_i(t_i) \in D_H$.*

If $t \in \mathfrak{g}(N)$, we see that $N \cdot ((t)-(0)) \in D_H$, hence $N \cdot ((t)-(0)) = \mathrm{div}(f)$ with a function f on E. Take a point t' on E so that $Nt' = t$. By (4.3.2), there exists a function g on E such that

$$\mathrm{div}(g)=\sum_{u\in\mathfrak{z}(N)}(t'+u)-\sum_{u\in\mathfrak{z}(N)}(u).$$

We see easily that the functions $f(Nx)$ and $g(x)^N$ $(x\in E)$ have the same divisor. Replacing f by a suitable constant multiple, we thus obtain two functions f and g which are characterized, up to constant factors, by the properties

$$\mathrm{div}(f)=N\cdot((t)-(0)),\qquad g(x)^N=f(Nx)\qquad (x\in E).$$

If $s\in\mathfrak{g}(N)$, we see that $g(x+s)^N=g(x)^N$, hence

$$g(x+s)=e_N(s,t)g(x)$$

with an N-th root of unity $e_N(s,t)$.

PROPOSITION 4.2. *Suppose that N is prime to the characteristic of the universal domain. Then the function $e_N(s,t)$ on $\mathfrak{g}(N)\times\mathfrak{g}(N)$ has the following properties:*

(1) $e_N(s_1+s_2,t)=e_N(s_1,t)e_N(s_2,t)$;

(2) $e_N(s,t_1+t_2)=e_N(s,t_1)e_N(s,t_2)$;

(3) $e_N(t,s)=e_N(s,t)^{-1}$;

(4) *$e_N(s,t)$ is non-degenerate, i.e., if $e_N(s,t)=1$ for all $s\in\mathfrak{g}(N)$, then $t=0$;*

(5) *if t is of order N, $e_N(s,t)$ is a primitive N-th root of unity for some $s\in\mathfrak{g}(N)$;*

(6) *for every automorphism σ of the universal domain over a field of definition for E, $e_N(s,t)^\sigma=e_N(s^\sigma,t^\sigma)$.*

PROOF. The first and last properties are obvious from our definition. To show (2), put $t_3=t_1+t_2$, and let f_i and g_i be functions with the above properties for t_i, for $i=1,2,3$. Since $t_1+t_2-t_3-0=0$, by (4.3.2), there exists a function h on E such that $\mathrm{div}(h)=(t_1)+(t_2)-(t_3)-(0)$. Then $\mathrm{div}(f_1f_2f_3^{-1})=\mathrm{div}(h^N)$, so that $f_1f_2f_3^{-1}=ch^N$ with a constant c. Therefore $(g_1g_2g_3^{-1})(x)=c'h(Nx)$ with a constant c', from which we obtain (2). To prove (3), observe that

$$\mathrm{div}\left(\prod_{i=0}^{N-1}f(x-it)\right)=N\cdot\sum_{i=0}^{N-1}((it+t)-(it))=0,$$

hence $\prod_{i=0}^{N-1}f(x-it)$ is a constant. Therefore, if $Nt'=t$, we see that $\prod_{i=0}^{N-1}g(x-it')$ must be a constant. Substituting $x-t'$ for x, we obtain

$$g(x)g(x-t')\cdots g(x-(N-1)t')=g(x-t')\cdots g(x-(N-1)t')g(x-t),$$

so that $g(x)=g(x-t)$, which implies $e_N(t,t)=1$, hence (3). If $e_N(s,t)=1$ for all $s\in\mathfrak{g}(N)$, then $g(x+s)=g(x)$ for all $s\in\mathfrak{g}(N)$. Therefore $g(x)=p(Nx)$ for some function p on E. It follows that $f(x)=p(x)^N$, hence $\mathrm{div}(p)=(t)-(0)$, which is possible only when $t=0$, since E is of genus one. This proves (4). Finally, to prove (5), let t be of order N, and let T_N be the group of all N-th roots of unity. Then $s \mapsto e_N(t, s)$ is a homomorphism of $\mathfrak{g}(N)$ into T_N. If this is not surjective, there exists a positive divisor M of N smaller than N such that $e_N(t, s)^M = 1$ for all $s \in \mathfrak{g}(N)$. This implies, by virtue of (4), that $Mt = 0$, which is a contradiction. This completes the proof.

4.4. Isogenies and endomorphisms of elliptic curves over C

Let E and E' be elliptic curves isomorphic to C/L and C/L', respectively, with lattices L and L' in C. Then every homomorphism of E into E' corresponds to a complex analytic homomorphism of C/L into C/L', and vice versa. Now every complex analytic homomorphism of C/L into C/L' is given by a linear map $u \mapsto \mu u$ with a complex number μ such that $\mu L \subset L'$. Therefore

$$\mathrm{Hom}(E, E') \cong \mathrm{Hom}(C/L, C/L') = \{\mu \in C \mid \mu L \subset L'\}.$$

Especially

$$\mathrm{End}(E) \cong \mathrm{End}(C/L) = \{\mu \in C \mid \mu L \subset L\},$$

$$\mathrm{End}_Q(E) \cong \mathrm{End}_Q(C/L) = \{\mu \in C \mid \mu \cdot (QL) \subset QL\}.$$

Here QL denotes the Q-linear span of L. We say that an elliptic curve E *has complex multiplications* if $\mathrm{End}(E) \neq Z$.

PROPOSITION 4.3. *Let* $L = Z\omega_1 + Z\omega_2$ *and* $L' = Z\omega_1' + Z\omega_2'$ *with* $z = \omega_1/\omega_2 \in \mathfrak{H}$, $z' = \omega_1'/\omega_2' \in \mathfrak{H}$. *Then* C/L *and* C/L' *are isogenous (resp. isomorphic) if and only if there exists an element* α *of* $GL_2^+(Q)$ *(resp.* $SL_2(Z)$*) such that* $\alpha(z') = z$, *where* $GL_2^+(Q) = \{\xi \in GL_2(Q) \mid \det(\xi) > 0\}$.

PROOF. If $0 \neq \mu \in C$ is such that $\mu L \subset L'$, we obtain an element $\alpha = \begin{bmatrix} a & b \\ c & d \end{bmatrix}$ of $M_2(Z) \cap GL_2(Q)$ such that

$$\mu \begin{bmatrix} \omega_1 \\ \omega_2 \end{bmatrix} = \begin{bmatrix} a & b \\ c & d \end{bmatrix} \begin{bmatrix} \omega_1' \\ \omega_2' \end{bmatrix},$$

hence $\det(\alpha) > 0$ and $z = \alpha(z')$. Conversely, if $\alpha(z') = z$ for $\alpha = \begin{bmatrix} a & b \\ c & d \end{bmatrix} \in M_2(Z) \cap GL_2^+(Q)$, put $\lambda = cz' + d$. Then we see that $\lambda \neq 0$, and

$$\lambda \begin{bmatrix} z \\ 1 \end{bmatrix} = \begin{bmatrix} a & b \\ c & d \end{bmatrix} \begin{bmatrix} z' \\ 1 \end{bmatrix}, \quad \text{or} \quad (\lambda\omega_2'/\omega_2) \begin{bmatrix} \omega_1 \\ \omega_2 \end{bmatrix} = \begin{bmatrix} a & b \\ c & d \end{bmatrix} \begin{bmatrix} \omega_1' \\ \omega_2' \end{bmatrix}, \tag{4.4.1}$$

hence $\mu L \subset L'$ with $\mu = \lambda\omega_2'/\omega_2$. Especially $\mu L = L'$ if and only if $\alpha \in SL_2(Z)$.

PROPOSITION 4.4. *Let* $L = Z\omega_1 + Z\omega_2$ *with* $z = \omega_1/\omega_2 \in \mathfrak{H}$. *Then* C/L *has complex multiplications if and only if there exists a non-scalar element* α *of* $GL_2^+(Q)$ *such that* $\alpha(z) = z$.

PROOF. Repeat the proof of Prop. 4.3 with $z=z'$ and $\omega_i=\omega_i'$. Then we see that every $\mu\neq 0$ satisfying $\mu L\subset L$ corresponds to an element $\alpha=\begin{bmatrix} a & b \\ c & d\end{bmatrix}$ of $M_2(\boldsymbol{Z})\cap GL_2^+(\boldsymbol{Q})$ through the relation (4.4.1) with $\mu=\lambda$ and $z=z'$. From (4.4.1), we obtain

(4.4.2) $$\begin{bmatrix} z & \bar{z} \\ 1 & 1\end{bmatrix}\begin{bmatrix} \mu & 0 \\ 0 & \bar{\mu}\end{bmatrix}=\begin{bmatrix} a & b \\ c & d\end{bmatrix}\begin{bmatrix} z & \bar{z} \\ 1 & 1\end{bmatrix}.$$

We see easily that $\mu\in\boldsymbol{Z}$ if and only if α is a scalar matrix, hence our assertion.

PROPOSITION 4.5. *Let L and z be as in Prop. 4.4. Then $\boldsymbol{C}/L$ has complex multiplications if and only if $\boldsymbol{Q}(z)$ is an imaginary quadratic field. If that is so,* $\mathrm{End}_{\boldsymbol{Q}}(\boldsymbol{C}/L)$ *is isomorphic to $\boldsymbol{Q}(z)$.*

PROOF. In the relation (4.4.2), if $\alpha=\begin{bmatrix} a & b \\ c & d\end{bmatrix}$ is not a scalar matrix, μ cannot be real; moreover μ and $\bar{\mu}$ are the characteristic roots of α, and therefore satisfy a quadratic equation over $\boldsymbol{Q}$. Since $\mu=cz+d$ and $c\neq 0$, we have $\boldsymbol{Q}(z)=\boldsymbol{Q}(\mu)$, so that $\boldsymbol{Q}(z)$ must be imaginary quadratic. Conversely, if $K=\boldsymbol{Q}(z)$ is imaginary quadratic, we have $\boldsymbol{Q}L=\omega_2\cdot(\boldsymbol{Q}z+\boldsymbol{Q})=\omega_2 K$, so that

(4.4.3) $$\mathrm{End}_{\boldsymbol{Q}}(\boldsymbol{C}/L)=\{\mu\in\boldsymbol{C}\mid \mu\boldsymbol{Q}L\subset\boldsymbol{Q}L\}=\{\mu\in\boldsymbol{C}\mid \mu K\subset K\}=K.$$

PROPOSITION 4.6. *Let L and z be as in Prop. 4.4. Suppose that $\boldsymbol{C}/L$ has complex multiplications, and let $K=\boldsymbol{Q}(z)$. Then there is an injective homomorphism (or simply, an embedding) q of K into $M_2(\boldsymbol{Q})$ such that*

(4.4.4) $$q(K^\times)=\{\alpha\in GL_2^+(\boldsymbol{Q})\mid \alpha(z)=z\}.$$

PROOF. In view of (4.4.2) and (4.4.3), we can define $q(\mu)$ for $\mu\in K$ by

(4.4.5) $$\mu\begin{bmatrix} z \\ 1\end{bmatrix}=q(\mu)\begin{bmatrix} z \\ 1\end{bmatrix}.$$

Then our assertion is obvious from (4.4.3) and what we said in the proof of Prop. 4.4.

PROPOSITION 4.7. *Let K be an imaginary qudratic field, and q an embedding of K into $M_2(\boldsymbol{Q})$. Then there exists a point z on $\mathfrak{H}$ for which the relation* (4.4.4) *holds.*

PROOF. Let $\lambda\in K-\boldsymbol{Q}$, and $\alpha=q(\lambda)$. Then $\det(\alpha)=N_{K/\boldsymbol{Q}}(\lambda)=\lambda\bar{\lambda}>0$, and α has λ and $\bar{\lambda}$ as its characteristic roots. Therefore, α, as a transformation on $\mathfrak{H}$, is elliptic, and has a fixed point z in $\mathfrak{H}$. If we write the relation (4.4.2) for the present α and z, then $\lambda=\mu$ or $\lambda=\bar{\mu}$. In any case $\boldsymbol{Q}(z)=\boldsymbol{Q}(\lambda)=K$. If q' denotes the embedding of K into $M_2(\boldsymbol{Q})$ defined by $\mu\begin{bmatrix} z \\ 1\end{bmatrix}=q'(\mu)\begin{bmatrix} z \\ 1\end{bmatrix}$, we see that $q(\lambda)=q'(\lambda)$ or $q(\lambda)=q'(\bar{\lambda})$. Since $K=\boldsymbol{Q}(\lambda)$, this implies either $q(\mu)=q'(\mu)$ for all $\mu\in K$, or $q(\mu)=q'(\bar{\mu})$ for all $\mu\in K$. Therefore we obtain our assertion from Prop. 4.6.

We have also seen that there are exactly two embeddings of K into $M_2(\boldsymbol{Q})$ with the property (4.4.4) for a fixed point z. We call an embedding q *normalized* if it is defined by (4.4.5). The other one is defined by (4.4.5) with $\bar{z}$ in place of z.

Let q and q' be arbitrary embeddings of the same K into $M_2(\boldsymbol{Q})$. Then there exists an element β of $GL_2(\boldsymbol{Q})$ such that $q'(\mu)=\beta q(\mu)\beta^{-1}$ for all $\mu\in K$. (This is well-known, and can be proved as follows. Through the embedding q (resp. q'), regard $\boldsymbol{Q}^2$ as a one-dimensional vector space V (resp. V') over K. Then V and V' must be isomorphic over K; this means the existence of a $\boldsymbol{Q}$-linear automorphism β of $\boldsymbol{Q}^2$ such that $q'(\mu)\beta=\beta q(\mu)$.) Let z (resp. z') be the fixed point of $q(K^\times)$ (resp. $q'(K^\times)$) on $\mathfrak{H}$. Then $\beta(z)=z'$ or $\bar{z}'$, since z' and $\bar{z}'$ are the only fixed points of $q'(K^\times)$ on $\boldsymbol{C}$. Therefore $\beta\begin{bmatrix} z \\ 1\end{bmatrix}=c\begin{bmatrix} z' \\ 1\end{bmatrix}$ or $=c\begin{bmatrix} \bar{z}' \\ 1\end{bmatrix}$ with a non-zero complex number c. It follows that *if both q and q' are normalized,* $\det(\beta)$ *must be positive.*

Let us now fix an imaginary quadratic field K (always considered as a subfield of $\boldsymbol{C}$), and determine all isomorphism classes of elliptic curves E such that $\mathrm{End}_{\boldsymbol{Q}}(E)$ is isomorphic to K. We first observe that $\mathrm{End}(E)$ is an order in $\mathrm{End}_{\boldsymbol{Q}}(E)$. In general, by an *order* in an algebraic number field F of finite degree, we mean a subring of F, containing $\boldsymbol{Z}$, which is a free $\boldsymbol{Z}$-module of rank $[F:\boldsymbol{Q}]$. Every order in F is contained in the ring of all algebraic integers in F, which is called *the maximal order in F.* By a *lattice* (or *$\boldsymbol{Z}$-lattice*) in F, we mean a free $\boldsymbol{Z}$-submodule of F of rank $[F:\boldsymbol{Q}]$. For a $\boldsymbol{Z}$-lattice $\mathfrak{a}$ in F, if we put $\mathfrak{o}=\{\mu\in F\mid \mu\mathfrak{a}\subset\mathfrak{a}\}$, then $\mathfrak{o}$ is an order in F. We call $\mathfrak{o}$ *the order of* $\mathfrak{a}$, and $\mathfrak{a}$ a *proper $\mathfrak{o}$-ideal.* We can classify all the proper $\mathfrak{o}$-ideals, for a fixed $\mathfrak{o}$, with respect to multiplication by the elements of $F^\times$, as we usually do for the fractional ideals in F. Coming back to the imaginary quadratic field K, let $\mathfrak{a}$ be a $\boldsymbol{Z}$-lattice in K. If we consider $\mathfrak{a}$ as a submodule of $\boldsymbol{C}$, it is a lattice in $\boldsymbol{C}$, so that $\boldsymbol{C}/\mathfrak{a}$ is a complex torus. Then we have

$$\mathrm{End}(\boldsymbol{C}/\mathfrak{a})=\{\mu\in\boldsymbol{C}\mid\mu\mathfrak{a}\subset\mathfrak{a}\}=\{\mu\in K\mid\mu\mathfrak{a}\subset\mathfrak{a}\}. \tag{4.4.6}$$

PROPOSITION 4.8. *Let E be an elliptic curve defined over $\boldsymbol{C}$ such that* $\mathrm{End}_{\boldsymbol{Q}}(E)$ *is isomorphic to K, and $\mathfrak{o}$ an order in K corresponding to* $\mathrm{End}(E)$. *Then E is isomorphic to $\boldsymbol{C}/\mathfrak{a}$ with a proper $\mathfrak{o}$-ideal $\mathfrak{a}$. Conversely, for any proper $\mathfrak{o}$-ideal $\mathfrak{a}$,* $\mathrm{End}(\boldsymbol{C}/\mathfrak{a})$ *is isomorphic to $\mathfrak{o}$. Moreover the class of proper $\mathfrak{o}$-ideals $\mathfrak{a}$ is uniquely determined by the isomorphism class of $\boldsymbol{C}/\mathfrak{a}$. In other words, $\boldsymbol{C}/\mathfrak{a}$ is isomorphic to $\boldsymbol{C}/\mathfrak{b}$ if and only if $\mu\mathfrak{a}=\mathfrak{b}$ for some $\mu\in K^\times$.*

PROOF. Since there are two isomorphisms of K onto $\mathrm{End}_{\boldsymbol{Q}}(E)$, $\mathfrak{o}$ may depend on the choice of isomorphism. But, if $a \in \mathfrak{o}$, we have $a+\bar{a} \in \boldsymbol{Z} \subset \mathfrak{o}$, so that $\bar{a} \in \mathfrak{o}$. This shows $\bar{\mathfrak{o}}=\mathfrak{o}$, hence $\mathfrak{o}$ *is independent of the choice of the isomorphism of K to* $\mathrm{End}_{\boldsymbol{Q}}(E)$. Now E is isomorphic to a torus of the form $\boldsymbol{C}/(\boldsymbol{Z}z+\boldsymbol{Z})$ with $z \in K$. Put $\mathfrak{a}=\boldsymbol{Z}z+\boldsymbol{Z}$. Then $\mathfrak{a}$ must be a proper $\mathfrak{o}$-ideal by (4.4.6). The converse part is just a restatement of (4.4.6). The last assertion can be verified in a straightforward way.

From this result, we obtain the following two propositions:

PROPOSITION 4.9. *Let E and E' be elliptic curves defined over $\boldsymbol{C}$. Suppose that E has complex multiplications. Then E' is isogenous to E if and only if* $\mathrm{End}_{\boldsymbol{Q}}(E')$ *is isomorphic to* $\mathrm{End}_{\boldsymbol{Q}}(E)$.

PROPOSITION 4.10. *For an order $\mathfrak{o}$ in K, the number of classes of proper $\mathfrak{o}$-ideals is exactly the number of isomorphism classes of elliptic curves E such that* $\mathrm{End}(E)$ *is isomorphic to $\mathfrak{o}$. Especially if $\mathfrak{o}$ is the maximal order in K, the number is nothing but the class number of K.*

PROPOSITION 4.11. *Let $\mathfrak{o}_K$ be the maximal order in K, and $\mathfrak{o}$ an order in K. Then there is a unique positive integer c such that $\mathfrak{o}=\boldsymbol{Z}+c\mathfrak{o}_K$. Further, for every proper $\mathfrak{o}$-ideal $\mathfrak{a}$, there exists an element μ of $K^\times$ such that $\mu\mathfrak{a}+c\mathfrak{o}=\mathfrak{o}$. Moreover, for two proper $\mathfrak{o}$-ideals $\mathfrak{a}$ and $\mathfrak{b}$, let $\mathfrak{ab}$ denote the $\boldsymbol{Z}$-module generated by the elements xy with $x \in \mathfrak{a}$ and $y \in \mathfrak{b}$. Then all the proper $\mathfrak{o}$-ideals form a group with respect to this law of multiplication, with $\mathfrak{o}$ as the identity element.*

PROOF. It is well-known that $\mathfrak{o}_K=\boldsymbol{Z}+\boldsymbol{Z}\lambda$ with an element λ. We can put $\mathfrak{o} \cap \boldsymbol{Z}\lambda=\boldsymbol{Z}c\lambda$ with a positive integer c. Then $\boldsymbol{Z}+c\mathfrak{o}_K=\boldsymbol{Z}+\boldsymbol{Z}c\lambda \subset \mathfrak{o}$. If $r+s\lambda \in \mathfrak{o}$ with r and s in $\boldsymbol{Z}$, then $s\lambda \in \mathfrak{o}$, so that $s \in c\boldsymbol{Z}$. Therefore $\mathfrak{o}=\boldsymbol{Z}+c\mathfrak{o}_K$. The uniqueness of c is obvious. For a $\boldsymbol{Z}$-lattice $\mathfrak{a}$ in K, put

$$\mathfrak{a}^*=\{\mu \in K \mid \mathrm{Tr}_{K/\boldsymbol{Q}}(\mu\mathfrak{a}) \subset \boldsymbol{Z}\}\ .$$

Then we see easily that $\mathfrak{a}^*$ is a $\boldsymbol{Z}$-lattice in K, $(\mathfrak{a}^*)^*=\mathfrak{a}$, and $\mathfrak{a}^* \subset \mathfrak{b}^*$ if $\mathfrak{b} \subset \mathfrak{a}$. Moreover, if $\mathfrak{o}\mathfrak{a} \subset \mathfrak{a}$, we have $\mathfrak{o}\mathfrak{a}^* \subset \mathfrak{a}^*$. Therefore if $\mathfrak{o}$ (resp. $\mathfrak{o}'$) is the order of $\mathfrak{a}$ (resp. $\mathfrak{a}^*$), we have $\mathfrak{o} \subset \mathfrak{o}'$, and $\mathfrak{o}' \subset \mathfrak{o}$ since $\mathfrak{a}^{**}=\mathfrak{a}$, so that $\mathfrak{o}=\mathfrak{o}'$. We can verify in a straightforward way that $\mathfrak{o}^*=g'(c\lambda)^{-1}\mathfrak{o}$ if $g(x)=0$ is the monic irreducible equation for $c\lambda$ over $\boldsymbol{Q}$. Let $\mathfrak{a}$ be a proper $\mathfrak{o}$-ideal. If $\xi \in (\mathfrak{a}\mathfrak{a}^*)^*$, then $\mathrm{Tr}_{K/\boldsymbol{Q}}(\xi\mathfrak{a}\mathfrak{a}^*) \subset \boldsymbol{Z}$, so that $\xi\mathfrak{a}^* \subset \mathfrak{a}^*$, hence $\xi \in \mathfrak{o}$. It follows that $(\mathfrak{a}\mathfrak{a}^*)^* \subset \mathfrak{o}$, hence $\mathfrak{o}^* \subset \mathfrak{a}\mathfrak{a}^*$. On the other hand, $\mathrm{Tr}(\mathfrak{a}\mathfrak{a}^*\mathfrak{o})=\mathrm{Tr}(\mathfrak{a}\mathfrak{a}^*) \subset \boldsymbol{Z}$, hence $\mathfrak{a}\mathfrak{a}^* \subset \mathfrak{o}^*$. Therefore we have $\mathfrak{a}\mathfrak{a}^*=\mathfrak{o}^*$, so that $\mathfrak{a} \cdot (g'(c\lambda)\mathfrak{a}^*)=\mathfrak{o}$. Thus we have shown the existence of inverse in the semi-group of proper $\mathfrak{o}$-ideals, hence the last assertion. Put $\mathfrak{b}=g'(c\lambda)\mathfrak{a}^*$. Define a $\boldsymbol{Q}$-linear map f of K into $\boldsymbol{Q}$ by $f(r+s\lambda)=r$ for r and s in $\boldsymbol{Q}$. Then $f(\mathfrak{b}\mathfrak{a})=f(\mathfrak{o})=\boldsymbol{Z}$. Therefore, for every rational prime p, there exists an element μ_p of $\mathfrak{b}$ such that $f(\mu_p\mathfrak{a})$ is not contained in $p\boldsymbol{Z}$. Then we can find an element μ of $\mathfrak{b}$ so that $\mu\equiv\mu_p \bmod p\mathfrak{b}$ for all prime factors p of c. Then $f(\mu\mathfrak{a})$ is not contained in $p\boldsymbol{Z}$ for all such p. Hence $f(\mu\mathfrak{a})=m\boldsymbol{Z}$ with a positive integer m prime to c. Then $f(\mu\mathfrak{a}+c\mathfrak{o}_K)=m\boldsymbol{Z}+c\boldsymbol{Z}=\boldsymbol{Z}$. If $\alpha\in\mathfrak{o}$, we have $f(\alpha)=f(\beta)$ for some $\beta\in\mu\mathfrak{a}+c\mathfrak{o}_K$. Then $\alpha-\beta\in\boldsymbol{Z}c\lambda\subset c\mathfrak{o}_K$, so that $\alpha=(\alpha-\beta)+\beta\in\mu\mathfrak{a}+c\mathfrak{o}_K$. This shows that $\mathfrak{o}=\mu\mathfrak{a}+c\mathfrak{o}_K$. Since both $\mu\mathfrak{a}$ and $c\mathfrak{o}_K$ are ideals of $\mathfrak{o}$, we have $\mathfrak{o}=\mathfrak{o}\mathfrak{o}=(\mu\mathfrak{a}+c\mathfrak{o}_K)(\mu\mathfrak{a}+c\mathfrak{o}_K)\subset\mu\mathfrak{a}+c^2\mathfrak{o}_K\subset\mu\mathfrak{a}+c\mathfrak{o}$, so that $\mathfrak{o}=\mu\mathfrak{a}+c\mathfrak{o}$.

The integer c (or the ideal $c\mathfrak{o}_K$) is called *the conductor of* $\mathfrak{o}$. It can easily be seen that $c\mathfrak{o}_K=\{\alpha\in K\mid \alpha\mathfrak{o}_K\subset\mathfrak{o}\}$. In (5.4.2), we shall show that every proper $\mathfrak{o}$-ideal is "locally principal".

As our argument shows, $\mathfrak{a}\mathfrak{a}^*=\mathfrak{o}^*$ holds for any proper $\mathfrak{o}$-ideal $\mathfrak{a}$ with any order $\mathfrak{o}$ in K, even if $[K:\boldsymbol{Q}]>2$. If $\mathfrak{o}=\boldsymbol{Z}[\pi]$ with an element π satisfying a monic irreducible equation $g(x)=0$ over $\boldsymbol{Q}$, then $\mathfrak{o}^*=g'(\pi)^{-1}\mathfrak{o}$, so that every proper $\mathfrak{o}$-ideal $\mathfrak{a}$ is invertible.

EXERCISE 4.12. Prove that the number of classes of proper $\mathfrak{o}$-ideals is given by

$$h\cdot c\cdot[\mathfrak{o}_K^\times:\mathfrak{o}^\times]^{-1}\cdot\prod_{p\mid c}\left[1-\left(\frac{K}{p}\right)p^{-1}\right],$$

where h is the class number of K; $\left(\frac{K}{p}\right)$ is 1, -1, or 0, according as the prime p decomposes in K, remains prime in K, or is ramified in K.

EXERCISE 4.13. Let F be an algebraic number field of finite degree, K a quadratic extension of F, and $\mathfrak{o}_F$ (resp. $\mathfrak{o}_K$) the maximal order in F (resp. K). Generalize Prop. 4.11 to the case of an order in K containing $\mathfrak{o}_F$. (Although this can be done globally, it may be easier to treat, at first, the corresponding problem for local fields. The assertion (5.4.2) can also be generalized.)

4.5. Automorphisms of an elliptic curve

Let $\mathrm{Aut}(E)$ denote the group of all automorphisms of an elliptic curve E defined over $\boldsymbol{C}$. If E has no complex multiplication, $\mathrm{Aut}(E)$ consists only of ±1. Therefore suppose that E has complex multiplications, and let $\mathfrak{o}$ and K be isomorphic to $\mathrm{End}(E)$ and $\mathrm{End}_{\boldsymbol{Q}}(E)$ as in Prop. 4.8. Then $\mathrm{Aut}(E)$ is isomorphic to $\mathfrak{o}^\times$. Since K is imaginary quadratic, as is well known, $\mathfrak{o}^\times$ contains more than ±1 only in the following two cases:

(A) $K=\boldsymbol{Q}(\sqrt{-1})$, $\mathfrak{o}=\boldsymbol{Z}[\sqrt{-1}]$, $\mathfrak{o}^\times=\{\pm1, \pm\sqrt{-1}\}$.

(B) $K=\boldsymbol{Q}(\zeta)$, $\zeta=e^{2\pi i/3}$, $\mathfrak{o}=\boldsymbol{Z}[\zeta]$, $\mathfrak{o}^\times=\{\pm1, \pm\zeta, \pm\zeta^2\}$.

In these two cases, $\mathfrak{o}$ is the maximal order in K, and the class number of K

is one, so that, by Prop. 4.10, in each case, there is one and only one elliptic curve E, up to isomorphisms over $\boldsymbol{C}$, such that $\mathrm{End}(E)$ is isomorphic to $\mathfrak{o}$.

Let E be defined by $y^2=4x^3-c_2x-c_3$ with c_2 and c_3 in $\boldsymbol{C}$. We observe that

(4.5.1) $$j_E=1 \Leftrightarrow c_3=0; \qquad j_E=0 \Leftrightarrow c_2=0.$$

Now, if $c_3=0$, $\mathrm{Aut}(E)$ contains at least 4 elements: $(x, y)\mapsto(x, \pm y)$, $(-x, \pm\sqrt{-1}\,y)$; if $c_2=0$, $\mathrm{Aut}(E)$ contains at least 6 elements: $(x, y)\mapsto(\zeta^\nu x, \pm y)$, $\nu=0, 1, 2$, with $\zeta=e^{2\pi i/3}$. Therefore, from (4.5.1), we obtain

(4.5.2) *E belongs to the case* (A) *resp.* (B) *if and only if $j_E=1$ resp. $j_E=0$.*

Moreover, we see that $\mathrm{Aut}(E)$ consists of those 4 or 6 elements.

Hereafter we denote by $\mathcal{E}$ the set of all elliptic curves E of the form $y^2=4x^3-c_2x-c_3$ with c_2 and c_3 in $\boldsymbol{C}$. We classify $\mathcal{E}$ into three classes $\mathcal{E}_i$ with $i=1, 2, 3$ according to the number $2i$ of automorphisms. Thus $\mathcal{E}_2$ and $\mathcal{E}_3$ consist of the members of $\mathcal{E}$ of the type (A) and (B) respectively, and $\mathcal{E}_1$ contains all the remaining members of $\mathcal{E}$.

For any elliptic curve E: $y^2=4x^3-c_2x-c_3$, we define three functions h_E^i on E by

$$h_E^1((x, y))=(c_2c_3/\Delta)\cdot x,$$
$$h_E^2((x, y))=(c_2^2/\Delta)\cdot x^2, \qquad (\Delta=c_2^3-27c_3^2),$$
$$h_E^3((x, y))=(c_3/\Delta)\cdot x^3.$$

They are obviously defined over any field of definition for E. If $E\in\mathcal{E}_2$, we have $h_E^1=h_E^3=0$ and $h_E^2((x, y))=c_2^{-1}x^2$; if $E\in\mathcal{E}_3$, we have $h_E^1=h_E^2=0$, and $h_E^3((x, y))=(-27c_3)^{-1}x^3$. By means of the explicit form of the elements of $\mathrm{Aut}(E)$ mentioned above, we can easily verify

(4.5.3) *When $E\in\mathcal{E}_i$, one has $h_E^i(t)=h_E^i(t')$ if and only if $t=\alpha t'$ for some $\alpha\in\mathrm{Aut}(E)$.*

(4.5.4) *Let E and E' be members of $\mathcal{E}$, and η an isomorphism of E to E'. Then $h_E^i=h_{E'}^i\circ\eta$ for $i=1, 2, 3$.*

In fact, if E is as above, and E' is defined by $y^2=4x^3-c_2'x-c_3'$, then, by Prop. 4.1, $\eta((x, y))=(\mu^2x, \mu^3y)$, $c_2'=\mu^4c_2$, $c_3'=\mu^6c_3$ with an element μ of $\boldsymbol{C}$. Therefore we obtain (4.5.4) from our definition of h_E^i.

4.6. Integrality properties of the invariant J

In Th. 2.9, we proved that the modular function

$$J(z)=12^3j(z)=12^3g_2(z)^3/\Delta(z)$$

has a Fourier expansion of the form

(4.6.1) $$J(z) = q^{-1}(1 + \sum_{n=1}^{\infty} c_n q^n), \qquad q = e^{2\pi i z}$$

with $c_n \in \boldsymbol{Z}$. Let us now prove

THEOREM 4.14. *If z belongs to an imaginary quadratic field and* $\mathrm{Im}(z) > 0$, *$J(z)$ is an algebraic integer.*

We shall give here an analytic proof of this fact, although a more intrinsic algebraic proof is now possible by virtue of the Néron minimal model [53], see Deuring [10], Serre and Tate [66].

The fact that $J(z)$ is an algebraic number can easily be seen as follows. Let $K = \boldsymbol{Q}(z)$, $L = \boldsymbol{Z}z + \boldsymbol{Z}$, and let E be an elliptic curve isomorphic to $\boldsymbol{C}/L$. Observe that, for any $\sigma \in \mathrm{Aut}(\boldsymbol{C})$, $\mathrm{End}_{\boldsymbol{Q}}(E^{\sigma})$ is isomorphic to K. Now there are only countably many isomorphism classes of elliptic curves whose endomorphism algebras are isomorphic to K. Since $j(E^{\sigma}) = j(E)^{\sigma}$, it follows that $\{j(E)^{\sigma} \mid \sigma \in \mathrm{Aut}(\boldsymbol{C})\}$ is a countable set, hence $j(E)$ must be algebraic.

The fact that $J(z)$ is integral is much deeper, and requires a more elaborate argument (whatever method one uses).

PROPOSITION 4.15. *Suppose that an equality*

$$\sum_{k=0}^{m} a_k J(z)^k = \sum_{n \geqq n_0} b_n q^n \qquad (q = e^{2\pi i z})$$

holds for all $z \in \mathfrak{H}$, with constants a_k and b_n in $\boldsymbol{C}$. Then the a_k belong to the ring generated by the b_n over $\boldsymbol{Z}$.

PROOF. Substitute the expression $q^{-1}(1 + \sum_{n=1}^{\infty} c_n q^n)$ for $J(z)$ in $\sum_{k=0}^{m} a_k J(z)^k$. Then we obtain

$$b_{-m} = a_m,$$
$$b_{1-m} = m a_m c_1 + a_{m-1},$$
$$b_{2-m} = (m(m-1)/2) \cdot a_m c_2 + (m-1) \cdot a_{m-1} c_1 + a_{m-2},$$
$$\cdots\cdots\cdots\cdots\cdots$$

Since $c_n \in \boldsymbol{Z}$, our assertion is obvious.

Let us call an element $\alpha = \begin{bmatrix} a & b \\ c & d \end{bmatrix}$ of $M_2(\boldsymbol{Z})$ *primitive* if a, b, c, d have no common divisors other than ± 1. If $\det(\alpha) = n > 0$, α is primitive if and only if $\alpha \in \Gamma \cdot \begin{bmatrix} n & 0 \\ 0 & 1 \end{bmatrix} \cdot \Gamma$, where $\Gamma = SL_2(\boldsymbol{Z})$. By Prop. 3.36, we have

$$\Gamma \cdot \begin{bmatrix} n & 0 \\ 0 & 1 \end{bmatrix} \cdot \Gamma = \bigcup_{\alpha \in A} \Gamma \alpha$$

with the set A of all the matrices $\alpha = \begin{bmatrix} a & b \\ 0 & d \end{bmatrix}$ under the conditions $d > 0$,

$ad=n,\ 0\leq b<d$, and $(a, b, d)=1$.

Now fix an integer $n>1$, and consider the polynomial

$$\prod_{\alpha\in A}(X-J\circ\alpha)=\sum_{m=0}^{M}s_m X^m$$

with an indeterminate X, where the s_m are the elementary symmetric functions in the $J\circ\alpha$, hence are holomorphic functions on $\mathfrak{H}$, which have Fourier expansions in $q^{1/n}$. For every $\gamma\in\Gamma$, we have $\bigcup_{\alpha\in A}\Gamma\alpha\gamma=\bigcup_{\alpha\in A}\Gamma\alpha$, so that

$$\{J\circ\alpha\circ\gamma\mid\alpha\in A\}=\{J\circ\alpha\mid\alpha\in A\}.$$

It follows that $s_m\circ\gamma=s_m$, hence s_m is a modular function of level 1. Since s_m is holomorphic on $\mathfrak{H}$, s_m is a polynomial in J, say $s_m=S_m(J)$.

For $\alpha=\begin{bmatrix} a & b\\ 0 & d\end{bmatrix}\in A$, the q-expansion of $J\circ\alpha$ is of the form

$$J(\alpha(z))=\zeta_d^{-b}q^{-a/d}\left[1+\sum_{m=1}^{\infty}c_m\zeta_d^{mb}q^{ma/d}\right],\qquad \zeta_d=e^{2\pi i/d}. \tag{4.6.2}$$

Thus the coefficients are algebraic integers of $\boldsymbol{Q}(\zeta_n)$. Let σ be an automorphism of $\boldsymbol{Q}(\zeta_n)$, such that $\zeta_n^\sigma=\zeta_n^t$ for some t with $(t, n)=1$. Transforming the coefficients of $J\circ\alpha$ by σ, we obtain $J\circ\beta$ with $\beta=\begin{bmatrix} a & b'\\ 0 & d\end{bmatrix}\in A$, $b'\equiv bt \bmod (d)$. Since $\alpha\mapsto\beta$ gives a permutation of the set A, we can conclude that the q-expansion of s_m has coefficients in $\boldsymbol{Z}$. Applying Prop. 4.15 to $S_m(J)$, we see that the polynomial S_m has integral coefficients. Thus we obtain a polynomial

$$F_n(X, J)=\prod_{\alpha\in A}(X-J\circ\alpha)=\sum_{m=0}^{M}S_m(J)X^m \tag{4.6.3}$$

belonging to $\boldsymbol{Z}[X, J]$.

PROPOSITION 4.16. *For any* $\xi\in GL_2(\boldsymbol{Q})$ *with* $\det(\xi)>0$, $J\circ\xi$ *is integral over* $\boldsymbol{Z}[J]$.

PROOF. Multiplying ξ by a suitable rational number, we may assume that ξ is a primitive element of $M_2(\boldsymbol{Z})$, since this does not change $J\circ\xi$. If $\det(\xi)=n>1$, $\xi\in\Gamma\cdot\begin{bmatrix} n & 0\\ 0 & 1\end{bmatrix}\cdot\Gamma$, so that $\Gamma\xi=\Gamma\alpha$ for some $\alpha\in A$. Then we have $J\circ\xi=J\circ\alpha$, so that $F_n(J\circ\xi, J)=0$, hence our assertion.

Let us now put

$$H_n(J)=F_n(J, J)=\prod_{\alpha\in A}(J-J\circ\alpha).$$

Then H_n is a polynomial in J with coefficients in $\boldsymbol{Z}$.

PROPOSITION 4.17. *If n is not a square, the highest coefficient of the polynomial $H_n(J)$ is ± 1.*

PROOF. If n is not a square, we have $a/d\neq 1$ in (4.6.2), hence the leading coefficient of the q-expansion of $J-J\circ\alpha$ is a root of unity, and so is the leading coefficient of the q-expansion of $H_n(J)$. This coefficient is equal to the highest coefficient of the polynomial H_n, which is rational, so that it must be ± 1.

Now suppose that $K=\boldsymbol{Q}(z)$ is imaginary quadratic, $L=\boldsymbol{Z}+\boldsymbol{Z}z$, and let $\mathfrak{o}$ be the order in K isomorphic to $\mathrm{End}(\boldsymbol{C}/L)$. First assume that $\mathfrak{o}$ is the maximal order in K. Then we can find an element μ of $\mathfrak{o}$ such that $N_{K/\boldsymbol{Q}}(\mu)$ is a square-free integer $n>1$. (In fact, if $K=\boldsymbol{Q}(\sqrt{-1})$, take $\mu=1+\sqrt{-1}$, and if $K=\boldsymbol{Q}(\sqrt{-m})$, $m>1$ and square-free, take $\mu=\sqrt{-m}$.) Define an element ξ of $M_2(\boldsymbol{Z})$ by

$$\mu\begin{bmatrix} z \\ 1 \end{bmatrix}=\xi\begin{bmatrix} z \\ 1 \end{bmatrix} \qquad (\xi=q(\mu) \text{ with the notation of (4.4.5))}.$$

Then $\det(\xi)=n$, and ξ is primitive, since n is square-free. Therefore $J\circ\xi=J\circ\alpha$ for some $\alpha\in A$ as in the proof of Prop. 4.16. Since $\xi(z)=z$, we have $J(z)=J(\xi(z))=J(\alpha(z))$, so that $H_n(J(z))=0$. By Prop. 4.17, this shows that $J(z)$ is an algebraic integer.

Next consider the case where $\mathfrak{o}$ is not the maximal order. By Prop. 4.3, there exists an element β of $GL_2^+(\boldsymbol{Q})$ such that $\mathrm{End}(\boldsymbol{C}/(\boldsymbol{Z}z'+\boldsymbol{Z}))$, with $z'=\beta(z)$, is the maximal order. By Prop. 4.16, $J(z)$ is integral over $\boldsymbol{Z}[J(z')]$. Since $J(z')$ is integral, this completes the proof of Th. 4.14.

Actually an arbitrary order $\mathfrak{o}$ in K contains an element μ such that $N_{K/\boldsymbol{Q}}(\mu)$ is a prime. In fact, take a positive integer h so that $h\mathfrak{o}_K\subset\mathfrak{o}$. By the generalized Dirichlet theorem, there is an element μ of K such that $\mu\equiv 1 \bmod h\mathfrak{o}_K$, and $N_{K/\boldsymbol{Q}}(\mu)$ is a prime. Then $\mu\in\mathfrak{o}$. Applying the above argument to this μ, we can show that $J(z)$ is integral, without reducing the question to the case of maximal order.

The equation $H_n(J)=0$ is called *the modular equation* for the degree n. For the classical treatment of this and other related topics, the reader is referred to Fricke [21], Hurwitz [32], and Weber [89].

CHAPTER 5
ABELIAN EXTENSIONS OF IMAGINARY QUADRATIC FIELDS AND COMPLEX MULTIPLICATION OF ELLIPTIC CURVES

The purpose of this chapter is to study the behavior of an elliptic curve E with complex multiplications under $\mathrm{Gal}(K_{ab}/K)$, where K is an imaginary quadratic field isomorphic to $\mathrm{End}_{\mathbf{Q}}(E)$, and K_{ab} the maximal abelian extension of K. The reader will be required to have some knowledge of class field theory. We shall state the main theorem 5.4 in the adelic language, and derive from it the classical result on the construction of K_{ab} by means of special values of elliptic or elliptic modular functions. This topic will be taken up again in § 6.8, in a different formulation without elliptic curves.

5.1. Preliminary considerations

There is a simple principle concerning the field of rationality, which we shall often make use of in this and next chapters. Let X be an algebro-geometric object defined in the universal domain $\mathbf{C}$, such that X^σ is meaningfully defined for every automorphism σ of $\mathbf{C}$. Thus X may be a variety, a rational map, or a differential form on a variety (see Appendix). Then our principle is as follows:

Let k be a subfield of $\mathbf{C}$. If $X^\sigma = X$ for all $\sigma \in \mathrm{Aut}(\mathbf{C}/k)$, then X is rational over k. Or equivalently, if X^σ, for $\sigma \in \mathrm{Aut}(\mathbf{C}/k)$, depends only on the restriction of σ to k, then X is rational over k.

This is not a completely rigorous statement if X is defined with respect to some other algebro-geometric objects. For example, if X is a rational map of a variety U into a variety V, it is better to assume that U and V are defined over k. The same remark applies to a differential form.

We can state a similar fact for two subfields of $\mathbf{C}$:

Let k and k' be subfields of $\mathbf{C}$ with countably many elements. Suppose that k' is stable under $\mathrm{Aut}(\mathbf{C}/k)$. Then the composite kk' is a (finite or an infinite) Galois extension of k. Moreover, if every element of $\mathrm{Aut}(\mathbf{C}/k)$ induces the identity map on k', then $k' \subset k$.

Now let us consider a projective non-singular curve V defined over a field k of any characteristic. We shall denote by $k(V)$ the field of all functions on V rational over k (see Appendix N° 4). Let W be a projective non-singular curve, and λ a rational map of V into W, both defined over k. Then it is well-known that λ is a morphism, i. e., defined everywhere on V. Suppose that λ is not a constant map. Then $f \mapsto f \circ \lambda$ defines an isomorphism of $k(W)$ into $k(V)$. Let $k(W) \circ \lambda$ denote the image of $k(W)$ by this isomorphism. We say that λ is *separable, inseparable,* or *purely inseparable,* according as $k(V)$ is separable, inseparable, or purely inseparable over $k(W) \circ \lambda$. Further we put

$$\deg(\lambda) = [k(V) : k(W) \circ \lambda],$$

and call it the *degree* of λ; this does not depend on the choice of k.

Let $\mathrm{Dif}(V)$ denote the set of all differential forms on V, and $\mathcal{D}(V)$ the set of all holomorphic elements of $\mathrm{Dif}(V)$, i. e., all differential forms of the first kind on V. Further let $\mathcal{D}(V; k)$ denote the set of all elements of $\mathcal{D}(V)$ rational over k (see Appendix N°s 8, 9). If λ and W are as above, for every $\omega = h \cdot df \in \mathrm{Dif}(W; k)$ with f and h in $k(W)$, we can define an element $\omega \circ \lambda$ of $\mathrm{Dif}(V; k)$ by

$$\omega \circ \lambda = (h \circ \lambda) \cdot d(f \circ \lambda).$$

If $\omega \in \mathcal{D}(W)$, then $\omega \circ \lambda \in \mathcal{D}(V)$.

PROPOSITION 5.1. *Let V, W, λ, and k be as above, and let $0 \neq \omega \in \mathrm{Dif}(W; k)$. Then $\omega \circ \lambda \neq 0$ if and only if λ is separable.*

PROOF. The differential form df has the property

(5.1.0) *$df \neq 0$ if and only if $k(W)$ is separably algebraic over $k(f)$.*

(See Appendix N°s 8, 9.) Put $\omega = h \cdot df$ with h and f in $k(W)$. Since $\omega \neq 0$, $k(W)$ is separable over $k(f)$, so that $k(W) \circ \lambda$ is separable over $k(f \circ \lambda)$. Applying (5.1.0) to $d(f \circ \lambda)$, we see that $k(V)$ is separable over $k(f \circ \lambda)$ if and only if $\omega \circ \lambda \neq 0$, hence our assertion.

PROPOSITION 5.2. *Let V, W, λ, and k be as above. If λ is purely inseparable and $q = \deg(\lambda)$, then there exists a biregular isomorphism μ of W to V^q, rational over k, such that $\mu \circ \lambda$ is the q-th power morphism of V to V^q, where V^q denotes the transform of V by the q-th power automorphism of the universal domain.*

PROOF. Let v be a generic point of V over k, and let $w = \lambda(v)$, $K = k(v)$, $L = k(w)$. Our assertion is equivalent to (or, at least follows from) the equality $L = k \cdot K^q$, where $K^q = \{a^q \mid a \in K\}$. In fact, if $k \cdot K^q = L$, we have $k(v^q) = k(w)$. Since v^q is a generic point of V^q over k, we can define a birational map μ of W to V^q by $\mu(w) = v^q$. Since W and V^q are projective non-singular, μ is biregular. Then $\mu(\lambda(v)) = v^q$, so that $\mu \circ \lambda$ is the q-th power morphism of V

to V^q. Thus our question is reduced to show that $k \cdot K^q = L$. By our assumption, K is purely inseparable over L, and $[K : L] = q$, so that $k \cdot K^q \subset L$. Therefore it is sufficient to show that $[K : k \cdot K^q] = q$. Since K is a regular extension of k, there exists an element x of K such that K is separably algebraic over $k(x)$. Then $k \cdot K^q$ is separable over $k(x^q)$. Now K is separable over $k(x)$, and purely inseparable over $k \cdot K^q$, so that K is the composite of $k(x)$ and $k \cdot K^q$. Since $k(x)$ is purely inseparable over $k(x^q)$, and $k \cdot K^q$ is separable over $k(x^q)$, we have

$$[K : k \cdot K^q] = [k(x) : k(x^q)] = q\,,$$

which completes the proof.

PROPOSITION 5.3. *Let E_1 and E_2 be elliptic curves defined over a subfield k of $\boldsymbol{C}$, and $\bar{k}$ the algebraic closure of k in $\boldsymbol{C}$. Then every element of* $\mathrm{Hom}(E_1, E_2)$ *is defined over $\bar{k}$. Moreover, if* $\mathrm{End}(E_1)$ *is isomorphic to $\boldsymbol{Z}$, and $\lambda \in \mathrm{Hom}(E_1, E_2)$, then $\lambda^\sigma = \pm\lambda$ for every automorphism σ of $\bar{k}$ over k.*

PROOF. If $\lambda \in \mathrm{Hom}(E_1, E_2)$ and σ is an automorphism of $\boldsymbol{C}$ over k, then $\lambda^\sigma \in \mathrm{Hom}(E_1, E_2)$. Since $\mathrm{Hom}(E_1, E_2)$ is at most a countable set, there are at most countably many λ^σ, so that λ must be defined over $\bar{k}$. If $\mathrm{End}(E_1)$ is isomorphic to $\boldsymbol{Z}$ and $\lambda \neq 0$, we see that $\mathrm{Hom}(E_1, E_2)$ is isomorphic to $\boldsymbol{Z}$, so that $m\lambda^\sigma = n\lambda$ with non-zero integers m and n. Then $m^2 \cdot \deg(\lambda^\sigma) = \deg(m\lambda^\sigma) = \deg(n\lambda) = n^2 \cdot \deg(\lambda)$. Since $\deg(\lambda^\sigma) = \deg(\lambda)$, we obtain $m = \pm n$, so that $\lambda^\sigma = \pm\lambda$.

Let us now consider an elliptic curve E over $\boldsymbol{C}$ such that $\mathrm{End}_{\boldsymbol{Q}}(E)$ is isomorphic to an imaginary quadratic field K. We shall now show a way of choosing a canonical one among the two isomorphisms of K onto $\mathrm{End}_{\boldsymbol{Q}}(E)$. First observe that the vector space $\mathfrak{D}(E)$ of holomorphic differential forms on E is one-dimensional over $\boldsymbol{C}$. Let $0 \neq \omega \in \mathfrak{D}(E)$. For every $\alpha \in \mathrm{End}(E)$, we have $\omega \circ \alpha \in \mathfrak{D}(E)$, so that $\omega \circ \alpha = \mu_\alpha \omega$ with an element μ_α of $\boldsymbol{C}$. If E is identified with a complex torus $\boldsymbol{C}/L$, with a lattice L in $\boldsymbol{C}$, and if u denotes the variable on $\boldsymbol{C}$, then $\omega = c \cdot du$ with $c \in \boldsymbol{C}$. Therefore, if α corresponds to the linear map $u \mapsto \mu u$ as in § 4.4, we have $\omega \circ \alpha = c \cdot d(\mu u) = c\mu \cdot du = \mu \cdot \omega$, so that $\mu = \mu_\alpha$. Thus we can choose an isomorphism θ of K onto $\mathrm{End}_{\boldsymbol{Q}}(E)$ which is completely characterized by the condition

$$(5.1.1) \qquad \omega \circ \theta(\mu) = \mu\omega \qquad (\mu \in K,\ \theta(\mu) \in \mathrm{End}(E))\,.$$

Observe that this condition does not depend on the choice of ω. We say that (E, θ) (or simply θ) is *normalized* if this condition is satisfied. If (E', θ') is another normalized couple with the same K, then every isogeny λ of E to E' satisfies

(5.1.2) $$\lambda \circ \theta(\mu) = \theta'(\mu) \circ \lambda \qquad (\mu \in K).$$

In fact, if ω (resp. ω') is a differential form on E (resp. E') as considered above, we have $\omega' \circ \lambda = b\omega$ with a constant b, so that $\omega' \circ \lambda \circ \theta(\mu) = b\mu\omega = \omega' \circ \theta'(\mu) \circ \lambda$, hence (5.1.2). As another application of this idea, we can prove

(5.1.3) *If E is defined over a field k, every element of* End(E) *is rational over kK.*

To show this, observe that we can take ω rational over k. Let $\sigma \in \mathrm{Aut}(\boldsymbol{C}/kK)$, $\mu \in K$, $\theta(\mu) \in \mathrm{End}(E)$. Since E, ω, and μ are invariant under σ, we have $\omega \circ \theta(\mu)^\sigma = (\omega \circ \theta(\mu))^\sigma = (\mu\omega)^\sigma = \mu\omega = \omega \circ \theta(\mu)$ (see Appendix N° 8), so that $\theta(\mu)^\sigma = \theta(\mu)$. This implies that $\theta(\mu)$ is rational over kK.

The couple (E, θ) and K being as above, suppose now that E is defined by

$$y^2 = 4x^3 - c_2 x - c_3$$

with c_2 and c_3 in an algebraic number field k of finite degree, containing K. (In view of Th. 4.14, we can always find such a model E among a given isomorphism class of curves.) Take a prime ideal $\mathfrak{p}$ in k, prime to 2 and 3, for which E has good reduction modulo $\mathfrak{p}$.[5] By this we mean that c_2 and c_3 are $\mathfrak{p}$-integers, and $c_2^3 - 27c_3^2$ is a $\mathfrak{p}$-unit. Then E modulo $\mathfrak{p}$ is, by definition, an elliptic curve

$$y^2 = 4x^3 - \tilde{c}_2 x - \tilde{c}_3,$$

where the tilde means the residue class modulo $\mathfrak{p}$. We denote this curve by $\mathfrak{p}(E)$, or $\tilde{E}$ when $\mathfrak{p}$ is fixed. Obviously $j(\tilde{E})$ is the residue class of $j(E)$ modulo $\mathfrak{p}$. For a point t on E rational over k, we can define $\mathfrak{p}(t) = \tilde{t} = (t$ modulo $\mathfrak{p})$ as a point on $\tilde{E}$ in a natural way. It can be shown that $t \mapsto \mathfrak{p}(t)$ is a homomorphism. Furthermore, we have

(5.1.4) *If $\mathfrak{p}(t) = 0$ and $Nt = 0$ with an integer N prime to $\mathfrak{p}$, then $t = 0$.*

An elementary proof is given in Lutz, "Sur l'équation $y^2 = x^3 - Ax - B$ dans les corps $\mathfrak{p}$-adiques," J. Reine Angew. Math. 177 (1937), 238-247. See also [81, § 11, Prop. 13], where a corresponding fact for higher dimensional abelian varieties is proved.

Now consider another elliptic curve E' defined over k which has also good reduction modulo $\mathfrak{p}$. Let λ be an element of Hom(E, E') rational over k.

---

5) As to the general theory of reduction modulo $\mathfrak{p}$ of algebraic varieties, especially abelian varieties, see Shimura [69], Shimura and Taniyama [81, Ch. III]. Néron [53] established a model of an abelian variety with the best behavior for reduction modulo $\mathfrak{p}$. For further study of this topic, especially a criterion for good reduction, see Serre and Tate [66].

Then we can define $\tilde{\lambda}=\mathfrak{p}(\lambda)$ in a natural way as an element of $\mathrm{Hom}(\tilde{E}', \tilde{E})$. It can be shown that $\lambda\mapsto\mathfrak{p}(\lambda)$ defines an injective homomorphism of $\mathrm{Hom}(E', E)$ into $\mathrm{Hom}(\tilde{E}', \tilde{E})$, and $\deg(\tilde{\lambda})=\deg(\lambda)$ (see [81, § 11.1, p. 94, Prop. 12]). Especially, when $E=E'$, we obtain an injective ring-homomorphism of $\mathrm{End}(E)$ into $\mathrm{End}(\tilde{E})$. Therefore we can define an injective map

$$\tilde{\theta}:\ K\to\mathrm{End}_{\boldsymbol{Q}}(\tilde{E})$$

by $\tilde{\theta}(\mu)=\mathfrak{p}(\theta(\mu))$ for $\mu\in K$, $\theta(\mu)\in\mathrm{End}(E)$. The image $\tilde{\theta}(K)$ does not necessarily coincide with $\mathrm{End}_{\boldsymbol{Q}}(\tilde{E})$. We have, however, the following assertion:

(5.1.5) *Every element of* $\mathrm{End}_{\boldsymbol{Q}}(\tilde{E})$ *commuting with all the elements of* $\tilde{\theta}(K)$ *belongs to* $\tilde{\theta}(K)$, *i.e., the commutor of* $\tilde{\theta}(K)$ *in* $\mathrm{End}_{\boldsymbol{Q}}(\tilde{E})$ *is* $\tilde{\theta}(K)$.

This follows immediately from the fact that $\mathrm{End}_{\boldsymbol{Q}}(\tilde{E})$ is either a quadratic field or a quaternion algebra over $\boldsymbol{Q}$. Another way is to consider an l-adic representation of $\mathrm{End}_{\boldsymbol{Q}}(E)$; this method is applicable to the higher dimensional case, see [81, § 5.1, Prop. 1].

If $\omega=dx/y$, we can define $\mathfrak{p}(\omega)=\tilde{\omega}$ in a natural way as a differential form on $\tilde{E}$, different from 0. If c is a $\mathfrak{p}$-integer, we put $\mathfrak{p}(c\omega)=\tilde{c}\tilde{\omega}$. We can then verify the formula $\mathfrak{p}(\omega\circ\lambda)=\tilde{\omega}\circ\tilde{\lambda}$ for every $\lambda\in\mathrm{Hom}(E', E)$ rational over k. (See [81, § 10.4].)

5.2. Class field theory in the adelic language

Before going further with (E, θ), we recall some elementary properties of the idele group of an algebraic number field and fundamental facts of class field theory.[6]

For an algebraic number field K of finite degree, we denote by $K_A^\times$ the idele group of K, by $K_\infty^\times$ the archimedean part of $K_A^\times$, and by $K_{\infty+}^\times$ the connected component of the identity element of $K_\infty^\times$. Further we denote by K_{ab} the maximal abelian extension of K. Then there is a canonical exact sequence

(5.2.1) $$1\longrightarrow\overline{K^\times K_{\infty+}^\times}\longrightarrow K_A^\times\longrightarrow\mathrm{Gal}(K_{ab}/K)\longrightarrow 1,$$

where $\overline{K^\times K_{\infty+}^\times}$ denotes the closure of $K^\times K_{\infty+}^\times$.[7] We shall denote by $[s, K]$ the element of $\mathrm{Gal}(K_{ab}/K)$ corresponding to an element s of $K_A^\times$. For an element x of $K_A^\times$ and for a finite prime $\mathfrak{p}$ of K, we denote by $x_\mathfrak{p}$ the $\mathfrak{p}$-component of x. Then we define a fractional ideal $il(x)$ in K by $il(x)_\mathfrak{p}=x_\mathfrak{p}\mathfrak{o}_\mathfrak{p}$ for all $\mathfrak{p}$, where $\mathfrak{o}_\mathfrak{p}$

---

6) As for these, we refer the reader to Cassels and Fröhlich [6] and Weil [99]. We follow, for the most part, the notation of the latter.

7) One can easily verify that, if K is either $\boldsymbol{Q}$ or an imaginary quadratic field, then $K^\times K_{\infty+}^\times$ itself is closed. This is because in both cases the group of units of K is finite.

denotes the maximal compact subring of the completion $K_\mathfrak{p}$ of K at $\mathfrak{p}$.

Put

$$U(1)=\{x\in K_A^\times \mid x_\mathfrak{p}\in \mathfrak{o}_\mathfrak{p}^\times \text{ for all finite primes } \mathfrak{p} \text{ of } K\}\,,$$

and for every integral ideal $\mathfrak{c}$ in K,

$$W(\mathfrak{c})=\{x\in K_A^\times \mid x_\mathfrak{p}-1\in \mathfrak{c}\mathfrak{o}_\mathfrak{p} \text{ for all } \mathfrak{p} \text{ dividing } \mathfrak{c}\}\,,$$

$$U(\mathfrak{c})=U(1)\cap W(\mathfrak{c})\,.$$

Since $K^\times U(\mathfrak{c})$ is an open subgroup of $K_A^\times$ containing $K^\times K_{\infty+}^\times$, there is a finite abelian extension $F_\mathfrak{c}$ of K characterized by

$$F_\mathfrak{c}=\{a\in K_{ab}\mid a^{[s,K]}=a \text{ for all } s\in U(\mathfrak{c})\}\,.$$

We call $F_\mathfrak{c}$ *the maximal ray class field modulo* $\mathfrak{c}$ *over* K. It is the maximal one among the class fields whose conductors divide $\mathfrak{c}$. Let $u\in W(\mathfrak{c})$. Then $il(u)$ is prime to $\mathfrak{c}$, and $[u, K]$ coincides with the Artin symbol $\left(\frac{F_\mathfrak{c}/K}{il(u)}\right)$ on $F_\mathfrak{c}$. In particular, if $\mathfrak{q}$ is a prime ideal in K prime to $\mathfrak{c}$, and if $u_\mathfrak{q}$ is a prime element of $\mathfrak{o}_\mathfrak{q}$ and $u_\mathfrak{p}=1$ for all finite $\mathfrak{p}\neq\mathfrak{q}$, then $[u, K]$ induces the Frobenius element of $\mathrm{Gal}(F_\mathfrak{c}/K)$ for $\mathfrak{q}$.

Let $\mathfrak{a}$ be an arbitrary $\boldsymbol{Z}$-lattice in K, which is not necessarily a fractional ideal. For each rational prime p, put $K_p=K\otimes_{\boldsymbol{Q}}\boldsymbol{Q}_p$, and $\mathfrak{a}_p=\mathfrak{a}\otimes_{\boldsymbol{Z}}\boldsymbol{Z}_p$. Then $\mathfrak{a}_p$ is a $\boldsymbol{Z}_p$-lattice in K_p. For every $x\in K_A^\times$, we can speak of the p-component x_p of x, belonging to $K_p^\times$, since $K_A=K\otimes_{\boldsymbol{Q}}\boldsymbol{A}$. Observe that $x_p\mathfrak{a}_p$ is a $\boldsymbol{Z}_p$-lattice in K_p. By a well-known principle, there exists a $\boldsymbol{Z}$-lattice $\mathfrak{b}$ in K such that $\mathfrak{b}_p=x_p\mathfrak{a}_p$ for all p. We denote $\mathfrak{b}$ simply by $x\mathfrak{a}$. In other words, $x\mathfrak{a}$ is a unique $\boldsymbol{Z}$-lattice in K characterized by the property $(x\mathfrak{a})_p=x_p\mathfrak{a}_p$ for all p. We can now associate with x an isomorphism of $K/\mathfrak{a}$ onto $K/x\mathfrak{a}$. To do this, first observe that $K/\mathfrak{a}$ is canonically isomorphic to the direct sum of $K_p/\mathfrak{a}_p$ for all p. (In fact, $\boldsymbol{Q}/\boldsymbol{Z}$ is the direct sum of $\boldsymbol{Q}_p/\boldsymbol{Z}_p$ for all p, and $K/\mathfrak{a}$ is isomorphic to $\boldsymbol{Q}^2/\boldsymbol{Z}^2$.) Then multiplication by x_p defines an isomorphism of $K_p/\mathfrak{a}_p$ onto $K_p/x_p\mathfrak{a}_p$. Combining these isomorphisms together for all p, we obtain an isomorphism of $K/\mathfrak{a}$ onto $K/x\mathfrak{a}$. We shall denote by xw the image of an element w of $K/\mathfrak{a}$ by this isomorphism. The situation is explained by the commutative diagram

(5.2.2)
$$\begin{array}{ccc} K_p/\mathfrak{a}_p & \xrightarrow{\quad x_p \quad} & K_p/x_p\mathfrak{a}_p \\ \downarrow & & \downarrow \\ K/\mathfrak{a} & \xrightarrow{\quad x \quad} & K/x\mathfrak{a} \end{array}$$

where the vertical arrows are canonical injections. In other words, if $u\in K$, we take an element v of K such that $v \equiv x_p u \bmod x_p\mathfrak{a}_p$ for all p, and put

$$x \cdot (u \bmod \mathfrak{a}) = v \bmod x\mathfrak{a}.$$

We shall write this element also $xu \bmod x\mathfrak{a}$. Although xu itself is meaningless, the notation may be justified, since the p-component of $x \cdot (u \bmod \mathfrak{a})$ in $K_p/x_p\mathfrak{a}_p$ is exactly $x_p u \bmod x_p\mathfrak{a}_p$. It should be remembered that we have been discussing the localization with respect to *rational primes*. However, *if* $\mathfrak{a}$ *is a fractional ideal,* $K/\mathfrak{a}$ is canonically isomorphic to the direct sum of the modules $K_\mathfrak{p}/\mathfrak{a}_\mathfrak{p}$ for all prime ideals $\mathfrak{p}$ in K. Therefore we can define, in such a case, the above homomorphism of $K/\mathfrak{a}$ to $K/x\mathfrak{a}$ by means of the commutative diagram similar to (5.2.2) with prime ideals $\mathfrak{p}$ in place of rational primes p.

5.3. Main theorem of complex multiplication of elliptic curves

Let us come back to a normalized couple (E, θ) and an imaginary quadratic field K. By Prop. 4.8, we can find a $\boldsymbol{Z}$-lattice $\mathfrak{a}$ in K so that $\boldsymbol{C}/\mathfrak{a}$ is isomorphic to E. Fix an isomorphism ξ of $\boldsymbol{C}/\mathfrak{a}$ to E. Since θ is normalized, we have $\xi(\alpha v) = \theta(\alpha)(\xi(v))$ for any α in K satisfying $\alpha\mathfrak{a} \subset \mathfrak{a}$. Observe that $\xi(K/\mathfrak{a})$ is the set of all points of E of finite order. We are now ready to state the main theorem of complex multiplication.

THEOREM 5.4.[8] *Let K, (E, θ), $\mathfrak{a}$, and ξ be as above. Let σ be an automorphism of $\boldsymbol{C}$ over K, and s an element of $K_A^\times$ such that $\sigma = [s, K]$ on K_{ab}. Then there is an isomorphism*

$$\xi' : \ \boldsymbol{C}/s^{-1}\mathfrak{a} \longrightarrow E^\sigma$$

such that $\xi(u)^\sigma = \xi'(s^{-1}u)$ for every $u \in K/\mathfrak{a}$, i.e., the following diagram is commutative.

$$\begin{array}{ccc} K/\mathfrak{a} & \xrightarrow{\ \xi\ } & E \\ {\scriptstyle s^{-1}}\big\downarrow & & \big\downarrow{\scriptstyle \sigma} \\ K/s^{-1}\mathfrak{a} & \xrightarrow{\ \xi'\ } & E^\sigma \end{array}$$

Obviously ξ' is uniquely determined by the above property, once ξ is fixed.

PROOF. In the above statement, we have not assumed that E is defined over an algebraic number field. Actually if we can prove the theorem for a

8) This theorem was originally given (in the lectures at Princeton University) in terms of a finite number of points on E, as in [80, Th. 4.3]. The present formulation for all points of E has been suggested by A. Robert.

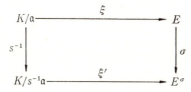

curve which is isomorphic to E (whether or not defined over an algebraic number field), then we can easily derive from it our assertion for E. Therefore it is sufficient to prove our assertion for a specially chosen curve in a given isomorphism class of elliptic curves.

In the next place, let us reduce the proof to the case $\mathrm{End}(E)=\theta(\mathfrak{o}_K)$ with the maximal order $\mathfrak{o}_K$ in K. Take any fractional ideal $\mathfrak{b}$ in K contained in $\mathfrak{a}$, and let E_1 be an elliptic curve with an isomorphism $\xi_1: \boldsymbol{C}/\mathfrak{b}\to E_1$. Let $\lambda: E_1\to E$ be the isogeny for which the diagram

$$\begin{array}{ccccc} \boldsymbol{C} & \longrightarrow & \boldsymbol{C}/\mathfrak{b} & \xrightarrow{\ \xi_1\ } & E_1 \\ {\scriptstyle \mathrm{id.}}\downarrow & & \downarrow & & \downarrow{\scriptstyle \lambda} \\ \boldsymbol{C} & \longrightarrow & \boldsymbol{C}/\mathfrak{a} & \xrightarrow{\ \xi\ } & E \end{array}$$

is commutative. Assuming our assertion to be true for E_1, we obtain an isomorphism $\xi_1': \boldsymbol{C}/s^{-1}\mathfrak{b}\to E_1^\sigma$ and a commutative diagram:

$$\begin{array}{ccc} K/\mathfrak{b} & \xrightarrow{\ \xi_1\ } & E_1 \\ {\scriptstyle s^{-1}}\downarrow & & \downarrow{\scriptstyle \sigma} \\ K/s^{-1}\mathfrak{b} & \xrightarrow{\ \xi_1'\ } & E_1^\sigma \end{array}$$

Now we have $\mathrm{Ker}(\lambda)=\xi_1(\mathfrak{a}/\mathfrak{b})$, so that $\mathrm{Ker}(\lambda^\sigma)=\mathrm{Ker}(\lambda)^\sigma=\xi_1(\mathfrak{a}/\mathfrak{b})^\sigma=\xi_1'(s^{-1}\mathfrak{a}/s^{-1}\mathfrak{b})$. Since $s^{-1}\mathfrak{b}\subset s^{-1}\mathfrak{a}$, we can find an elliptic curve E' and an isogeny λ' of E_1^σ to E' such that the diagram

$$\begin{array}{ccccc} \boldsymbol{C} & \longrightarrow & \boldsymbol{C}/s^{-1}\mathfrak{b} & \xrightarrow{\ \xi_1'\ } & E_1^\sigma \\ {\scriptstyle \mathrm{id.}}\downarrow & & \downarrow & & \downarrow{\scriptstyle \lambda'} \\ \boldsymbol{C} & \longrightarrow & \boldsymbol{C}/s^{-1}\mathfrak{a} & \xrightarrow{\ \eta\ } & E' \end{array}$$

is commutative. Then $\mathrm{Ker}(\lambda')=\xi_1'(s^{-1}\mathfrak{a}/s^{-1}\mathfrak{b})=\mathrm{Ker}(\lambda^\sigma)$. Therefore we can find an isomorphism ε of E' to E^σ so that $\varepsilon\circ\lambda'=\lambda^\sigma$. Putting $\xi'=\varepsilon\circ\eta$, we obtain a commutative diagram:

$$\begin{array}{ccccc} \boldsymbol{C} & \longrightarrow & \boldsymbol{C}/s^{-1}\mathfrak{b} & \xrightarrow{\ \xi_1'\ } & E_1^\sigma \\ {\scriptstyle \mathrm{id.}}\downarrow & & \downarrow & & \downarrow{\scriptstyle \lambda^\sigma} \\ \boldsymbol{C} & \longrightarrow & \boldsymbol{C}/s^{-1}\mathfrak{a} & \xrightarrow{\ \xi'\ } & E^\sigma \end{array}$$

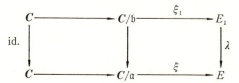

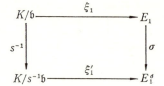

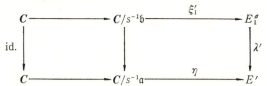

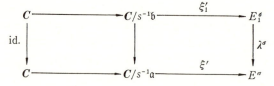

Then we have, for $u\in K$,

$$\begin{aligned}\xi(u \bmod \mathfrak{a})^\sigma &= \lambda^\sigma(\xi_1(u \bmod \mathfrak{b})^\sigma)\\ &= \lambda^\sigma(\xi_1'(s^{-1}u \bmod s^{-1}\mathfrak{b})) = \xi'(s^{-1}u \bmod \mathfrak{a}),\end{aligned}$$

which proves our assertion for E.

Thus we may assume that $\mathfrak{a}$ is a fractional ideal in K, so that $\theta(\mathfrak{o}_K) = \mathrm{End}(E)$. Furthermore, as is remarked at the beginning of our proof, we may take E to be defined over $\boldsymbol{Q}(j_E)$. Now let h be the class number of K, and $\{j_1, \cdots, j_h\}$ be the set of all invariants of elliptic curves whose endomorphism rings are isomorphic to $\mathfrak{o}_K$ (see Prop. 4.10). For each j_i, we take an elliptic curve E_i such that $j(E_i)=j_i$ defined over $\boldsymbol{Q}(j_i)$ (see § 4.1). We put $E=E_1$. Take any positive integer $m>2$, which we shall make large afterwards. Since $\mathfrak{o}_K^\times$ is a finite group, if $\zeta\in\mathfrak{o}_K^\times$ and $\zeta\equiv 1 \bmod m\mathfrak{o}_K$, then $\zeta=1$. Define an abelian extension F_m of K as in § 5.2 with $m\mathfrak{o}_K$ as $\mathfrak{c}$. We can find a finite Galois extension L of K so that $F_m\subset L$, $j_1, \cdots, j_h\in L$, and every point of order m on E is rational over L. Further, for the given automorphism σ of $\boldsymbol{C}$ over K, we take a prime ideal $\mathfrak{P}$ in L so that the following conditions (i—v) are satisfied:

(i) *The restriction of* σ *to* L *is a Frobenius element of* $\mathrm{Gal}(L/K)$ *for* $\mathfrak{P}$.

(ii) *If* $\mathfrak{p}=\mathfrak{P}\cap K$, *then* $N(\mathfrak{p})$ *is a rational prime, and* $\mathfrak{p}$ *is unramified in* L.

(iii) $\mathfrak{P}$ *does not divide* $6m$.

(iv) *The curves* E_i^τ *have good reduction modulo* $\mathfrak{P}$ *for every* $\tau\in\mathrm{Gal}(L/K)$.

(v) *The residue classes of* $j_1, \cdots, j_h$ *modulo* $\mathfrak{P}$ *are different from each other.*

The existence of such a $\mathfrak{P}$ is ensured by the Tchebotarev density theorem. Note also that the conditions (iii—v) exclude only finitely many primes.

Put $p=N(\mathfrak{p})$. Now $\boldsymbol{C}/\mathfrak{p}^{-1}\mathfrak{a}$ is isomorphic to E_i for a unique i. Fix an isomorphism η of $\boldsymbol{C}/\mathfrak{p}^{-1}\mathfrak{a}$ to E_i. Take an integral ideal $\mathfrak{x}$ in K prime to p so that $\mathfrak{x}\mathfrak{p}=\alpha\mathfrak{o}_K$ with $\alpha\in\mathfrak{o}_K$. Since $\mathfrak{a}\subset\mathfrak{p}^{-1}\mathfrak{a}$ and $\alpha\mathfrak{p}^{-1}\mathfrak{a}\subset\mathfrak{a}$, we obtain a commutative diagram

$$(*)\qquad \begin{array}{ccccc} \boldsymbol{C} & \longrightarrow & \boldsymbol{C}/\mathfrak{a} & \xrightarrow{\ \xi\ } & E \\ {\scriptstyle \mathrm{id.}}\downarrow & & \downarrow & & \downarrow{\scriptstyle \lambda} \\ \boldsymbol{C} & \longrightarrow & \boldsymbol{C}/\mathfrak{p}^{-1}\mathfrak{a} & \xrightarrow{\ \eta\ } & E_i \\ {\scriptstyle \alpha}\downarrow & & \downarrow & & \downarrow{\scriptstyle \mu} \\ \boldsymbol{C} & \longrightarrow & \boldsymbol{C}/\mathfrak{a} & \xrightarrow{\ \xi\ } & E \end{array}$$

with isogenies λ and μ. Then we have $\mu\circ\lambda=\theta(\alpha)$. By Prop. 5.3, λ and μ are defined over a finite algebraic extension L' of L. Take a prime ideal $\mathfrak{O}$

in L' which divides $\mathfrak{P}$. We shall now consider reduction modulo $\mathfrak{Q}$ and indicate the reduced objects by putting tildes (see § 5.1). Take a holomorphic differential form ω on E rational over L for which $\tilde{\omega} \neq 0$, as mentioned in § 5.1. Then $\tilde{\omega} \circ \tilde{\mu} \circ \tilde{\lambda} = \mathfrak{Q}(\omega \circ \theta(\alpha)) = \mathfrak{Q}(\alpha\omega) = \tilde{\alpha}\tilde{\omega} = 0$, since $\alpha \in \mathfrak{p}$. It follows that the isogeny $\tilde{\mu} \circ \tilde{\lambda}$ is inseparable, by Prop. 5.1. The above diagram shows that $\operatorname{Ker}(\mu) = \eta(\alpha^{-1}\mathfrak{a}/\mathfrak{p}^{-1}\mathfrak{a}) = \eta(\mathfrak{x}^{-1}\mathfrak{p}^{-1}\mathfrak{a}/\mathfrak{p}^{-1}\mathfrak{a})$, which is of order $N(\mathfrak{x})$. Since $\mathfrak{x}$ is prime to p, this shows that $\tilde{\mu}$ is separable, hence $\tilde{\lambda}$ must be inseparable. Since $\operatorname{Ker}(\lambda) = \xi(\mathfrak{p}^{-1}\mathfrak{a}/\mathfrak{a})$ is of order $N(\mathfrak{p}) = p$, we see that $\deg(\tilde{\lambda}) = \deg(\lambda) = p$, and $\tilde{\lambda}$ is purely inseparable. Let φ denote the p-th power automorphism of the universal domain of characteristic p, and π the p-th power morphism of $\tilde{E}$ to $\tilde{E}^{\varphi}$. By Prop. 5.2, there is an isomorphism ε of $\tilde{E}_i$ to $\tilde{E}^{\varphi}$ such that $\varepsilon \circ \tilde{\lambda} = \pi$. It follows especially that $\tilde{E}_i$ and $\tilde{E}^{\varphi}$ have the same invariant. Therefore we have $\tilde{j}_i = \tilde{j}^p = \mathfrak{P}(j^{\sigma})$, in view of the condition (i). Now both j_i and j^{σ} belong to $\{j_1, \cdots, j_h\}$. By (v), we have $j_i = j^{\sigma}$, so that E_i is isomorphic to E^{σ}. Therefore we can replace E_i by E^{σ} in the above diagram, and repeat the above reasoning, (possibly changing L' and $\mathfrak{Q}$). Since $\mathfrak{P}(E^{\sigma}) = \tilde{E}^{\varphi}$, both $\tilde{\lambda}$ and π are isogenies of $\tilde{E}$ to $\tilde{E}^{\varphi}$, so that ε is an automorphism of $\tilde{E}^{\varphi}$. Since $\sigma = \text{id.}$ on K, we have $\omega^{\sigma} \circ \theta(a)^{\sigma} = (\omega \circ \theta(a))^{\sigma} = (a\omega)^{\sigma} = a\omega^{\sigma}$ for every $a \in \mathfrak{o}_K$, so that $(E^{\sigma}, \theta^{\sigma})$ is normalized in the sense of § 5.1. Therefore, by (5.1.2), we have $\lambda \circ \theta(a) = \theta^{\sigma}(a) \circ \lambda$ so that $\tilde{\lambda} \circ \tilde{\theta}(a) = \tilde{\theta}(a)^{\varphi} \circ \tilde{\lambda}$ for all $a \in \mathfrak{o}_K$. Now the isogeny π has the same property $\pi \circ \tilde{\theta}(a) = \tilde{\theta}(a)^{\varphi} \circ \pi$ (see Appendix (7.1)), hence $\varepsilon \circ \tilde{\theta}(a)^{\varphi} = \tilde{\theta}(a)^{\varphi} \circ \varepsilon$ for all $a \in \mathfrak{o}_K$. By (5.1.5), $\varepsilon = \tilde{\theta}(\gamma)^{\varphi}$ with an element γ of $\mathfrak{o}_K$, which must be a unit of $\mathfrak{o}_K$, since ε is an automorphism. Put $\kappa = \theta(\gamma)^{\sigma} \circ \lambda$, $\xi^* = \theta(\gamma)^{\sigma} \circ \eta$. Then κ is an isogeny of E to E^{σ}, and $\tilde{\kappa} = \pi$. Now by replacing λ, η by κ, ξ^*, the upper part of our diagram $(*)$ becomes as follows:

$$(**) \qquad \begin{array}{ccccc} \boldsymbol{C} & \longrightarrow & \boldsymbol{C}/\mathfrak{a} & \xrightarrow{\ \xi\ } & E \\ {\scriptstyle \text{id.}}\downarrow & & \downarrow & & \downarrow{\scriptstyle \kappa} \\ \boldsymbol{C} & \longrightarrow & \boldsymbol{C}/\mathfrak{p}^{-1}\mathfrak{a} & \xrightarrow{\ \xi^*\ } & E^{\sigma} \end{array}$$

This is still commutative.

Let t be an element of E such that $mt = 0$. Then $\mathfrak{P}(t^{\sigma}) = \pi\tilde{t} = \mathfrak{Q}(\kappa t)$. Since m is prime to p, we have $t^{\sigma} = \kappa t$ by (5.1.4). For $u \in m^{-1}\mathfrak{a}$, put $u_1 = u \bmod \mathfrak{a}$, and $u_2 = u \bmod \mathfrak{p}^{-1}\mathfrak{a}$. Then $\xi(u_1)^{\sigma} = \kappa(\xi(u_1)) = \xi^*(u_2)$. Now let c be an element of $K_A^{\times}$ such that $c_{\mathfrak{p}}$ is a prime element of $K_{\mathfrak{p}}$ and $c_{\mathfrak{q}} = 1$ for all $\mathfrak{q} \neq \mathfrak{p}$. Then the restriction of σ to F_m is $[s, K] = [c, K]$, so that $c = sde$ with some $d \in K^{\times}$ and $e \in U(m\mathfrak{o}_K)$, where $U(m\mathfrak{o}_K)$ is as in § 5.2 (with $m\mathfrak{o}_K$ as $\mathfrak{c}$). Since $\mathfrak{p}^{-1}\mathfrak{a} = c^{-1}\mathfrak{a} = d^{-1}s^{-1}\mathfrak{a}$, we can extend $(**)$ to a commutative diagram

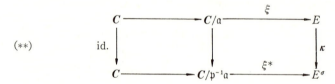

$$
(***)\qquad
\begin{array}{ccccc}
\boldsymbol{C} & \longrightarrow & \boldsymbol{C}/\mathfrak{a} & \xrightarrow{\ \xi\ } & E \\
{\scriptstyle \mathrm{id.}}\downarrow & & \downarrow & & \downarrow{\scriptstyle \kappa} \\
\boldsymbol{C} & \longrightarrow & \boldsymbol{C}/\mathfrak{p}^{-1}\mathfrak{a} & \xrightarrow{\ \xi^*\ } & E^{\sigma} \\
{\scriptstyle d}\downarrow & & \downarrow & & \downarrow{\scriptstyle \mathrm{id.}} \\
\boldsymbol{C} & \longrightarrow & \boldsymbol{C}/s^{-1}\mathfrak{a} & \xrightarrow{\ \xi'\ } & E^{\sigma}
\end{array}
$$

with a suitable choice of an isomorphism ξ'. Then, for u, u_1, and u_2 as above, we have $\xi(u_1)^\sigma = \xi^*(u_2) = \xi'$ $(du \bmod s^{-1}\mathfrak{a})$. We have $mu \in \mathfrak{a}$, $e \in U(m\mathfrak{o}_K)$, and $d = s^{-1}ce^{-1}$. Let $\mathfrak{q}$ be a prime ideal in K. If $\mathfrak{q} \neq \mathfrak{p}$, we have $c_\mathfrak{q} = 1$, so that

$$du = s_\mathfrak{q}^{-1}e_\mathfrak{q}^{-1}u \equiv s_\mathfrak{q}^{-1}u \mod s_\mathfrak{q}^{-1}\mathfrak{a}_\mathfrak{q}\ ;$$

if $\mathfrak{q} = \mathfrak{p}$, we have $u \in \mathfrak{a}_\mathfrak{p}$, so that

$$du = s_\mathfrak{p}^{-1}c_\mathfrak{p}e_\mathfrak{p}^{-1}u \in s_\mathfrak{p}^{-1}c_\mathfrak{p}\mathfrak{a}_\mathfrak{p} = s_\mathfrak{p}^{-1}(\mathfrak{p}\mathfrak{a})_\mathfrak{p}\,.$$

These relations imply that

$$du \bmod s^{-1}\mathfrak{a} = s^{-1}u \bmod s^{-1}\mathfrak{a}\,.$$

Therefore we obtain

$$\xi(u \bmod \mathfrak{a})^\sigma = \xi'(s^{-1}u \bmod s^{-1}\mathfrak{a})$$

for every $u \in m^{-1}\mathfrak{a}$.

Now taking any multiple n of m in place of m, we obtain an isomorphism ξ'' of $\boldsymbol{C}/s^{-1}\mathfrak{a}$ to E^σ such that $\xi(v)^\sigma = \xi''(s^{-1}v)$ for every $v \in n^{-1}\mathfrak{a}/\mathfrak{a}$. Since $\xi'' \circ \xi'^{-1}$ is an automorphism of E^σ, we have $\xi'' = \theta^\sigma(\zeta)\circ\xi'$ with a unit ζ of $\mathfrak{o}_K$ satisfying $\zeta\mathfrak{a} \subset \mathfrak{a}$. Then, for every $v \in m^{-1}\mathfrak{a}/\mathfrak{a}$, we have

$$\xi(\zeta v)^\sigma = \theta(\zeta)^\sigma(\xi(v)^\sigma) = \theta(\zeta)^\sigma(\xi'(s^{-1}v)) = \xi''(s^{-1}v) = \xi(v)^\sigma\,,$$

so that $\zeta v = v$ for every $v \in m^{-1}\mathfrak{a}/\mathfrak{a}$. It follows that $\zeta \equiv 1 \bmod m\mathfrak{o}_K$. Since $m > 2$, as is remarked above, we have $\zeta = 1$, hence $\xi' = \xi''$. This implies that $\xi(v)^\sigma = \xi'(s^{-1}v)$ for every $v \in n^{-1}\mathfrak{a}/\mathfrak{a}$, for every multiple n of m. Thus ξ' has the required property, and the proof is completed.

5.4. Construction of class fields over an imaginary quadratic field

Let us now derive from the above theorem a few classical results of complex multiplication due to Kronecker, Weber, Takagi, and Hasse. First let us denote by $j(\mathfrak{a})$ the invariant of an elliptic curve isomorphic to $\boldsymbol{C}/\mathfrak{a}$ for a $\boldsymbol{Z}$-lattice $\mathfrak{a}$ in K. Then σ and s being as in Th. 5.4, we have $j(\mathfrak{a})^\sigma = j(s^{-1}\mathfrak{a})$. This means that $j(\mathfrak{a})^\sigma$ depends only on the restriction of σ to K_{ab}. Thus we obtain

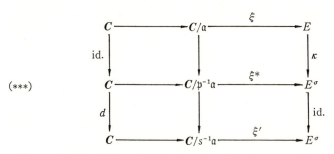

(5.4.1) *For every* $\boldsymbol{Z}$*-lattice* $\mathfrak{a}$ *in* K, *one has* $j(\mathfrak{a}) \in K_{ab}$, *and* $j(\mathfrak{a})^{[s,K]} = j(s^{-1}\mathfrak{a})$ *for all* $s \in K_A^{\times}$.

We shall now prove

(5.4.2) *For an order* $\mathfrak{o}$ *in* K, *a* $\boldsymbol{Z}$*-lattice* $\mathfrak{a}$ *is a proper* $\mathfrak{o}$*-ideal if and only if* $\mathfrak{a} = x\mathfrak{o}$ *for some* x *of* $K_A^{\times}$.

The "if"-part is obvious. To prove the converse, let c be the conductor of $\mathfrak{o}$. If $\mathfrak{a}$ is a proper $\mathfrak{o}$-ideal, there exists, by Prop. 4.11, an element μ of K such that $\mu\mathfrak{a} + c\mathfrak{o} = \mathfrak{o}$. Let p be a rational prime. If $p \nmid c$, we have $\mathfrak{o}_p = (\mathfrak{o}_K)_p$, so that $\mathfrak{a}_p$ is a principal $\mathfrak{o}_p$-ideal. If $p \mid c$, we have $c\mathfrak{o}_p \subset p\mathfrak{o}_p$, so that $\mu\mathfrak{a}_p + p\mathfrak{o}_p = \mathfrak{o}_p$. Then $\mathfrak{o}_p = \mu\mathfrak{a}_p + p(\mu\mathfrak{a}_p + p\mathfrak{o}_p) = \mu\mathfrak{a}_p + p^2\mathfrak{o}_p$. Similarly, by induction, we can show that $\mathfrak{o}_p = \mu\mathfrak{a}_p + p^m\mathfrak{o}_p$ for every positive integer m. But we can find m so that $p^m\mathfrak{o}_p \subset \mu\mathfrak{a}_p$. Therefore $\mu\mathfrak{a}_p = \mathfrak{o}_p$. Thus $\mathfrak{a}_p$ is a principal $\mathfrak{o}_p$-ideal for all p, hence (5.4.2).

THEOREM 5.5. *Let* K, E, $\mathfrak{a}$, *and* ξ *be as in Th.* 5.4, *and* h_E^i *the function on* E *defined in* § 4.5. *Let* u *be an element of* $K/\mathfrak{a}$, *and*

$$W = \{s \in K_A^{\times} \mid s\mathfrak{a} = \mathfrak{a},\ su = u\}.$$

Suppose that E *belongs to* $\mathcal{E}_i$. *Then the field* $K(j_E, h_E^i(\xi(u)))$ *is the subfield of* K_{ab} *corresponding to the subgroup* $K^{\times}W$ *of* $K_A^{\times}$.

PROOF. Observe that W is an open subgroup of $K_A^{\times}$ containing $K_\infty^{\times}$. Let F denote the subfield of K_{ab} corresponding to $K^{\times}W$. Let $\sigma \in \mathrm{Aut}(\boldsymbol{C}/K)$. Take $s \in K_A^{\times}$ so that $\sigma = [s, K]$ on K_{ab}, and take ξ' as in Th. 5.4. Put $t = \xi(u)$.

(I) Suppose that σ is the identity map on F. Then we can take s from W, so that $s\mathfrak{a} = \mathfrak{a}$. It follows that E^σ is isomorphic to E, hence $j_E^\sigma = j_E$. Further we can find an isomorphism ε of E^σ to E so that $\varepsilon \circ \xi' = \xi$. By (4.5.4), we have $h_E^i(\varepsilon t^\sigma) = h_{E^\sigma}^i(t^\sigma) = h_E^i(t)^\sigma$. Since $\varepsilon t^\sigma = \varepsilon(\xi(u)^\sigma) = \varepsilon(\xi'(s^{-1}u)) = \xi(u) = t$, we have $h_E^i(t) = h_E^i(t)^\sigma$. This means that σ is the identity map on $K(j_E, h_E^i(t))$, hence $K(j_E, h_E^i(t)) \subset F$.

(II) Conversely, suppose that $\sigma = \mathrm{id.}$ on $K(j_E, h_E^i(t))$. Then $j(E) = j(E)^\sigma = j(E^\sigma)$, so that there exists an isomorphism δ of E^σ to E. By Prop. 4.8, there exists an element μ of $K^{\times}$ such that $\mu s^{-1}\mathfrak{a} = \mathfrak{a}$. Choosing δ suitably, we obtain a commutative diagram:

$$\begin{array}{ccccc}
\boldsymbol{C} & \longrightarrow & \boldsymbol{C}/s^{-1}\mathfrak{a} & \xrightarrow{\ \xi'\ } & E^\sigma \\
\Big\downarrow \mu & & \Big\downarrow & & \Big\downarrow \delta \\
\boldsymbol{C} & \longrightarrow & \boldsymbol{C}/\mathfrak{a} & \xrightarrow{\ \xi\ } & E
\end{array}$$

By (4.5.4), we have $h_E^i(\delta t^\sigma)=h_{E^\sigma}^i(t^\sigma)=h_E^i(t)^\sigma=h_E^i(t)$. Hence, by (4.5.3), there exists an element ζ of K such that $\zeta\mathfrak{a}=\mathfrak{a}$ and $\theta(\zeta)\delta t^\sigma=t$. On the other hand, $\delta t^\sigma=\delta(\xi(u)^\sigma)=\delta(\xi'(s^{-1}u))=\xi(\mu s^{-1}u)$, hence $\xi(u)=\xi(\zeta\mu s^{-1}u)$. Putting $\zeta\mu s^{-1}=s'$, we see that $s'\mathfrak{a}=\mathfrak{a}$ and $s'u=u$, hence $s'\in W$, and $s\in K^\times W$. Therefore $\sigma=\mathrm{id}.$ on F. This shows that $F\subset K(j_E, h_E^i(t))$, and our proof is completed.

COROLLARY 5.6. *Let E be an elliptic curve belonging to $\mathcal{E}_i$. Then K_{ab} is generated over K by j_E and the values $h_E^i(t)$ for all points t of finite order on E.*

This follows immediately from the easy fact that $K^\times K_\infty^\times$ (closed by itself) is the intersection of the $K^\times W$, with the subgroups W of the type described in Th. 5.5, for all choices of u.

THEOREM 5.7. *Let $\mathfrak{o}$ be an order in K, and $\mathfrak{a}$ a proper $\mathfrak{o}$-ideal. Then the following assertions hold.*

(i) *$\mathrm{Gal}(K(j(\mathfrak{a}))/K)$ is isomorphic to the group of all classes of proper $\mathfrak{o}$-ideals, through the correspondence $\sigma\mapsto\mathfrak{b}$ such that $j(\mathfrak{a})^\sigma=j(\mathfrak{b}^{-1}\mathfrak{a})$.*

(ii) $[K(j(\mathfrak{a})):K]=[\boldsymbol{Q}(j(\mathfrak{a})):\boldsymbol{Q}]$.

(iii) *If $\mathfrak{a}_1,\cdots,\mathfrak{a}_n$ are representatives for the classes of proper $\mathfrak{o}$-ideals, then $j(\mathfrak{a}_1),\cdots,j(\mathfrak{a}_n)$ form a complete set of conjugates of $j(\mathfrak{a})$ over $\boldsymbol{Q}$, and over K.*

(iv) *If $\mathfrak{o}=\mathfrak{o}_K$, and hence $\mathfrak{a}$ is a fractional ideal in K, then $K(j(\mathfrak{a}))$ is the maximal unramified abelian extension of K, and $j(\mathfrak{a})^\sigma=j(\mathfrak{b}^{-1}\mathfrak{a})$ for $\sigma=\left(\frac{K(j(\mathfrak{a}))/K}{\mathfrak{b}}\right)$, $\mathfrak{b}$ any fractional ideal in K.*

PROOF. The notation being as in Th. 5.5, put $u=0$ (or disregard u). Then $W=K_\infty^\times\cdot\prod_p\mathfrak{o}_p^\times$. On account of (5.4.2), we see easily that $K_A^\times\ni s\mapsto s\mathfrak{o}$ gives an isomorphism of $K_A^\times/K^\times W$ onto the group of classes of proper $\mathfrak{o}$-ideals. Therefore we obtain (i) from Th. 5.5 and (5.4.1). If $\mathfrak{o}=\mathfrak{o}_K$, the class field F over K corresponding to $K^\times W$ is the maximal unramified abelian extension of K. Further, if $\mathfrak{b}=s\mathfrak{o}_K$, we have $[s,K]=\left(\frac{F/K}{\mathfrak{b}}\right)$ on F. Therefore we obtain (iv). The assertion of (iii) with the basic field K follows from (i). Let E be an elliptic curve isomorphic to $\boldsymbol{C}/\mathfrak{a}$, and let $\sigma\in\mathrm{Aut}(\boldsymbol{C}/\boldsymbol{Q})$. Then $\mathrm{End}(E^\sigma)$ is isomorphic to $\mathrm{End}(E)$, and hence to $\mathfrak{o}$. By Prop. 4.8, E^σ is isomorphic to $\boldsymbol{C}/\mathfrak{a}_\nu$ for some ν. Then $j(\mathfrak{a})^\sigma=j(E^\sigma)=j(\mathfrak{a}_\nu)$. This shows that $[\boldsymbol{Q}(j(\mathfrak{a})):\boldsymbol{Q}]\leqq n=[K(j(\mathfrak{a})):K]$. Since the inequality in the opposite direction is obvious, we obtain (ii) and the assertion of (iii) over $\boldsymbol{Q}$.

Since the Fourier expansion of $j(z)$ (see (4.6.1) and Th. 2.9) has rational Fourier coefficients, we see that $j(-\bar{z})=\overline{j(z)}$ for all $z\in\mathfrak{H}$. Therefore, if $\mathfrak{a}=\boldsymbol{Z}\omega_1+\boldsymbol{Z}\omega_2$ and $\omega_1/\omega_2\in\mathfrak{H}$, we have $\bar{\mathfrak{a}}=\boldsymbol{Z}\cdot(-\bar{\omega}_1)+\boldsymbol{Z}\bar{\omega}_2$, so that $j(\bar{\mathfrak{a}})=j(-\bar{\omega}_1/\bar{\omega}_2)=\overline{j(\omega_1/\omega_2)}=\overline{j(\mathfrak{a})}$. This implies that $j(\mathfrak{a})$ is real if and only if $\mathfrak{a}$ and $\bar{\mathfrak{a}}$ belong to the same class of proper $\mathfrak{o}$-ideals. On account of (5.4.2), we can easily show that $\mathfrak{a}\bar{\mathfrak{a}}$ is a principal $\mathfrak{o}$-ideal. Therefore we obtain

(5.4.3) *Let $\mathfrak{o}$ be an order in K, and $\mathfrak{a}$ a proper $\mathfrak{o}$-ideal. Then $j(\mathfrak{a})$ is real if and only if $\mathfrak{a}^2$ is a principal $\mathfrak{o}$-ideal.*

EXERCISE 5.8. For $\mathfrak{o}$ and $\mathfrak{a}$ as above, prove that $K(j(\mathfrak{a}))$ is normal over $\boldsymbol{Q}$, and study the structure of $\mathrm{Gal}(K(j(\mathfrak{a}))/\boldsymbol{Q})$. Further prove that the following three statements are equivalent to each other: (i) $\boldsymbol{Q}(j(\mathfrak{a}))$ is normal over $\boldsymbol{Q}$; (ii) $\boldsymbol{Q}(j(\mathfrak{a}))$ is totally real; (iii) the group of all classes of proper $\mathfrak{o}$-ideals is a product of cyclic groups of order 2.

EXERCISE 5.9. Let F' be the subfield generated over K by the values $j(z)$ for all $z \in K$ such that $\mathrm{Im}(z) > 0$. Prove that F' is the subfield of K_{ab} corresponding to $\boldsymbol{Q}_A^\times K^\times K_\infty^\times$. (Observe that $\boldsymbol{Q}_A^\times K^\times K_\infty^\times = \prod_p \boldsymbol{Z}_p^\times K^\times K_\infty^\times$.)

EXERCISE 5.10. Let E be an elliptic curve belonging to $\mathcal{E}_i$ such that $\mathrm{End}(E)$ is isomorphic to the maximal order $\mathfrak{o}_K$. Prove the following assertions:

(1) *For any integral ideal $\mathfrak{c}$ in K, there exists a point t on E such that*

$$\alpha \in \mathfrak{o}_K, \quad \theta(\alpha)t = 0 \quad \Leftrightarrow \quad \alpha \in \mathfrak{c},$$

where θ is the normalized isomorphism of K onto $\mathrm{End}_{\boldsymbol{Q}}(E)$.

(2) *For any such point t, the field $K(j_E, h_E^i(t))$ is the maximal ray class field modulo $\mathfrak{c}$ over K, defined in* § 5.2.

Complex multiplication of elliptic functions may be a fascinating subject of the history of mathematics. But we refrain from making any historical comments, and mention only a few classical and modern works: Weber [89], Hasse [25], Deuring [11], [13], Ramachandra [59]. Further references can be found in these articles. In § 6.8, we shall discuss another formulation of complex multiplication in terms of modular functions of arbitrary level.

5.5. Complex multiplication of abelian varieties of higher dimension

We shall now briefly explain how the results of the previous section can be generalized to the higher dimensional case. Here we must assume that the reader is familiar with abelian varieties (over the complex number field). For the terminology and notation, see Appendix. Except for the notion of a CM-field (see below), the results of this section will be used only in § 7.8.

A. Algebraic preliminaries

In this section we denote by x^ρ the complex conjugate of a complex number x. By an *algebraic number field,* we always mean a subfield of $\boldsymbol{C}$ algebraic over $\boldsymbol{Q}$ of finite degree. By a *CM-field,* we understand a totally imaginary quadratic extension of a totally real algebraic number field.

PROPOSITION 5.11. *An algebraic number field K is a CM-field if and only if the following two conditions are satisfied.*

(1) *ρ induces a non-trivial automorphism of K.*

(2) *$\rho\tau = \tau\rho$ for every isomorphism τ of K into $\boldsymbol{C}$.*

The proof is straightforward, and left to the reader as an exercise. As an application, we obtain

PROPOSITION 5.12. *The composite of a finite number of CM-fields is a CM-field. If K is a CM-field, then every conjugate of K over $\boldsymbol{Q}$ and the smallest Galois extension of $\boldsymbol{Q}$ containing K are CM-fields.*

Let K be a *CM*-field, and Φ an absolute equivalence class of $\boldsymbol{Q}$-linear representations of K by complex matrices. We shall often denote by the same letter Φ any representation of K in the class Φ. We call (K, Φ) a *CM*-type if the following condition is satisfied:

(5.5.1) *The direct sum of Φ and its complex conjugate is the equivalence class of regular representations of K over $\boldsymbol{Q}$.*

Under this assumption, if $[K:\boldsymbol{Q}] = 2n$, Φ is the direct sum of n isomorphisms $\varphi_1, \cdots, \varphi_n$ of K into $\boldsymbol{C}$ such that

(5.5.2) *$\{\varphi_1, \cdots, \varphi_n, \varphi_1\rho, \cdots, \varphi_n\rho\}$ is the set of all isomorphisms of K into $\boldsymbol{C}$; in other words, $\varphi_1, \cdots, \varphi_n$ correspond to all distinct archimedean valuations of K.*

We write then $\Phi = \sum_{i=1}^n \varphi_i$, and

$$\det \Phi(x) = \prod_{i=1}^n x^{\varphi_i}, \qquad \operatorname{tr} \Phi(x) = \sum_{i=1}^n x^{\varphi_i} \qquad (x \in K).$$

Let us now construct another *CM*-type (K^*, Φ^*) from a given *CM*-type (K, Φ). First let K^* be the field generated by $\operatorname{tr}\Phi(x)$ over $\boldsymbol{Q}$ for all $x \in K$. Then, for any $\sigma \in \operatorname{Aut}(\bar{\boldsymbol{Q}})$, we have, by Prop. 5.11,

$$\operatorname{tr}\Phi(x)^{\sigma\rho} = \sum_{i=1}^n x^{\varphi_i\sigma\rho} = \sum_{i=1}^n x^{\rho\varphi_i\sigma} = \sum_{i=1}^n x^{\varphi_i\rho\sigma} = \operatorname{tr}\Phi(x)^{\rho\sigma},$$

so that $\sigma\rho = \rho\sigma$ on K^*, i.e., K^* satisfies (2) of Prop. 5.11. Since $\operatorname{tr}\Phi(x)^\rho = \operatorname{tr}\Phi(x^\rho)$, ρ induces an automorphism of K^*. If $\rho = \mathrm{id.}$ on K^*, we have $\operatorname{tr}\Phi(x) = \operatorname{tr}\Phi(x)^\rho$ for all $x \in K$, so that Φ is equivalent to Φ^ρ, a contradiction. Therefore, by Prop. 5.11, K^* is a *CM*-field.

Let F be the smallest Galois extension of $\boldsymbol{Q}$ containing K, and let $G = \operatorname{Gal}(F/\boldsymbol{Q})$. Denote by H (resp. H^*) the subgroup of G corresponding to K (resp. K^*). Extend φ_i to an element of G, and denote it again by φ_i. Put $S = \bigcup_{i=1}^n H\varphi_i$. Then we see easily that

$$H^* = \{\gamma \in G \mid S\gamma = S\}.$$

Therefore $S^{-1}=\{\sigma^{-1} \mid \sigma \in S\}$ is a union of cosets with respect to H^*. We have thus $S^{-1}=\bigcup_{j=1}^{m} H^*\psi_j$ with elements ψ_j of G. By Prop. 5.12, F is a *CM*-field, so that by Prop. 5.11, the restriction of ρ to F belongs to the center of G. In view of (5.5.2), we have $G=S\cup S\rho$, so that $G=S^{-1}\cup S^{-1}\rho$, which shows that $[K^*:\boldsymbol{Q}]=[G:H^*]=2m$, and $\{\psi_1, \cdots, \psi_m\}$ satisfies (5.5.2). Therefore we obtain a *CM*-type (K^*, Φ^*) with $\Phi^*=\sum_{i=1}^{m}\psi_i$. We call (K^*, Φ^*) the *reflex* of (K, Φ).[9] Since $S^{-1}\gamma=S^{-1}$ for $\gamma\in H$, we see that $\det \Phi^*(x)\in K$ for every $x\in K^*$. Consider the idele groups $K_A^\times$ and $K_A^{*\times}$ of K and K^*. Then the map

$$\det \Phi^*: \quad K^{*\times} \longrightarrow K^\times$$

can be extended to a continuous homomorphism of $K_A^{*\times}$ to $K_A^\times$. For simplicity, we put

(5.5.3) $$\eta(x)=\det \Phi^*(x) \qquad (x\in K_A^{*\times}).$$

B. Abelian varieties with many complex multiplications

Let A be an abelian variety of dimension n, defined over (a subfield of) $\boldsymbol{C}$. Take a complex torus $\boldsymbol{C}^n/L$ with a lattice L in $\boldsymbol{C}^n$, isomorphic to A, or rather, consider an exact sequence

(5.5.4) $$0 \longrightarrow L \longrightarrow \boldsymbol{C}^n \xrightarrow{\ \xi\ } A \longrightarrow 0$$

with a holomorphic map ξ. Then every element of $\mathrm{End}_{\boldsymbol{Q}}(A)$ corresponds to a $\boldsymbol{C}$-linear transformation of $\boldsymbol{C}^n$. Thus we obtain a $\boldsymbol{Q}$-linear isomorphism Φ_1 of $\mathrm{End}_{\boldsymbol{Q}}(A)$ into $M_n(\boldsymbol{C})$ by $\xi\circ\Phi_1(\lambda)=\lambda\circ\xi$ for $\lambda\in\mathrm{End}(A)$. Observe that

(5.5.5) $\Phi_1(\lambda)$ *maps* $\boldsymbol{Q}L$ *to* $\boldsymbol{Q}L$ *for every* $\lambda\in\mathrm{End}_{\boldsymbol{Q}}(A)$.

Since $\boldsymbol{R}L=\boldsymbol{C}^n$, one can easily show that

(5.5.6) *The direct sum of* Φ_1 *and its complex conjugate is equivalent to a rational representation of* $\mathrm{End}_{\boldsymbol{Q}}(A)$ (see Appendix N° 11).

Now we impose the condition that $\mathrm{End}_{\boldsymbol{Q}}(A)$ has a subalgebra isomorphic to an algebraic number field K of degree $2n$. This is a generalization of an elliptic curve with complex multiplications. It is convenient to discuss a couple (A, θ) with a fixed isomorphism θ of K into $\mathrm{End}_{\boldsymbol{Q}}(A)$ for the following two reasons: (i) there may be many isomorphisms of K into $\mathrm{End}_{\boldsymbol{Q}}(A)$; (ii) one has to deal with various A's with the same K. It can be shown that θ

9) In [81, §8.3], we called (K^*, Φ^*) the *dual* of (K, Φ). The notion of reflex can be defined for a couple (K, Φ) with any algebraic number field K and any representation class Φ. For details, see [75], [77]. A more intrinsic definition of reflex without the extension F is given in [80].

maps the identity element of K to the identity element of $\mathrm{End}_{\boldsymbol{Q}}(A)$ [81, p. 39, Prop. 1]. Put $\Phi=\Phi_1\circ\theta$. Then Φ is a $\boldsymbol{Q}$-linear isomorphism of K into $M_n(\boldsymbol{C})$, and $\Phi(1)=1_n$. Therefore, we can find n isomorphisms $\varphi_1, \cdots, \varphi_n$ of K into $\boldsymbol{C}$ such that Φ is equivalent to the direct sum of $\varphi_1, \cdots, \varphi_n$. We say that (A, θ) is of type (K, Φ) or $(K, \{\varphi_i\})$. From (5.5.6), we see that Φ satisfies (5.5.1).

In view of (5.5.5), we can consider $\boldsymbol{Q}L$ as a K-module, through Φ. Since $[\boldsymbol{Q}L:\boldsymbol{Q}]=2n=[K:\boldsymbol{Q}]$, we can find an element w of $\boldsymbol{C}^n$ such that $\boldsymbol{Q}L=\Phi(K)w$. Changing the coordinate system of $\boldsymbol{C}^n$, we may assume

(5.5.7) $$\Phi(a)=\begin{bmatrix} a^{\varphi_1} & & \\ & \ddots & \\ & & a^{\varphi_n}\end{bmatrix} \qquad (a\in K).$$

If $w=\begin{bmatrix} w_1\\ \vdots \\ w_n\end{bmatrix}$, we have

$$\boldsymbol{Q}L=\left\{\begin{bmatrix} a^{\varphi_1}w_1\\ \vdots \\ a^{\varphi_n}w_n\end{bmatrix} \,\middle|\, a\in K\right\}.$$

Since $\boldsymbol{R}L=\boldsymbol{C}^n$, none of the w_i can be 0. Therefore, changing again the coordinate system by the matrix $\begin{bmatrix} w_1 & & \\ & \ddots & \\ & & w_n\end{bmatrix}$, and putting

(5.5.8) $$u(a)=\begin{bmatrix} a^{\varphi_1}\\ \vdots \\ a^{\varphi_n}\end{bmatrix} \qquad (a\in K),$$

we see that u is an isomorphism of K onto $\boldsymbol{Q}L$, and can be extended to an $\boldsymbol{R}$-linear isomorphism of $K_{\boldsymbol{R}}=K\otimes_{\boldsymbol{Q}}\boldsymbol{R}$ onto $\boldsymbol{R}L=\boldsymbol{C}^n$, which we write again u. Put $\mathfrak{a}=u^{-1}(L)$. Then we obtain a commutative diagram

(5.5.9) $$\begin{array}{ccccccccc} 0 & \longrightarrow & \mathfrak{a} & \longrightarrow & K_{\boldsymbol{R}} & \longrightarrow & K_{\boldsymbol{R}}/\mathfrak{a} & \longrightarrow & 0 \quad \text{(exact)}\\ & & \downarrow & & \downarrow u & & \downarrow & & \\ 0 & \longrightarrow & L & \longrightarrow & \boldsymbol{C}^n & \xrightarrow{\ \xi\ } & A & \longrightarrow & 0 \quad \text{(exact)}. \end{array}$$

In other words, A is obtained as $K_{\boldsymbol{R}}/\mathfrak{a}$ with a $\boldsymbol{Z}$-lattice $\mathfrak{a}$ in K; the complex structure of A is determined by u; and $\theta: K\to\mathrm{End}_{\boldsymbol{Q}}(A)$ is obtained by (5.5.7). This implies especially

PROPOSITION 5.13. *Any two (A, θ) of the same type (K, Φ) are isogenous.*

Let us now take a polarization $\mathcal{C}$ of A and consider a triple $(A, \mathcal{C}, \theta)$.

Let γ denote the involution of $\mathrm{End}_{\boldsymbol{Q}}(A)$ determined by $\mathcal{C}$ (see Appendix N° 13). We now impose the following condition on $(A, \mathcal{C}, \theta)$.

(5.5.10) $$\theta(K)^{\gamma} = \theta(K).$$

This holds whenever A is simple, since $\theta(K) = \mathrm{End}_{\boldsymbol{Q}}(A)$ if A is simple (see [81, p. 42, Prop. 6]). Under the assumption (5.5.10), it can be shown that K is a CM-field, so that (K, Φ) is a CM-type. The condition (5.5.10) implies

(5.5.11) $$\theta(a^{\rho}) = \theta(a)^{\gamma} \quad \textit{for every} \quad a \in K.$$

Now take a basic polar divisor in $\mathcal{C}$, and consider its Riemann form $E(x, y)$ on $\boldsymbol{C}^n$ with respect to (5.5.4) (see Appendix N°s 11–13). Then (5.5.11) is equivalent to

(5.5.12) $$E(\Phi(a)x, y) = E(x, \Phi(a^{\rho})y).$$

Put $f(a) = E(u(a), u(1))$ for $a \in K$. Then f is a $\boldsymbol{Q}$-linear map of K into $\boldsymbol{Q}$, so that $f(a) = \mathrm{Tr}_{K/\boldsymbol{Q}}(\zeta a)$ with an element ζ of K. Then we have

$$\begin{aligned} E(u(a), u(b)) &= E(u(a), \Phi(b)u(1)) \\ &= E(\Phi(b^{\rho})u(a), u(1)) = E(u(b^{\rho}a), u(1)), \end{aligned}$$

so that we obtain

(5.5.13) $$E(u(a), u(b)) = \mathrm{Tr}_{K/\boldsymbol{Q}}(\zeta a b^{\rho}) \qquad (a \in K,\ b \in K).$$

Since E is alternating, we have

(5.5.14) $$\zeta^{\rho} = -\zeta.$$

Now we can show

(5.5.15) $$E(z, w) = \textstyle\sum_{\nu=1}^{n} \zeta^{\varphi_\nu} z_\nu \bar{w}_\nu \qquad \textit{for} \quad z \in \boldsymbol{C}^n,\ w \in \boldsymbol{C}^n,$$

where z_ν and w_ν denote the components of z and w, respectively. In fact, (5.5.13) shows that (5.5.15) is true for $z, w \in u(K)$. Since $u(K)$ is dense in $\boldsymbol{C}^n$, we obtain (5.5.15). Now E, being a Riemann form of a positive non-degenerate divisor, has the property that $E(z, \sqrt{-1}\,w)$ is symmetric and positive definite. This holds if and only if

(5.5.16) $$\mathrm{Im}(\zeta^{\varphi_\nu}) > 0 \qquad \textit{for} \quad \nu = 1, \cdots, n.$$

Thus, from a given $(A, \mathcal{C}, \theta)$, we have obtained a CM-type (K, Φ), a $\boldsymbol{Z}$-lattice $\mathfrak{a}$ in K, and an element ζ of K satisfying (5.5.14) and (5.5.16).

Conversely, we can construct $(A, \mathcal{C}, \theta)$ from these data. In fact, let (K, Φ) be a CM-type, and $\mathfrak{a}$ a $\boldsymbol{Z}$-lattice in K. Then we define u by (5.5.8), and form a complex torus $A = \boldsymbol{C}^n/L$ so that (5.5.9) holds. Define $\theta(a)$ for $a \in K$ by $\theta(a) \circ \xi = \xi \circ \Phi(a)$. Take an element ζ satisfying (5.5.14) and (5.5.16). (The existence of such a ζ is clear.) Define E by (5.5.15). Then it can easily be verified that E is a Riemann form so that A has a structure of an abelian variety with a specified polarization. This shows also that the isomorphism class of (A, C, θ) is completely determined by the data $(K, \Phi; \mathfrak{a}, \zeta)$. We say that (A, C, θ) is of type $(K, \Phi; \mathfrak{a}, \zeta)$ (with respect to ξ) in this situation. Observe that $(\mathfrak{a}, \zeta)$ depends on the choice of the map ξ of (5.5.9).

C. Main theorem

Let (A, C, θ) be as above, and let $\sigma \in \mathrm{Aut}(\boldsymbol{C})$. Then C^σ is naturally defined as a polarization of A^σ. We define $\theta^\sigma : K \to \mathrm{End}_{\boldsymbol{Q}}(A^\sigma)$ by $\theta^\sigma(a) = \theta(a)^\sigma$ for $a \in K$, $\theta(a) \in \mathrm{End}(A)$. By our definition, if (A, θ) is of type $(K, \{\varphi_\nu\})$, we can find n linearly independent holomorphic differential forms $\omega_1, \cdots, \omega_n$ of degree 1 on A such that

$$\omega_\nu \circ \theta(a) = a^{\varphi_\nu} \omega_\nu \qquad (a \in K,\ \theta(a) \in \mathrm{End}(A),\ \nu = 1, \cdots, n).$$

Then we have $\omega_\nu^\sigma \circ \theta^\sigma(a) = a^{\varphi_\nu \sigma} \omega_\nu^\sigma$, so that

(5.5.17) *$(A^\sigma, \theta^\sigma)$ is of type (K, Φ^σ).*

PROPOSITION 5.14. *Let (K^*, Φ^*) be the reflex of (K, Φ). If $\sigma = \mathrm{id.}$ on K^*, $(A^\sigma, \theta^\sigma)$ is of type (K, Φ), and isogenous to (A, θ).*

This follows immediately from the definition of K^* and Prop. 5.13.

Now the relation of (A, C, θ) with $(A^\sigma, C^\sigma, \theta^\sigma)$ is given by the following main theorem, which is a generalization of Th. 5.4.

THEOREM 5.15. *Let (K, Φ) be a CM-type, (K^*, Φ^*) the reflex of (K, Φ), $\mathfrak{a}$ a $\boldsymbol{Z}$-lattice in K, and ζ an element of K satisfying* (5.5.14, 16). *Let (A, C, θ) be of type $(K, \Phi; \mathfrak{a}, \zeta)$, u the map defined by* (5.5.8), *and ξ a map such that* (5.5.9) *holds and ζ corresponds to C through ξ. Further let σ be an element of* $\mathrm{Aut}(\boldsymbol{C}/K^*)$, *and s an element of $K_A^{*\times}$ such that $\sigma = [s, K^*]$ on K_{ab}^*. Define η by* (5.5.3). *Then there is an exact sequence*

$$0 \longrightarrow u(\eta(s)^{-1}\mathfrak{a}) \longrightarrow \boldsymbol{C}^n \xrightarrow{\ \xi'\ } A^\sigma \longrightarrow 0$$

with the following properties:

(i) *$(A^\sigma, C^\sigma, \theta^\sigma)$ is of type $(K, \Phi; \eta(s)^{-1}\mathfrak{a}, \zeta')$ with respect to ξ', where $\zeta' = N(il(s))\zeta$. (For the symbol $il(s)$, see* § 5.2.)

(ii) *$\xi(u(a))^\sigma = \xi'(u(\eta(s)^{-1}a))$ for all $a \in K/\mathfrak{a}$.*

A proof in a more general setting is given in [80, 4.3]. If the reader is familiar with the results of [81], especially with the prime ideal decomposition of the Frobenius endomorphism [81, § 13, Th. 1], then he will be able to give a proof exactly in the same manner as has been done for Th. 5.4.

In the above theorem, we have imposed no condition upon the field of definition for (A, C, θ). Actually there exists a model of (A, C, θ) defined over an algebraic number field, see [81, p. 109, Prop. 26].

Let $t_1, \cdots, t_r$ be points of A (of finite or infinite order). One can prove that there exists a subfield k of $\boldsymbol{C}$ which is uniquely characterized by the following condition:

(5.5.18) *An automorphism σ of $\boldsymbol{C}$ is the identity map on k if and only if there is an isomorphism λ of A to A^σ such that $\lambda(C)=C^\sigma$, $\lambda t_i = t_i^\sigma$ for $i=1,\cdots,r$, and $\lambda\circ\theta(a)=\theta^\sigma(a)\circ\lambda$ for all $a \in K$.* (Such a λ is called an *isomorphism of $(A, C, \theta; t_1, \cdots, t_r)$ to $(A^\sigma, C^\sigma, \theta^\sigma; t_1^\sigma, \cdots, t_r^\sigma)$*.)

We call k *the field of moduli of $(A, C, \theta; t_1, \cdots, t_r)$*. (For the proof of the existence of k, see [72], [75, II].) With this concept, the following result can easily be derived from the above theorem:

COROLLARY 5.16. *The notation and the assumption being as in Th.* 5.14, *let $v_1, \cdots, v_r$ be elements of $K/\mathfrak{a}$, and let T be the set of all the elements s of $K_A^{*\times}$ such that*

$$qq^\rho N(il(s))=1\,,\quad q\eta(s)\mathfrak{a}=\mathfrak{a}\,,\quad q\eta(s)v_i=v_i \qquad (i=1,\cdots,r)$$

for some $q\in K^\times$. Then the field of moduli of $(A, C, \theta; \xi(u(v_1)), \cdots, \xi(u(v_r)))$ is the subfield of K_{ab}^* corresponding to the subgroup T of $K_A^{*\times}$.

Let k_1 be the field of moduli of (A, C, θ), and G the group of all automorphisms of (A, C, θ). Then G is isomorphic to the group of all units of K, and one can construct a quotient variety W of A by G and a projection map $p: A\rightarrow W$ satisfying the following conditions:

(i) *W is defined over k_1.*

(ii) *If $\sigma\in \mathrm{Aut}(\boldsymbol{C}/k_1)$, and f is an isomorphism of (A, C, θ) to $(A^\sigma, C^\sigma, \theta^\sigma)$, then $p=p^\sigma\circ f$. (Observe that such an f exists for any $\sigma\in \mathrm{Aut}(\boldsymbol{C}/k_1)$, on account of the definition of the field of moduli.)*

Then one can easily show that

(5.5.19) *For every point t of A, the field of moduli of $(A, C, \theta; t)$ is $k_1(p(t_1))$.*

It may be worth while noting that, if A is simple, G coincides with the group of all automorphisms of (A, C).

If A is an elliptic curve E, we see easily that $k_1=\boldsymbol{Q}(j_E)$, and $G=\mathrm{Aut}(E)$. Thus the map p is a generalization of h_E^i, and hence the combination of (5.5.19) with Cor. 5.16 yields a generalization of Th. 5.5. There remains the question of finding a generalization of the function $j(z)$ in the higher dimensional case. But this is settled in the following way.

The polarized abelian variety (A, C) determines a point z in the Siegel upper half space $\mathfrak{H}_n$ of degree n, modulo a certain discrete subgroup Γ of $Sp(n, \boldsymbol{R})$ commensurable with $Sp(n, \boldsymbol{Z})$. (Γ depends on the type of C.) There exists a Γ-invariant holomorphic map φ of $\mathfrak{H}_n$ into a complex projective space, such that $\boldsymbol{Q}(\varphi(z))$ is the field of moduli of (A, C) for any (A, C) with a polarization C whose type determines Γ. One can also formulate a similar result by using the Hilbert modular group instead of the Siegel modular group. For details, see [77], [78], [80].

Finally let us make a few remarks about the relation of the field of moduli of (A, C, θ) and that of (A, C, θ'), where θ' is the restriction of θ to any subfield F of K. *The field of moduli of* (A, C, θ') is a unique subfield k of $\boldsymbol{C}$ satisfying (5.5.18) with the following modification: the points t_i are disregarded; $\lambda\circ\theta(a)=\theta^\sigma(a)\circ\lambda$ is required only for $a\in F$. If $F=\boldsymbol{Q}$, k is the field of moduli of (A, C).

PROPOSITION 5.17. *Let K, K^*, and (A, C, θ) be as in Th.* 5.15, *F a subfield of K, θ' the restriction of θ to F, and k_0 the field of moduli of (A, C, θ'). Suppose that A is simple. Then the following assertions hold:*

(1) *k_0K^* is the field of moduli of (A, C, θ).*

(2) *K^* is normal over $k_0\cap K^*$.*

(3) *k_0K^* is normal over k_0.*

(4) $\mathrm{Gal}(k_0K^*/k_0)$ *is isomorphic to a subgroup of* $\mathrm{Aut}(K/F)$.

(5) *k_0 contains the smallest subfield of K^* over which K^* is normal.*

PROOF. Let $\sigma\in\mathrm{Aut}(\boldsymbol{C})$. If there exists an isomorphism of (A, C, θ) to $(A^\sigma, C^\sigma, \theta^\sigma)$, we see, by (5.5.17), that Φ^σ is equivalent to Φ, and hence σ is the identity on K^*. This shows that k_0K^* is contained in the field of moduli of (A, C, θ). Let $\tau\in\mathrm{Aut}(\boldsymbol{C}/k_0)$. Then there exists an isomorphism f of (A, C, θ') to $(A^\tau, C^\tau, \theta'^\tau)$. Since A is simple, we have $\theta(K)=\mathrm{End}_{\boldsymbol{Q}}(A)$ by [81, § 5.1, p. 42, Prop. 6]. Therefore we can define an element μ of $\mathrm{Aut}(K/F)$ such that $\theta^\tau(a)\circ f=f\circ\theta(a^\mu)$ for all $a\in K$. Then $\Phi(a)^\tau$ and $\Phi(a^\mu)$ have the same set of characteristic roots, so that $\{\varphi_1\tau, \cdots, \varphi_n\tau\}$ coincides with $\{\mu\varphi_1, \cdots, \mu\varphi_n\}$ as a whole. Therefore we have $(\sum_i a^{\varphi_i})^\tau=\sum_i a^{\mu\varphi_i}$ for all $a\in K$, which shows that $K^{*\tau}=K^*$. This proves (3). If

$$M=\{x\in K^*\mid x^\sigma=x \text{ for all } \sigma\in\mathrm{Aut}(K^*)\}\ ,$$

we have $\tau=\mathrm{id.}$ on M for every $\tau\in\mathrm{Aut}(\boldsymbol{C}/k_0)$, so that $M\subset k_0$. This proves (5) and (2). Now, if $\tau=\mathrm{id.}$ on K^*, $\{\mu\varphi_1, \cdots, \mu\varphi_n\}$ coincides with $\{\varphi_1, \cdots, \varphi_n\}$ as a whole. Let F, G, H, H^*, and S be as in the definition of (K^*, Φ^*) in the paragraph A. Take elements of G which coincide with $\mu, \varphi_1, \cdots, \varphi_n$, and denote them again by the same letters. Then $\mu H=H\mu$, and $\mu S=\bigcup_i\mu H\varphi_i$ $=\bigcup_i H\mu\varphi_i=\bigcup_i H\varphi_i=S$. By [81, § 8.2, p. 69, Prop. 26], we have $\mu\in H$, so that $\mu=\text{id.}$ on K, hence f is an isomorphism of $(A, \mathcal{C}, \theta)$ to $(A^\tau, \mathcal{C}^\tau, \theta^\tau)$. This shows that k_0K^* contains the field of moduli of $(A, \mathcal{C}, \theta)$, and hence (1). We have also seen that $\tau=\text{id.}$ on k_0K^* if and only if $\mu=\text{id.}$ on K. Therefore, assigning μ to τ, we obtain an isomorphism of $\mathrm{Gal}(k_0K^*/k_0)$ into $\mathrm{Aut}(K/F)$.

For example, if K is not normal over $\boldsymbol{Q}$ and $[K:\boldsymbol{Q}]=4$, then K^* is also a field of the same type, and A is simple (see [81, § 8.4, (2), c), p. 74]). Therefore, in this case, taking F to be $\boldsymbol{Q}$, we know, by (5), that *the field of moduli of* $(A, \mathcal{C})$ *contains the real quadratic subfield of* K^*.

CHAPTER 6
MODULAR FUNCTIONS OF HIGHER LEVEL

6.1. Modular functions of level N obtained by division of elliptic curves

A. The functions $f_a^i(z)$

Let N be a positive integer, and $\Gamma_N = \Gamma(N)$ the principal congruence subgroup of $\Gamma_1 = SL_2(\boldsymbol{Z})$ of level N, which is defined by

$$\Gamma_N = \{\gamma \in SL_2(\boldsymbol{Z}) \mid \gamma \equiv 1_2 \bmod (N)\} \qquad (\text{see } \S\, 1.6).$$

We shall now construct some functions which generate the field of all modular functions of level N, and which behave nicely under the transformations of Γ_1. The main idea is to consider the points of finite order on the elliptic curve

(6.1.1) $$E_L:\ y^2 = 4x^3 - g_2(L)x - g_3(L)$$

with variable L. If $L = \boldsymbol{Z}\omega_1 + \boldsymbol{Z}\omega_2$, we see that every point of finite order on E_L can be written as

$$\left(\wp\left(a\begin{bmatrix}\omega_1\\ \omega_2\end{bmatrix};\ L\right),\ \wp'\left(a\begin{bmatrix}\omega_1\\ \omega_2\end{bmatrix};\ L\right)\right)$$

with $a \in \boldsymbol{Q}^2$, and conversely such a point is of finite order for any $a \in \boldsymbol{Q}^2$. Here we consider a as a row vector. In view of the definition of $\wp$, we see that $\wp\left(a\begin{bmatrix}\omega_1\\ \omega_2\end{bmatrix};\ \omega_1, \omega_2\right)$ is a homogeneous function of degree -2 in ω_1, ω_2. Therefore we can define three types of functions $f_a = f_a^1,\ f_a^2,\ f_a^3$ on $\mathfrak{H}$ by

$$f_a(z) = f_a^1(z) = \frac{g_2(\omega_1, \omega_2)g_3(\omega_1, \omega_2)}{\Delta(\omega_1, \omega_2)}\,\wp\left(a\begin{bmatrix}\omega_1\\ \omega_2\end{bmatrix};\ \omega_1, \omega_2\right),$$

$$f_a^2(z) = \frac{g_2(\omega_1, \omega_2)^2}{\Delta(\omega_1, \omega_2)}\,\wp\left(a\begin{bmatrix}\omega_1\\ \omega_2\end{bmatrix};\ \omega_1, \omega_2\right)^2,$$

$$f_a^3(z) = \frac{g_3(\omega_1, \omega_2)}{\Delta(\omega_1, \omega_2)}\,\wp\left(a\begin{bmatrix}\omega_1\\ \omega_2\end{bmatrix};\ \omega_1, \omega_2\right)^3$$

$$(z = \omega_1/\omega_2 \in \mathfrak{H};\ a \in \boldsymbol{Q}^2,\ \notin \boldsymbol{Z}^2).$$

Especially we can substitute $(z, 1)$ for (ω_1, ω_2). We see then that these functions are holomorphic on $\mathfrak{H}$. Since $j(z) = g_2^3/\Delta$, we have $j(z) - 1 = 27 \cdot g_3^2/\Delta$, so that

$$(6.1.2)\qquad f_a^2(z)=27(j(z)-1)^{-1}f_a(z)^2,$$
$$f_a^3(z)=27j(z)^{-1}(j(z)-1)^{-1}f_a(z)^3.$$

The functions f_a^2 and f_a^3 are rather auxiliary, and will be put to use only in § 6.8.

Let

$$\gamma\in\Gamma_1,\quad z=\omega_1/\omega_2,\quad \begin{bmatrix}\omega_1'\\ \omega_2'\end{bmatrix}=\gamma\begin{bmatrix}\omega_1\\ \omega_2\end{bmatrix},\quad z'=\omega_1'/\omega_2',\quad \text{and}\quad a\gamma=a'.$$

Then

$$a'\begin{bmatrix}\omega_1\\ \omega_2\end{bmatrix}=a\begin{bmatrix}\omega_1'\\ \omega_2'\end{bmatrix},\quad z'=\gamma(z),\quad \text{and}\quad \boldsymbol{Z}\omega_1+\boldsymbol{Z}\omega_2=\boldsymbol{Z}\omega_1'+\boldsymbol{Z}\omega_2'.$$

Therefore substituting z' for z in $f_a^i(z)$, we obtain

(6.1.3) $f_a^i\circ\gamma=f_{a\gamma}^i$ *for every* $\gamma\in\Gamma_1$ *and every* $a\in\boldsymbol{Q}^2$, $\notin\boldsymbol{Z}^2$.

Since $\wp(u\,;L)=\wp(v\,;L)$ if and only if $u\equiv\pm v \bmod L$, we see that

$$(6.1.4)\qquad a\equiv\pm b \bmod \boldsymbol{Z}^2 \quad\Rightarrow\quad f_a^i=f_b^i.$$

$$(6.1.5)\qquad f_a=f_b \quad\Leftrightarrow\quad a\equiv\pm b \bmod \boldsymbol{Z}^2.$$

From (6.1.3, 4) we obtain

(6.1.6) *If* $Na\in\boldsymbol{Z}^2$, *then* $f_a^i\circ\gamma=f_a^i$ *for all* $\gamma\in\Gamma_N\cdot\{\pm1\}$.

Therefore, in order to ensure that f_a^i, with $a\in N^{-1}\boldsymbol{Z}^2$, is a modular function of level N, it is sufficient, by virtue of Prop. 2.7 and (6.1.2), to show that f_a is algebraic over $\boldsymbol{C}(j)$. We shall actually prove in Th. 6.6 that f_a is algebraic over $\boldsymbol{Q}(j)$. (Also, the Fourier expansion of f_a will be explicitly given in the proof of Prop. 6.9.) Assuming this result, we obtain

PROPOSITION 6.1. *For every positive integer* N, $\boldsymbol{C}(j,f_a\,|\,a\in N^{-1}\boldsymbol{Z}^2,\ \notin\boldsymbol{Z}^2)$ *is the field of all modular functions of level* N.

PROOF. Let $\mathfrak{K}_N$ denote the field of all modular functions of level N. Then

$$\boldsymbol{C}(j)\subset\boldsymbol{C}(j,f_a\,|\,a\in N^{-1}\boldsymbol{Z}^2,\ \notin\boldsymbol{Z}^2)\subset\mathfrak{K}_N.$$

Now $\mathfrak{K}_N$ is a Galois extension of $\boldsymbol{C}(j)$, whose Galois group is $\Gamma_1/\Gamma_N\cdot\{\pm1\}$. Therefore, to prove our proposition, it is sufficient to show that if $\gamma\in\Gamma_1$ and $f_a\circ\gamma=f_a$ for all $a\in N^{-1}\boldsymbol{Z}^2$, $\notin\boldsymbol{Z}^2$, then $\gamma\in\Gamma_N\cdot\{\pm1\}$. But this follows immediately from (6.1.3, 5) and the following

LEMMA 6.2. *Let* α *be an automorphism of the module* $(\boldsymbol{Z}/N\boldsymbol{Z})^2$ *such that* $\alpha u=\varepsilon_u u$ *for every* $u\in(\boldsymbol{Z}/N\boldsymbol{Z})^2$ *with* $\varepsilon_u=\pm1$. *Then* $\alpha=\pm1$.

The proof is very easy and may therefore be left to the reader.

B. The field generated by the points of finite order on an elliptic curve

Let us now discuss the points of finite order on an elliptic curve in a more intrinsic way, without any reference to complex tori or $\mathfrak{H}$. Consider an elliptic curve

$$E:\ y^2 = 4x^3 - c_2 x - c_3$$

with c_2 and c_3 in $\boldsymbol{C}$, such that $\mathrm{Aut}(E) = \{\pm 1\}$, and functions h_E^i, for $i = 1, 2, 3$, defined in § 4.5. For simplicity we write h for h_E^1. For a positive integer N, we put

$$\mathfrak{g}_N = \{t \in E \mid Nt = 0\}\ ,$$

and consider the field

$$F_N = \boldsymbol{Q}(j_E,\ h(t) \mid t \in \mathfrak{g}_N)\ .$$

In view of (4.5.4), we see that the field F_N depends only on N and the isomorphism class of E. Therefore, to study the structure of F_N, we can assume that E is defined over $\boldsymbol{Q}(j_E)$, by changing E for a suitable curve isomorphic to E. Assuming this, let σ be an automorphism of $\boldsymbol{C}$ over $\boldsymbol{Q}(j_E)$. Then $E^\sigma = E$, and $t \mapsto t^\sigma$ gives an automorphism of the module $\mathfrak{g}_N$. Since $\mathfrak{g}_N$ is isomorphic to $(\boldsymbol{Z}/N\boldsymbol{Z})^2$, the group of all automorphisms of $\mathfrak{g}_N$ is isomorphic to $GL_2(\boldsymbol{Z}/N\boldsymbol{Z})$. Since h is rational over $\boldsymbol{Q}(j_E)$, we have $h(t)^\sigma = h(t^\sigma)$, so that F_N is stable under σ. Therefore F_N is a Galois extension of $\boldsymbol{Q}(j_E)$. If $\sigma = \mathrm{id.}$ on F_N, we have $h(t^\sigma) = h(t)$, so that by (4.5.3), $t^\sigma = \varepsilon_t t$ with $\varepsilon_t = \pm 1$. By Lemma 6.2, ε_t is independent of t. Thus σ induces an automorphism ± 1 on $\mathfrak{g}_N$ if $\sigma = \mathrm{id.}$ on F_N. Therefore we obtain an injective homomorphism

$$\text{(6.1.7)} \qquad \mathrm{Gal}(F_N/\boldsymbol{Q}(j_E)) \longrightarrow GL_2(\boldsymbol{Z}/N\boldsymbol{Z})/\{\pm 1_2\}\ .$$

More explicitly, take two elements t_1 and t_2 of $\mathfrak{g}_N$ so that $\mathfrak{g}_N = \boldsymbol{Z}t_1 + \boldsymbol{Z}t_2$. For an automorphism σ of $\boldsymbol{C}$ over $\boldsymbol{Q}(j_E)$, put

$$\text{(6.1.8)} \qquad \begin{aligned} t_1^\sigma &= pt_1 + qt_2\ , \\ t_2^\sigma &= rt_1 + st_2\ , \end{aligned}$$

with an element $\beta = \begin{bmatrix} p & q \\ r & s \end{bmatrix}$ of $M_2(\boldsymbol{Z})$. Then $\det(\beta)$ is prime to N, and the restriction of σ to F_N corresponds to $\pm\beta \bmod (N)$. We have clearly

$$\text{(6.1.9)} \qquad h(at_1 + bt_2)^\sigma = h(a't_1 + b't_2) \quad \text{if} \quad (a\ \ b)\beta = (a'\ \ b')\,.$$

PROPOSITION 6.3. *The notation being as above, the following assertions hold:*

(1) *F_N contains a primitive N-th root of unity, say ζ.*

(2) *If an element* τ *of* $\mathrm{Gal}(F_N/\boldsymbol{Q}(j_E))$ *corresponds to an element* α *of* $GL_2(\boldsymbol{Z}/N\boldsymbol{Z})$, *then* $\zeta^\tau=\zeta^{\det(\alpha)}$. *(Note that* $\zeta^{\det(\alpha)}$ *is meaningful.)*

(3) *If* λ *is an isogeny of* E *onto an elliptic curve* E' *such that* $\mathrm{Ker}(\lambda)\subset\mathfrak{g}_N$, *then* $j(E')\in F_N$. *Moreover, if* $\mathrm{End}(E)=\boldsymbol{Z}$, *then, for* $\sigma\in\mathrm{Aut}(\boldsymbol{C}/\boldsymbol{Q}(j_E))$, *one has* $j(E')^\sigma=j(E')$ *if and only if* $\mathrm{Ker}(\lambda)^\sigma=\mathrm{Ker}(\lambda)$.

PROOF. Consider the symbol $e_N(s,t)$ of § 4.3. For an automorphism σ of $\boldsymbol{C}$ over $\boldsymbol{Q}(j_E)$, define $\beta=\begin{bmatrix} p & q \\ r & s\end{bmatrix}$ as above. Put $\zeta=e_N(t_1,t_2)$. By Prop. 4.2, we have

$$\zeta^\sigma=e_N(t_1^\sigma,t_2^\sigma)=e_N(pt_1+qt_2,rt_1+st_2)=e_N(t_1,t_2)^{ps-qr}=\zeta^{\det(\beta)}.$$

For every u and v in $\boldsymbol{Z}$, we have

$$e_N(t_1,ut_1+vt_2)=e_N(t_1,t_2)^v,$$

so that, by (5) of Prop. 4.2, $\zeta=e_N(t_1,t_2)$ must be a primitive N-th root of unity. If $\sigma=\mathrm{id.}$ on F_N, we have $\beta\equiv\pm1 \bmod (N)$, so that $\zeta^\sigma=\zeta$. This shows that $\zeta\in F_N$, hence (1) and (2). Let λ and E' be as in (3), and again σ an automorphism of $\boldsymbol{C}$ over $\boldsymbol{Q}(j_E)$. Then λ^σ is an isogeny of E onto E'^σ. If $\mathrm{Ker}(\lambda)^\sigma=\mathrm{Ker}(\lambda)$, E'^σ is isomorphic to E', so that $j(E')^\sigma=j(E')$. This is so especially if $\sigma=\mathrm{id.}$ on F_N, since one has then $t^\sigma=\pm t$ for all $t\in\mathfrak{g}_N$. This proves that $j(E')\in F_N$. Suppose conversely that $j(E')^\sigma=j(E')$, and further $\mathrm{End}(E)=\boldsymbol{Z}$. Then there exists an isomorphism μ of E' onto E'^σ. Observe that $\mu\circ\lambda$ and λ^σ are elements of $\mathrm{Hom}(E,E'^\sigma)$ of the same degree (cf. § 5.1). Since $\mathrm{Hom}(E,E'^\sigma)$ is isomorphic to $\boldsymbol{Z}$, we have $\mu\circ\lambda=\pm\lambda^\sigma$, so that $\mathrm{Ker}(\lambda)=\mathrm{Ker}(\lambda^\sigma)=\mathrm{Ker}(\lambda)^\sigma$. This completes the proof of (3).

6.2. The field of modular functions of level N rational over $\boldsymbol{Q}(e^{2\pi i/N})$

We are going to connect together the results of Parts A and B of the preceding section, by means of the following two lemmas.

LEMMA 6.4. *Let* $L=\boldsymbol{Z}\omega_1+\boldsymbol{Z}\omega_2$, *and let* E *be a member of* $\mathcal{E}$ *isomorphic to* $\boldsymbol{C}/L$ *(see* § 4.5*). Then, for any isomorphism* ξ *of* $\boldsymbol{C}/L$ *onto* E, *we have*[10]

$$h_E^i\left(\xi\left(a\begin{bmatrix}\omega_1\\ \omega_2\end{bmatrix}\right)\right)=f_a^i(\omega_1/\omega_2) \qquad (a\in\boldsymbol{Q}^2,\ \notin\boldsymbol{Z}^2;\ i=1,2,3).$$

PROOF. Let E' be defined by $y^2=4x^3-g_2(L)x-g_3(L)$, and let ξ' be an isomorphism of $\boldsymbol{C}/L$ to E' defined by

10) One should actually write $\xi(u \bmod L)$ instead of $\xi(u)$ for $u\in\boldsymbol{C}$. But we shall hereafter use the abbreviated form $\xi(u)$ if there is no fear of confusion.

$$\xi'(u) = (\wp(u\,;L), \wp'(u\,;L)).$$

Put $\eta = \xi' \circ \xi^{-1}$. Since η is an isomorphism of E onto E', we have $h_E^i(\xi(u)) = h_{E'}^i(\xi'(u))$ by (4.5.4). From our definition of h_E^i and f_a^i, we obtain $h_{E'}^i(\xi'(u)) = f_a^i(\omega_1/\omega_2)$ if $u = a\begin{bmatrix}\omega_1\\ \omega_2\end{bmatrix}$, hence our assertion.

LEMMA 6.5. *Let $\{f_\alpha \mid \alpha \in A\}$ be a set of meromorphic functions in a connected open subset D of $\boldsymbol{C}^d$, indexed by an at most countable set A. Let k be a subfield of $\boldsymbol{C}$ with only countably many elements. Then there exists a point z_0 of D such that the specialization $\{f_\alpha\}_{\alpha\in A} \to \{f_\alpha(z_0)\}_{\alpha\in A}$ defines an isomorphism of the field $k(f_\alpha \mid \alpha \in A)$ onto $k(f_\alpha(z_0) \mid \alpha \in A)$ over k.*

We call such a point z_0 *generic over k for the functions f_α*. Actually we need this lemma only in the special case $d=1$, where the proof is much simpler.

PROOF. We may assume that $A = \{1, 2, 3, \cdots\}$ (finite or not). By induction, we see that there exists a subset $B = \{\nu_1, \nu_2, \cdots\}$ of A such that: (i) $\nu_1 < \nu_2 < \cdots$; (ii) $f_{\nu_1}, f_{\nu_2}, \cdots$ are algebraically independent over k; and (iii) $f_1, \cdots, f_n$ are algebraic over $k(f_\nu \mid \nu \in B,\ \nu \leqq n)$. Let S_m be the set of all polynomials $P(X_1, \cdots, X_m) \neq 0$ in m indeterminates with coefficients in k, and W_ν the set of the points of D where f_ν is not holomorphic. Put, for each $P \in S_m$,

$$F_P = \{z \in D - \textstyle\bigcup_{i=1}^m W_{\nu_i} \mid P(f_{\nu_1}(z), \cdots, f_{\nu_m}(z)) = 0\}.$$

The closure of F_P in D has no interior point of D. Now observe that S_m has only countably many elements. By Lemma 1.2, there exists a point z_0 of D not belonging to the countable union $(\bigcup_{\nu\in A} W_\nu) \cup (\bigcup_{m=1}^\infty \bigcup_{P\in S_m} F_P)$. Then, by virtue of our construction, $k(f_1, \cdots, f_n)$ has the same transcendence degree as $k(f_1(z_0), \cdots, f_n(z_0))$ over k for every n. Therefore the specialization $f_\alpha \to f_\alpha(z_0)$ over k defines an isomorphism of these fields, hence our assertion.

Now let us put, for a positive integer N,

$$\mathfrak{F}_N = \boldsymbol{Q}(j, f_a \mid a \in N^{-1}\boldsymbol{Z}^2, \notin \boldsymbol{Z}^2).$$

We have seen in Prop. 6.1 that $\boldsymbol{C}\cdot\mathfrak{F}_N$ is the field of all modular functions of level N. We call (by abuse of language) an element of $\mathfrak{F}_N$ a *modular function of level N rational over $\boldsymbol{Q}(e^{2\pi i/N})$*. The following theorem will justify this definition.

THEOREM 6.6. *The field $\mathfrak{F}_N$ has the following properties.*

(1) *$\mathfrak{F}_N$ is a Galois extension of $\boldsymbol{Q}(j)$.*

(2) *For every $\beta \in GL_2(\boldsymbol{Z}/N\boldsymbol{Z})$, $f_a \mapsto f_{a\beta}$ gives an element of* $\mathrm{Gal}(\mathfrak{F}_N/\boldsymbol{Q}(j))$, *which we write $\tau(\beta)$. Then $\beta \mapsto \tau(\beta)$ gives an isomorphism of $GL_2(\boldsymbol{Z}/N\boldsymbol{Z})/\{\pm 1\}$*

to Gal$(\mathfrak{F}_N/\boldsymbol{Q}(j))$.

(3) *If ζ is a primitive N-th root of unity, then $\zeta\in\mathfrak{F}_N$, and* $\zeta^{\tau(\beta)}=\zeta^{\det(\beta)}$.

(4) $\boldsymbol{Q}(\zeta)$ *is algebraically closed in* $\mathfrak{F}_N$.

(5) $\mathfrak{F}_N$ *contains the functions $j\circ\alpha$ for all $\alpha\in M_2(\boldsymbol{Z})$ such that* $\det(\alpha)=N$.

PROOF. By Lemma 6.5, we can find a point z_0 of $\mathfrak{H}$ generic for the functions j, f_a, $j\circ\alpha$ for all $a\in N^{-1}\boldsymbol{Z}^2$, $\notin\boldsymbol{Z}^2$, and for all $\alpha\in M_2(\boldsymbol{Z})$ such that $\det(\alpha)=N$. Since the substitution of z_0 for z gives an isomorphism, it is sufficient to prove our assertions for $j(z_0)$, $f_a(z_0)$, $j(\alpha(z_0))$ instead of j, f_a, $j\circ\alpha$. Obviously $j(z_0)$ is transcendental. Take $c\in\boldsymbol{C}$ so that $c/(c-27)=j(z_0)$, and consider an elliptic curve $E: y^2=4x^3-cx-c$. Then $j_E=j(z_0)$, so that there exists an isomorphism ξ of $\boldsymbol{C}/(\boldsymbol{Z}z_0+\boldsymbol{Z})$ onto E. Consider $\mathfrak{g}_N$, $h=h_E^1$, and F_N of § 6.1, Part B, with respect to the present elliptic curve E. Put $\eta(a)=\xi\left(a\begin{bmatrix} z_0\\ 1\end{bmatrix}\right)$ for $a\in\boldsymbol{Q}^2$. By Lemma 6.4, we have $h(t)=f_a(z_0)$ if $t=\eta(a)$, so that

$$F_N=\boldsymbol{Q}(j(z_0), f_a(z_0)\mid a\in N^{-1}\boldsymbol{Z}^2,\ \notin\boldsymbol{Z}^2).$$

Then the assertion (1) follows from the fact that F_N is a Galois extension of $\boldsymbol{Q}(j_E)$. Put $t_1=\eta((N^{-1},0))$, $t_2=\eta((0,N^{-1}))$. If σ and $\beta=\begin{bmatrix} p & q\\ r & s\end{bmatrix}$ are defined with respect to these t_1 and t_2 as in (6.1.8, 9), then $\eta(a)^\sigma=\eta(a\beta)$ for all $a\in N^{-1}\boldsymbol{Z}^2$, so that $f_a(z_0)^\sigma=h(\eta(a))^\sigma=h(\eta(a\beta))=f_{a\beta}(z_0)$. Therefore we obtain (2) and (3) from Prop. 6.3, if we could prove the surjectivity of the map (6.1.7) in the present case. Let A be the image of the map (6.1.7). Let $\gamma\in SL_2(\boldsymbol{Z})$. Since $f_{a\gamma}=f_a\circ\gamma$ by (6.1.3), we see that $f_a\mapsto f_{a\gamma}$ defines an automorphism of $\mathfrak{F}_N$ over $\boldsymbol{Q}(j)$. Transferring this result to F_N, we can conclude that $SL_2(\boldsymbol{Z}/N\boldsymbol{Z})/\{\pm1\}\subset A$. Identifying A with Gal$(F_N/\boldsymbol{Q}(j_E))$, let B denote the subgroup of A corresponding to $\boldsymbol{Q}(\zeta, j_E)$. By Galois theory, we obtain

$$[A:B]=[\boldsymbol{Q}(\zeta, j_E):\boldsymbol{Q}(j_E)]=[(\boldsymbol{Z}/N\boldsymbol{Z})^\times:1].$$

By (2) of Prop. 6.3, we have $SL_2(\boldsymbol{Z}/N\boldsymbol{Z})/\{\pm1\}\subset B$, so that $A=GL_2(\boldsymbol{Z}/N\boldsymbol{Z})/\{\pm1\}$, and $B=SL_2(\boldsymbol{Z}/N\boldsymbol{Z})/\{\pm1\}$. To prove (4), put $k=\boldsymbol{C}\cap\mathfrak{F}_N$. Then every element of k is invariant under $SL_2(\boldsymbol{Z}/N\boldsymbol{Z})$, since, as is shown above, the action of $SL_2(\boldsymbol{Z}/N\boldsymbol{Z})$ is obtained from the substitution $z\mapsto\gamma(z)$ with $\gamma\in SL_2(\boldsymbol{Z})$. Moreover, we have seen that $\boldsymbol{Q}(\zeta, j)$ is the subfield of $\mathfrak{F}_N$ corresponding to $SL_2(\boldsymbol{Z}/N\boldsymbol{Z})$. Therefore $k\subset\boldsymbol{Q}(\zeta, j)$, so that $k\subset\boldsymbol{Q}(\zeta)$. This proves (4). To prove (5), let $\alpha\in M_2(\boldsymbol{Z})$, $\det(\alpha)=N$, $\alpha\begin{bmatrix} z_0\\ 1\end{bmatrix}=\begin{bmatrix}\omega_1'\\ \omega_2'\end{bmatrix}$, and let E' be an elliptic curve isomorphic to $\boldsymbol{C}/(\boldsymbol{Z}\omega_1'+\boldsymbol{Z}\omega_2')$. Since $N\alpha^{-1}\in M_2(\boldsymbol{Z})$, we see that $N(\boldsymbol{Z}z_0+\boldsymbol{Z})\subset\boldsymbol{Z}\omega_1'+\boldsymbol{Z}\omega_2'$. Therefore we obtain an isogeny λ of E onto E' such that $\lambda(\xi(u))=\xi'(Nu)$ for $u\in\boldsymbol{C}$, where ξ' is an isomorphism of $\boldsymbol{C}/(\boldsymbol{Z}\omega_1'+\boldsymbol{Z}\omega_2')$ onto E'. Then $\mathrm{Ker}(\lambda)=\xi(N^{-1}(\boldsymbol{Z}\omega_1'+\boldsymbol{Z}\omega_2'))\subset\xi(N^{-1}(\boldsymbol{Z}z_0+\boldsymbol{Z}))\subset\mathfrak{g}_N$. Now we have

$j(\alpha(z_0))=j(\omega_1'/\omega_2')=j(E')$, and $j(E')\in F_N$ by (3) of Prop. 6.3. This proves (5).

The Galois theoretical correspondence between fields and groups in the above theorem can best be described by the following diagram, in which we put $k_N=\boldsymbol{Q}(e^{2\pi i/N})$. $\mathfrak{K}_N$ denotes the field of all modular functions of level N.

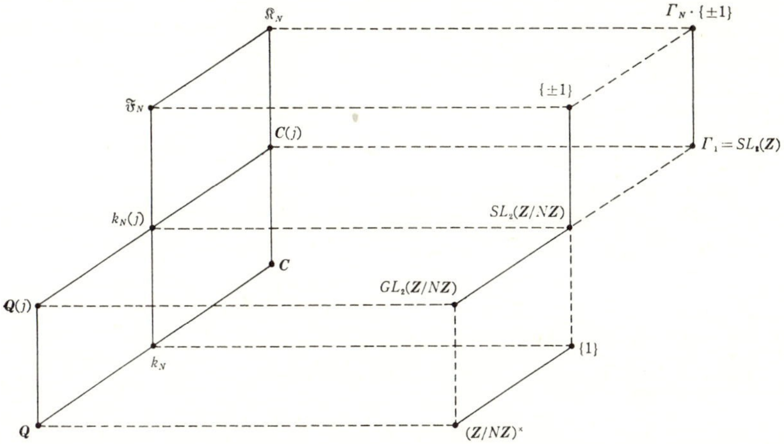

REMARK 6.7. The notation E and $\mathfrak{g}_N$ being as above, we see easily that $\boldsymbol{Q}(j_E, t\mid t\in\mathfrak{g}_N)$ is a Galois extension of $\boldsymbol{Q}(j_E)$ whose Galois group is isomorphic to a subgroup H of $GL_2(\boldsymbol{Z}/N\boldsymbol{Z})$. The above result implies that $H\cdot\{\pm1\}=GL_2(\boldsymbol{Z}/N\boldsymbol{Z})$. Take an element γ of $SL_2(\boldsymbol{Z}/N\boldsymbol{Z})$ so that $\gamma^2=-1$, say $\gamma=\begin{bmatrix}0&-1\\1&0\end{bmatrix}$. Then either γ or $-\gamma$ is contained in H, hence $-1=\gamma^2\in H$. Therefore we have $H=GL_2(\boldsymbol{Z}/N\boldsymbol{Z})$. We shall make no use of this result in the rest of the book.

PROPOSITION 6.8. *Let $\mathfrak{D}_N$ denote the field generated over $\boldsymbol{Q}(j)$ by the functions of the form $j\circ\alpha$ with $\alpha\in M_2(\boldsymbol{Z})$, $\det(\alpha)=N$ for a fixed N. Then $\mathfrak{D}_N$ is the subfield of $\mathfrak{F}_N$ corresponding to the subgroup*

$$\left\{\begin{bmatrix}a&0\\0&a\end{bmatrix}\,\middle|\, a\in(\boldsymbol{Z}/N\boldsymbol{Z})^\times\right\}/\{\pm1\}$$

of $GL_2(\boldsymbol{Z}/N\boldsymbol{Z})/\{\pm1\}$.

PROOF. Let z_0, E, $\mathfrak{g}_N$, t_1, t_2, and F_N be as in the proof of Th. 6.6. As is shown in the proof, every $\alpha\in M_2(\boldsymbol{Z})$ such that $\det(\alpha)=N$ corresponds to an isogeny λ of E to an elliptic curve E' such that $\operatorname{Ker}(\lambda)\subset\mathfrak{g}_N$. Especially, $\operatorname{Ker}(\lambda)=\boldsymbol{Z}t_1$, $\boldsymbol{Z}t_2$, or $\boldsymbol{Z}(t_1+t_2)$ according as $\alpha=\begin{bmatrix}1&0\\0&N\end{bmatrix}$, $\begin{bmatrix}N&0\\0&1\end{bmatrix}$, or $\begin{bmatrix}1&1\\0&N\end{bmatrix}$. We then have $j(E')=j(z_0/N)$, $j(Nz_0)$, or $j((z_0+1)/N)$, accordingly. Let σ be an automorphism of $\boldsymbol{C}$ over $\boldsymbol{Q}(j_E)$ whose restriction to F_N corresponds to an element β of $GL_2(\boldsymbol{Z}/N\boldsymbol{Z})$. By (3) of Prop. 6.3, if σ leaves $j(\alpha(z_0))=j(E')$ invariant for all such α, then $\operatorname{Ker}(\lambda)$ for all corresponding λ must be stable under σ. Especially $\boldsymbol{Z}t_1$, $\boldsymbol{Z}t_2$, and $\boldsymbol{Z}(t_1+t_2)$ must be stable under σ. Then we see easily that β is of the form $\beta=\begin{bmatrix} a & 0 \\ 0 & a \end{bmatrix}$. Conversely, if $\beta=\begin{bmatrix} a & 0 \\ 0 & a \end{bmatrix}$, then every subgroup of $\mathfrak{g}_N$ is stable under σ, so that, by (3) of Prop. 6.3, $j(E')^\sigma=j(E')$ for any E' as above, hence $\sigma=\mathrm{id.}$ on $\mathfrak{D}_N$.

REMARK. As the above proof shows, the conclusion of Prop. 6.8 is true even if we restrict α to the elements whose elementary divisors are 1 and N.

PROPOSITION 6.9. (1) *$\mathfrak{F}_N$ coincides with the field of all the modular functions of level N whose Fourier expansions with respect to $e^{2\pi iz/N}$ have coefficients in $k_N=\boldsymbol{Q}(e^{2\pi i/N})$.*

(2) *The field $\boldsymbol{Q}(j(z), j(Nz), f_{a_1}(z))$, with $a_1=(N^{-1}, 0)$, coincides with the field of all the modular functions of level N whose Fourier expansions with respect to $e^{2\pi iz/N}$ have rational coefficients.*

(3) *The field of (2) corresponds to the subgroup*

$$\left\{\begin{bmatrix} \pm 1 & 0 \\ 0 & x \end{bmatrix} \,\middle|\, x\in(\boldsymbol{Z}/N\boldsymbol{Z})^\times\right\}/\{\pm 1\}$$

of $GL_2(\boldsymbol{Z}/N\boldsymbol{Z})/\{\pm 1\}$.

These results will be needed only in the proof of Prop. 6.35 and Ex. 6.26.

PROOF. To prove (3), let z_0, E, and F_N be as in the proofs of Th. 6.6 and Prop. 6.8. We have seen that there exists an isogeny λ of E onto E' such that $j(Nz_0)=j(E')$, and $\operatorname{Ker}(\lambda)=\boldsymbol{Z}t_2$. Let σ be an automorphism of $\boldsymbol{C}$ over $\boldsymbol{Q}(j(z_0))$, and let $\beta=\begin{bmatrix} p & q \\ r & s \end{bmatrix}$ be an element of $GL_2(\boldsymbol{Z}/N\boldsymbol{Z})$ corresponding to the restriction of σ to F_N. Then $\sigma=\mathrm{id.}$ on $\boldsymbol{Q}(j(z_0), j(Nz_0), f_{a_1}(z_0))$ if and only if $\operatorname{Ker}(\lambda)^\sigma=\operatorname{Ker}(\lambda)$ and $a_1\beta\equiv\pm a_1 \bmod \boldsymbol{Z}^2$. This is so if and only if $\beta=\begin{bmatrix} \pm 1 & 0 \\ 0 & s \end{bmatrix}$, hence (3).

To prove (1) and (2), we consider the Fourier expansion of f_a. Putting $v=u/\omega_1$ and $z=\omega_1/\omega_2$, we have

$$\begin{aligned} \omega_2^2\cdot\wp(u\,;\omega_1,\omega_2) &= v^{-2}+\textstyle\sum'\left[(v-mz-n)^{-2}-(mz+n)^{-2}\right] \qquad ((m,n)\neq(0,0)) \\ &= -2\textstyle\sum_{n=1}^\infty n^{-2}-2\sum_{m=1}^\infty\sum_{n=-\infty}^\infty(mz+n)^{-2}+\sum_{n=-\infty}^\infty(v+n)^{-2} \\ &\quad +\textstyle\sum_{m=1}^\infty\sum_{n=-\infty}^\infty\left[(v+mz+n)^{-2}+(-v+mz+n)^{-2}\right]. \end{aligned}$$

By virtue of (2.2.3), this is equal to

$$-\pi^2/3+8\pi^2 \sum_{m=1}^{\infty} \sum_{n=1}^{\infty} n \cdot e^{2\pi imnz} - 4\pi^2 \sum_{n=1}^{\infty} n \cdot e^{2\pi inv}$$
$$-4\pi^2 \sum_{m=1}^{\infty} \sum_{n=1}^{\infty} n \cdot [e^{2\pi in(v+mz)} + e^{2\pi in(-v+mz)}].$$

Therefore, putting $u=(r\omega_1+s\omega_2)/N$ with integers r and s, $\zeta=e^{2\pi i/N}$, $q=e^{2\pi iz}$, and $q_N=e^{2\pi iz/N}$, we obtain

$$(6.2.1)\qquad (\omega_2/2\pi)^2\wp((r\omega_1+s\omega_2)/N;\ \omega_1,\omega_2)$$
$$=-(1/12)+2\sum_{n=1}^{\infty} nq^n/(1-q^n)$$
$$-\zeta^s q_N^r/(1-\zeta^s q_N^r)^2 - \sum_{n=1}^{\infty} (\zeta^{ns} q_N^{nr} + \zeta^{-ns} q_N^{-nr}) \cdot nq^n/(1-q^n),$$
$$(0 \leq r < N,\ (r,s) \notin N\boldsymbol{Z}^2).$$

This, together with the results of § 2.2, shows that the Fourier coefficients of f_a belong to k_N for every $a \in N^{-1}\boldsymbol{Z}^2$, $\notin \boldsymbol{Z}^2$. Let X (resp. X') denote the field of all the modular functions of level N whose Fourier coefficients with respect to q_N belong to $\boldsymbol{Q}$ (resp. k_N). Then X (resp. X') and $\boldsymbol{C}$ are linearly disjoint over $\boldsymbol{Q}$ (resp. k_N). In fact, let $\mu_1, \cdots, \mu_m$ be elements of $\boldsymbol{C}$ linearly independent over $\boldsymbol{Q}$. Suppose $\sum_{i=1}^m \mu_i g_i = 0$ with g_i in X. Let $g_i = \sum_n c_{in} q_N^n$ with $c_{in} \in \boldsymbol{Q}$. Then $\sum_i \mu_i c_{in} = 0$ for every n, so that $c_{in}=0$ for all i and n, hence $g_1 = \cdots = g_m = 0$. The same argument applies to X' and k_N. Now, since $\mathfrak{F}_N \subset X' \subset \boldsymbol{C}\mathfrak{F}_N$, we obtain, from the linear disjointness, $\mathfrak{F}_N = X'$. To prove (2), put $Y=\boldsymbol{Q}(j(z), j(Nz), f_{a_1}(z))$. From the above formula (6.2.1), we see that $f_{a_1} \in X$, so that $Y \subset X$. By our assertion (3) which is already proved, and by (3) of Th. 6.6, we see that only the identity element of $\mathrm{Gal}(\mathfrak{F}_N/\boldsymbol{Q}(j))$ can leave the elements of $Y(\zeta)$ invariant, hence $\mathfrak{F}_N = Y(\zeta)$. Thus $Y \subset X \subset Y(\zeta)$. From the linear disjointness of X and $\boldsymbol{Q}(\zeta)$ over $\boldsymbol{Q}$, we obtain $Y = X$. This completes the proof.

6.3. A generalization of Galois theory

Let k be a field, and K an *arbitrary* extension of k. We shall now make a few elementary observations about the Galois-like correspondence between the subgroups of $\mathrm{Aut}(K/k)$ and the subfields of K. In later sections, our results will be applied to the field of all modular functions rational over cyclotomic fields, i. e., the composite of the $\mathfrak{F}_N$ for all N. In this section, for simplicity, we fix the fields k and K, and put $\mathfrak{A} = \mathrm{Aut}(K/k)$. For a subfield F of K containing k, we put

$$\mathfrak{g}(F) = \mathrm{Aut}(K/F) = \{\sigma \in \mathfrak{A} \mid x^\sigma = x \text{ for all } x \in F\},$$

and for every subgroup S of $\mathfrak{A}$,

$$\mathfrak{f}(S) = \{x \in K \mid x^\sigma = x \text{ for all } \sigma \in S\}.$$

We can make $\mathfrak{A}$ a Hausdorff topological group by taking, as a basis of neighborhoods of the identity element, all subgroups of the form

$$\{\sigma \in \mathfrak{A} \mid x_1^\sigma = x_1, \cdots, x_n^\sigma = x_n\}$$

for any finite set $\{x_1, \cdots, x_n\}$ of elements of K. We observe that the topology of $\mathrm{Aut}(K/F) = \mathfrak{g}(F)$ is the same as that induced from the topology of $\mathfrak{A}$. The following proposition is fundamental and well-known.

PROPOSITION 6.10. *If K is a (finite or an infinite) Galois extension of k, then $\mathfrak{A}$ is compact, $\mathfrak{g}(\mathfrak{f}(S)) = S$ for every closed subgroup S of $\mathfrak{A}$, and $\mathfrak{f}(\mathfrak{g}(F)) = F$ for every subfield F of K containing k. (In this case, of course $\mathfrak{A} = \mathrm{Gal}(K/k)$.)*

In a more general case, we have

PROPOSITION 6.11. *Let Σ denote the set of all compact subgroups of $\mathfrak{A}$, and Φ the set of all subfields of K containing k, over which K is a (finite or an infinite) Galois extension. Then $\mathfrak{g}(\mathfrak{f}(S)) = S$ and $\mathfrak{f}(S) \in \Phi$ for every $S \in \Sigma$*[11]*; $\mathfrak{f}(\mathfrak{g}(F)) = F$ and $\mathfrak{g}(F) \in \Sigma$ for every $F \in \Phi$. Thus there is a one-to-one correspondence between Σ and Φ.*

PROOF. The fact that $\mathfrak{f}(\mathfrak{g}(F)) = F$ and $\mathfrak{g}(F) \in \Sigma$ for every $F \in \Phi$ follows immediately from Prop. 6.10. To prove the remaining part, let $S \in \Sigma$, and $a \in K$. Obviously $S = \bigcup_{b \in K} \{\sigma \in S \mid a^\sigma = b\}$. Since S is compact, S is covered by a finite number of the sets of the form $\{\sigma \in S \mid a^\sigma = b\}$. This shows that $\{a^\sigma \mid \sigma \in S\}$ is a finite set, say $\{a_1, \cdots, a_n\}$. Then the polynomial $\prod_{i=1}^n (X - a_i)$ has coefficients in $\mathfrak{f}(S)$. This shows that every element a of K is algebraic over $\mathfrak{f}(S)$, and an irreducible equation for a over $\mathfrak{f}(S)$ splits completely in K. Therefore K is a Galois extension of $\mathfrak{f}(S)$. Now S is a closed subgroup of $\mathfrak{g}(\mathfrak{f}(S)) = \mathrm{Gal}(K/\mathfrak{f}(S))$. Applying Prop. 6.10 to S, we obtain $S = \mathfrak{g}(\mathfrak{f}(S))$.

PROPOSITION 6.12. *The notation being as in Prop. 6.11, let Σ' be the set of all open compact subgroups of $\mathfrak{A}$, and Φ' the subset of Φ consisting of all $F \in \Phi$ which are finitely generated over $\mathfrak{f}(\mathfrak{A})$. Suppose that Φ' is not empty. Then $\mathfrak{A}$ is locally compact, and the one-to-one correspondence between Σ and Φ induces a one-to-one correspondence between Σ' and Φ'.*

PROOF. Put $k_0 = \mathfrak{f}(\mathfrak{A})$. Suppose that a member M of Φ is generated by a finite number of elements $x_1, \cdots, x_n$ over k_0. Then

$$\mathfrak{g}(M) = \{\sigma \in \mathfrak{A} \mid x_1^\sigma = x_1, \cdots, x_n^\sigma = x_n\}.$$

---

11) The fact that every compact subgroup S corresponds to a member of Φ is mentioned in N. Jacobson, Lectures in abstract algebra, vol. III (1964), p. 151, Ex. 5. See also Pjateckii-Shapiro and Shafarevic [58] and Ihara [34].

Therefore $\mathfrak{g}(M)$ is open, hence $\mathfrak{g}(M)\in\Sigma'$. It follows that $\mathfrak{A}$ is locally compact. Conversely, let $S\in\Sigma'$, and $F=\mathfrak{f}(S)$. Then we have $\mathfrak{g}(MF)=\mathfrak{g}(M)\cap\mathfrak{g}(F)$, which is open and compact, hence $[MF:M]=[\mathfrak{g}(M):\mathfrak{g}(MF)]<\infty$. It follows that $MF\in\Phi'$, hence $F\in\Phi'$.

PROPOSITION 6.13. *Let S be a subgroup of $\mathfrak{A}$, $F=\mathfrak{f}(S)$, and F_1 the algebraic closure of F in K. Then F_1 is a Galois extension of F. If moreover $\mathfrak{g}(F)=S$, then $\mathfrak{g}(F_1)$ is a normal subgroup of S, and $S/\mathfrak{g}(F_1)$, as an abstract group, is canonically isomorphic to a dense subgroup of* $\mathrm{Gal}(F_1/F)$.

PROOF. If $u\in F_1$, $\{u^\sigma\mid\sigma\in S\}$ is obviously a finite set, say $\{u_1,\cdots,u_n\}$. Then $\prod_{i=1}^n(X-u_i)$ has coefficients in F, so that F_1 is Galois over F. If $\mathfrak{g}(F)=S$, then $\mathfrak{g}(F_1)\subset S$, and $F_1^\sigma=F_1$ for every $\sigma\in S$, hence $\mathfrak{g}(F_1)=\mathfrak{g}(F_1^\sigma)=\sigma^{-1}\mathfrak{g}(F_1)\sigma$ for every $\sigma\in S$. Now $S/\mathfrak{g}(F_1)$ can be identified with a subgroup of $\mathrm{Gal}(F_1/F)$ in a natural way. Since F is the fixed subfield of F_1 for this subgroup, we obtain the last assertion.

PROPOSITION 6.14. *If $\mathfrak{f}(\mathfrak{g}(F))=F$ for a subfield F of K containing k, then $\mathfrak{f}(\mathfrak{g}(M))=M$ for every finite algebraic extension M of F contained in K.*

PROOF. Put $S=\mathfrak{g}(F)$ and $T=\mathfrak{g}(M)$. Let F_1 be the algebraic closure of F in K. Considering the restriction of the elements of S to M, we find $[S:T]\leqq[M:F]$. If $\mathfrak{f}(S)=F$, we see from Prop. 6.13 that every isomorphism of M into F_1 over F can be obtained from an element of S. Therefore $[S:T]=[M:F]$. Let $S=\bigcup_{\sigma\in R}T\sigma$ be a disjoint union. We see that, for every $v\in\mathfrak{f}(T)$, $\prod_{\sigma\in R}(X-v^\sigma)$ has coefficients in F, hence $\mathfrak{f}(T)\subset F_1$. Now for every finite extension M' of M contained in $\mathfrak{f}(T)$, we have $\mathfrak{g}(M')=T$. Taking M' in place of M, we obtain $[S:T]=[M':F]$, so that $M=M'$. This proves $M=\mathfrak{f}(T)$, q. e. d.

6.4. The adelization of GL_2

Throughout the rest of this chapter, we denote by G the group GL_2, viewed as an algebraic group defined over $\boldsymbol{Q}$. We are going to define the adelization G_A of G, the suffix A denoting the adeles of $\boldsymbol{Q}$.[12] First put

$$G_p=GL_2(\boldsymbol{Q}_p)\qquad (p:\text{ rational prime}),$$
$$G_\infty=GL_2(\boldsymbol{R}),$$
$$G_{\infty+}=\{x\in G_\infty\mid\det(x)>0\}.$$

Then G_A is, by definition, the group consisting of all elements $x=(\cdots,x_p,\cdots,x_\infty)$

12) For the general theory of adelization of algebraic groups, see Weil [96].

of $\prod_p G_p \times G_\infty$ such that $x_p \in GL_2(\boldsymbol{Z}_p)$ for all except a finite number of p. $G_{\boldsymbol{A}}$ can be identified with $GL_2(\boldsymbol{A})$. Put

$$U = \prod_p GL_2(\boldsymbol{Z}_p) \times G_{\infty+} .$$

Then U is a subgroup of $G_{\boldsymbol{A}}$, locally compact with respect to the usual product topology. We define the topology of $G_{\boldsymbol{A}}$ by taking U to be an open subgroup of $G_{\boldsymbol{A}}$.

Put $G_{\boldsymbol{Q}} = GL_2(\boldsymbol{Q})$, and consider it a subgroup of $G_{\boldsymbol{A}}$ by the diagonal embedding $\alpha \mapsto (\alpha, \alpha, \alpha, \cdots) \in G_{\boldsymbol{A}}$. We denote by G_0 the non-archimedean part of $G_{\boldsymbol{A}}$, i.e., the set of all elements of $G_{\boldsymbol{A}}$ whose ∞-component is 1. Then we put

$$G_{\boldsymbol{A}+} = G_0 G_{\infty+} ,$$

$$G_{\boldsymbol{Q}+} = G_{\boldsymbol{Q}} \cap G_{\boldsymbol{A}+} = \{\alpha \in GL_2(\boldsymbol{Q}) \mid \det(\alpha) > 0\} .$$

Observe that the map $x \mapsto \det(x)$ defines a continuous homomorphism of $G_{\boldsymbol{A}}$ into $\boldsymbol{Q}_{\boldsymbol{A}}^\times$. We define a homomorphism

(6.4.1) $$\sigma : G_{\boldsymbol{A}} \to \mathrm{Gal}(\boldsymbol{Q}_{ab}/\boldsymbol{Q})$$
$$\text{by} \quad \sigma(x) = [\det(x)^{-1}, \boldsymbol{Q}] \qquad (x \in G_{\boldsymbol{A}}).$$

(For the notation $[s, \boldsymbol{Q}]$ with $s \in \boldsymbol{Q}_{\boldsymbol{A}}^\times$, see § 5.2.) Note that $\sigma(x)=1$ if $x \in G_{\boldsymbol{Q}} G_{\infty+}$.

For any positive integer N, we put

(6.4.2) $$U_N = \{x = (x_p) \in U \mid x_p \equiv 1 \bmod N \cdot M_2(\boldsymbol{Z}_p)\} .$$

Obviously $U = U_1$, and U_N is an open subgroup of $G_{\boldsymbol{A}}$. We also observe:

(6.4.3) *Every open subgroup of $G_{\boldsymbol{A}}$ contains U_N for some N.*

For every open subgroup S of $G_{\boldsymbol{A}}$, we see that $\det(S)$ is open in $\boldsymbol{Q}_{\boldsymbol{A}}^\times$. Therefore the subgroup $\boldsymbol{Q}^\times \cdot \det(S)$ of $\boldsymbol{Q}_{\boldsymbol{A}}^\times$ corresponds to a finite abelian extension of $\boldsymbol{Q}$, which we write $k_S = k(S)$. It is easy to see that $k(U_N) = k_N = \boldsymbol{Q}(e^{2\pi i/N})$, and

(6.4.4) $$k(S) = k(xSx^{-1}) \textit{ for every } x \in G_{\boldsymbol{A}} ,$$

(6.4.5) $$S \subset T \quad \Rightarrow \quad k_T \subset k_S .$$

Let $\mathfrak{x}$ be a $\boldsymbol{Z}$-lattice in $\boldsymbol{Q}^2$. We can then define the action of an element of $G_{\boldsymbol{A}}$ on $\boldsymbol{Q}^2/\mathfrak{x}$ in the same manner as in § 5.2. To do this, first let $\mathfrak{x}_p$ denote the closure of $\mathfrak{x}$ in $\boldsymbol{Q}_p^2$, and identify $\boldsymbol{Q}^2/\mathfrak{x}$ with the direct sum of the modules $\boldsymbol{Q}_p^2/\mathfrak{x}_p$ for all p. For every $c = (c_p) \in G_{\boldsymbol{A}}$, we define $\mathfrak{x}c$ to be the $\boldsymbol{Z}$-lattice in $\boldsymbol{Q}^2$ characterized by the property $(\mathfrak{x}c)_p = \mathfrak{x}_p c_p$. Then right multiplication by c_p defines an isomorphism of $\boldsymbol{Q}_p^2/\mathfrak{x}_p$ onto $\boldsymbol{Q}_p^2/\mathfrak{x}_p c_p$, hence an isomorphism of $\boldsymbol{Q}^2/\mathfrak{x}$ to $\boldsymbol{Q}^2/\mathfrak{x}c$. We shall denote by wc the image of an element w of $\boldsymbol{Q}^2/\mathfrak{x}$ by this isomorphism. The situation is explained by the commutative diagram

$$\begin{array}{ccc} \boldsymbol{Q}_p^2/\mathfrak{x}_p & \xrightarrow{\;c_p\;} & \boldsymbol{Q}^2/\mathfrak{x}_p c_p \\ \downarrow & & \downarrow \\ \boldsymbol{Q}^2/\mathfrak{x} & \xrightarrow{\;c\;} & \boldsymbol{Q}^2/\mathfrak{x}c \end{array}$$

where the vertical arrows are canonical injections. In particular, every element of U gives an automorphism of $\boldsymbol{Q}^2/\boldsymbol{Z}^2$. We also note that

$$U = \{c \in G_{A+} \mid \boldsymbol{Z}^2 c = \boldsymbol{Z}^2\}\,. \tag{6.4.6}$$

Let us now prove a few useful lemmas. Put

$$SL_2(\boldsymbol{A}) = \{x \in G_{\boldsymbol{A}} \mid \det(x) = 1\}\,.$$

LEMMA 6.15. *For every open subgroup S of $G_{\boldsymbol{A}}$, one has*

$$SL_2(\boldsymbol{A}) = SL_2(\boldsymbol{Q}) \cdot (S \cap SL_2(\boldsymbol{A})) = (S \cap SL_2(\boldsymbol{A})) \cdot SL_2(\boldsymbol{Q})\,.$$

PROOF. This is the simplest case of the "strong approximation theorem" for semi-simple algebraic groups. In the present case, it is merely a reformulation of Lemma 1.38. Let $\mathfrak{x} = \boldsymbol{Z}^2$, and let $c \in G_{\boldsymbol{A}}$. Then we can find an element α of $G_{\boldsymbol{Q}}$ such that $\mathfrak{x}c = \mathfrak{x}\alpha$. By (6.4.6), we see that $\alpha c^{-1} \in UG_\infty$ (which proves the equality $G_{\boldsymbol{A}} = U \cdot G_{\boldsymbol{Q}}$). If $c \in SL_2(\boldsymbol{A})$, we have $\det(\alpha) \in \det(UG_\infty) \cap \boldsymbol{Q}^\times = \{\pm 1\}$. Take an element ε of $G_{\boldsymbol{Q}}$ such that $\mathfrak{x}\varepsilon = \mathfrak{x}$ and $\det(\varepsilon) = \det(\alpha)$. Then $\mathfrak{x}c = \mathfrak{x}\varepsilon\alpha$, so that $c \cdot (\varepsilon\alpha)^{-1}$ belongs to $U \cap SL_2(\boldsymbol{A})$. This proves

$$SL_2(\boldsymbol{A}) = (U \cap SL_2(\boldsymbol{A})) \cdot SL_2(\boldsymbol{Q})\,. \tag{6.4.7}$$

In view of (6.4.3), it is sufficient to prove our lemma in the special case $S = U_N$. By virtue of (6.4.7), the question is reduced to showing that

$$U \cap SL_2(\boldsymbol{A}) \subset (U_N \cap SL_2(\boldsymbol{A})) \cdot SL_2(\boldsymbol{Z})\,. \tag{6.4.8}$$

Let $v \in U \cap SL_2(\boldsymbol{A})$. We can find an element β of $M_2(\boldsymbol{Z})$ such that $\beta \equiv v_p \bmod N \cdot M_2(\boldsymbol{Z}_p)$ for all p. Then $\det(\beta) \equiv 1 \bmod (N)$. By Lemma 1.38, there exists an element γ of $SL_2(\boldsymbol{Z})$ such that $\gamma \equiv \beta \bmod (N)$. Then $v\gamma^{-1} \in U_N \cap SL_2(\boldsymbol{A})$, hence (6.4.8), q. e. d.

LEMMA 6.16. *The restriction of σ to G_{A+} is surjective.*

PROOF. Since $G_{\boldsymbol{A}} = G_{A+}G_{\boldsymbol{Q}}$, we have $\sigma(G_{A+}) = \sigma(G_{\boldsymbol{A}}) = [\det(G_{\boldsymbol{A}}), \boldsymbol{Q}]$. It is easy to see that $\det(G_{\boldsymbol{A}}) = \boldsymbol{Q}_{\boldsymbol{A}}^\times$, hence our assertion.

LEMMA 6.17. *Let S be an open subgroup of G_{A+}. Then*

(i) $SG_{\boldsymbol{Q}+} = G_{\boldsymbol{Q}+}S = \{x \in G_{A+} \mid \sigma(x) = \text{id. } on\ k_S\}$,

(ii) *for* $y \in G_{A+}$, *one has* $SG_{Q+}y = \{x \in G_{A+} \mid \sigma(x) = \sigma(y)$ *on* $k_S\}$; *the product* $SG_{Q+}y$ *can be taken in any order of* S, G_{Q+}, *and* y.

PROOF. (i) By our definition of k_S, $\sigma(s) = \text{id.}$ on k_S for $s \in S$. Therefore it is sufficient to show that if $\sigma(x) = \text{id.}$ on k_S for $x \in G_{A+}$, then $x \in SG_{Q+}$ and $x \in G_{Q+}S$. But the hypothesis implies that $\det(x) \in \mathbf{Q}^\times \cdot \det(S)$, hence $\det(x) = \det(\alpha)\det(s)$ for some $\alpha \in G_Q$ and $s \in S$. Then $\det(\alpha) > 0$, and $\det(\alpha^{-1}xs^{-1}) = 1$. By Lemma 6.15, $\alpha^{-1}xs^{-1} = \beta t$ with $\beta \in SL_2(\mathbf{Q})$ and $t \in S$, hence $x = \alpha\beta \cdot ts \in G_{Q+}S$, and similarly $x \in SG_{Q+}$.

(ii) This is only an obvious generalization of (i). Indeed, from (i), we obtain

$$SG_{Q+}y = ySG_{Q+} = G_{Q+}Sy = yG_{Q+}S = \{x \in G_{A+} \mid \sigma(x) = \sigma(y) \text{ on } k_S\}.$$

Further, since $k_T = k_S$ if $T = y^{-1}Sy$, we have $y^{-1}SyG_{Q+} = SG_{Q+}$ by (i), so that $SyG_{Q+} = ySG_{Q+}$. Similarly $G_{Q+}yS = G_{Q+}Sy$.

LEMMA 6.18. *Let* S *be an open subgroup of* G_{A+}. *Then* σ *induces an isomorphism of* G_{A+}/SG_{Q+} *onto* $\mathrm{Gal}(k_S/\mathbf{Q})$, *and*

$$[G_{A+} : SG_{Q+}] = [k_S : \mathbf{Q}] = [\mathbf{Q}_A^\times : \mathbf{Q}^\times \cdot \det(S)].$$

This is an immediate consequence of Lemmas 6.16 and 6.17.

LEMMA 6.19. $G_{A+} = G_{Q+}U = UG_{Q+}$.

This follows immediately from Lemma 6.17, since $k_U = \mathbf{Q}$. More directly, in the proof of Lemma 6.16, we have seen that $G_A = UG_Q$, hence $G_{A+} = G_{Q+}U$.

EXERCISE 6.20. Prove:
(i) The normalizer of U_N in G_{A+} is $U\mathbf{Q}_A^\times$, and $U\mathbf{Q}_A^\times = U\mathbf{Q}^\times$;
(ii) If G^* denotes the closure of $G_{Q+}G_{\infty+}$, then

$$G^* = G_{Q+}G_{\infty+}SL_2(A) = \{x \in G_{A+} \mid \det(x) \in \mathbf{Q}^\times\mathbf{Q}_{\infty+}^\times\}.$$

6.5. The action of U on $\mathfrak{F}$

Let us now come back to the field $\mathfrak{F}_N$ defined in § 6.2. We see easily that $\mathfrak{F}_N \subset \mathfrak{F}_M$ if M is a multiple of N. Therefore, if we put

$$\mathfrak{F} = \bigcup_{N=1}^{\infty} \mathfrak{F}_N,$$

then $\mathfrak{F}$ is a Galois extension of $\mathfrak{F}_1$, and $\mathbf{C} \cdot \mathfrak{F}$ is the field of all modular functions of all levels. Further we see that $\mathbf{Q}_{ab}$ is the algebraic closure of $\mathbf{Q}$ in $\mathfrak{F}$, so that $\mathfrak{F}$ and $\mathbf{C}$ are linearly disjoint over $\mathbf{Q}_{ab}$. Our next goal is the determination of $\mathrm{Aut}(\mathfrak{F})$ (Th. 6.23). In this section, we study the part of $\mathrm{Aut}(\mathfrak{F})$ obtained from the elements of U, and its relation with the substitution

$z \mapsto \alpha(z)$ for any $\alpha \in G_{Q+}$. For convenience, we understand that the suffix a in the notation f_a indicates also an element of Q^2/Z^2, since f_a depends only on the class of a mod Z^2. (It is also convenient to put $f_0 = j$. But we shall not do this for fear of confusion.)

PROPOSITION 6.21. *For every $u \in U$, one can define an element $\tau(u)$ of* $\mathrm{Gal}(\mathfrak{F}/\mathfrak{F}_1)$ *by $f_a^{\tau(u)} = f_{au}$ for all $a \in Q^2/Z^2$, $\neq 0$. Moreover, $\tau(u)$ has the following properties:*

(1) *The sequence $1 \to \{\pm 1\} \cdot G_{\infty+} \to U \to \mathrm{Gal}(\mathfrak{F}/\mathfrak{F}_1) \to 1$ is exact.*

(2) *$\tau(u) = \sigma(u)$ on Q_{ab}.*

(3) *$h^{\tau(\gamma)} = h \circ \gamma$ for all $h \in \mathfrak{F}$ and $\gamma \in SL_2(Z)$.*

PROOF. For every $u \in U$ and every N, there exists an element α of $M_2(Z) \cap G_{Q+}$ such that $u_p \equiv \alpha \bmod N \cdot M_2(Z_p)$ for all p. Then $au = a\alpha$ for every $a \in N^{-1}Z^2/Z^2$. Therefore, by Th. 6.6, $f_a \mapsto f_{au}$ defines an element of $\mathrm{Gal}(\mathfrak{F}_N/\mathfrak{F}_1)$, hence an element of $\mathrm{Gal}(\mathfrak{F}/\mathfrak{F}_1)$. Call it $\tau(u)$. By (2) of Th. 6.6, we see that the restriction of $\tau(u)$ to $\mathfrak{F}_N$ defines an exact sequence

$$1 \longrightarrow \{\pm 1\} \cdot U_N \longrightarrow U \longrightarrow \mathrm{Gal}(\mathfrak{F}_N/\mathfrak{F}_1) \longrightarrow 1. \tag{6.5.1}$$

Therefore $\mathrm{Ker}(\tau) = \bigcap_{N=1}^{\infty} \{\pm 1\} \cdot U_N = \{\pm 1\} \cdot G_{\infty+}$; τ is a continuous homomorphism of U to $\mathrm{Gal}(\mathfrak{F}/\mathfrak{F}_1)$; and $\tau(U)$ is dense in $\mathrm{Gal}(\mathfrak{F}/\mathfrak{F}_1)$. Since $U/G_{\infty+}$ is compact, we obtain the assertion (1). To see (2), let u and α be as above. Define two elements c and c' of $Q_A^\times$ by

$$c_p = \begin{cases} \det(\alpha) & \text{for } p \mid N \\ 1 & \text{for } p \nmid N \text{ or } p = \infty, \end{cases}$$

and $cc' = \det(\alpha)$. Then we have, by (3) of Th. 6.6,

$$\begin{aligned} \tau(u) = \left(\frac{k_N/Q}{\det(\alpha)}\right) = [c', Q] = [\det(\alpha)^{-1}c', Q] = [c^{-1}, Q] \\ = [\det(u)^{-1}, Q] = \sigma(u) \qquad \text{on} \quad k_N, \end{aligned}$$

so that $\tau(u) = \sigma(u)$ on k_N for every N, hence (2). If $u = \gamma \in SL_2(Z)$, we can take γ as the above α, so that $f_a^{\tau(u)} = f_{a\gamma} = f_a \circ \gamma$ by (6.1.3), hence (3).

PROPOSITION 6.22. (1) *For every $\alpha \in G_{Q+}$ and for every $h \in \mathfrak{F}$, the function $h \circ \alpha$ belongs to $\mathfrak{F}$.*

(2) *If $\alpha \in G_{Q+}$, $\beta \in G_{Q+}$, $u \in U$, $v \in U$, and $\alpha u = v\beta$, then $(j \circ \alpha)^{\tau(u)} = j \circ \beta$, and $(f_a \circ \alpha)^{\tau(u)} = f_{av} \circ \beta$ for every $a \in Q^2/Z^2$, $\neq 0$.*

PROOF. Let $\mathfrak{F}'$ be the field generated over Q by the functions $h \circ \alpha$ for all $h \in \mathfrak{F}$ and all $\alpha \in G_{Q+}$. By Lemma 6.5, there exists a point z_0 of $\mathfrak{H}$ such that $g \mapsto g(z_0)$ defines an isomorphism of $\mathfrak{F}'$ onto the subfield $\mathfrak{F}_0' =$

$\boldsymbol{Q}(h(\alpha(z_0)) \mid \alpha \in G_{\boldsymbol{Q}+},\, h \in \mathfrak{F})$ of $\boldsymbol{C}$. Therefore, it is sufficient to prove the assertions corresponding to (1) and (2) on the field $\mathfrak{F}_0'$. To prove (1) and (2), taking suitable scalar multiples of α and β instead of α and β, we may assume that α^{-1} and β^{-1} belong to $M_2(\boldsymbol{Z})$. Now, for every $z \in \mathfrak{H}$, put $L(z) = \boldsymbol{Z}z + \boldsymbol{Z}$. Define c and E by

$$E:\ y^2 = 4x^3 - cx - c\,, \qquad c/(c-27) = j(z_0)\,.$$

Take any isomorphism ξ of $\boldsymbol{C}/L(z_0)$ to E, and put

$$t(a) = \xi\left(a\begin{bmatrix} z_0 \\ 1 \end{bmatrix}\right) \qquad (a \in \boldsymbol{Q}^2/\boldsymbol{Z}^2)\,.$$

By Lemma 6.4, $f_a(z_0) = h_E^1(t(a))$. To simplify our notation, put $\alpha = \alpha_1$, $\beta = \alpha_2$, and $w_i = \alpha_i(z_0)$ for $i = 1, 2$. Then there exist $\mu_i \in \boldsymbol{C}^\times$ such that

$$\begin{bmatrix} z_0 \\ 1 \end{bmatrix}\mu_i = \alpha_i^{-1}\begin{bmatrix} w_i \\ 1 \end{bmatrix} \qquad (i = 1, 2)\,,$$

and multiplication by μ_i defines an isogeny of $\boldsymbol{C}/L(z_0)$ to $\boldsymbol{C}/L(w_i)$. Define c_i and E_i, for $i = 1, 2$, by

$$E_i:\ y^2 = 4x^3 - c_i x - c_i\,, \qquad c_i/(c_i - 27) = j(w_i)\,.$$

Let η_i be an isomorphism of $\boldsymbol{C}/L(w_i)$ to E_i, and put

$$s_i(a) = \eta_i\left(a\begin{bmatrix} w_i \\ 1 \end{bmatrix}\right) \qquad (a \in \boldsymbol{Q}^2/\boldsymbol{Z}^2;\ i = 1, 2)\,.$$

Then there exists an isogeny λ_i of E onto E_i such that the following diagram is commutative.

$$\begin{array}{ccccc} \boldsymbol{C} & \longrightarrow & \boldsymbol{C}/L(z_0) & \xrightarrow{\ \xi\ } & E \\ {\scriptstyle \mu_i}\downarrow & & \downarrow & & \downarrow{\scriptstyle \lambda_i} \\ \boldsymbol{C} & \longrightarrow & \boldsymbol{C}/L(w_i) & \xrightarrow{\ \eta_i\ } & E_i \end{array}$$

Then

$$\lambda_i(t(a)) = \lambda_i\left(\xi\left(a\begin{bmatrix} z_0 \\ 1 \end{bmatrix}\right)\right) = \eta_i\left(a\begin{bmatrix} z_0 \\ 1 \end{bmatrix}\mu_i\right) = \eta_i\left(a\alpha_i^{-1}\begin{bmatrix} w_i \\ 1 \end{bmatrix}\right) = s_i(a\alpha_i^{-1})$$

for every $a \in \boldsymbol{Q}^2/\boldsymbol{Z}^2$. Therefore $\mathrm{Ker}\,(\lambda_i) = t(\boldsymbol{Z}^2\alpha_i/\boldsymbol{Z}^2)$. Now consider the automorphism σ of $\boldsymbol{Q}(h(z_0) \mid h \in \mathfrak{F})$ over $\boldsymbol{Q}(c)$ such that $f_a(z_0)^\sigma = f_{au}(z_0)$ for all $a \in \boldsymbol{Q}^2/\boldsymbol{Z}^2$, $\neq 0$ (this corresponds to $\tau(u)$). Extend it to an automorphism of $\boldsymbol{C}$, and denote it again by σ. Then we see that $E^\sigma = E$, and by (4.5.3), $t(a)^\sigma = \pm t(au)$, since $f_a(z_0) = h_E^1(t(a))$. It follows that

$$\mathrm{Ker}\,(\lambda_1^\sigma) = \mathrm{Ker}\,(\lambda_1)^\sigma = t(\boldsymbol{Z}^2\alpha u/\boldsymbol{Z}^2) = t(\boldsymbol{Z}^2 v\beta/\boldsymbol{Z}^2) = t(\boldsymbol{Z}^2\beta/\boldsymbol{Z}^2) = \mathrm{Ker}\,(\lambda_2)\,.$$

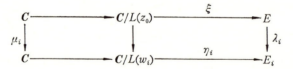

Therefore E_1^σ is isomorphic to E_2, so that $j(w_1)^\sigma = j(w_2)$, hence $c_1^\sigma = c_2$, and $E_1^\sigma = E_2$. Both λ_1^σ and λ_2 are isogenies of E onto E_2 with the same kernel, so that $\lambda_1^\sigma = \varepsilon\lambda_2$ with an automorphism ε of E_2. Since $j(z_0)$ is transcendental, E, E_1, E_2 have no complex multiplication, hence $\varepsilon = \pm 1$, so that $\lambda_1^\sigma = \pm\lambda_2$. Therefore, for every $a \in \boldsymbol{Q}^2/\boldsymbol{Z}^2$, we have

$$s_1(a\alpha^{-1})^\sigma = (\lambda_1(t(a)))^\sigma = \lambda_1^\sigma(t(a)^\sigma) = \pm\lambda_2(\pm t(au)) = \pm\lambda_2(t(au)) = \pm s_2(au\beta^{-1}).$$

Let $b = a\alpha^{-1}$. Then $au\beta^{-1} = b\alpha u\beta^{-1} = bv$. (In fact, let $\bar{a}$ be an element of $\boldsymbol{Q}^2$ which represents a, and $\bar{b} = \bar{a}\alpha^{-1}$. Then $\bar{a}u_p\beta^{-1} = \bar{b}\alpha u_p\beta^{-1} = \bar{b}v_p$, which shows $au\beta^{-1} = bv$.) Since $a \mapsto a\alpha^{-1} = b$ is a surjective endomorphism of $\boldsymbol{Q}^2/\boldsymbol{Z}^2$, we obtain $s_1(b)^\sigma = \pm s_2(bv)$ for every $b \in \boldsymbol{Q}^2/\boldsymbol{Z}^2$. By Lemma 6.4, we have, for every $b \in \boldsymbol{Q}^2/\boldsymbol{Z}^2$,

$$f_b(w_1)^\sigma = h_{E_1}^1(s_1(b))^\sigma = h_{E_2}^1(s_2(bv)) = f_{bv}(w_2).$$

Thus we have proved

$$(6.5.2) \qquad j(\alpha(z_0))^\sigma = j(\beta(z_0)), \qquad f_b(\alpha(z_0))^\sigma = f_{bv}(\beta(z_0)) \qquad (b \in \boldsymbol{Q}^2/\boldsymbol{Z}^2).$$

This applies to any automorphism σ of $\boldsymbol{C}$ over $\boldsymbol{Q}(j(z_0))$ such that $f_a(z_0)^\sigma = f_{au}(z_0)$. Suppose especially that σ is the identity on $\boldsymbol{Q}(h(z_0) \mid h \in \mathfrak{F})$. Then by (1) of Prop. 6.21, $u \in \{\pm 1\} \cdot G_{\infty+}$, and we can apply the formula (6.5.2) to the case $\alpha = \beta$, $v = \alpha u \alpha^{-1} \in \{\pm 1\} \cdot G_{\infty+}$. Then we see that σ leaves the elements $j(\alpha(z_0))$ and $f_b(\alpha(z_0))$ invariant. It follows that these elements belong to the field $\boldsymbol{Q}(h(z_0) \mid h \in \mathfrak{F})$. By virtue of our choice of z_0, this proves (1) of our proposition. Then the assertion (2) follows from (6.5.2).

6.6. The structure of Aut $(\mathfrak{F})$

We shall now define a homomorphism

$$\tau: \ G_{A+} \longrightarrow \text{Aut}(\mathfrak{F}).$$

By Lemma 6.19, we have $G_{A+} = UG_{Q+} = G_{Q+}U$. For $u \in U$, we define $\tau(u)$ to be the same as the element $\tau(u)$ of Gal $(\mathfrak{F}/\mathfrak{F}_1)$ defined in Prop. 6.21. As for $\alpha \in G_{Q+}$, we define $\tau(\alpha)$ by

$$(6.6.1) \qquad h^{\tau(\alpha)} = h \circ \alpha \qquad \text{for all} \quad h \in \mathfrak{F}.$$

Obviously this defines a homomorphism of G_{Q+} into Aut $(\mathfrak{F})$. Thus the symbol τ is defined on $SL_2(\boldsymbol{Z}) = U \cap G_{Q+}$ in two different ways, but, both definitions coincide by virtue of (3) of Prop. 6.21. Now, for $x = u\alpha \in G_{A+}$ with $u \in U$ and $\alpha \in G_{Q+}$, we put $\tau(x) = \tau(u)\tau(\alpha)$ so that

$$j^{\tau(x)} = j \circ \alpha, \qquad f_a^{\tau(x)} = f_{au} \circ \alpha.$$

If $x=u'\alpha'$ is another expression with $u'\in U$ and $\alpha'\in G_{\mathbf{Q}+}$, then $u^{-1}u'=\alpha\alpha'^{-1}\in SL_2(\mathbf{Z})$. Therefore, putting $\delta=u^{-1}u'$, we have

$$\tau(u')\tau(\alpha')=\tau(u\delta)\tau(\delta^{-1}\alpha)=\tau(u)\tau(\delta)\tau(\delta)^{-1}\tau(\alpha)=\tau(u)\tau(\alpha),$$

since τ is multiplicative on U and on $G_{\mathbf{Q}+}$. Thus the symbol $\tau(x)$ is well defined independently of the choice of u and α. We have to show that τ is actually a homomorphism. To see this, let $x=u\alpha$ and $y=v\beta$ with $u\in U$, $v\in U$, $\alpha\in G_{\mathbf{Q}+}$, $\beta\in G_{\mathbf{Q}+}$. Since $G_{\mathbf{A}+}=UG_{\mathbf{Q}+}$, there exist elements $w\in U$ and $\gamma\in G_{\mathbf{Q}+}$ such that $\alpha v=w\gamma$. By our definition,

$$\tau(xy)=\tau(uw)\tau(\gamma\beta)=\tau(u)\tau(w)\tau(\gamma)\tau(\beta) \quad \text{and} \quad \tau(x)\tau(y)=\tau(u)\tau(\alpha)\tau(v)\tau(\beta).$$

Therefore it is sufficient to show that $\tau(w)\tau(\gamma)=\tau(\alpha)\tau(v)$. But this is nothing but the assertion (2) of Prop. 6.22.

Since both $\tau(\alpha)$ and $\sigma(\alpha)$ are trivial on $\mathbf{Q}_{ab}$ if $\alpha\in G_{\mathbf{Q}+}$, we obtain, from (2) of Prop. 6.21,

(6.6.2) $$\tau(x)=\sigma(x) \text{ on } \mathbf{Q}_{ab} \text{ for every } x\in G_{\mathbf{A}+}.$$

Let us now prove

(6.6.3) $$\mathbf{Q}^\times U_N=\{x\in G_{\mathbf{A}+} \mid \tau(x)=\text{id. on } \mathfrak{F}_N\}.$$

The inclusion $\subset$ is obvious in view of (6.5.1). Let $x\in G_{\mathbf{A}+}$, and suppose that $\tau(x)=\text{id.}$ on $\mathfrak{F}_N$. By (6.6.2), $\sigma(x)=\text{id.}$ on k_N. Therefore, by Lemma 6.17, $x=u\alpha$ with $u\in U_N$ and $\alpha\in G_{\mathbf{Q}+}$. Then $\tau(\alpha)=\text{id.}$ on $\mathfrak{F}_N$, hence $\alpha\in\mathbf{Q}^\times\Gamma_N$, so that $x\in\mathbf{Q}^\times U_N$, which completes the proof of (6.6.3).

From (6.6.3), we obtain

$$\operatorname{Ker}(\tau)=\textstyle\bigcap_{N=1}^{\infty}\mathbf{Q}^\times U_N=\text{the closure of } \mathbf{Q}^\times G_{\infty+}=\mathbf{Q}^\times G_{\infty+}$$

(since $\mathbf{Q}^\times\mathbf{Q}_{\infty+}^\times$ is closed in $\mathbf{Q}_{\mathbf{A}}^\times$).

The relation (6.6.3) shows also that τ is continuous, and moreover, τ induces an open map of $G_{\mathbf{A}+}/\mathbf{Q}^\times G_{\infty+}$ to $\tau(G_{\mathbf{A}+})$. Therefore τ induces a topological isomorphism of $G_{\mathbf{A}+}/\mathbf{Q}^\times G_{\infty+}$ onto $\tau(G_{\mathbf{A}+})$. By (1) of Prop. 6.21, $\tau(U)=\operatorname{Gal}(\mathfrak{F}/\mathfrak{F}_1)$. Since $\operatorname{Gal}(\mathfrak{F}/\mathfrak{F}_1)$ is open in $\operatorname{Aut}(\mathfrak{F})$, it follows that $\tau(G_{\mathbf{A}+})$ is open, and hence closed, in $\operatorname{Aut}(\mathfrak{F})$.[13] Now we have one of the main theorems of our theory:

THEOREM 6.23. *The sequence*

$$1\longrightarrow\mathbf{Q}^\times G_{\infty+}\longrightarrow G_{\mathbf{A}+}\xrightarrow{\ \tau\ }\operatorname{Aut}(\mathfrak{F})\longrightarrow 1$$

is exact, so that $\operatorname{Aut}(\mathfrak{F})$ *is isomorphic to* $G_{\mathbf{A}+}/\mathbf{Q}^\times G_{\infty+}$ *as a topological group.*

13) That $\tau(G_{\mathbf{A}+})$ is closed can be shown also as follows. Since $\tau(G_{\mathbf{A}+})$ is homeomorphic to $G_{\mathbf{A}+}/\mathbf{Q}^\times G_{\infty+}$, it is locally compact, and hence closed in $\operatorname{Aut}(\mathfrak{F})$, by virtue of Prop. 1.4.

PROOF. Since $\tau(G_{A+})$ is closed in Aut $(\mathfrak{F})$, it is sufficient to show that $\tau(G_{A+})$ is dense in Aut $(\mathfrak{F})$. Let $\zeta \in$ Aut $(\mathfrak{F})$. By Lemma 6.16, there exists an element y of G_{A+} such that $\sigma(y)=\zeta$ on $\boldsymbol{Q}_{ab}$. Put $\pi=\zeta\cdot\tau(y)^{-1}$. Then π is the identity map on $\boldsymbol{Q}_{ab}$. Since $\mathfrak{F}$ and $\boldsymbol{C}$ are linearly disjoint over $\boldsymbol{Q}_{ab}$, we can extend π to an automorphism of $\boldsymbol{C}\mathfrak{F}$ over $\boldsymbol{C}$, which we denote again by π. Take and fix any positive integer $N>2$. We can find two positive integers M and M' such that $N<M<M'$, $\mathfrak{F}_N^{\pi^{-1}}\subset\mathfrak{F}_M$, and $\mathfrak{F}_M^{\pi}\subset\mathfrak{F}_{M'}$. Then we have $\boldsymbol{C}\mathfrak{F}_N\subset\boldsymbol{C}\mathfrak{F}_M^{\pi}\subset\boldsymbol{C}\mathfrak{F}_{M'}$, so that there exists a subgroup $\varDelta$ of $\varGamma_N$ containing $\varGamma_{M'}$ such that $\boldsymbol{C}\mathfrak{F}_M^{\pi}$ is the field of all modular functions with respect to $\varDelta$.

Let $\mathfrak{H}^*$ be the union of $\mathfrak{H}$ and the cusps of $\varGamma_1$. Put $V=\mathfrak{H}^*/\varGamma_M$ and $V'=\mathfrak{H}^*/\varDelta$, and denote by φ (resp. φ') the projection map of $\mathfrak{H}^*$ to V (resp. V'). Then V and V' are compact Riemann surfaces, and $\boldsymbol{C}\mathfrak{F}_M$ (resp. $\boldsymbol{C}\mathfrak{F}_M^{\pi}$) can be identified with the field $\boldsymbol{C}(V)$ (resp. $\boldsymbol{C}(V')$) of all meromorphic functions on V (resp. V'), through the map $\boldsymbol{C}(V)\ni f\mapsto f\circ\varphi$ (resp. $\boldsymbol{C}(V')\ni f\mapsto f\circ\varphi'$). Since π is an isomorphism of $\boldsymbol{C}\mathfrak{F}_M$ onto $\boldsymbol{C}\mathfrak{F}_M^{\pi}$ over $\boldsymbol{C}$, there exists a biregular isomorphism η of V' onto V such that $(f\circ\varphi)^{\pi}=f\circ\eta\circ\varphi'$ for every $f\in\boldsymbol{C}(V)$. Put $V_0=\varphi(\mathfrak{H})$, and $V_0'=\varphi'(\mathfrak{H})$. We are going to show that $\eta(V_0')=V_0$. Let $p\in V_0'$, and assume that $\eta(p)\notin V_0$, i.e., $\eta(p)=\varphi(s)$ with a cusp s of $\varGamma_M$. If v is the discrete valuation of $\boldsymbol{C}\mathfrak{F}_M^{\pi}$ corresponding to the point p, then v is unramified in $\boldsymbol{C}\mathfrak{F}$, since $p=\varphi'(z)$ with a point z on $\mathfrak{H}$ which is not elliptic. (Here observe that neither $\varGamma_M$ nor $\varDelta$ has elliptic elements, since $N>2$.) Now define a valuation v^* of $\boldsymbol{C}\mathfrak{F}_M$ by $v^*(h)=v(h^{\pi})$ for $h\in\boldsymbol{C}\mathfrak{F}_M$. Since π is an automorphism of $\boldsymbol{C}\mathfrak{F}$, v^* must be unramified in $\boldsymbol{C}\mathfrak{F}$. On the other hand, v^* is the discrete valuation of $\boldsymbol{C}\mathfrak{F}_M$ corresponding to the point $\eta(p)=\varphi(s)$. Since s is a cusp, v^* is ramified in $\boldsymbol{C}\mathfrak{F}$. (In fact, if L is a multiple of M, the ramification index of v^* in $\boldsymbol{C}\mathfrak{F}_L$ is L/M, see Prop. 1.37 and § 1.6.) Thus we get a contradiction, hence $\eta(p)$ must be contained in V_0. Similarly we can prove that η^{-1} maps V_0 into V_0', hence η gives a biregular isomorphism of V_0' onto V_0. Since $V_0'=\mathfrak{H}/\varDelta$, $V_0=\mathfrak{H}/\varGamma_M$, and neither $\varDelta$ nor $\varGamma_M$ has elliptic elements, we can find an element β of $SL_2(\boldsymbol{R})$ such that $\varphi\circ\beta=\eta\circ\varphi'$, and $\beta^{-1}(\{\pm 1\}\cdot\varGamma_M)\beta=\{\pm 1\}\cdot\varDelta$. Observe that $\varGamma_d$ spans $M_2(\boldsymbol{Q})$ over $\boldsymbol{Q}$, for every positive integer d. (In fact, four elements $\begin{bmatrix}1 & d\\ 0 & 1\end{bmatrix}$, $\begin{bmatrix}1 & 0\\ d & 1\end{bmatrix}$, $\begin{bmatrix}d^2+1 & d\\ d & 1\end{bmatrix}$, $\begin{bmatrix}1 & d\\ d & d^2+1\end{bmatrix}$ of $\varGamma_d$ are linearly independent over $\boldsymbol{Q}$.) Therefore we have $\beta^{-1}M_2(\boldsymbol{Q})\beta=M_2(\boldsymbol{Q})$, so that $x\mapsto\beta^{-1}x\beta$ is an automorphism of $M_2(\boldsymbol{Q})$. By a well known theorem, there exists an element α of $GL_2(\boldsymbol{Q})$ such that $\beta^{-1}x\beta=\alpha^{-1}x\alpha$ for all $x\in M_2(\boldsymbol{Q})$. Then $\alpha\beta^{-1}x=x\alpha\beta^{-1}$ for all $x\in M_2(\boldsymbol{Q})$, so that $\alpha\beta^{-1}=c\cdot 1_2$ with $c\in\boldsymbol{R}^{\times}$. It follows that $\alpha=c\beta$, $\det(\alpha)=c^2>0$, and $\varphi\circ\alpha=\varphi\circ\beta=\eta\circ\varphi'$, hence $(f\circ\varphi)^{\pi}=f\circ\eta\circ\varphi'=f\circ\varphi\circ\alpha$ for every $f\in\boldsymbol{C}(V)$, i.e., $h^{\pi}=h\circ\alpha$ for every $h\in\boldsymbol{C}\mathfrak{F}_M$. Therefore we have $\pi=\tau(\alpha)$ on $\mathfrak{F}_M$, so that $\zeta=\pi\cdot\tau(y)=\tau(\alpha y)$ on $\mathfrak{F}_M$.

Since M can be taken arbitrarily large, this shows that $\tau(G_{A+})$ is dense in Aut $(\mathfrak{F})$, and completes the proof.

There is an obvious analogy between the above theorem and the fundamental exact sequence (5.2.1) of class field theory. Actually they are not only analogous, but also closely connected with each other by a certain explicit formula, which describes the behavior of the values of the functions of $\mathfrak{F}$ at special points belonging to imaginary quadratic fields. We shall discuss this in § 6.8.

EXERCISE 6.24. Let $\mathfrak{F}'$ be the subfield of $\mathfrak{F}$ generated over $\boldsymbol{Q}$ by the functions $j\circ\alpha$ for all $\alpha\in G_{\boldsymbol{Q}+}$. Prove: (i) The subgroup of G_{A+} corresponding to $\mathfrak{F}'$ (in the sense of Prop. 6.11) is $\boldsymbol{Q}_A^\times\cdot G_{\infty+}$; (ii) $\boldsymbol{Q}_{ab}\cap\mathfrak{F}'$ is the composite of all quadratic extensions of $\boldsymbol{Q}$; (iii) The subgroup of G_{A+} corresponding to $\boldsymbol{Q}_{ab}\mathfrak{F}'$ is $\{x\in\boldsymbol{Q}_A^\times G_{\infty+}\mid \det(x)\in\boldsymbol{Q}^\times\boldsymbol{Q}_{\infty+}^\times\}$; (iv) Every element of Aut $(\mathfrak{F}')$ is extensible to an element of Aut $(\mathfrak{F})$; (v) Aut $(\mathfrak{F}')$ is (canonically) isomorphic to $G_{A+}/\boldsymbol{Q}_A^\times G_{\infty+}$ (cf. Prop. 6.8).

EXERCISE 6.25. Show that every automorphism of $\mathfrak{F}_N$ extensible to an element of Aut $(\mathfrak{F})$ must belong to Gal $(\mathfrak{F}_N/\mathfrak{F}_1)$, i. e., it is the restriction of an element of $\tau(U)$ to $\mathfrak{F}_N$. Especially, no automorphism of $\mathfrak{F}_1$ other than the identity map is extensible to an automorphism of $\mathfrak{F}$.

EXERCISE 6.26. Let $\mathfrak{F}_0$ be the subfield of $\mathfrak{F}$ consisting of the elements invariant under $\tau(x)$ for all $x=\begin{bmatrix}1 & 0\\ 0 & d\end{bmatrix}\in U$ (with $d\in\boldsymbol{Q}_A^\times$). Prove: (i) $\mathfrak{F}=\boldsymbol{Q}_{ab}\mathfrak{F}_0$; (ii) $\mathfrak{F}_0\cap\boldsymbol{Q}_{ab}=\boldsymbol{Q}$; (iii) $\mathfrak{F}_0$ is generated over $\boldsymbol{Q}$ by the functions $j(Nz)$ and f_a with $a=(1/N, 0)$ for all positive integers N; (iv) $\mathfrak{F}_0$ is the field of all modular functions (of any level) with rational Fourier coefficients at ∞ (with respect to $e^{2\pi iz/N}$ for some N) (cf. Prop. 6.9).

6.7. The canonical system of models of $\Gamma\backslash\mathfrak{H}^*$ for all congruence subgroups Γ of $GL_2(\boldsymbol{Q})$

Before discussing the main topic of this section, let us first introduce the notion of a model of $\Gamma\backslash\mathfrak{H}^*$, where Γ is a Fuchsian group of the first kind, and $\mathfrak{H}^*$ is the union of $\mathfrak{H}$ and the cusps of Γ. (Γ may be a subgroup of $SL_2(\boldsymbol{R})$, $SL_2(\boldsymbol{R})/\{\pm 1\}$, $G_{\infty+}$, or $G_{\infty+}/\boldsymbol{R}^\times$.) Since $\Gamma\backslash\mathfrak{H}^*$ is a compact Riemann surface, as shown in § 1.5, there exists a projective non-singular algebraic curve V, defined over (a subfield of) $\boldsymbol{C}$, biregularly isomorphic to $\Gamma\backslash\mathfrak{H}^*$. It is often convenient to specify a Γ-invariant holomorphic map φ of $\mathfrak{H}^*$ to V which gives a biregular isomorphism of $\Gamma\backslash\mathfrak{H}^*$ to V. If V and φ are in that situation, we call (V, φ) a *model* of $\Gamma\backslash\mathfrak{H}^*$. For example, if $\Gamma=SL_2(\boldsymbol{Z})$ and

P^1 denotes the projective line, (P^1, j) is a model of $\Gamma\backslash\mathfrak{H}^*$.

Coming back to the general case, let Γ' be another Fuchsian group of the first kind, $\mathfrak{H}^{*\prime}$ the union of $\mathfrak{H}$ and the cusps of Γ', and (V', φ') a model of $\Gamma'\backslash\mathfrak{H}^{*\prime}$. Suppose that $\alpha\Gamma\alpha^{-1}\subset\Gamma'$ with an element α of $G_{\infty+}$. Then, as is shown in § 2.1, we can define a rational map T of V to V' by $T(\varphi(z))=\varphi'(\alpha(z))$, i. e., by the following commutative diagram:

$$\begin{array}{ccc} \mathfrak{H}^* & \xrightarrow{\ \alpha\ } & \mathfrak{H}^{*\prime} \\ \varphi\downarrow & & \downarrow\varphi' \\ V & \xrightarrow{\ T\ } & V' \end{array}$$

This includes, as special cases, the following two types of maps:

CASE a: $\alpha=1$, hence $\Gamma\subset\Gamma'$. Then T is the usual projection map.

CASE b: $\alpha\Gamma\alpha^{-1}=\Gamma'$. Then T is a biregular isomorphism of V to V'.

Now the purpose of this section is to discuss the following question, which is actually somewhat too naive a problem setting, though, so that a modification will be made afterwards.

To any Fuchsian group Γ which is contained in G_{Q+} and contains Γ_N for some N, associate, once for all, a model $(V_\Gamma, \varphi_\Gamma)$ of $\Gamma\backslash\mathfrak{H}^*$, and an algebraic number field k_Γ in such a way that the following conditions are satisfied:

(1) *V_Γ is defined over k_Γ.*

(2) *If $\alpha\in G_{Q+}$ is such that $\alpha\Gamma\alpha^{-1}\subset\Delta$, then $k_\Delta\subset k_\Gamma$, and the rational map T of V_Γ to V_Δ defined by $T\circ\varphi_\Gamma=\varphi_\Delta\circ\alpha$ is rational over k_Γ.*

Here and henceforth $\mathfrak{H}^*$ means of course $\mathfrak{H}\cup\boldsymbol{Q}\cup\{\infty\}$.

Suppose we could find such a system of $(V_\Gamma, \varphi_\Gamma)$ and k_Γ. Then consider a field

$$\mathfrak{F}_\Gamma=\{f\circ\varphi_\Gamma \mid f\in k_\Gamma(V_\Gamma)\},$$

where $k_\Gamma(V_\Gamma)$ denotes the field of functions on V_Γ rational over k_Γ, see Appendix N° 4. It is natural to assume that $\mathfrak{F}_\Delta=\mathfrak{F}_N$ if $\Delta=\Gamma_N$. By our assumption, Γ contains Γ_N for some N. In view of the condition (2), we see that $k_\Gamma\subset k_N$, and $\mathfrak{F}_\Gamma\subset\mathfrak{F}_N$. Therefore $\mathfrak{F}_\Gamma$ is a subfield of $\mathfrak{F}$. Then, (assuming that $\mathfrak{F}$ is a Galois extension of $\mathfrak{F}_\Gamma$,) $\mathfrak{F}_\Gamma$ corresponds to an open compact subgroup of Aut $(\mathfrak{F})$ by Prop. 6.12. Now Aut $(\mathfrak{F})$ is isomorphic to $G_{A+}/\boldsymbol{Q}^\times G_{\infty+}$. Therefore, it seems reasonable to consider, instead of the family of Γ, the family of all open compact subgroups of $G_{A+}/\boldsymbol{Q}^\times G_{\infty+}$, or the subgroups of G_{A+} corresponding to them.

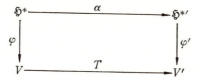

Thus we are led to consider the set $\mathfrak{Z}$ of all open subgroups S of G_{A+} containing $\boldsymbol{Q}^\times G_{\infty+}$ such that $S/\boldsymbol{Q}^\times G_{\infty+}$ is compact. We see easily that $\mathfrak{Z}$ has the following properties:

(6.7.1) *If* $S \in \mathfrak{Z}$ *and* $T \in \mathfrak{Z}$, *then* S *and* T *are commensurable, and* $S \cap T \in \mathfrak{Z}$.

(6.7.2) *If* $S \in \mathfrak{Z}$ *and* $x \in G_A$, *then* $xSx^{-1} \in \mathfrak{Z}$.

Put, for each $S \in \mathfrak{Z}$,

$$\Gamma_S = S \cap G_{Q+},$$

$$\mathfrak{F}_S = \{h \in \mathfrak{F} \mid h^{\tau(x)} = h \text{ for all } x \in S\}.$$

By Prop. 6.12, $\mathfrak{F}_S$ is finitely generated over $\boldsymbol{Q}$, $\mathfrak{F}$ is a Galois extension of $\mathfrak{F}_S$, and

(6.7.3) $$S = \{x \in G_{A+} \mid \tau(x) = \text{id. on } \mathfrak{F}_S\}, \quad \text{i.e.,} \quad \tau(S) = \mathrm{Gal}(\mathfrak{F}/\mathfrak{F}_S).$$

For example, if $S = \boldsymbol{Q}^\times U_N$, we have $\Gamma_S = (\boldsymbol{Q}^\times U_N) \cap G_{Q+} = \boldsymbol{Q}^\times (U_N \cap G_{Q+}) = \boldsymbol{Q}^\times \Gamma_N$, so that the group Γ_S (or rather $\Gamma_S/\boldsymbol{Q}^\times$), as a transformation group on $\mathfrak{H}$, is the same as Γ_N. Moreover, $\mathfrak{F}_S = \mathfrak{F}_N$, by (6.6.3) and Prop. 6.11. In general, we have the following

PROPOSITION 6.27. *For any* $S \in \mathfrak{Z}$, Γ_S *is commensurable with* $\boldsymbol{Q}^\times \Gamma_1$, *(so that* $\Gamma_S/\boldsymbol{Q}^\times$ *is a Fuchsian group of the first kind commensurable with* $\Gamma_1/\{\pm 1\}$*), and* $\boldsymbol{C}\mathfrak{F}_S$ *is the field of all automorphic functions with respect to* Γ_S. *Furthermore,* k_S *is algebraically closed in* $\mathfrak{F}_S$, *where* k_S *is as in* § 6.4, p. 144.

PROOF. By (6.4.3), S contains $\boldsymbol{Q}^\times U_N$ for some N. Put $T = \boldsymbol{Q}^\times U_N$. By (6.7.1), we have $[S:T] < \infty$, so that $[\Gamma_S : \Gamma_T] < \infty$. Since $\Gamma_T = \boldsymbol{Q}^\times \Gamma_N$, it follows that Γ_S is commensurable with $\boldsymbol{Q}^\times \Gamma_1$. By Lemmas 6.16 and 6.17, every element of $\mathrm{Gal}(\boldsymbol{Q}_{ab}/k_S)$ can be written as $\sigma(s)$ with some $s \in S$. Since every element of $\mathfrak{F}_S \cap \boldsymbol{Q}_{ab}$ is invariant under $\tau(s)$, we obtain $\mathfrak{F}_S \cap \boldsymbol{Q}_{ab} \subset k_S$, so that $\mathfrak{F}_S \cap \boldsymbol{Q}_{ab} = k_S$. Since $\boldsymbol{Q}_{ab}$ is algebraically closed in $\mathfrak{F}$, this implies that k_S is algebraically closed in $\mathfrak{F}_S$. By the definition of $\mathfrak{F}_S$, we observe that $\mathrm{Gal}(\mathfrak{F}/\mathfrak{F}_S)$ is isomorphic to $S/\boldsymbol{Q}^\times G_{\infty+}$ under τ, and $T/\boldsymbol{Q}^\times G_{\infty+}$ corresponds to $\mathfrak{F}_T = \mathfrak{F}_N$. Put

$$R = \{x \in G_{A+} \mid \tau(x) = \text{id. on } k_T\mathfrak{F}_S\}.$$

Obviously $\Gamma_S T \subset R$. Conversely, by Lemma 6.17 and (6.7.3), we obtain

$$R \subset S \cap (G_{Q+}T) = (S \cap G_{Q+})T = \Gamma_S T,$$

so that $\Gamma_S T = R$. Therefore $k_T\mathfrak{F}_S$ corresponds to $\Gamma_S T$, hence

$$[\mathfrak{F}_T : k_T\mathfrak{F}_S] = [\Gamma_S T : T] = [\Gamma_S : \Gamma_T].$$

Since $\boldsymbol{C}$ and $\mathfrak{F}_T$ are linearly disjoint over k_T, we have

$$[\boldsymbol{C}\mathfrak{F}_T : \boldsymbol{C}\mathfrak{F}_S] = [\mathfrak{F}_T : k_T\mathfrak{F}_S] = [\Gamma_S : \Gamma_T].$$

Let $\mathfrak{M}_S$ be the field of all automorphic functions with respect to Γ_S. Then $\boldsymbol{C}\mathfrak{F}_S\subset\mathfrak{M}_S$, and $\boldsymbol{C}\mathfrak{F}_T=\mathfrak{M}_T$, hence

$$[\boldsymbol{C}\mathfrak{F}_T:\mathfrak{M}_S]=[\mathfrak{M}_T:\mathfrak{M}_S]=[\Gamma_S:\Gamma_T]=[\boldsymbol{C}\mathfrak{F}_T:\boldsymbol{C}\mathfrak{F}_S].$$

This proves that $\mathfrak{M}_S=\boldsymbol{C}\mathfrak{F}_S$.

REMARK 6.28. It can happen that $S\neq T$ even if $\Gamma_S=\Gamma_T$ and $k_S=k_T$. Take for example

$$S=\boldsymbol{Q}^\times\cdot\left\{x\in U\,\middle|\,x_p\equiv\begin{bmatrix}a&0\\0&1\end{bmatrix}\ \text{mod}\ \ N\cdot M_2(\boldsymbol{Z}_p)\quad(a\in\boldsymbol{Z}_p^\times)\right\},$$

$$T=\boldsymbol{Q}^\times\cdot\left\{x\in U\,\middle|\,x_p\equiv\begin{bmatrix}1&0\\0&d\end{bmatrix}\ \text{mod}\ \ N\cdot M_2(\boldsymbol{Z}_p)\quad(d\in\boldsymbol{Z}_p^\times)\right\}.$$

Then $\Gamma_S=\Gamma_T=\boldsymbol{Q}^\times\Gamma_N$, $k_S=k_T=\boldsymbol{Q}$, but $S\neq T$ if $N>2$.

Nevertheless, we have:

LEMMA 6.29. *Let $S\in\mathfrak{Z}$, $T\in\mathfrak{Z}$. If $\Gamma_S=\Gamma_T$, $k_S=k_T$, and $S\subset T$, then $S=T$.*

PROOF. By (i) of Lemma 6.17, the assumption $k_S=k_T$ implies $G_{\boldsymbol{Q}+}S=G_{\boldsymbol{Q}+}T$, so that if $S\subset T$, we have $T\subset(G_{\boldsymbol{Q}+}\cap T)S=\Gamma_TS$. Therefore $\Gamma_T=\Gamma_S\subset S$ furnishes the opposite inclusion $T\subset S$, so that $T=S$.

PROPOSITION 6.30. *Let Γ' be a discrete subgroup of $G_{\infty+}/\boldsymbol{R}^\times$ commensurable with $\boldsymbol{Q}^\times\Gamma_1/\boldsymbol{Q}^\times$, and containing Γ_N for some N. Then $\Gamma'=\Gamma_S/\boldsymbol{Q}^\times$ for some $S\in\mathfrak{Z}$.*

PROOF. Let β be an element of $G_{\infty+}$ which represents an element of Γ', and let $\Gamma''=\Gamma_1\cap\beta\Gamma_1\beta^{-1}$. Since $[\Gamma_1:\Gamma'']<\infty$, we see easily that Γ'' spans $M_2(\boldsymbol{Q})$ over $\boldsymbol{Q}$, so that $\beta M_2(\boldsymbol{Q})\beta^{-1}=M_2(\boldsymbol{Q})$. By the same argument as in the proof of Th. 6.23, we have $\beta=c\alpha$ with $c\in\boldsymbol{R}^\times$ and $\alpha\in G_{\boldsymbol{Q}+}$. Therefore we may assume that $\Gamma'=\Delta/\boldsymbol{Q}^\times$ with a subgroup Δ of $G_{\boldsymbol{Q}+}$. Take N so that $\Gamma_N\subset\Delta$. We can find a finite number of elements $\alpha_1,\cdots,\alpha_d$ such that $\Delta=\bigcup_{i=1}^d\boldsymbol{Q}^\times\Gamma_N\alpha_i$. Put $W=\bigcap_{\alpha\in\Delta}\alpha^{-1}U_N\alpha=\bigcap_{i=1}^d\alpha_i^{-1}U_N\alpha_i$, and $S=\Delta W$. Then W is an open subgroup of G_{A+}, $W/G_{\infty+}$ is compact, and $\alpha^{-1}W\alpha=W$ for every $\alpha\in\Delta$. Therefore S is an open subgroup of G_{A+}, and $S/\boldsymbol{Q}^\times G_{\infty+}$ is compact, so that $S\in\mathfrak{Z}$. We have then $\Gamma_S=\Delta\cdot(W\cap G_{\boldsymbol{Q}+})=\Delta$, since $W\cap G_{\boldsymbol{Q}+}\subset U_N\cap G_{\boldsymbol{Q}+}=\Gamma_N\subset\Delta$. This completes the proof.

By virtue of Prop. 6.27, we can find a model (V_S,φ_S) of $\Gamma_S\backslash\mathfrak{H}^*$, which is characterized by the following properties:

(6.7.4) *V_S is defined over k_S,*

(6.7.5) $\mathfrak{F}_S=\{f\circ\varphi_S\mid f\in k_S(V_S)\}$.

We fix (V_S, φ_S) for each $S \in \mathfrak{Z}$ once for all. Let $S \in \mathfrak{Z}$, $T \in \mathfrak{Z}$, and $x \in G_{A+}$. Suppose that $xSx^{-1} \subset T$. Then $\tau(x)$ gives an isomorphism of $\mathfrak{F}_T$ to a subfield of $\mathfrak{F}_S$. Replacing $\mathfrak{F}_S$ and $\mathfrak{F}_T$ by $k_S(V_S)$ and $k_T(V_T)$, we obtain an isomorphism $\tau'(x)$ of $k_T(V_T)$ into $k_S(V_S)$ such that $f^{\tau'(x)} \circ \varphi_S = (f \circ \varphi_T)^{\tau(x)}$ for $f \in k_T(V_T)$. Therefore by Appendix N° 6, we find a unique biregular morphism $J_{TS}(x)$ of V_S to $V_T^{\sigma(x)}$ such that $f^{\sigma(x)} \circ J_{TS}(x) = f^{\tau'(x)}$ for $f \in k_T(V_T)$, i.e.,

(6.7.6) $\quad f^{\sigma(x)} \circ J_{TS}(x) \circ \varphi_S = (f \circ \varphi_T)^{\tau(x)} \qquad$ *for* $f \in k_T(V_T)$.

It can easily be verified that $J_{TS}(x)$ has the following properties:

(6.7.7) $\quad J_{TS}(x)$ *is rational over* k_S;

(6.7.8) $\quad J_{TS}(x)^{\sigma(y)} \circ J_{SR}(y) = J_{TR}(xy)$;

(6.7.9) $\quad J_{SS}(x) = \mathrm{id.}$ *if* $x \in S$;

(6.7.10) $\quad J_{TS}(\alpha)[\varphi_S(z)] = \varphi_T(\alpha(z))$ *if* $\alpha \in G_{Q+}$ *and* $T = \alpha S \alpha^{-1}$.

Especially, if $S \subset T$, $J_{TS}(1)$ is defined, and

$$J_{TS}(1)[\varphi_S(z)] = \varphi_T(z).$$

Therefore $J_{TS}(1)$ corresponds to the natural projection map of $\Gamma_S \backslash \mathfrak{H}^*$ to $\Gamma_T \backslash \mathfrak{H}^*$. If $xSx^{-1} = T$, both $J_{TS}(x)$ and $J_{ST}(x^{-1})$ are meaningful, and $J_{ST}(x^{-1})^{\sigma(x)} \circ J_{TS}(x) = \mathrm{id.}$, so that $J_{TS}(x)$ is a biregular isomorphism of V_S to $V_T^{\sigma(x)}$. In the most general situation $xSx^{-1} \subset T$, we have

$$\begin{aligned} J_{TS}(x) &= J_{TR}(1)^{\sigma(x)} \circ J_{RS}(x) \qquad (R = xSx^{-1}), \\ &= J_{TP}(x) \circ J_{PS}(1) \qquad (P = x^{-1}Tx), \end{aligned}$$

so that $J_{TS}(x)$ is a composed map of a biregular isomorphism and a projection map, in either order.

As an illustration, fix a positive integer N, and let us consider a member S of $\mathfrak{Z}$ defined as follows:

$$\begin{aligned} S &= \boldsymbol{Q}^\times U', \\ U' &= \{x \in U \mid x_p \in U'_p \text{ for all finite } p\}, \\ U'_p &= \left\{ \begin{bmatrix} a & b \\ c & d \end{bmatrix} \in GL_2(\boldsymbol{Z}_p) \,\middle|\, c \equiv 0 \bmod N\boldsymbol{Z}_p \right\}. \end{aligned}$$

Put $\alpha = \begin{bmatrix} N & 0 \\ 0 & 1 \end{bmatrix}$. Then we see easily that $\Gamma_S = \boldsymbol{Q}^\times(U' \cap G_{\boldsymbol{Q}}) = \boldsymbol{Q}^\times \Gamma_0(N)$, where $\Gamma_0(N)$ is as in (1.6.5), and $\boldsymbol{Q}^\times \cdot \det(S) = \boldsymbol{Q}^\times \cdot \det(U) = \boldsymbol{Q}_A^\times$, so that $k_S = \boldsymbol{Q}$. Further we have $\boldsymbol{Q}^\times U_N \subset S = \boldsymbol{Q}^\times U \cap \boldsymbol{Q}^\times \alpha^{-1} U \alpha$, so that the functions j and $j \circ \alpha$ are contained in the field $\mathfrak{F}_S$, and $\mathfrak{F}_S \subset \mathfrak{F}_N$. Observe that $j(\alpha(z)) = j(Nz)$. By

Prop. 6.27 and Prop. 2.10, we have $C\mathfrak{F}_S = C(j(z), j(Nz))$. Since $Q(j(z), j(Nz)) \subset \mathfrak{F}_S$, and C is linearly disjoint with $\mathfrak{F}_S$ over $k_S = Q$, we obtain

$$\mathfrak{F}_S = Q(j(z), j(Nz)).$$

Consider, as another example, the groups S and T of Remark 6.28. Since $\Gamma_S = \Gamma_T = Q^\times \Gamma_N$, we have $C\mathfrak{F}_S = C\mathfrak{F}_T = \mathfrak{F}_N$ by Prop. 6.27. Thus V_S and V_T are models of $\Gamma_N \backslash \mathfrak{H}^*$, defined over Q, but an obvious biregular map $Y: V_T \to V_S$ defined by $Y \circ \varphi_T = \varphi_S$ is not rational over Q if $N > 2$. It can be shown that Y is defined over $Q(\zeta_N + \zeta_N^{-1})$, where $\zeta_N = e^{2\pi i/N}$.

6.8. An explicit reciprocity-law at the fixed points of G_{Q+} on $\mathfrak{H}$

Let K be an imaginary quadratic field, and q a normalized embedding of K into $M_2(Q)$ in the sense of § 4.4, and z the fixed point of $q(K^\times)$ on $\mathfrak{H}$ (see Prop. 4.6 and Prop. 4.7). In § 4.4, we have shown that every non-trivial fixed point of an element of G_{Q+} on $\mathfrak{H}$ is obtained as such a point z. The purpose of this section is to study the nature of the values of functions in the field $\mathfrak{F}$ at z. First we observe that the embedding q defines a continuous homomorphism of $K_A^\times$ into G_{A+}; we denote it again by q.

THEOREM 6.31. *The symbols K, q, and z being as above, the following assertions hold.*

(i) *For every $h \in \mathfrak{F}$, defined and finite at z, the value $h(z)$ belongs to K_{ab}, and*

$$h(z)^{[s,K]} = h^{\tau(q(s)^{-1})}(z)$$

for every $s \in K_A^\times$.

(ii) *For any $S \in \mathfrak{Z}$, the point $\varphi_S(z)$ is rational over K_{ab}, and for every $s \in K_A^\times$, one has*

$$\varphi_S(z)^{[s,K]} = J_{ST}(q(s)^{-1})(\varphi_T(z)),$$

where $T = q(s)Sq(s)^{-1}$.

One may notice that the relation (i) explains the deep arithmetic meaning of the map τ, exactly similar to the fact that the canonical map of $K_A^\times$ to $\mathrm{Gal}(K_{ab}/K)$ is defined locally by the Frobenius automorphisms. Thus our two theorems 6.23 and 6.31 provide an analogue of class field theory for the field $\mathfrak{F}$ which is of Kroneckerian dimension 2. It should also be observed that the relations (i) and (ii) are generalizations of (5.4.1).

PROOF. As is seen in § 4.4, z belongs to K. Define a Q-linear isomorphism $\iota_z: Q^2 \to K$ by $\iota_z(a) = a\begin{bmatrix} z \\ 1 \end{bmatrix}$ for $a \in Q^2$ (row vector!). Since $q(\mu)\begin{bmatrix} z \\ 1 \end{bmatrix} = \begin{bmatrix} \mu z \\ \mu \end{bmatrix}$ for $\mu \in K^\times$ (see (4.4.5)), the diagram

$$\begin{array}{ccc} \boldsymbol{Q}^2 & \xrightarrow{\iota_z} & K^\times \\ {\scriptstyle q(\mu)}\downarrow & & \downarrow{\scriptstyle \mu} \\ \boldsymbol{Q}^2 & \xrightarrow{\iota_z} & K^\times \end{array}$$

is commutative. If we put $\mathfrak{a}_z = \boldsymbol{Z}z + \boldsymbol{Z}$, then ι_z induces an isomorphism of $\boldsymbol{Q}^2/\boldsymbol{Z}^2$ onto $K/\mathfrak{a}_z$, which we also denote by ι_z. Let ξ be an isomorphism of $\boldsymbol{C}/\mathfrak{a}_z$ to an elliptic curve $E \in \mathcal{E}$. Let σ be an element of $\mathrm{Aut}(\boldsymbol{C}/K)$, and s an element of $K_A^\times$ such that $\sigma = [s, K]$ on K_{ab}. Take an isomorphism ξ' of $\boldsymbol{C}/s^{-1}\mathfrak{a}_z$ onto E^σ as in Th. 5.4 for these σ and s, so that

(1) $$\xi(x)^\sigma = \xi'(s^{-1}x) \qquad (x \in K/\mathfrak{a}_z).$$

By Lemma 6.19, we can find an element y of U and an element α of $G_{\boldsymbol{Q}+}$ so that $q(s)^{-1} = y\alpha^{-1}$. Then $\boldsymbol{Z}^2 q(s)^{-1} = \boldsymbol{Z}^2\alpha^{-1}$. Put $w = \alpha^{-1}(z)$. Then we find an element λ of $K^\times$ such that

$$\alpha^{-1}\begin{bmatrix} z \\ 1 \end{bmatrix} = \begin{bmatrix} \lambda w \\ \lambda \end{bmatrix}.$$

Observe that

$$\mathfrak{a}_z = \iota_z(\boldsymbol{Z}^2), \qquad s^{-1}\mathfrak{a}_z = \iota_z(\boldsymbol{Z}^2 q(s)^{-1}) = \iota_z(\boldsymbol{Z}^2\alpha^{-1}) = \lambda \cdot \iota_w(\boldsymbol{Z}^2) = \lambda\mathfrak{a}_w .$$

Therefore we have a commutative diagram

$$\begin{array}{ccccc} \boldsymbol{C} & \longrightarrow & \boldsymbol{C}/\mathfrak{a}_w & \xrightarrow{\xi''} & E^\sigma \\ {\scriptstyle \lambda}\downarrow & & \downarrow & & \downarrow{\scriptstyle \mathrm{id.}} \\ \boldsymbol{C} & \longrightarrow & \boldsymbol{C}/s^{-1}\mathfrak{a}_z & \xrightarrow{\xi'} & E^\sigma \end{array}$$

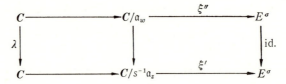

with a suitable choice of an isomorphism ξ''. Let $a \in \boldsymbol{Q}^2/\boldsymbol{Z}^2$, and $u = \iota_z(a)$. By Lemma 6.4, we have

(2) $$f_a^i(z) = h_E^i(\xi(u)) \qquad (i = 1, 2, 3),$$

and $s^{-1}u = s^{-1} \cdot \iota_z(a) = \iota_z(a \cdot q(s)^{-1}) = \iota_z(ay\alpha^{-1}) = \lambda \cdot \iota_w(ay) \pmod{s^{-1}\mathfrak{a}_z = \lambda\mathfrak{a}_w}$. Therefore we have $\xi'(s^{-1}u) = \xi''(\iota_w(ay))$, so that

$$h_{E^\sigma}^i(\xi'(s^{-1}u)) = f_{ay}^i(w) \qquad (i = 1, 2, 3)$$

by Lemma 6.4, hence from (1) and (2), we obtain

(3) $$f_a^i(z)^\sigma = h_{E^\sigma}^i(\xi(u)^\sigma) = f_{ay}^i(\alpha^{-1}(z)) \qquad (i = 1, 2, 3).$$

Also we have

(4) $$j(z)^\sigma = j(E^\sigma) = j(w) = j(\alpha^{-1}(z)).$$

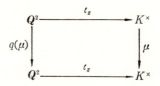

Now fix a positive integer $N>2$, and let $N^{-1}\boldsymbol{Z}^2/\boldsymbol{Z}^2-\{0\}=\{a, b, \cdots\}$. Further let V'_N be the locus of

$$\varphi'(\mathfrak{z})=(j(\mathfrak{z}), f_a^1(\mathfrak{z}), f_b^1(\mathfrak{z}), \cdots, f_a^2(\mathfrak{z}), f_b^2(\mathfrak{z}), \cdots, f_a^3(\mathfrak{z}), f_b^3(\mathfrak{z}), \cdots)$$

in the affine space of dimension $3(N^2-1)+1$, where $\mathfrak{z}$ denotes the variable on $\mathfrak{H}$. If $P=\boldsymbol{Q}^\times U_N$, V_P is birationally equivalent to V'_N, and there exists a birational map X of V_P to V'_N, rational over $k_P=k_N$, such that $X\circ\varphi_P=\varphi'$. Since V_P is non-singular, X is defined at every point of $\varphi_P(\mathfrak{H})$; X is not biregular, but we see that X is one-to-one in the following sense:

(5) *If* $z_1\in\mathfrak{H}$, $z_2\in\mathfrak{H}$, *and* $\varphi'(z_1)=\varphi'(z_2)$, *then* $\varphi_P(z_1)=\varphi_P(z_2)$.

In fact, if $\varphi'(z_1)=\varphi'(z_2)$, we have $j(z_1)=j(z_2)$, so that there exists an element γ of Γ_1 such that $\gamma(z_1)=z_2$. Put $L_1=\boldsymbol{Z}z_1+\boldsymbol{Z}$, and $\iota(a)=a\begin{bmatrix} z_1 \\ 1 \end{bmatrix}$ for $a\in\boldsymbol{R}^2$. Denote by the same letter ι the map of $\boldsymbol{R}^2/\boldsymbol{Z}^2$ onto $\boldsymbol{C}/L_1$ obtained from ι. Let ξ_1 be an isomorphism of $\boldsymbol{C}/L_1$ to an elliptic curve $E_1\in\mathcal{E}$. If $x=\iota(a)$, $a\in N^{-1}\boldsymbol{Z}^2/\boldsymbol{Z}^2-\{0\}$, and $y=\iota(a\gamma)$, then we have, by (6.1.3) and Lemma 6.4,

$$h^i_{E_1}(\xi_1(x))=f^i_a(z_1)=f^i_a(z_2)=f^i_a(\gamma(z_1))=f^i_{a\gamma}(z_1)=h^i_{E_1}(\xi_1(y)) \qquad (i=1, 2, 3).$$

Therefore, by (4.5.3), $\varepsilon_a(\xi_1(x))=\xi_1(y)$ with an automorphism ε_a of E_1. If $E_1\in\mathcal{E}_1$, we have $\varepsilon_a=\pm1$, and $\varepsilon_a a=a\gamma$ for all $a\in N^{-1}\boldsymbol{Z}^2/\boldsymbol{Z}^2-\{0\}$. By Lemma 6.2, we have $\gamma\in\Gamma_N\cdot\{\pm1\}$, so that $\varphi_P(z_2)=\varphi_P(\gamma(z_1))=\varphi_P(z_1)$, which proves (5) in the case $E_1\in\mathcal{E}_1$. Suppose that $E_1\in\mathcal{E}_2$; then L_1 is a fractional ideal in $K=\boldsymbol{Q}(\sqrt{-1})$, and ε_a is identified with (multiplication by) one of the four units ±1, $\pm\sqrt{-1}$. For $\varepsilon\in\{\pm1, \pm\sqrt{-1}\}$, we can define an element ε^* of Γ_1 by $\varepsilon^*\begin{bmatrix} z_1 \\ 1 \end{bmatrix}=\begin{bmatrix} \varepsilon z_1 \\ \varepsilon \end{bmatrix}$, so that $\iota(b\varepsilon^*)=\varepsilon\cdot\iota(b)$ for $b\in\boldsymbol{Q}^2/\boldsymbol{Z}^2$. Then we have $a\gamma=a\varepsilon^*_a$ for every $a\in N^{-1}\boldsymbol{Z}^2/\boldsymbol{Z}^2-\{0\}$. Now we need the following

LEMMA 6.32. *Let N be a positive integer >2, γ an element of Γ_1, z_1 an elliptic point of Γ_1, and $\Delta=\{\delta\in\Gamma_1 \mid \delta(z_1)=z_1\}$. Suppose that, for every $u\in\boldsymbol{Z}^2$, there exists an element δ_u of Δ such that $u\gamma\equiv u\delta_u$* mod (N). *Then $\gamma\in\Delta\Gamma_N$.*

PROOF. Since every elliptic point of Γ_1 is Γ_1-equivalent to $\sqrt{-1}$ or $e^{2\pi i/3}$, it is sufficient to prove our assertion in the cases $z_1=\sqrt{-1}$ and $z_1=e^{2\pi i/3}$. If $z_1=e^{2\pi i/3}$, we have, by the result of § 1.4,

$$\Delta=\left\{\pm1_2, \pm\begin{bmatrix} 0 & 1 \\ -1 & 1 \end{bmatrix}, \pm\begin{bmatrix} -1 & 1 \\ -1 & 0 \end{bmatrix}\right\}.$$

Put $b=(1, 0)$, $c=(0, 1)$, and $\gamma'=\gamma\delta_b^{-1}$. Then $b\gamma'\equiv b$ mod (N), hence $\gamma'\equiv\begin{bmatrix} 1 & 0 \\ p & q \end{bmatrix}$ mod (N) with some integers p and q. Since $\det(\gamma')=1$, we have $q\equiv1$ mod (N), so that $\gamma'\equiv\begin{bmatrix} 1 & 0 \\ p & 1 \end{bmatrix}$ mod (N). On the other hand we have $(p, 1)\equiv c\gamma'\equiv c\delta$

mod (N) for some $\delta \in \Delta$. Looking at the elements of Δ, we see that $\delta = 1_2$ or $= \begin{bmatrix} 0 & 1 \\ -1 & 1 \end{bmatrix}$. But if $\delta = \begin{bmatrix} 0 & 1 \\ -1 & 1 \end{bmatrix}$, we have $\gamma' \equiv \begin{bmatrix} 1 & 0 \\ -1 & 1 \end{bmatrix}$ mod (N), so that $(b+c)\gamma' \equiv (0, 1) \not\equiv (b+c)\varepsilon$ mod (N) for any element ε of Δ, a contradiction. Therefore $\delta = 1_2$, so that $\gamma' \in \Gamma_N$, hence $\gamma \in \Gamma_N \Delta$. The case $z_1 = \sqrt{-1}$ can be treated in a similar and simpler way.

Applying this lemma to the present situation, we see that $\gamma\delta \in \Gamma_N$ for some δ of Γ_1 such that $\delta(z_1) = z_1$. Then $\varphi_P(z_2) = \varphi_P(\gamma\delta(z_1)) = \varphi_P(z_1)$, which proves (5) in the case $E \in \mathcal{E}_2$. The remaining case $E \in \mathcal{E}_3$ can be treated by the same argument, on account of Lemma 6.32.

Coming back to the original z, s, σ, α, y and $P = \boldsymbol{Q}^\times U_N$, we see that $\sigma(y) = \sigma(q(s)^{-1}) = [s, K] = \sigma$ on $\boldsymbol{Q}_{ab}$. Now $f_a^i \mapsto f_{ay}^i$ defines an automorphism of $\mathfrak{F}_N$ over $\mathfrak{F}_1$ which induces a birational map J' of V'_N to $V_N'^{\sigma(y)}$, which is obviously defined everywhere on V'_N, and satisfies $J' \circ X = X^\sigma \circ J_{PP}(y)$. From (3) and (4) we obtain $\varphi'(z)^\sigma = J'[\varphi'(\alpha^{-1}(z))]$, so that

$$X^\sigma[\varphi_P(z)^\sigma] = J'[X[\varphi_P(\alpha^{-1}(z))]] = X^\sigma[J_{PP}(y)[\varphi_P(\alpha^{-1}(z))]].$$

By (5), we have $\varphi_P(z)^\sigma = J_{PP}(y)[\varphi_P(\alpha^{-1}(z))]$. Putting $R = \alpha P \alpha^{-1} = q(s)Pq(s)^{-1}$, we obtain

$$\varphi_P(z)^\sigma = J_{PP}(y)[J_{PR}(\alpha^{-1})[\varphi_R(z)]] = J_{PR}(q(s)^{-1})[\varphi_R(z)].$$

This formula holds for $P = \boldsymbol{Q}^\times U_N$ with any $N > 2$. Now for every $S \in \mathfrak{Z}$, we can find a positive integer $N > 2$ such that $\boldsymbol{Q}^\times U_N \subset S$. Therefore, putting $P = \boldsymbol{Q}^\times U_N$ and $T = q(s)Sq(s)^{-1}$, we have

(6) $$\varphi_S(z)^\sigma = J_{SP}(1)^\sigma[\varphi_P(z)^\sigma] = J_{SP}(1)^\sigma[J_{PR}(q(s)^{-1})[\varphi_R(z)]]$$
$$= J_{SR}(q(s)^{-1})[\varphi_R(z)] = J_{ST}(q(s)^{-1})[J_{TR}(1)[\varphi_R(z)]] = J_{ST}(q(s)^{-1})[\varphi_T(z)].$$

Let h be an element of $\mathfrak{F}$ defined and finite at z. Then $h = f \circ \varphi_S$ for some $S \in \mathfrak{Z}$ and some function f on V_S, which is rational over k_S, and defined at $\varphi_S(z)$. Therefore, by (6.7.6),

(7) $$h(z)^\sigma = f^\sigma(\varphi_S(z)^\sigma) = f^\sigma(J_{ST}(q(s)^{-1})[\varphi_T(z)]) = f^{\tau(q(s)^{-1})}(z).$$

In both formulas (6) and (7), we observe that $\varphi_S(z)^\sigma$ and $h(z)^\sigma$ depend only on s, i.e., they depend only on the restriction of σ to K_{ab}. Therefore $\varphi_S(z)$ and $h(z)$ are rational over K_{ab}, so that we can replace σ by $[s, K]$ in (6) and (7), and thus obtain (ii) and (i) of our theorem.

PROPOSITION 6.33. *The notation being as in Th.* 6.31, *let* $S \in \mathfrak{Z}$, *and*

$$W = \{s \in K_A^\times \mid q(s) \in S\}.$$

Then $K \cdot k_S(\varphi_S(z))$ *is the subfield of* K_{ab} *corresponding to the subgroup* $K^\times W$ *of* $K_A^\times$.

PROOF. Let $s \in K_A^\times$, and $\pi = [s, K]$. Then $\pi = \sigma(q(s)^{-1})$ on $\boldsymbol{Q}_{ab}$. If $s \in W$, we see that $\pi = \text{id.}$ on k_S. By (6.7.9) and (ii) of Th. 6.31, we have $\varphi_S(z)^\pi = \varphi_S(z)$, so that $\pi = \text{id.}$ on $k_S(\varphi_S(z))$. Conversely suppose that $\pi = \text{id.}$ on $k_S(\varphi_S(z))$. By (i) of Lemma 6.17, we have $q(s)^{-1} = t\alpha$ with $t \in S$ and $\alpha \in G_{\boldsymbol{Q}+}$. Putting $T = q(s)Sq(s)^{-1}$, by (ii) of Th. 6.31, and (6.7.10), we have

$$\varphi_S(z) = \varphi_S(z)^\pi = J_{ST}(t\alpha)[\varphi_T(z)] = \varphi_S(\alpha(z)) ,$$

so that $z = \gamma\alpha(z)$ with $\gamma \in \Gamma_S$. By (4.4.4), $\gamma\alpha = q(b)$ with $b \in K^\times$. Then $q(bs)^{-1} = t\gamma^{-1} \in S$, so that $s \in K^\times W$, which completes the proof.

We now specialize the formula (i) of Th. 6.31 by taking h to be a more explicitly given function. First we take the function f_a^i as h. Although the result in this case is essentially the same as (3) of the proof of Th. 6.31, we formulate it in a somewhat different way.

PROPOSITION 6.34. *Let $\mathfrak{a}$ be a fractional ideal in K, and $\{\omega_1, \omega_2\}$ be a basis of $\mathfrak{a}$ over $\boldsymbol{Z}$, with $z_0 = \omega_1/\omega_2 \in \mathfrak{H}$. Further let N be a positive integer, C_N the maximal ray class field over K modulo N, and $\mathfrak{b}$ a fractional ideal in K, prime to N. Then, for every $a \in N^{-1}\boldsymbol{Z}^2$, $\notin \boldsymbol{Z}^2$, the value $f_a^i(z_0)$ belongs to C_N. Moreover, if*

$$\sigma = \left(\frac{C_N/K}{\mathfrak{b}}\right), \quad \mathfrak{a}\mathfrak{b}^{-1} = \boldsymbol{Z}\omega_1' + \boldsymbol{Z}\omega_2', \quad \omega_1'/\omega_2' = z_0' \in \mathfrak{H}, \quad \begin{bmatrix}\omega_1\\ \omega_2\end{bmatrix} = \xi\begin{bmatrix}\omega_1'\\ \omega_2'\end{bmatrix}$$

with $\xi \in G_{\boldsymbol{Q}+}$, then one has

$$f_a^i(z_0)^\sigma = f_b^i(z_0') \qquad (i = 1, 2, 3) ,$$

where b is an element of $N^{-1}\boldsymbol{Z}^2$ such that $b \equiv a\xi \bmod \boldsymbol{Z}_p^2$ for all prime factors p of N.

PROOF. From Prop. 6.33, we see that $h(z_0) \in C_N$ for every $h \in \mathfrak{F}_N$, which proves the first assertion. To prove the second one, let s be an element of $K_A^\times$ such that $s\mathfrak{o}_K = \mathfrak{b}$. We can take s so that $s_p = 1$ for all prime factors p of N. Then $[s, K] = \sigma$ on C_N. Define an embedding $q : K \to M_2(\boldsymbol{Q})$ by $\begin{bmatrix}\mu\omega_1\\ \mu\omega_2\end{bmatrix} = q(\mu)\begin{bmatrix}\omega_1\\ \omega_2\end{bmatrix}$ for $\mu \in K$. For every rational prime p, we have

$$\boldsymbol{Z}_p^2\begin{bmatrix}\omega_1'\\ \omega_2'\end{bmatrix} = (\mathfrak{a}\mathfrak{b}^{-1})_p = \mathfrak{a}_p s_p^{-1} = \boldsymbol{Z}_p^2\begin{bmatrix}\omega_1 s_p^{-1}\\ \omega_2 s_p^{-1}\end{bmatrix} = \boldsymbol{Z}_p^2 q(s_p^{-1})\begin{bmatrix}\omega_1\\ \omega_2\end{bmatrix} = \boldsymbol{Z}_p^2 q(s_p^{-1})\xi\begin{bmatrix}\omega_1'\\ \omega_2'\end{bmatrix}.$$

(Each term is a lattice in $K_p = K \otimes_{\boldsymbol{Q}} \boldsymbol{Q}_p = \boldsymbol{Q}_p\omega_1 + \boldsymbol{Q}_p\omega_2$; the p-component s_p of s is an element of $K_p^\times$.) Therefore $\boldsymbol{Z}_p^2 = \boldsymbol{Z}_p^2 q(s_p^{-1})\xi$ for all p, so that $q(s^{-1})\xi = t$ with an element t of U. By (i) of Th. 6.31, we have

$$f_a^i(z_0)^\sigma = (f_a^i)^{\tau(t)}(\xi^{-1}(z_0)) = (f_a^i)^{\tau(t)}(z_0') .$$

Since $s_p=1$ for all p dividing N, we see that $at\equiv a\xi \bmod \boldsymbol{Z}_p^2$ for all p dividing N. Since $\mathfrak{b}$ is prime to N, we see that $\xi\in GL_2(\boldsymbol{Z}_p)$ for all such p. Therefore we have $(f_a^i)^{\tau(t)}=f_b^i$ with b as described above. This completes the proof.

REMARK. If $\mathfrak{b}$ is an integral ideal, we have $\mathfrak{a}\subset\mathfrak{a}\mathfrak{b}^{-1}$, so that $\xi\in M_2(\boldsymbol{Z})$. In this case we can put $b=a\xi$, so that the formula becomes

$$f_a^i(z_0)^\sigma=f_{a\xi}^i(\xi^{-1}(z_0))\,. \tag{6.8.1}$$

Next we consider a modular function which is obtained from automorphic forms with rational Fourier coefficients:

PROPOSITION 6.35. *Let g_1 and g_2 be automorphic forms of weight k with respect to $\Gamma=SL_2(\boldsymbol{Z})$, other than* 0, *and α an element of $G_{\boldsymbol{Q}+}$. Put $S=\boldsymbol{Q}^\times(\alpha^{-1}U\alpha\cap U)$, and $h=(g_1\,|\,[\alpha]_k)/g_2$, where*

$$g_1\,|\,[\alpha]_k=\det(\alpha)^{k/2}g_1(\alpha(z))j(\alpha,z)^{-k}$$

(see §2.1), *and U is as in* §6.4. *Suppose that the Fourier expansions of g_1 and g_2 with respect to $e^{2\pi iz}$ have rational coefficients. Then $h\in\mathfrak{F}_S$.*

Note that the weight k must be even, since there is no non-zero automorphic form of an odd weight with respect to Γ (see §2.1).

PROOF. We can find two elements γ and δ of Γ so that $\alpha=\gamma\beta\delta$, $\beta=\begin{bmatrix} rm & 0\\ 0 & r\end{bmatrix}$ with $r\in\boldsymbol{Q}$, $m\in\boldsymbol{Z}$. Then we have $(g_1\,|\,[\beta]_k)/g_2=h\circ\delta^{-1}$, and $\boldsymbol{Q}^\times(\beta^{-1}U\beta\cap U)=\delta S\delta^{-1}$. Therefore it is sufficient to prove our assertion for β. In other words, we may assume that $\alpha=\begin{bmatrix} rm & 0\\ 0 & r\end{bmatrix}$. Then $\alpha^{-1}\Gamma\alpha\cap\Gamma=\Gamma_0(m)$, and $h(z)=m^{k/2}g_1(mz)/g_2(z)$. Therefore h is invariant under $\Gamma_0(m)$, and has rational Fourier coefficients, so that h belongs to the field $\mathfrak{F}'_m=\boldsymbol{Q}(j,j(mz),f_a)$ considered in (2) of Prop. 6.9. By virtue of that proposition, and through the isomorphism of U/U_m onto $GL_2(\boldsymbol{Z}/m\boldsymbol{Z})$, we see that $\mathfrak{F}'_m=\mathfrak{F}_T$ with

$$T=\boldsymbol{Q}^\times\cdot\left\{x\in U\,\middle|\,x_p\equiv\begin{bmatrix}1 & 0\\ 0 & d\end{bmatrix}\bmod m\cdot M_2(\boldsymbol{Z}_p)\quad(d\in\boldsymbol{Z}_p^\times)\right\}.$$

Now it can easily be verified that $\alpha^{-1}U\alpha\cap U\subset\Gamma_0(m)\cdot T$. Since h is invariant under both $\Gamma_0(m)$ and T, we have $h\in\mathfrak{F}_S$, q. e. d.

PROPOSITION 6.36. *Let g_1, g_2, α, and h be as in Prop.* 6.35, *and N, $\mathfrak{a}$, $\mathfrak{b}$, ω_1, ω_2, z_0, σ, and C_N be as in Prop.* 6.34. *Suppose that* $\det(\alpha)=N$ *and $\alpha\in M_2(\boldsymbol{Z})$. Then $h(z_0)\in C_N$. Moreover there exists an element η of $G_{\boldsymbol{Q}+}$ satisfying the following condition:*

(∗) $\eta\begin{bmatrix}\omega_1\\ \omega_2\end{bmatrix}$ *is a basis of $\mathfrak{a}\mathfrak{b}^{-1}$ over $\boldsymbol{Z}$, and $\alpha\eta\alpha^{-1}\in GL_2(\boldsymbol{Z}_p)$ for all p dividing N.*

If η *satisfies* $(*)$, *then* $h(z_0)^\sigma = h(\eta(z_0))$.

PROOF. Put $S = \boldsymbol{Q}^\times(\alpha^{-1}U\alpha \cap U)$. By Prop. 6.35, $h \in \mathfrak{F}_S$. Observe that $U_N \subset \alpha^{-1}U\alpha \cap U$, hence $h \in \mathfrak{F}_N$. Therefore $h(z_0) \in C_N$, as is remarked at the beginning of the proof of Prop. 6.34. Take ω_1', ω_2', ξ, s, and t as in Prop. 6.34 and its proof. Put $L = \boldsymbol{Z}^2$. Since $t \in U$, we see that $L/L\alpha t$ is isomorphic to $L/L\alpha$, so that $L\alpha t = L\alpha\gamma$ for some $\gamma \in \Gamma$ by Lemma 3.12. Then $\alpha\gamma t^{-1}\alpha^{-1} \in U$. Put $\eta = \gamma\xi^{-1}$. Then $\eta\begin{bmatrix}\omega_1\\ \omega_2\end{bmatrix} = \gamma\begin{bmatrix}\omega_1'\\ \omega_2'\end{bmatrix}$. Since $t_p = \xi$ for all p dividing N, we have $\alpha\eta\alpha^{-1} \in GL_2(\boldsymbol{Z}_p)$ for all such p. Thus we have shown the existence of η satisfying $(*)$.

Next let η be any element satisfying $(*)$. Take η^{-1} as ξ in the proof of Prop. 6.34. Then we have $q(s^{-1})\eta^{-1} = t$ with $t \in U$, as is proved there. Since $s_p = 1$ for all p dividing N, we have $\eta^{-1} = t_p$ for all such p, so that $\alpha t_p \alpha^{-1} \in GL_2(\boldsymbol{Z}_p)$. The last inclusion holds also for all p not dividing N, since $\det(\alpha) = N$ and $\alpha \in M_2(\boldsymbol{Z})$. Therefore $\alpha t \alpha^{-1} \in U$, hence $t \in \alpha^{-1}U\alpha \cap U \subset S$. By (i) of Th. 6.31, we have

$$h(z_0)^\sigma = h^{\tau(t\eta)}(z_0) = h^{\tau(t)}(\eta(z_0)) = h(\eta(z_0)),$$

since $h \in \mathfrak{F}_S$. This completes the proof.

EXERCISE 6.37. Generalize Propositions 6.34 and 6.36 to the case where the order of $\mathfrak{a} = \boldsymbol{Z}\omega_1 + \boldsymbol{Z}\omega_2$ is not necessarily maximal (cf. Prop. 4.11, (5.4.2)).

6.9. The action of an element of G_Q with negative determinant

For every $x \in G_A$, let x_0 denote its projection to G_0. If $\alpha \in G_{Q+}$, the element $\tau(\alpha) = \tau(\alpha_0)$ is defined by $h^{\tau(\alpha)} = h \circ \alpha$ for $h \in \mathfrak{F}$. If $\alpha \in G_Q$ and $\det(\alpha) < 0$, $\tau(\alpha_0)$ is meaningful since $\alpha_0 \in G_{A+}$, while $\tau(\alpha)$ is not defined. Therefore, it is a natural question to ask the nature of $\tau(\alpha_0)$. The answer is given by the following

THEOREM 6.38. *Let* α *be an element of* G_Q *such that* $\det(\alpha) < 0$, *and* α_0 *the projection of* α *to the non-archimedean part* G_0 *of* G_A. *Then*

(i) $h^{\tau(\alpha_0)}(z) = \overline{h(\alpha(\bar{z}))}$ *for all* $h \in \mathfrak{F}$ *and all* $z \in \mathfrak{H}$;

(ii) *if* $S \in \mathfrak{Z}$ *and* $S' = \alpha_0 S \alpha_0^{-1}$, *then* $J_{S'S}(\alpha_0)[\varphi_S(z)] = \overline{\varphi_{S'}(\alpha(\bar{z}))}$ *for all* $z \in \mathfrak{H}$. (Here a bar means the complex conjugation.)

PROOF. Let $z \in \mathfrak{H}$, and $L = \boldsymbol{Z}z + \boldsymbol{Z}$. Let ξ be an isomorphism of $\boldsymbol{C}/L$ to an elliptic curve $E \in \mathcal{E}$. Then we can define an isomorphism of $\boldsymbol{C}/\bar{L}$ to $\bar{E}$ by $\xi'(u) = \overline{\xi(\bar{u})}$. Put $\delta = \begin{bmatrix}0 & 1\\ 1 & 0\end{bmatrix}$. Then $\delta\begin{bmatrix}\bar{z}\\ 1\end{bmatrix} = \begin{bmatrix}1\\ \bar{z}\end{bmatrix}$, $\delta(\bar{z}) = 1/\bar{z} \in \mathfrak{H}$, and $\bar{L} = \boldsymbol{Z} + \boldsymbol{Z}\bar{z}$. Therefore

(1) $$\overline{j(z)}=\overline{j(E)}=j(\bar{E})=j(1/\bar{z})=j(\delta(\bar{z}))\,.$$

For every $a\in \boldsymbol{Q}^2$ we have

$$\xi'\left(a\begin{bmatrix}1\\ \bar{z}\end{bmatrix}\right)=\overline{\xi\left(a\delta\begin{bmatrix}z\\ 1\end{bmatrix}\right)},$$

so that, by Lemma 6.4,

(2) $$f_a^i(\delta(\bar{z}))=h_{\bar{E}}^i\left(\xi'\left(a\begin{bmatrix}1\\ \bar{z}\end{bmatrix}\right)\right)=\overline{h_E^i}\left(\overline{\xi\left(a\delta\begin{bmatrix}z\\ 1\end{bmatrix}\right)}\right)=\overline{f_{a\delta}^i(z)}\,.$$

Since $\delta_0\in U$, we obtain, from (1),

$$j^{\tau(\delta_0)}(z)=j(z)=\overline{j(\delta(\bar{z}))}\,,$$

which, together with (2), implies

$$h^{\tau(\delta_0)}(z)=\overline{h(\delta(\bar{z}))}\qquad \text{for all}\quad h\in\mathfrak{F}\,.$$

If α and α_0 are as in our theorem, we have $\alpha\delta^{-1}\in G_{\boldsymbol{Q}+}$, so that, putting $h'=h^{\tau(\alpha\delta^{-1})}=h\circ\alpha\delta^{-1}$, we have

$$h^{\tau(\alpha_0)}(z)=h^{\tau(\alpha\delta^{-1})\tau(\delta_0)}(z)=h'^{\tau(\delta_0)}(z)=\overline{h'(\delta(\bar{z}))}=\overline{h(\alpha(\bar{z}))}\,,$$

which proves the assertion (i). The second assertion follows immediately from (i) and (6.7.6).

COROLLARY 6.39. *Let K be an imaginary quadratic field, q a normalized embedding of K into $M_2(\boldsymbol{Q})$, and z the fixed point of $q(K^\times)$ on $\mathfrak{H}$. Further let $\mathfrak{N}$ be the normalizer of $q(K^\times)$ in $G_{\boldsymbol{Q}}$. Then*

(1) $[\mathfrak{N}:q(K^\times)]=2$;

(2) $\det(\alpha)<0$ *and* $\alpha(z)=\bar{z}$ *for every* $\alpha\in\mathfrak{N}-q(K^\times)$;

(3) $\overline{h(z)}=h^{\tau(\alpha_0)}(z)$ *for every* $h\in\mathfrak{F}$ *and every* $\alpha\in\mathfrak{N}-q(K^\times)$.

PROOF. By the discussion of § 4.4, there exists an element β of $G_{\boldsymbol{Q}}$ such that $\det(\beta)<0$, $\beta(z)=\bar{z}$, and $q(\bar{a})=\beta^{-1}q(a)\beta$ for all $a\in K$. Then $\beta\in\mathfrak{N}-q(K^\times)$. Let $\alpha\in\mathfrak{N}$. Since $\alpha^{-1}q(K)\alpha=q(K)$, we can define an automorphism σ of K by $q(a^\sigma)=\alpha^{-1}q(a)\alpha$ for $a\in K$. If $\sigma=\text{id.}$, α must be contained in $q(K)$, since $q(K)$ is the commutor of $q(K)$ itself in $M_2(\boldsymbol{Q})$. Therefore, if $\alpha\notin q(K)$, we have $a^\sigma=\bar{a}$ for all $a\in K$, so that $\alpha\beta^{-1}q(a)=q(a)\alpha\beta^{-1}$ for all $a\in K$. Then $\alpha\beta^{-1}\in q(K)$. This shows that $\mathfrak{N}=q(K^\times)\cup q(K^\times)\beta$, hence the assertions (1) and (2). The last assertion follows from (i) of Th. 6.38 and (2).

REMARK 6.40. Since $G_A/\boldsymbol{Q}^\times G_\infty$ is naturally isomorphic to $G_{A+}/\boldsymbol{Q}^\times G_{\infty+}$, we can define a homomorphism τ' of G_A to $\mathrm{Aut}(\mathfrak{F})$ with kernel $\boldsymbol{Q}^\times G_\infty$ so that $\tau=\tau'$ on G_{A+}. However, such an extension of τ does not keep one of the fundamental properties (6.6.2). To see this, take α and α_0 as in Th. 6.38, and put $\alpha=\alpha_0\alpha_\infty$. By (i) of Th. 6.38, $\tau(\alpha_0)$ coincides with the complex conjugation on $\boldsymbol{Q}_{ab}$. Since $\sigma(\alpha)=\text{id.}$, we have

$$\sigma(\alpha_\infty)=\sigma(\alpha_0)^{-1}=\tau(\alpha_0)^{-1}=\text{complex conjugation} \qquad (\text{on } \boldsymbol{Q}_{ab}).$$

On the other hand, $\tau'(\alpha_\infty)=\text{id.}$ by our definition, so that $\tau'(\alpha_\infty)\neq\sigma(\alpha_\infty)$ on $\boldsymbol{Q}_{ab}$.

Therefore, in order to discuss the whole G_A, it is necessary and natural to consider more functions than those of $\mathfrak{F}$. This can be done in the following way. Let $\mathfrak{H}^-$ denote the lower half complex plane, i.e.,

$$\mathfrak{H}^-=\{z\in\boldsymbol{C}\mid \operatorname{Im}(z)<0\}.$$

For every complex valued function f defined either in $\mathfrak{H}$ or in $\mathfrak{H}^-$, define f^* by $f^*(z)=\overline{f(\bar{z})}$. Put

$$\mathfrak{F}^*=\{f^*\mid f\in\mathfrak{F}\},$$

$$\mathfrak{R}\ =\mathfrak{F}\oplus\mathfrak{F}^*=\{(f,g)\mid f\in\mathfrak{F},\ g\in\mathfrak{F}^*\}.$$

Then $\mathfrak{F}^*$ is a field of meromorphic functions on $\mathfrak{H}^-$, and $\mathfrak{R}$ can be regarded as a ring of meromorphic functions on $\mathfrak{H}\cup\mathfrak{H}^-$. Let $\operatorname{Aut}(\mathfrak{R})$ denote the group of all automorphisms of the ring $\mathfrak{R}$. Define

$$\lambda:\ G_A\rightarrow\operatorname{Aut}(\mathfrak{R})$$

as follows:

$$(f,h^*)^{\lambda(x)}=(f^{\tau(x)},(h^{\tau(x)})^*) \qquad (x\in G_{A+};\ f\in\mathfrak{F},\ h\in\mathfrak{F}),$$

$$(f,h^*)^{\lambda(x)}=(h^{\tau(x_0)},(f^{\tau(x_0)})^*) \qquad (x\in G_A-G_{A+};\ f\in\mathfrak{F},\ h\in\mathfrak{F}).$$

Then it can easily be verified that λ is a homomorphism, and

(6.9.1) $\qquad \operatorname{Ker}(\lambda)=\boldsymbol{Q}^\times G_{\infty+}$,

(6.9.2) $\qquad (a,a)^{\lambda(x)}=(a^{\sigma(x)},a^{\sigma(x)}) \qquad (x\in G_A,\ a\in\boldsymbol{Q}_{ab})$,

(6.9.3) $\qquad r^{\lambda(\alpha)}=r\circ\alpha \qquad (r\in\mathfrak{R};\ \alpha\in G_{\boldsymbol{Q}})$.

The last formula follows from (6.6.1) and (i) of Th. 6.38. If we define an injection $\iota:\mathfrak{F}\rightarrow\mathfrak{R}$ by $\iota(f)=(f,f^*)$, then

(6.9.4) $\qquad \iota(f^{\tau(x)})=\iota(f)^{\lambda(x)} \qquad (f\in\mathfrak{F},\ x\in G_{A+})$.

Further, by a straightforward argument, we can show

(6.9.5) $\qquad \lambda(G_A)$ *is the commutor of* $\lambda(G_\infty)$ *in* $\operatorname{Aut}(\mathfrak{R})$.

REMARK 6.41. Let K, q, z, and $\mathfrak{R}$ be as in Cor. 6.39. We see that K_{ab} is a Galois extension of $\boldsymbol{Q}$, and $\operatorname{Gal}(K_{ab}/\boldsymbol{Q})$ is a non-abelian group with $\operatorname{Gal}(K_{ab}/K)$ as a subgroup of index 2. Put

$$\mathfrak{M} = q(K_A^\times)\mathfrak{N} = q(K_A^\times) \cup q(K_A^\times)\beta$$

with an element β of $\mathfrak{N} - q(K^\times)$. Then we can define a map

$$\rho : \mathfrak{M} \to \mathrm{Gal}\,(K_{ab}/\boldsymbol{Q})$$

as follows:

$$\rho(q(s)) = [s^{-1}, K] \qquad \text{for} \quad s \in K_A^\times,$$

$$\rho(\beta) = \text{complex conjugation},$$

$$\rho(x\beta) = \rho(x)\rho(\beta) \qquad \text{for} \quad x \in q(K_A^\times).$$

By (i) of Th. 6.31, (3) of Cor. 6.39, and (6.9.3, 4), we have, with the fixed point z,

$$r(z)^{\rho(y)} = r^{\lambda(y)}(z) \qquad (r \in \iota(\mathfrak{F}),\ y \in \mathfrak{M}).$$

It follows that ρ is a homomorphism. We then obtain a commutative diagram:

$$\begin{array}{ccccccccc}
1 & \longrightarrow & q(K_A^\times) & \longrightarrow & \mathfrak{M} & \longrightarrow & \mathfrak{M}/q(K_A^\times) & \longrightarrow & 1 \quad \text{(exact)} \\
 & & \downarrow & & \downarrow{\scriptstyle \rho} & & \downarrow & & \\
1 & \longrightarrow & \mathrm{Gal}\,(K_{ab}/K) & \longrightarrow & \mathrm{Gal}\,(K_{ab}/\boldsymbol{Q}) & \longrightarrow & \mathrm{Gal}\,(K/\boldsymbol{Q}) & \longrightarrow & 1 \quad \text{(exact)}
\end{array}.$$

CHAPTER 7
ZETA-FUNCTIONS OF ALGEBRAIC CURVES AND ABELIAN VARIETIES

7.1. Definition of the zeta-functions of algebraic curves and abelian varieties; the aim of this chapter

Let V be a projective non-singular curve of genus g, defined over an algebraic number field k of finite degree. For every prime ideal $\mathfrak{p}$ in k, let $\mathfrak{p}(V)$ denote the curve obtained from V by reduction modulo $\mathfrak{p}$. There exists a finite set $\mathfrak{B}$ of prime ideals in k such that $\mathfrak{p}(V)$ is a non-singular curve (of multiplicity one) if $\mathfrak{p} \notin \mathfrak{B}$. It can be shown that $\mathfrak{p}(V)$ is of genus g for such a $\mathfrak{p}$ [81, § 10.4, Prop. 11]. Now the zeta-function $Z(u\,;\mathfrak{p}(V))$ of $\mathfrak{p}(V)$ over the residue field $\kappa_{\mathfrak{p}}$ of $\mathfrak{p}$ has the following form:

$$Z(u\,;\mathfrak{p}(V)) = F_{\mathfrak{p}}(u)/[(1-u)(1-N(\mathfrak{p})u)]\,.$$

Here u is an indeterminate, $N(\mathfrak{p})$ is the number of elements of $\kappa_{\mathfrak{p}}$, and $F_{\mathfrak{p}}$ is a polynomial of degree $2g$ of which the constant term is 1. Then *the zeta-function of V over k* is defined (formally) as an infinite product[14]

$$\zeta(s\,;V/k) = \textstyle\prod_{\mathfrak{p}\notin\mathfrak{B}} F_{\mathfrak{p}}(N(\mathfrak{p})^{-s})^{-1}\,,$$

with a complex variable s. One can actually define in a similar way the zeta-function(s) of an arbitrary (non-singular projective) algebraic variety over k. But we shall consider here only the zeta-functions of curves and abelian varieties. To define the zeta-function of an abelian variety A defined over k, we first observe that there exists a finite set $\mathfrak{B}'$ of prime ideals in k such that, for every $\mathfrak{p} \notin \mathfrak{B}'$, A has good reduction modulo $\mathfrak{p}$ in the sense of [66], or equivalently, A has no defect for $\mathfrak{p}$ in the sense of [81, § 11]. Let $\mathfrak{p}(A)$ denote the abelian variety obtained from A by reduction modulo $\mathfrak{p}$, $\pi_{\mathfrak{p}}$ the Frobenius endomorphism of $\mathfrak{p}(A)$ of degree $N(\mathfrak{p})$, and R_l an l-adic representation of End $(\mathfrak{p}(A))$ for a rational prime l which is prime to $\mathfrak{p}$. Then (the one-dimensional part of) the zeta-function of $\mathfrak{p}(A)$ over $\kappa_{\mathfrak{p}}$ is given by

$$F'_{\mathfrak{p}}(u) = \det\,[1 - R_l(\pi_{\mathfrak{p}})u]\,.$$

Therefore we define the zeta-function of A over k to be an infinite product

---

14) Although we are neglecting the "bad" primes $\mathfrak{p}$ in our discussion, it is actually important to consider the Euler factors also for them. See § 7.9, C.

$$\zeta(s\,;A/k)=\prod_{\mathfrak{p}\notin\mathfrak{B}'}F'_{\mathfrak{p}}(N(\mathfrak{p})^{-s})^{-1}.$$

If A is the jacobian variety of the curve V, then A can be defined over the same field k of definition for V. Moreover, as Igusa has shown, we can choose a model of A so that $\mathfrak{B}'\subset\mathfrak{B}$. Then, for every $\mathfrak{p}\notin\mathfrak{B}$, we have $F_{\mathfrak{p}}=F'_{\mathfrak{p}}$ by Weil [92], so that $\zeta(s\,;A/k)$ is essentially the same as $\zeta(s\,;V/k)$.

Coming back to the general case, we can now state, in a somewhat specialized form,

THE CONJECTURE OF HASSE AND WEIL. *Each of the functions $\zeta(s\,;V/k)$ and $\zeta(s\,;A/k)$ can be holomorphically continued to the whole complex s-plane, and satisfies a functional equation.*

The zeta-function of a variety has been determined, and hence the conjecture has been verified, in the following cases:

($\mathrm{I_a}$) Algebraic curves of type $\alpha x^m+\beta y^n+\gamma=0$ (Weil [93]).

($\mathrm{I_b}$) Elliptic curves with complex multiplications (Deuring [12]).

($\mathrm{I_c}$) Abelian varieties with sufficiently many complex multiplications (Taniyama [87]).

($\mathrm{II_a}$) Algebraic curves isomorphic to $\Gamma\backslash\mathfrak{H}^*$ with certain congruence subgroups Γ of $SL_2(\boldsymbol{Z})$ (Eichler [16], Shimura [70]).

($\mathrm{II_b}$) Algebraic curves isomorphic to $\Gamma\backslash\mathfrak{H}^*$ with arithmetic Fuchsian groups Γ obtained from quaternion algebras (Shimura [73], [77]).

($\mathrm{II_c}$) Certain fibre varieties of which the base is a curve of type ($\mathrm{II_{a,b}}$), and the fibres are abelian varieties (especially elliptic curves) (Kuga and Shimura [42], Ihara [33], Deligne [9]).

The result of ($\mathrm{I_c}$) generalizes that of ($\mathrm{I_b}$), and in essence, of ($\mathrm{I_a}$). Similarly ($\mathrm{II_b}$) includes ($\mathrm{II_a}$) as a special case. The zeta-function in the cases ($\mathrm{I_{a,b,c}}$) is a product of several Hecke L-functions with Größen-characters of totally imaginary fields. On the other hand, the zeta-function in the cases ($\mathrm{II_{a,b,c}}$) is a product of Dirichlet series of the type of Ch. 3, or their generalizations. In this chapter, we shall discuss the cases ($\mathrm{I_c}$) and ($\mathrm{II_a}$), with more stress on the latter case than the former. More specifically, we shall verify the above conjecture for the curves V_S defined in § 6.7, and also for the abelian varieties of CM-type considered in § 5.5. Further in § 7.7, we shall investigate some class fields over real quadratic fields which are closely connected with the zeta-functions of V_S.

7.2. Algebraic correspondences on algebraic curves

Let us first recall some elementary properties of algebraic correspondences on curves. For a systematic treatment of this topic, the reader is referred to Weil [90, Ch. VIII] and [91]. Let U and V be projective non-singular curves defined over a field k. By an *algebraic 1-cycle*, simply a *1-cycle*, or an *algebraic correspondence*, on $U \times V$, we understand a formal finite sum $X = \sum_i n_i D_i$ with $n_i \in \boldsymbol{Z}$ and one-dimensional subvarieties D_i of $U \times V$. We denote by ${}^tX$ the 1-cycle on $V \times U$, which is the transform of X by the map $(u, v) \mapsto (v, u)$ of $U \times V$ to $V \times U$. By a *0-cycle*, or a *divisor* on U, we understand a formal finite sum $c = \sum_i m_i b_i$ with $m_i \in \boldsymbol{Z}$ and $b_i \in U$. We put $\deg(c) = \sum_i m_i$. For such X and c, we can define a 0-cycle $X[c]$ on V by

$$X[c] = pr_V[X \cdot (c \times V)],$$

where pr_V denotes the projection of $U \times V$ to V, and $X \cdot (c \times V)$ the intersection product of X and $c \times V$. Define two integers $d(X)$ and $d'(X)$ by

$$d(X)U = pr_U(X), \qquad d'(X)U = pr_V(X).$$

(See [91]. Roughly speaking, $d(X)$ (resp. $d'(X)$) is the number of sheets of X, viewed as a covering of U (resp. V).) Then we have

$$\deg(X[c]) = d(X) \cdot \deg(c).$$

Let W be another projective non-singular curve, and Y a 1-cycle on $V \times W$. Then we can define a 1-cycle $Z = Y \circ X$ on $U \times W$ by

$$Z = pr_{U \times W}[(X \times W) \cdot (U \times Y)].$$

Z is uniquely characterized by the property

(7.2.1) $Z(b) = Y[X[b]]$ *for every* $b \in U$, *and* ${}^tZ[c] = {}^tX[{}^tY[c]]$ *for every* $c \in W$.

Therefore ${}^tZ = {}^tX \circ {}^tY$.

We call X *proper*, if X has no component of the form $a \times V$ with $a \in U$ or $U \times b$ with $b \in V$. We see easily that $Y \circ X$ is proper if X and Y are proper. Moreover, if X and X' are proper 1-cycles on $U \times V$ and $X(a) = X'(a)$ for a generic point a of U over a field of rationality for U, V, X, and X', then $X = X'$.

Let A_U (resp. A_V) be the jacobian variety of U (resp. V), and f_U (resp. f_V) a canonical map of U into A_U (resp. V into A_V). With every 1-cycle X on $U \times V$, we can associate an element ξ of $\mathrm{Hom}(A_U, A_V)$ such that, if $X[u] = \sum_i v_i$ with $u \in U$ and $v_i \in V$, then

$$\xi(f_U(u)) = \sum_i f_V(v_i) + c$$

with a point c of A_V independent of u. If k is a field of rationality for U, V, and X, then A_U and A_V can be chosen so as to be rational over k. The maps f_U and f_V may not be rational over k, but it can easily be shown that ξ is rational over k.

For a projective variety Z, let $\mathcal{D}(Z)$ denote the vector space of holomorphic differential forms of degree one on Z. If U, V, and X are as above, we can associate with X a linear map δX of $\mathcal{D}(V)$ into $\mathcal{D}(U)$ as follows. By linearity, it is sufficient to consider the case where X is irreducible. Let k be a field of rationality for U, V, and X. Take a generic point u of U over k, and put $X[u]=\sum_{i=1}^{e} v_i$ with $v_i \in V$. Let W be a projective non-singular curve with a generic point w over the algebraic closure k_1 of k such that $k_1(w) = k_1(u, v_1, \cdots, v_e)$. Let p (resp. q_i) be the morphism of W into U (resp. V) defined by $p(w)=u$ (resp. $q_i(w)=v_i$) over k_1. For $\varepsilon \in \mathcal{D}(V)$, one can show the unique existence of an element $\varepsilon \circ X$ (also denoted by $\delta X(\varepsilon)$) of $\mathcal{D}(U)$ such that

$$(\varepsilon \circ X) \circ p = \textstyle\sum_{i=1}^{e} \varepsilon \circ q_i .$$

(For the notation $\varepsilon \circ q_i$, see § 5.1 and Appendix N° 8.) If A_U, f_U, A_V, f_V are as above, the map

$$\mathcal{D}(A_V) \ni \omega \mapsto \omega \circ f_V \in \mathcal{D}(V)$$

is an isomorphism, and

$$(\omega \circ f_V) \circ X = (\omega \circ \xi) \circ f_U$$

with the element ξ of $\mathrm{Hom}\,(A_U, A_V)$ associated with X. (For details of the proof of these facts, see [81, § 2.9, Prop. 9].) In other words, the diagram

(7.2.2) $$\begin{array}{ccc} \mathcal{D}(A_V) & \xrightarrow{\ \delta\xi\ } & \mathcal{D}(A_U) \\ {\scriptstyle \delta f_V}\downarrow & & \downarrow{\scriptstyle \delta f_U} \\ \mathcal{D}(V) & \xrightarrow[\ \delta X\ ]{} & \mathcal{D}(U) \end{array}$$

is commutative, where δ indicates the action of the map (or correspondence) on differential forms (see Appendix N° 8).

We shall now discuss a special type of correspondence for the curves which are models of the upper half plane modulo Fuchsian groups of the first kind. We fix a family $\mathcal{G}=\{\Gamma_\lambda \mid \lambda \in \Lambda\}$ of mutually commensurable subgroups of $SL_2(\boldsymbol{R})$ which are Fuchsian groups of the first kind, and denote by $\tilde{\Gamma}$ the set of all elements α of $GL_2^+(\boldsymbol{R})$ such that $\alpha\Gamma\alpha^{-1}$ is commensurable with Γ for a member Γ of $\mathcal{G}$ (see § 3.1). Note that $\tilde{\Gamma}$ does not depend on the choice of Γ, and all members of $\mathcal{G}$ have the same set of cusps (see Prop. 1.30). Let $\mathfrak{H}^*$ denote the union of $\mathfrak{H}$ and the cusps. For each $\Gamma_\lambda \in \mathcal{G}$, fix a model $(V_\lambda, \varphi_\lambda)$ of $\Gamma_\lambda \backslash \mathfrak{H}^*$ in the sense of § 6.7. Now, for $\Gamma_\lambda, \Gamma_\mu \in \mathcal{G}$, and $\alpha \in \tilde{\Gamma}$, put

(7.2.3) $$X = X(\Gamma_\lambda \alpha \Gamma_\mu) = \{\varphi_\mu(z) \times \varphi_\lambda(\alpha(z)) \mid z \in \mathfrak{H}^*\} \qquad (\subset V_\mu \times V_\lambda).$$

It can easily be verified that $X(\Gamma_\lambda \alpha \Gamma_\mu)$ is a proper 1-cycle, and actually an absolutely irreducible curve, on $V_\mu \times V_\lambda$; it depends only on the coset $\Gamma_\lambda \alpha \Gamma_\mu$, and not on the choice of α. If $\Gamma_\lambda \alpha \Gamma_\mu = \bigcup_{i=1}^{e} \Gamma_\lambda \alpha_i$ is a disjoint union, and $\Gamma_\lambda \cap \{\pm 1\} = \Gamma_\mu \cap \{\pm 1\}$, then

$$X[\varphi_\mu(z)] = \textstyle\sum_{i=1}^{e} \varphi_\lambda(\alpha_i(z)) . \tag{7.2.4}$$

Therefore, if we define $\deg(\Gamma_\lambda \alpha \Gamma_\mu)$ as in § 3.1, then

$$d(X(\Gamma_\lambda \alpha \Gamma_\mu)) = e = \deg(\Gamma_\lambda \alpha \Gamma_\mu) . \tag{7.2.5}$$

Further we see easily that

$${}^t X(\Gamma_\lambda \alpha \Gamma_\mu) = X(\Gamma_\mu \alpha^{-1} \Gamma_\lambda) = X(\Gamma_\mu \alpha^\iota \Gamma_\lambda) ,$$

where ι denotes the main involution of $M_2(\boldsymbol{R})$ (see § 3.3).

PROPOSITION 7.1. *Suppose that*

$$\Gamma_\lambda \cap \{\pm 1\} = \Gamma_\mu \cap \{\pm 1\} = \Gamma_\nu \cap \{\pm 1\} ,$$

and

$$(\Gamma_\lambda \alpha \Gamma_\mu) \cdot (\Gamma_\mu \beta \Gamma_\nu) = \sum c_\xi \cdot \Gamma_\lambda \xi \Gamma_\nu$$

with $c_\xi \in \boldsymbol{Z}$ in the sense of the multiplication-law defined in § 3.1. *Then*

$$X(\Gamma_\lambda \alpha \Gamma_\mu) \circ X(\Gamma_\mu \beta \Gamma_\nu) = \textstyle\sum_\xi c_\xi \cdot X(\Gamma_\lambda \xi \Gamma_\nu) .$$

This can easily be verified by applying the correspondences on $\varphi_\nu(z)$, on account of (7.2.1) and (7.2.4).

Let $S_2(\Gamma_\lambda)$ be the vector space of all cusp forms of weight 2 with respect to Γ_λ (see § 2.1). In Cor. 2.17, we have seen that the map $f(z) \mapsto f(z)dz$ is an isomorphism of $S_2(\Gamma_\lambda)$ onto $\mathfrak{D}(\Gamma_\lambda \backslash \mathfrak{H}^*)$. More precisely, if we distinguish V_λ from $\Gamma_\lambda \backslash \mathfrak{H}^*$, the isomorphism $S_2(\Gamma_\lambda) \ni f \mapsto \varepsilon \in \mathfrak{D}(V_\lambda)$ is obtained by the relation $f(z)dz = \varepsilon \circ \varphi_\lambda$. Now let us show that

$$\begin{array}{ccc} S_2(\Gamma_\lambda) & \xrightarrow{\quad [\Gamma_\lambda \alpha \Gamma_\mu]_2 \quad} & S_2(\Gamma_\mu) \\ \Big\downarrow & & \Big\downarrow \\ \mathfrak{D}(V_\lambda) & \xrightarrow{\quad \delta X(\Gamma_\lambda \alpha \Gamma_\mu) \quad} & \mathfrak{D}(V_\mu) \end{array} \tag{7.2.6}$$

is a commutative diagram, where $[\Gamma_\lambda \alpha \Gamma_\mu]_2$ is the map defined in § 3.4. Put $\Gamma = \bigcap_{i=1}^{e} \alpha_i^{-1} \Gamma_\lambda \alpha_i \cap \Gamma_\mu$ with the elements α_i as above. Let (W, ψ) be a model of $\Gamma \backslash \mathfrak{H}^*$. We can define morphisms $p : W \to V_\mu$ and $q_i : W \to V_\lambda$ by $p \circ \psi = \varphi_\mu$ and $q_i \circ \psi = \varphi_\lambda \circ \alpha_i$. If we put $u = \varphi_\mu(z)$ and $v_i = \varphi_\lambda(\alpha_i(z))$, then we see that the present symbols are exactly in the same situation as in the definition of δX. Therefore if $f \in S_2(\Gamma_\lambda)$ and $\varepsilon \in \mathfrak{D}(V_\lambda)$ are such that $f(z)dz = \varepsilon \circ \varphi_\lambda$, then

$$\begin{aligned}\varepsilon\circ X\circ\varphi_\mu &= (\varepsilon\circ X\circ p)\circ\psi\\ &= \textstyle\sum_{i=1}^e \varepsilon\circ q_i\circ\psi = \sum_{i=1}^e \varepsilon\circ\varphi_\lambda\circ\alpha_i\\ &= \textstyle\sum_{i=1}^e (f(z)dz)\circ\alpha_i = \sum_{i=1}^e f(\alpha_i(z))\det(\alpha_i)j(\alpha_i, z)^{-2}\\ &= f\,|\,[\Gamma_\lambda\alpha\Gamma_\mu]_2,\end{aligned}$$

which proves the commutativity of (7.2.6). Especially if $\lambda=\mu$, by virtue of (7.2.6) and (7.2.2), we see that the eigen-values of $[\Gamma_\lambda\alpha\Gamma_\lambda]_2$ coincide with those of the endomorphism ξ of A_λ associated with $X(\Gamma_\lambda\alpha\Gamma_\lambda)$. It follows that

(7.2.7) *The eigen-values of* $[\Gamma_\lambda\alpha\Gamma_\lambda]_2$ *are algebraic integers.*

7.3. Modular correspondences on the curves V_S

We shall now specialize our discussion to the groups Γ_S defined in § 6.7. Let $\mathfrak{Z}$ be as in § 6.7, and let $\Gamma'_S=\boldsymbol{R}^\times\Gamma_S\cap SL_2(\boldsymbol{R})$. Then the transformation group $\Gamma_S/\boldsymbol{Q}^\times$ on $\mathfrak{H}$ can be identified with $\Gamma'_S/\{\pm 1\}$. Let $\{V_S, \varphi_S, J_{TS}(x), (S, T\in\mathfrak{Z};\ x\in G_{A+})\}$ be as in § 6.7. We can consider $J_{TS}(x)$ as a proper 1-cycle on $V_S\times V_T^{\sigma(x)}$ rational over k_S.

Let $S\in\mathfrak{Z}$, $T\in\mathfrak{Z}$, and $x\in G_{A+}$. Put $W=S\cap x^{-1}Tx$. Then we can define a proper 1-cycle $X_{TS}(x)$ on $V_S\times V_T^{\sigma(x)}$ by

(7.3.1) $$X_{TS}(x)=J_{TW}(x)\circ {}^tJ_{SW}(1)\,.$$

Then we see that

(7.3.2) $$d(X_{TS}(x))=[\Gamma_S:\Gamma_W]\,,\qquad d'(X_{TS}(x))=[\Gamma_{x^{-1}Tx}:\Gamma_W]\,.$$

Note also that $X_{TS}(x)$ is the image of V_W by the map $(J_{SW}(1), J_{TW}(x))$, i.e., the locus of $J_{SW}(1)(v)\times J_{TW}(x)(v)$ with a generic point v of V_W.

PROPOSITION 7.2. *The cycles $X_{TS}(x)$ have the following properties.*

(1) *$X_{TS}(x)$ is absolutely irreducible, and rational over k_W, where $W=S\cap x^{-1}Tx$.*
(2) *$X_{TS}(x)=J_{TS}(x)$ if $S\subset x^{-1}Tx$.*
(3) *$X_{TS}(x)={}^tJ_{ST}(x^{-1})^{\sigma(x)}$ if $x^{-1}Tx\subset S$.*
(4) *$X_{TS}(x)$ depends only on $Tx\Gamma_S$.*
(5) *If $k_S=k_W$ for $W=S\cap x^{-1}Tx$, $X_{TS}(x)$ depends only on TxS.*
(6) *$X_{TS}(\alpha)=X(\Gamma'_T\alpha\Gamma'_S)$ if $\alpha\in G_{\boldsymbol{Q}+}$.*
(7) *$X_{TR}(xy)=X_{TS}(x)^{\sigma(y)}\circ J_{SR}(y)$ if $y\in G_{A+}$ and $R=y^{-1}Sy$.*
(8) *$X_{RS}(yx)=J_{RT}(y)^{\sigma(x)}\circ X_{TS}(x)$ if $y\in G_{A+}$ and $R=yTy^{-1}$.*
(9) *${}^tX_{TS}(x)=X_{ST}(x^{-1})^{\sigma(x)}$.*

PROOF. The assertions (1), (2), (6), (8) follow from our definition in a straightforward way. To show (9), put $W=S\cap x^{-1}Tx$ and $P=xWx^{-1}=T\cap xSx^{-1}$.

Then

$$\begin{aligned} X_{ST}(x^{-1})^{\sigma(x)} &= J_{SP}(x^{-1})^{\sigma(x)} \circ {}^tJ_{TP}(1)^{\sigma(x)} \\ &= J_{SW}(1) \circ J_{WP}(x^{-1})^{\sigma(x)} \circ {}^tJ_{TP}(1)^{\sigma(x)} \\ &= J_{SW}(1) \circ {}^tJ_{PW}(x) \circ {}^tJ_{TP}(1)^{\sigma(x)} \\ &= J_{SW}(1) \circ {}^tJ_{TW}(x) \\ &= {}^tX_{TS}(x) . \end{aligned}$$

We obtain (7) from (8) and (9). Then (4) and (5) follow from (7) and (8); (3) from (2) and (9).

PROPOSITION 7.3. *Let $S \in \mathfrak{Z}$, $T \in \mathfrak{Z}$, $x \in G_{A+}$, and $W = S \cap x^{-1}Tx$. Then the following three conditions are equivalent to each other.*

(1) $k_W = k_S$.

(2) $S = W\Gamma_S$.

(3) $TxS = Tx\Gamma_S$.

Moreover, if these conditions are satisfied, $d(X_{TS}(x)) = [S : W] = [\Gamma_S : \Gamma_W]$.

PROOF. By Lemma 6.17, we have $k_W = k_S$ if and only if $WG_{Q+} = SG_{Q+}$. Therefore the first two conditions are equivalent. Next, if $S = W\Gamma_S$, we have $S \subset x^{-1}Tx\Gamma_S$, so that $x^{-1}TxS = x^{-1}Tx\Gamma_S$, hence $TxS = Tx\Gamma_S$. Conversely, if $TxS = Tx\Gamma_S$, we have $x^{-1}TxS = x^{-1}Tx\Gamma_S$, so that $S \subset S \cap (x^{-1}Tx\Gamma_S) = W\Gamma_S$, hence $S = W\Gamma_S$. Since $\Gamma_W = \Gamma_S \cap W$, we have $[\Gamma_S : \Gamma_W] = [S : W]$ if $S = W\Gamma_S$. Therefore we obtain the last assertion from (7.3.2).

PROPOSITION 7.4. *Let $R, S, T \in \mathfrak{Z}$, and $x, y \in G_{A+}$. Suppose that $TxS = Tx\Gamma_S$, $SyR = Sy\Gamma_R$, and $TwR = Tw\Gamma_R$ for every $w \in TxSyR$. Let $(TxS)\cdot(SyR) = \sum c_w \cdot (TwR)$ with $c_w \in \boldsymbol{Z}$ in the sense of the multiplication-law of* § 3.1. *Then*

$$X_{TS}(x)^{\sigma(y)} \circ X_{SR}(y) = \sum c_w \cdot X_{TR}(w) .$$

PROOF. Put

$$Q = y^{-1}Sy\,, \quad P = y^{-1}x^{-1}Txy\,, \quad M = x^{-1}Tx\,, \quad R \cap Q = W\,, \quad Q \cap P = Z\,.$$

By Prop. 7.3, $k_R = k_W$, and $k_Q = k_Z$, hence $QR = Q\Gamma_R$, and $PQ = P\Gamma_Q$. Therefore we have $(PQ)\cdot(QR) = \sum c_\gamma \cdot (P\gamma R)$ with $c_\gamma \in \boldsymbol{Z}$ and elements γ of Γ_Q. Then we can show, in a straightforward way, that

$$(TxS)\cdot(SyR) = \sum c_\gamma \cdot Txy\gamma R\,.$$

By (8) of Prop. 7.2, we have

$$X_{TS}(x) = J_{TM}(x) \circ X_{MS}(1) \quad \text{and} \quad X_{SR}(y) = J_{SQ}(y) \circ X_{QR}(1)\,,$$

hence

$$\begin{aligned} X_{TS}(x)^{\sigma(y)} \circ X_{SR}(y) &= J_{TM}(x)^{\sigma(y)} \circ X_{MQ}(y) \circ X_{QR}(1) \\ &= J_{TM}(x)^{\sigma(y)} \circ J_{MP}(y) \circ X_{PQ}(1) \circ X_{QR}(1) \\ &= J_{TP}(xy) \circ X_{PQ}(1) \circ X_{QR}(1). \end{aligned}$$

Now let $\Gamma_R = \bigcup_j \Gamma_W \beta_j$ and $\Gamma_Q = \bigcup_i \Gamma_Z \alpha_i$ be disjoint unions. Then $QR = \bigcup_j Q\beta_j$ and $PQ = \bigcup_i P\alpha_i$ are disjoint unions. Let $z \in \mathfrak{H}$. Since $\Gamma_R \cap \boldsymbol{R}^\times = \Gamma_W \cap \boldsymbol{R}^\times = \boldsymbol{Q}^\times$, we have ${}^t J_{RW}(1)[\varphi_R(z)] = \sum_j \varphi_W(\beta_j(z))$, so that $X_{QR}(1)[\varphi_R(z)] = \sum_j \varphi_Q(\beta_j(z))$. For the same reason, we have $X_{PQ}(1)[X_{QR}(1)[\varphi_R(z)]] = \sum_{i,j} \varphi_P(\alpha_i \beta_j(z))$. From our definition of $(PQ) \cdot (QR)$ (see § 3.1), it follows that $X_{PQ}(1) \circ X_{QR}(1) = \sum c_\gamma \cdot X_{PR}(\gamma)$, hence

$$X_{TS}(x)^{\sigma(y)} \circ X_{SR}(y) = \sum c_\gamma \cdot J_{TP}(xy) \circ X_{PR}(\gamma) = \sum c_\gamma \cdot X_{TR}(xy\gamma).$$

By our assumption and (5) of Prop. 7.2, $X_{TR}(w)$ depends only on TwR, for every $w \in TxSyR$. Therefore we obtain our proposition.

PROPOSITION 7.5. *Let* $W = W_0 G_{\infty+}$ *and* $W' = W_0' G_{\infty+}$ *with compact subgroups* W_0 *and* W_0' *of* G_0. *Then*

$$\boldsymbol{Q}^\times W \cap \boldsymbol{Q}^\times W' = \boldsymbol{Q}^\times(\{\pm 1\} W \cap \{\pm 1\} W').$$

PROOF. Let $ax = by$ with $a \in \boldsymbol{Q}$, $b \in \boldsymbol{Q}$, $x \in W$, and $y \in W'$. Then

$$a^2/b^2 = \det(yx^{-1}) \in \boldsymbol{Q}^\times \cap \det(W_0 W_0' G_{\infty+}) = \{\pm 1\},$$

so that $a = \pm b$, hence $x = \pm y$. This proves our proposition.

Put $U_p = GL_2(\boldsymbol{Z}_p)$ for each rational prime p, and

$$(7.3.3) \qquad \begin{aligned} U &= G_{\infty+} \times \textstyle\prod_p U_p, \\ \mathfrak{g} &= \boldsymbol{R} \times \textstyle\prod_p \boldsymbol{Z}_p, \\ \mathfrak{g}^\times &= \boldsymbol{R}^\times \times \textstyle\prod_p \boldsymbol{Z}_p^\times, \\ \mathfrak{g}_+^\times &= \boldsymbol{R}_+^\times \times \textstyle\prod_p \boldsymbol{Z}_p^\times \qquad (\boldsymbol{R}_+^\times = \{x \in \boldsymbol{R}^\times \mid x > 0\}). \end{aligned}$$

For elements $a = (a_p)$ and $b = (b_p)$ of $\mathfrak{g}$, and for a positive integer s, we write $a \equiv b \bmod (s)$ if $a_p - b_p \in s\boldsymbol{Z}_p$ for all p.

Let us fix a positive integer N, a positive divisor t of N, and a subgroup $\mathfrak{h}^*$ of $\mathfrak{g}^\times$ such that

$$(7.3.4) \qquad \{a \in \mathfrak{g}^\times \mid a \equiv 1 \bmod (N)\} \subset \mathfrak{h}^*.$$

Since the left side is open in $\mathfrak{g}^\times$, so is $\mathfrak{h}^*$, and $\mathfrak{h}^* = \boldsymbol{R}^\times \cdot \mathfrak{h}_0^*$ with an open subgroup $\mathfrak{h}_0^*$ of $\prod_p \boldsymbol{Z}_p^\times$. Given N, t, and $\mathfrak{h}^*$, define U' and S by

$$(7.3.5) \qquad U' = \left\{ \begin{bmatrix} a & b \\ c & d \end{bmatrix} \in U \,\middle|\, d \in \mathfrak{h}^*,\ c \equiv 0 \bmod (N),\ b \equiv 0 \bmod (t) \right\},$$

$$S = \boldsymbol{Q}^\times U'.$$

Then $S \in \mathfrak{Z}$, $\det(U') = \mathfrak{g}_+^\times$, and $Q^\times \cdot \det(S) = Q^\times \cdot \mathfrak{g}_+^\times = Q_A^\times$, so that $k_S = Q$. Put $\Gamma' = G_Q \cap U'$. Then

(7.3.6) $$\Gamma' = \left\{ \begin{bmatrix} a & b \\ c & d \end{bmatrix} \in SL_2(Z) \,\middle|\, a \in \mathfrak{h}^*,\ d \in \mathfrak{h}^*,\ c \equiv 0 \bmod (N),\ b \equiv 0 \bmod (t) \right\},$$
$$\Gamma_S = Q^\times \Gamma'.$$

Therefore Γ' is exactly the group defined by (3.3.2). Note that the present $\mathfrak{h}^*$ corresponds uniquely to a subgroup of $(Z/NZ)^\times$, which we wrote $\mathfrak{h}$ in (3.3.2). We consider also a semi-group

(7.3.7) $$\Delta' = \left\{ \begin{bmatrix} a & b \\ c & d \end{bmatrix} \in M_2(Z) \cap G_{Q+} \,\middle|\, a \in \mathfrak{h}^*,\ c \equiv 0 \bmod (N),\ b \equiv 0 \bmod (t) \right\},$$

which is the same as (3.3.3).

LEMMA 7.6. *$\det(U_p \cap x^{-1} U_p x) = Z_p^\times$ for every $x \in GL_2(Q_p)$.*

PROOF. We can find elements y and z of U_p so that $yxz = a \cdot \begin{bmatrix} 1 & 0 \\ 0 & b \end{bmatrix}$ with $a \in Q_p^\times$ and $b \in Z_p$. Put $v = \begin{bmatrix} 1 & 0 \\ 0 & b \end{bmatrix}$. Then

$$\begin{bmatrix} 1 & 0 \\ 0 & c \end{bmatrix} \in U_p \cap v^{-1} U_p v = z^{-1}(U_p \cap x^{-1} U_p x) z$$

for every $c \in Z_p^\times$, q. e. d.

PROPOSITION 7.7. *The notation being as above, $X_{SS}(\alpha)$ is rational over Q for every $\alpha \in \Delta'$.*

PROOF. By (3) of Prop. 3.32, if $\alpha \in \Delta'$, we have

$$\Gamma' \alpha \Gamma' = (\Gamma' \xi \Gamma')(\Gamma' \eta \Gamma'),$$
$$\xi \equiv \begin{bmatrix} 1 & 0 \\ 0 & q \end{bmatrix} \bmod (N), \qquad \eta = \begin{bmatrix} 1 & 0 \\ 0 & m \end{bmatrix}, \qquad (q, N) = 1,$$

with positive integers q and m. By Prop. 7.1 and (6) of Prop. 7.2, we have $X_{SS}(\alpha) = X_{SS}(\xi) \circ X_{SS}(\eta)$. Therefore, it is sufficient to prove our assertion for ξ and η. As for η, observe that $\begin{bmatrix} a & 0 \\ 0 & 1 \end{bmatrix} \in U' \cap \eta^{-1} U' \eta$ for every $a \in \mathfrak{g}_+^\times$, hence $Q^\times \cdot \det(S \cap \eta^{-1} S \eta) = Q_A^\times$. By (1) of Prop. 7.2, $X_{SS}(\eta)$ is defined over Q. As for ξ, if p does not divide N, then $\det(U_p \cap \xi^{-1} U_p \xi) = Z_p^\times$ by Lemma 7.6. If p divides N, and U'_p denotes the projection of U' to G_p, then $U'_p \cap \xi^{-1} U'_p \xi$ contains $\begin{bmatrix} a & 0 \\ 0 & 1 \end{bmatrix}$ for every $a \in Z_p^\times$. Therefore $\det(U' \cap \xi^{-1} U' \xi) = \mathfrak{g}_+^\times$, so that $Q^\times \cdot \det(S \cap \xi^{-1} S \xi) = Q_A^\times$. By (1) of Prop. 7.2, $X_{SS}(\xi)$ is rational over Q, which completes the proof.

Let q be an integer prime to N, and σ_q be an element of $SL_2(\boldsymbol{Z})$ such that

(7.3.8) $$q\sigma_q \equiv \begin{bmatrix} 1 & 0 \\ 0 & q^2 \end{bmatrix} \bmod (N) \qquad \text{(see (3.3.10))}.$$

We see that $\sigma_q S \sigma_q^{-1} = S$, so that $J_{SS}(\sigma_q)$ is meaningful, and rational over $k_S = \boldsymbol{Q}$. Note also that $J_{SS}(\sigma_q)$ depends only on the residue class of q modulo (N), and not on the choice of σ_q.

PROPOSITION 7.8. *Let* $\tau = \begin{bmatrix} 0 & -t \\ N & 0 \end{bmatrix}$, *and* $\mathfrak{h}_0' = (\{\pm 1\} \cdot \mathfrak{h}^*) \cap (\prod_p \boldsymbol{Z}_p^\times)$. *Let* k *denote the subfield of* $\boldsymbol{Q}_{ab}$ *corresponding to the subgroup* $\boldsymbol{Q}^\times \boldsymbol{R}_+^\times \mathfrak{h}_0'$ *of* $\boldsymbol{Q}_A^\times$. *Then* $X_{SS}(\tau)$ *is a birational automorphism of* V_S *rational over* k. *Moreover, if* $k_N = \boldsymbol{Q}(e^{2\pi i/N})$, $\zeta = e^{2\pi i/N}$, *and* $\rho(q)$ *is the element of* Gal$(k_N/\boldsymbol{Q})$ *such that* $\zeta^{\rho(q)} = \zeta^q$, *then*

$$X_{SS}(\tau) = X_{SS}(\tau)^{\rho(q)} \circ J_{SS}(\sigma_q).$$

PROOF. Put $\mathfrak{h}' = \boldsymbol{R}^\times \mathfrak{h}_0'$, and

$$W = \left\{ \begin{bmatrix} a & b \\ c & d \end{bmatrix} \in U \,\middle|\, a \in \mathfrak{h}',\ d \in \mathfrak{h}',\ b \equiv 0 \bmod (t),\ c \equiv 0 \bmod (N) \right\}.$$

By Prop. 7.5, we have

$$S \cap \tau^{-1} S \tau = \boldsymbol{Q}^\times (U' \{\pm 1\} \cap \tau^{-1} U' \tau \{\pm 1\}) = \boldsymbol{Q}^\times W.$$

Since $\det(W) = \boldsymbol{R}_+^\times \mathfrak{h}_0'$, $X_{SS}(\tau)$ is rational over k, on account of (1) of Prop. 7.2. Further we see that $\tau^{-1} S \tau \cap G_{\boldsymbol{Q}+} = \boldsymbol{Q}^\times \Gamma'$, and $\tau^{-1} \Gamma' \tau = \Gamma'$. Therefore $X_{SS}(\tau) = X(\Gamma' \tau \Gamma')$ is a birational automorphism of V_S. Let $y = (y_p)$ be an element of G_0 such that $y_p = \begin{bmatrix} 1 & 0 \\ 0 & q \end{bmatrix}$ or 1 according as p divides N or not. Then we see that $\sigma_q^{-1} y \in U'$, so that $J_{SS}(\sigma_q) = J_{SS}(y)$. Now $\sigma(y) = [\det(y)^{-1}, \boldsymbol{Q}] = \rho(q)$ on k_N, $\tau y = y^\iota \tau$ and $y^\iota \in U'$, so that, by (4) and (7) of Prop. 7.2,

$$X_{SS}(\tau) = X_{SS}(\tau y) = X_{SS}(\tau)^{\rho(q)} \circ J_{SS}(y) = X_{SS}(\tau)^{\rho(q)} \circ J_{SS}(\sigma_q).$$

The algebraic correspondence $X_{SS}(\alpha)$ with $\alpha \in G_{\boldsymbol{Q}+}$ is often called a *modular correspondence* of level N. If $S = U$ (so that the level is 1), and α is a primitive element of $M_2(\boldsymbol{Z})$ of determinant n in the sense of § 4.6, the modular correspondence $X_{UU}(\alpha)$ can be represented by the equation $F_n(X, J) = 0$, with the polynomial F_n of (4.6.3).

7.4. Congruence relations for modular correspondences

Let p be a rational prime, and $\mathfrak{P}$ a prime divisor of $\bar{\boldsymbol{Q}}$ which divides p. If X is a variety, or a cycle etc., rational over $\bar{\boldsymbol{Q}}$, we shall denote by $\tilde{X}$ or $\mathfrak{P}(X)$ the object obtained from X by reduction modulo $\mathfrak{P}$. Let U, V, W be projective non-singular curves, X a proper positive 1-cycle on $U\times V$, and Y a proper positive 1-cycle on $V\times W$, all rational over $\bar{\boldsymbol{Q}}$. Suppose that $\tilde{U}$, $\tilde{V}$, $\tilde{W}$ are non-singular curves, and $\tilde{X}$, $\tilde{Y}$ are proper. Then we have

$$\mathfrak{P}(X\circ Y)=\tilde{X}\circ\tilde{Y}.$$

(For a general theory of reduction modulo $\mathfrak{P}$, we refer the reader to [69], [81, Ch. III].)

Let us now fix a member S of $\mathfrak{Z}$ of the form $S=\boldsymbol{Q}^{\times}U'$ with an open subgroup U' of U, where U is as in (7.2.3). We note that there is a finite set $\mathfrak{B}_S$ of rational primes such that the following statements hold for every p not contained in $\mathfrak{B}_S$.

(7.4.1) $U_p\subset U'$.

(7.4.2) $\mathfrak{P}(V_S^{\sigma})$ *is non-singular for every* $\sigma\in\mathrm{Gal}(k_S/\boldsymbol{Q})$, *and every prime divisor* $\mathfrak{P}$ *of* $\bar{\boldsymbol{Q}}$ *which divides* p.

(7.4.3) $\mathfrak{P}(J_{SS}(c))$ *is a biregular isomorphism of* $\mathfrak{P}(V_S)$ *to* $\mathfrak{P}(V_S^{g(c)})$ *for every* $c\in\boldsymbol{Q}_A^{\times}$, *and every prime divisor of* $\bar{\boldsymbol{Q}}$ *which divides* p.

(Note that there are only finitely many $J_{SS}(c)$, since $\boldsymbol{Q}_A^{\times}/(\boldsymbol{Q}_A^{\times}\cap S)$ is finite.)

(7.4.4) *If* $S_1=\boldsymbol{Q}^{\times}U$, $\mathfrak{P}(J_{S_1S}(1))$ *is a surjective morphism of* $\mathfrak{P}(V_{S_1})$ *to* $\mathfrak{P}(V_S)$ *for every* $\mathfrak{P}$ *which divides* p.

Here we take V_{S_1} to be the projective straight line, and φ_{S_1} to be the modular function J of Th. 2.9.

Now fix a rational prime p not contained in $\mathfrak{B}_S$, and a prime divisor $\mathfrak{P}$ of $\bar{\boldsymbol{Q}}$ which divides p. Let π denote the p-th power automorphism of the universal domain containing the residue field of $\bar{\boldsymbol{Q}}$ modulo $\mathfrak{P}$. We denote by Φ_S the Frobenius correspondence on $\tilde{V}_S\times\tilde{V}_S^{\pi}$, i.e., the locus of $a\times a^{\pi}$ on $\tilde{V}_S\times\tilde{V}_S^{\pi}$ with $a\in\tilde{V}_S$.

THEOREM 7.9. *The notation and assumptions being as above, let* w_p *be an element of* $M_2(\boldsymbol{Z}_p)$ *such that* $\det(w_p)=p$, *and* w *the element of* G_A *of which the* p*-component is* w_p, *and other components are all equal to* 1. *If* $p\notin\mathfrak{B}_S$, *then* $X_{SS}(w^{-1})$ *is rational over* k_S, *and one has*

$$\tilde{X}_{SS}(w^{-1})=\Phi_S+{}^t\Phi_S^{\pi}\circ\tilde{J}_{SS}(\det(w)^{-1}).$$

PROOF. Let U'' be the projection of U' to $\prod_{l\neq p}U_l$. Then $U'=U_pU''$, so that $wU'w^{-1}\cap U'=(w_pU_pw_p^{-1}\cap U_p)U''$. On account of Lemma 7.6, $\det(wU'w^{-1}\cap U')=\det(U')$. It follows that $k_S=k_Y$ if $Y=wSw^{-1}\cap S$. By (1) of Prop. 7.2, $X_{SS}(w^{-1})$ is rational over k_S. Now we can find infinitely many imaginary quadratic fields K such that p decomposes in K. Take such a K, and a normalized embedding q of K into $M_2(\boldsymbol{Q})$ such that $q(\mathfrak{o}_K)\subset M_2(\boldsymbol{Z})$, where $\mathfrak{o}_K$ denotes the ring of algebraic integers in K. Let z be the fixed point of $q(K^\times)$ on $\mathfrak{H}$. Let $\mathfrak{p}=\mathfrak{P}\cap K$. By Prop. 6.33, $K\cdot k_S(\varphi_S(z))$ is the subfield of K_{ab} corresponding to $K^\times\cdot\{s\in K_A^\times \mid q(s)\in S\}$. In view of (7.4.1), we see that p is unramified in $K\cdot k_S(\varphi_S(z))$. Let $K_\mathfrak{p}$ be the completion of K at $\mathfrak{p}$, $u_\mathfrak{p}$ a prime element of $K_\mathfrak{p}$, and u an element of $K_A^\times$ of which the $\mathfrak{p}$-component is $u_\mathfrak{p}$, and other components are all equal to 1. Let μ be a Frobenius automorphism of $\bar{\boldsymbol{Q}}$ over K with respect to $\mathfrak{P}$ and $\mathfrak{p}$. Since $\mu=[u, K]$ on $K\cdot k_S(\varphi_S(z))$, we have, by (ii) of Th. 6.31, $\varphi_S(z)^\mu=J_{ST}(q(u)^{-1})[\varphi_T(z)]$, where $T=q(u)Sq(u)^{-1}$. Since $q(\mathfrak{o}_K)\subset M_2(\boldsymbol{Z})$, we see that $q(u)_p$ is contained in $M_2(\boldsymbol{Z}_p)$, and has the same elementary divisors as w_p. Therefore, by (7.4.1), we have $Sq(u)^{-1}S=Sw^{-1}S$. We have seen in the above that $k_S=k_Y$ if $Y=S\cap wSw^{-1}$. By (5) of Prop. 7.2, this implies that $X_{SS}(w^{-1})$ depends only on $Sw^{-1}S$. Therefore, putting $R=S\cap T$, we have

$$\begin{aligned} X_{SS}(w^{-1}) &= X_{SS}(q(u)^{-1}) \\ &= J_{SR}(q(u)^{-1})\circ {}^tJ_{SR}(1) \\ &= J_{ST}(q(u)^{-1})\circ J_{TR}(1)\circ {}^tJ_{SR}(1)\,. \end{aligned}$$

Since $\varphi_T(z)\times\varphi_S(z)^\mu\in J_{ST}(q(u)^{-1})$ as shown above, we see that $\varphi_S(z)\times\varphi_S(z)^\mu \in X_{SS}(w^{-1})$. Put $a=\mathfrak{P}(\varphi_S(z))$. Then $a\times a^\pi\in \tilde{X}_{SS}(w^{-1})$. Now we have $J_{S_1S}(1)(a) =\varphi_{S_1}(z)=J(z)$. If E is an elliptic curve isomorphic to $\boldsymbol{C}/(\boldsymbol{Z}z+\boldsymbol{Z})$, $\widetilde{J(z)}$ is the invariant of $\tilde{E}$. Since p decomposes in K, $\mathrm{End}_{\boldsymbol{Q}}(\tilde{E})$ must be isomorphic to K. (This result is due to Deuring. For the proof, see [10] or [81, p. 114, Th. 2].) Taking infinitely many distinct fields K, we obtain infinitely many distinct $\widetilde{J(z)}$, and hence in view of (7.4.4), infinitely many distinct points a on $\tilde{V}_S$ such that $a\times a^\pi\in\tilde{X}_{SS}(w^{-1})$. It follows that $\Phi_S\subset\tilde{X}_{SS}(w^{-1})$. By (9) of Prop. 7.2, we have $X_{SS}(w)={}^tX_{SS}(w^{-1})^{\sigma(w)}$, hence $\tilde{X}_{SS}(w)^\pi={}^t\tilde{X}_{SS}(w^{-1})$. Put $c=\det(w)$, $w^\iota=cw^{-1}$. Since w_p^ι and w_p have the same elementary divisors, we have $Sw^\iota S=SwS$, so that $Sw^{-1}S=Sw^\iota c^{-1}S=Swc^{-1}S$, hence $X_{SS}(w^{-1})=X_{SS}(w)^{\sigma(c^{-1})}\circ J_{SS}(c^{-1})$ by (7) of Prop. 7.2. Since $\sigma(c^{-1})=[c^2, \boldsymbol{Q}]=\mu^2$ on k_S, we have

$$\begin{aligned} \tilde{X}_{SS}(w^{-1}) &= \tilde{X}_{SS}(w)^{\pi^2}\circ J_{SS}(c^{-1}) \\ &= {}^t\tilde{X}_{SS}(w^{-1})^\pi\circ\tilde{J}_{SS}(c^{-1})\supset {}^t\Phi_S^\pi\circ\tilde{J}_{SS}(c^{-1})\,. \end{aligned}$$

It can easily be seen that $d(\Phi_S)=d'({}^t\Phi_S^\pi\circ\tilde{J}_{SS}(c^{-1}))=1$, and $d'(\Phi_S)=d({}^t\Phi_S^\pi\circ J_{SS}(c^{-1})) =p$. Since Φ_S and ${}^t\Phi_S^\pi\circ J_{SS}(c^{-1})$ are irreducible and distinct, we obtain

$$(*)\qquad \Phi_S+{}^t\Phi_S^\pi\circ J_{SS}(c^{-1})\subset\tilde{X}_{SS}(w^{-1})\,.$$

On the other hand, we see that

$$[S: wSw^{-1}\cap S] \leqq [U_p : w_pU_pw_p^{-1}\cap U_p] = p+1,$$

hence, by Prop. 7.3, $d(\tilde{X}_{SS}(w^{-1})) = d(X_{SS}(w^{-1})) \leqq p+1$. Comparing d and d' of both sides of (∗), we conclude that the inclusion must actually be an equality. This completes the proof.

COROLLARY 7.10. *Let S and U' be defined by* (7.3.5). *Let σ_p be an element of $SL_2(\boldsymbol{Z})$ satisfying* (7.3.8), *and let* $\alpha = \begin{bmatrix} 1 & 0 \\ 0 & p \end{bmatrix}$, $\tau = \begin{bmatrix} 0 & -t \\ N & 0 \end{bmatrix}$. *If $p \notin \mathfrak{B}_S$, one has*

$$\tilde{X}_{SS}(\alpha) = \Phi_S + {}^t\Phi_S \circ \tilde{J}_{SS}(\sigma_p), \tag{1}$$

$${}^t\Phi_S \circ \tilde{J}_{SS}(\sigma_p) = {}^t\tilde{X}_{SS}(\tau) \circ {}^t\Phi_S \circ \tilde{X}_{SS}(\tau). \tag{2}$$

PROOF. We first observe that $U_p \subset U'$ if and only if p does not divide N. Therefore, if $p \notin \mathfrak{B}_S$, p does not divide N. Let y be the element of G_A of which the p-component is 1, and other components are all equal to α. Put $\alpha = wy$, $c = \det(w)$. Then $y^{\iota} \in U'$, $y^{-1}Sy = S$, and $\sigma(y) = \sigma(w^{-1})$. By (7) and (9) of Prop. 7.2, we have

$$X_{SS}(\alpha) = X_{SS}(w)^{\sigma(y)} \circ J_{SS}(y) = {}^tX_{SS}(w^{-1}) \circ J_{SS}(y). \tag{∗}$$

Now we see that $y\sigma_p^{-1} \in U'$, and $p = \det(\alpha) = cyy^{\iota}$. Therefore

$$J_{SS}(c^{-1}) = J_{SS}(yy^{\iota}) = J_{SS}(y) = J_{SS}(\sigma_p).$$

From the above theorem and (∗), we obtain

$$\tilde{X}_{SS}(\alpha) = {}^t\tilde{X}_{SS}(w^{-1}) \circ J_{SS}(y) = {}^t\Phi_S \circ \tilde{J}_{SS}(\sigma_p) + {}^t\tilde{J}_{SS}(\sigma_p) \circ \Phi_S \circ \tilde{J}_{SS}(\sigma_p).$$

Here note that $k_S = \boldsymbol{Q}$, hence $\Phi_S^{\pi} = \Phi_S$. Since $J_{SS}(\sigma_p)$ is rational over $k_S = \boldsymbol{Q}$, Φ_S commutes with $\tilde{J}_{SS}(\sigma_p)$, by virtue of (7.1) of Appendix. Thus we obtain (1). The formula (2) follows immediately from Prop. 7.8 and (7.1) of Appendix.

7.5. Zeta-functions of V_S and the factors of the jacobian variety of V_S

We shall now determine the zeta-functions of the curves V_S with members S of $\mathfrak{Z}$ of the type (7.3.5), and the zeta-functions of some abelian varieties occurring as factors of the jacobian variety of V_S. The main idea is to connect the Frobenius morphism with Hecke operators by means of the congruence relations of Th. 7.9, or Cor. 7.10.

THEOREM 7.11. *Let U', S, Γ' be as in* (7.3.5) *and* (7.3.6), *and $\mathfrak{B}_S$ be as in* § 7.4. *Let σ_p, for a prime $p \notin \mathfrak{B}_S$, be an element of $SL_2(\boldsymbol{Z})$ satisfying* (7.3.8) *(see also* (3.3.10)), *and* $\alpha_p = \begin{bmatrix} 1 & 0 \\ 0 & p \end{bmatrix}$. *Further let $S_2(\Gamma')$ be the vector space of* *all cusp forms of weight* 2 *with respect to* Γ', *and* $[\Gamma'\alpha\Gamma']_2$ *the action of* $\Gamma'\alpha\Gamma'$ *on* $S_2(\Gamma')$ *defined by* (3.4.1). *Then, for every* p *not contained in* $\mathfrak{B}_S$, *the zeta-function of* $p(V_S)$ *over the prime field is given by*

$$Z(u\,;p(V_S))=[(1-u)(1-pu)]^{-1}\cdot\det(1-[\Gamma'\alpha_p\Gamma']_2u+p\cdot[\Gamma'\sigma_p\Gamma']_2u^2)\,.$$

PROOF. Fix a prime $p\notin\mathfrak{B}_S$, and a prime divisor $\mathfrak{P}$ of $\bar{\boldsymbol{Q}}$ which divides p. Denote, as before, by $\tilde{X}$ or $\mathfrak{P}(X)$ the object obtained from X by reduction modulo $\mathfrak{P}$. Let A_S be the jacobian variety of V_S, and R_l (resp. R_l') an l-adic representation of $\mathrm{End}(A_S)$ (resp. $\mathrm{End}(\tilde{A}_S)$) for a rational prime l different from p. By [81, § 11, Prop. 14], we can assume $R_l(\lambda)=R_l'(\tilde{\lambda})$ for every $\lambda\in\mathrm{End}(A_S)$. Let π_p denote the p-th power endomorphism of $\tilde{A}_S$. Then there exists an element π_p^* of $\mathrm{End}(\tilde{A}_S)$ which is associated with ${}^t\Phi_S$ and satisfies $\pi_p\pi_p^*=p$. Let ξ_p, η_p, and β be the elements of $\mathrm{End}(A_S)$ associated with $X_{SS}(\alpha_p)$, $J_{SS}(\sigma_p)$, and $X_{SS}(\tau)$, respectively. From Cor. 7.10, we obtain

$$\tilde{\xi}_p=\pi_p+\pi_p^*\tilde{\eta}_p\,, \tag{7.5.1}$$

$$\pi_p^*\tilde{\eta}_p=\tilde{\beta}^{-1}\pi_p^*\tilde{\beta}\,. \tag{7.5.2}$$

Therefore, if u is an indeterminate, we have

$$\begin{aligned}[1-u\cdot R_l'(\pi_p)][1-u\cdot R_l'(\tilde{\beta}^{-1}\pi_p^*\tilde{\beta})]&=1-u\cdot R_l'(\tilde{\xi}_p)+pu^2\cdot R_l'(\tilde{\eta}_p)\\&=1-u\cdot R_l(\xi_p)+pu^2\cdot R_l(\eta_p)\,.\end{aligned}$$

Since π_p and π_p^* have the same characteristic polynomial, we have

$$\det[1-u\cdot R_l'(\pi_p)]^2=\det[1-u\cdot R_l(\xi_p)+pu^2\cdot R_l(\eta_p)]\,.$$

Now the representation R_l is equivalent to the representation R^0 of $\mathrm{End}_{\boldsymbol{Q}}(A_S)$ on the first cohomology group of A_S. If R denotes the representation of $\mathrm{End}_{\boldsymbol{Q}}(A_S)$ on $\mathfrak{D}(A_S)$, then R^0 is equivalent to the direct sum of R and its complex conjugate (see Appendix N° 11). In § 7.3, we have shown that $X_{SS}(\alpha_p)$ and $J_{SS}(\sigma_p)$ are rational over $\boldsymbol{Q}$, so that ξ_p and η_p are rational over $\boldsymbol{Q}$. Therefore, taking a basis of $\mathfrak{D}(A_S)$ over $\boldsymbol{Q}$, we can assume that $R(\xi_p)$ and $R(\eta_p)$ are rational matrices. Therefore we obtain, through those two equivalences of representations,

$$\det[1-u\cdot R_l'(\pi_p)]^2=\det[1-u\cdot R(\xi_p)+pu^2\cdot R(\eta_p)]^2\,,$$

so that

$$\det[1-u\cdot R_l'(\pi_p)]=\det[1-u\cdot R(\xi_p)+pu^2\cdot R(\eta_p)]\,.$$

By virtue of the commutativity of the diagrams (7.2.2) and (7.2.6), we may put $R(\xi_p)=[\Gamma'\alpha_p\Gamma']_2$, $R(\eta_p)=[\Gamma'\sigma_p\Gamma']_2$. This completes the proof.

THEOREM 7.12. *Let* $T'(n)_{k,\psi}$ *be as in* § 3.5. *Then, for almost all primes* p, *every eigen-value* λ_p *of* $T'(p)_{2,\psi}$ *satisfies* $|\lambda_p| \leqq 2p^{1/2}$.

PROOF. Take $\mathfrak{h}^*$ to be $\{a \in \mathfrak{g}^\times \mid a \equiv 1 \bmod (N)\}$. Then Γ' coincides with the group Γ'' of (3.5.1′). By (3.5.6), $T'(p)_{2,\psi}$ is the restriction of $[\Gamma''\alpha_p\Gamma'']_2$ to $S_2(\Gamma'_0, \psi)$. Since π_p commutes with $\pi_p^*\eta_p$, we see, from (7.5.1) and (7.5.2), that any characteristic root of $R_l(\xi_p) = R'_l(\tilde{\xi}_p)$ is of the form $\mu+\mu'$ with a characteristic root μ of $R'_l(\pi_p)$ and a characteristic root μ' of $R'_l(\pi_p^*)$. By the Weil theorem, $|\mu| = |\mu'| = p^{1/2}$. Since $R(\xi_p) = [\Gamma''\alpha_p\Gamma'']_2$, we obtain our assertion for all p not contained in $\mathfrak{B}_S$.

A prototype of Th. 7.9 is already in the works of Kronecker. The relation (7.5.1), in the present formulation, was first proved by Eichler [16] for $\Gamma_0(N)$ and its subgroups Γ' of index 2; he then obtained Th. 7.12 for such groups and Th. 7.11 for $\Gamma_0(N)$. The generalization to the present form and the formula (7.5.2) were given in [70]. Actually in [70], Th. 7.9, or rather Cor. 7.10, was proved by means of the congruence relations for an elliptic curve with a variable modulus. This method is simpler than the above proof of Th. 7.9 in the sense that it does not require any result of complex multiplication. It was shown by Igusa [36] that the set of primes $\mathfrak{B}_S$ is contained in the set of all prime factors of N. But we shall not discuss this point in the present book.

Let Δ' be defined by (7.3.7). Let $T'(n)$ and $T'(a, d)$ be the elements of $R(\Gamma', \Delta')$ as in § 3.3 (see especially Th. 3.34). Denote by $T'(n)_2$ and $T'(a, d)_2$ the action of $T'(n)$ and $T'(a, d)$ on $S_2(\Gamma')$, respectively (see §§ 3.4–5). Then we have $[\Gamma'\alpha_p\Gamma']_2 = T'(p)_2$, and $[\Gamma'\sigma_p\Gamma']_2 = T'(p, p)_2$ by (3.4.4) and (3.3.11). Therefore, by virtue of Th. 7.11, the zeta-function of V_S over $\boldsymbol{Q}$ has the form

$$\zeta(s\,;\,V/\boldsymbol{Q}) = \textstyle\prod_{p \notin \mathfrak{B}_S} \det\,[1 - T'(p)_2 p^{-s} + T'(p, p)_2 p^{1-2s}]^{-1}.$$

This is, up to finitely many Euler factors, exactly a Dirichlet series of the type discussed in §§ 3.3, 3.5, 3.6. More precisely, let $S_k(\Gamma'_0, \psi)$ and $T'(n)_{k,\psi}$ be as in § 3.5. Then $S_k(\Gamma')$ is the direct sum of all the $S_k(\Gamma'_0, \psi)$ such that $\psi(\mathfrak{h}) = 1$. Put

$$D_{k,\psi}(s) = \textstyle\sum_{n=1}^{\infty} T'(n)_{k,\psi} \cdot n^{-s}.$$

Then $\zeta(s\,;\,V/\boldsymbol{Q})$ coincides, up to a finite number of Euler factors, with the product

(7.5.3) $$D(s) = \textstyle\prod_{\psi(\mathfrak{h})=1} \det\,[D_{2,\psi}(s)].$$

For each element $f(z) = \sum_{n=1}^{\infty} a_n e^{2\pi i n z/t}$ of $S_k(\Gamma'')$, put

(7.5.4) $$L(s, f) = \textstyle\sum_{n=1}^{\infty} a_n n^{-s}.$$

By virtue of the discussion of § 3.5, especially of Prop. 3.47, we can find a set of elements $\{h_1, \cdots, h_\kappa\}$ of $S_k(\Gamma'_0, \psi)$ such that

$$(7.5.4') \qquad h_\nu(z) = \textstyle\sum_{n=1}^{\infty} \lambda_\nu(n) e^{2\pi i n z/t}, \quad h_\nu | T'(n)_{k,\psi} = \lambda_\nu(n) h_\nu \qquad (\nu = 1, \cdots, \kappa),$$

$$\det(D_{k,\psi}(s)) = \textstyle\prod_{\nu=1}^{\kappa} L(s, h_\nu).$$

($\{h_1, \cdots, h_\kappa\}$ is not necessarily a basis of $S_k(\Gamma'_0, \psi)$.) Therefore we obtain, from Th. 3.66 and Remark 3.58,

THEOREM 7.13. *The zeta-function* $\zeta(s; V_S/\boldsymbol{Q})$ *of* V_S *over* $\boldsymbol{Q}$ *is an entire function, and satisfies a functional equation.*

By Remark 3.58 and Th. 3.66, the functional equation of $L(s, f)$ is given by

$$R(s, f) = (tN)^{s/2}(2\pi)^{-s}\Gamma(s)L(s, f) = i^k \cdot R(k-s, f | [\tau]_k),$$

where $\tau = \begin{bmatrix} 0 & -t \\ N & 0 \end{bmatrix}$. The above h_ν may not be an eigen-function of $[\tau]_k$. However, on account of Prop. 3.57 and the result of Hecke mentioned in Remark 3.60, if N is a prime and $t=1$, we see that $h_\nu | [\tau]_2 = \varepsilon_\nu h_\nu$ for the basis $\{h_1, \cdots, h_\kappa\}$ of $S_2(\Gamma'_0, \psi)$ with $\varepsilon_\nu = \pm 1$, for a real character ψ; if ψ is not real, $[\tau]_2$ sends a common eigen-function of the $T'(n)_{2,\psi}$ to a common eigen-function of the $T'(n)_{2,\bar{\psi}}$. Therefore the Dirichlet series $D(s)$ of (7.5.3) satisfies the functional equation

$$(7.5.5) \qquad R(s) = [N^{s/2}(2\pi)^{-s}\Gamma(s)]^g D(s) = \mu \cdot R(2-s)$$

with $\mu = \pm 1$, where g is the genus of V_S.

For example, assume $t=1$, and $\mathfrak{h} = \mathfrak{g}^\times$, so that $\Gamma' = \Gamma'_0 = \Gamma_0(N)$. By Prop. 1.40 and 1.43, V_S is of genus 1 for the following 12 values of N:

(7.5.6) 11, 14, 15, 17, 19, 20, 21, 24, 27, 32, 36, 49.

In these cases, the zeta-function of V_S is (up to possible bad factors)

$$\zeta(s; V/\boldsymbol{Q}) = L(s, h) = \textstyle\sum_{n=1}^{\infty} c_n n^{-s}$$
$$= \textstyle\prod_{p|N} (1 - c_p p^{-s})^{-1} \cdot \prod_{p\nmid N} (1 - c_p p^{-s} + p^{1-2s})^{-1},$$

with an element $h(z) = \sum_{n=1}^{\infty} c_n e^{2\pi i n z}$ of $S_2(\Gamma_0(N))$. It can be shown that $h | [\tau]_2 = -h$, and hence the functional equation has the form

$$R(s, h) = N^{s/2}(2\pi)^{-s}\Gamma(s)L(s, h) = R(2-s, h).$$

As Fricke [21] observed, the elliptic curve $\Gamma_0(N)\backslash\mathfrak{H}^*$ has no complex multiplications for the first 8 values of (7.5.6). For further examples, see Ex. 7.26 below.

In the next place, we shall consider the zeta-functions of abelian varieties which occur as "factors" of the jacobian A_S of V_S. We denote by ξ_n, for each positive integer n, the element of $\mathrm{End}(A_S)$ corresponding to the sum of $X_{SS}(\alpha)$ for all $\Gamma'\alpha\Gamma'$ such that $\alpha \in \Delta'$ and $\det(\alpha)=n$. Then we see, from Prop. 7.7, that ξ_n is rational over $\boldsymbol{Q}$. Moreover, through the diagrams (7.2.2) and (7.2.6), ξ_n corresponds to $T'(n)_2$. For simplicity, we assume hereafter

(7.5.7) $$t=1,$$

without losing much generality, in view of Remark 3.58.

THEOREM 7.14. *Let $f(z)$ be an element of $S_2(\Gamma')$ which is a common eigenfunction of $T'(n)_2$ for all n, and let $f \mid T'(n)_2 = a_n f$. Let K be the subfield of $\boldsymbol{C}$ generated over $\boldsymbol{Q}$ by the complex numbers a_n for all n. Then there exists an abelian subvariety A of A_S and an isomorphism θ of K into $\mathrm{End}_{\boldsymbol{Q}}(A)$ with the following properties:*

(1) $\dim(A)=[K:\boldsymbol{Q}]$;

(2) *$\theta(a_n)$ is the restriction of ξ_n to A for all n;*

(3) *A is defined over $\boldsymbol{Q}$.*

The couple (A, θ) is uniquely determined by (1) *and* (2). *Moreover, for every isomorphism σ of K into $\boldsymbol{C}$, there exists an element f_σ of $S_2(\Gamma')$ such that $f_\sigma \mid T'(n)_2 = a_n^\sigma f_\sigma$ for all n, and $f_\sigma(z)=\sum_{n=1}^{\infty} a_n^\sigma e^{2\pi i n z}$.*

We may and do assume, for simplicity, $f(z)=\sum_{n=1}^{\infty} a_n e^{2\pi i n z}$ (see Th. 3.43).

THEOREM 7.15. *The notation being as in Th. 7.14, suppose that $\mathfrak{h}^* = \mathfrak{g}^\times$ in the definition* (7.3.5) *of U' and S. (This implies that*

$$\Gamma' = \Gamma_0(N) = \left\{ \begin{bmatrix} a & b \\ c & d \end{bmatrix} \in SL_2(\boldsymbol{Z}) \,\middle|\, c \equiv 0 \bmod (N) \right\}.)$$

Then the zeta-function of A over $\boldsymbol{Q}$ coincides, up to a finite number of Euler factors, with the product

$$\textstyle\prod_\sigma L(s, f_\sigma) = \prod_\sigma \left(\sum_{n=1}^{\infty} a_n^\sigma n^{-s}\right),$$

where the product is taken over all the isomorphisms σ of K into $\boldsymbol{C}$.

PROOF of Th. 7.14 and Th. 7.15. Let $\mathfrak{T}$ be the subalgebra of $\mathrm{End}_{\boldsymbol{Q}}(A_S)$ generated by the ξ_n for all n. If A_S is of dimension g, $\mathfrak{T}$ is a commutative algebra of rank g over $\boldsymbol{Q}$, by Th. 3.51. Let $\mathfrak{R}$ be the radical of $\mathfrak{T}$. By a theorem of Wedderburn, there exists a semi-simple subalgebra $\mathfrak{S}$ of $\mathfrak{T}$ such that $\mathfrak{T} = \mathfrak{S} \oplus \mathfrak{R}$. Let $\mathfrak{K}_1, \cdots, \mathfrak{K}_r$ be the simple components of $\mathfrak{S}$. Now the map $\xi_n \mapsto a_n$ defines a homomorphism ρ of $\mathfrak{T}$ onto K. Therefore $\rho(\mathfrak{R})=\{0\}$, and $\rho(\mathfrak{K}_i) \neq \{0\}$ for one and only one $\mathfrak{K}_i$, say $\mathfrak{K}_1$. Then ρ gives an isomorphism of $\mathfrak{K}_1$ to K. Denote by ρ' the inverse map of this isomorphism. Then $\rho' \circ \rho$ is the projection map of $\mathfrak{T}$ to $\mathfrak{K}_1$, so that $\rho'(a_n)$ is the projection of ξ_n to $\mathfrak{K}_1$.

Take an integer $q \geqq 0$ so that $\mathfrak{K}_1\mathfrak{R}^q \neq \{0\}$ and $\mathfrak{K}_1\mathfrak{R}^{q+1} = \{0\}$, where we understand that $\mathfrak{R}^0 = \mathfrak{T}$. Let $\mathfrak{M}$ be an irreducible $\mathfrak{K}_1$-submodule of $\mathfrak{K}_1\mathfrak{R}^q$. Then $\mathfrak{M}$ is a minimal ideal of $\mathfrak{T}$. Put $\mathfrak{M}_0 = \mathfrak{M} \cap \mathrm{End}(A_S)$, and $A = \mathfrak{M}_0 A_S$. Since every element of $\mathfrak{M}_0$ is defined over $\boldsymbol{Q}$, A is an abelian subvariety of A_S defined over $\boldsymbol{Q}$. Since the action of $\mathfrak{T}$ on $S_2(\Gamma')$ is equivalent to a regular representation of $\mathfrak{T}$ (see Th. 3.51), we see that $\dim(A) = [\mathfrak{M} : \boldsymbol{Q}] = [\mathfrak{K}_1 : \boldsymbol{Q}] = [K : \boldsymbol{Q}]$. For each $a \in K$ such that $\rho'(a) \in \mathrm{End}(A_S)$, denote by $\theta(a)$ the restriction of $\rho'(a)$ to A. Then θ can be extended to an isomorphism of K into $\mathrm{End}_{\boldsymbol{Q}}(A)$. It is clear that $\theta(a_n)$ is the restriction of ξ_n to A. To prove the uniqueness of (A, θ), let (A', θ') be a couple satisfying (1) and (2). Consider A_S as a complex torus $\boldsymbol{C}^g/L$ with a lattice L in $\boldsymbol{C}^g$. We may assume that $\boldsymbol{C}^g = S_2(\Gamma')$ and ξ_n is represented by the operator $T'(n)_2$ on $S_2(\Gamma')$. Let W be the subspace of $S_2(\Gamma')$ corresponding to A'. Since $\theta'(a_n)$ is the restriction of ξ_n to A', $\theta'(\rho(\xi))$ is the restriction of ξ to A' for every $\xi \in \mathfrak{T}$. Therefore, A' (and hence W) is annihilated by $\mathfrak{R}$ and $\mathfrak{K}_i$ for $i > 1$. Consider W as a module over $\mathfrak{K}_1 \otimes_{\boldsymbol{Q}} \boldsymbol{C}$, or over $K \otimes_{\boldsymbol{Q}} \boldsymbol{C}$. Then we find a basis $\{f_1, \cdots, f_m\}$ of W over $\boldsymbol{C}$ such that f_ν is sent to $a^{\sigma_\nu} f_\nu$ by $\theta'(a)$ for every $a \in K$, where σ_ν is an isomorphism of K into $\boldsymbol{C}$ for each ν. Here $m = \dim(A') = [K : \boldsymbol{Q}]$. Then

$$(*) \qquad f_\nu \mid T'(n)_2 = a_n^{\sigma_\nu} f_\nu \qquad (1 \leqq \nu \leqq m\,;\ 1 \leqq n < \infty)\,.$$

By Prop. 3.53 and Cor. 3.44, f_ν is uniquely determined by $(*)$ up to constant factors. Therefore we see that $\sigma_1, \cdots, \sigma_m$ are distinct, and W is uniquely determined by f. This implies that A' is unique, and hence $A = A'$. This completes the proof of Th. 7.14.

Suppose that $\mathfrak{h}^* = \mathfrak{g}^\times$. Let p be a rational prime not contained in $\mathfrak{B}_S$. By [40] or [66], A has no defect for p. Since $\sigma_p \in \Gamma'$, we have $\eta_p = 1$, so that the relation (7.5.1) becomes $\tilde{\xi}_p = \pi_p + \pi_p^*$. Let R_l'' denote the l-adic representation of $\mathrm{End}(\tilde{A})$, and π_p^A the Frobenius endomorphism of $\tilde{A}$ of degree p. By the same reasoning as in the proof of Th. 7.11, we have

$$\det[1 - u \cdot R_l''(\pi_p^A)] = \det[1 - T^W(p)_2 u + p u^2]\,,$$

where $T^W(p)_2$ means the restriction of $T'(p)_2$ to W. Therefore, in view of $(*)$, we obtain Th. 7.15.

To obtain further information, particularly in the case $\mathfrak{h}^* \neq \mathfrak{g}^\times$, we first observe, in view of Prop. 3.53, that f belongs to $S_2(\Gamma_0', \psi)$ with a unique character ψ of $(\boldsymbol{Z}/N\boldsymbol{Z})^\times$ such that $\psi(\mathfrak{h}) = 1$, where $\mathfrak{h}$ is the subgroup of $(\boldsymbol{Z}/N\boldsymbol{Z})^\times$ corresponding to $\mathfrak{h}^*$. The values $\psi(n)$ belong to the number field K, on account of (3.5.8) and (5) of Th. 3.34. Let $\mathfrak{J}$ denote the set of all isomorphisms of K into $\boldsymbol{C}$. Then, for every $\sigma \in \mathfrak{J}$, f_σ belongs to $S_2(\Gamma_0', \psi^\sigma)$, again on account of (3.5.8) and the equality $pT'(p, p) = T'(p)^2 - T'(p^2)$. Now suppose that the following condition is satisfied:

(7.5.8) *For every* $\sigma \in \mathfrak{Z}$, *all the* $T'(n)_{2,\psi^\sigma}$ *belong to the algebra generated over* $\boldsymbol{Q}$ *by the* $T'(n)_{2,\psi^\sigma}$ *for all* n *prime to* N.

We obtain then

(7.5.9) K *is generated by the* a_n *over* $\boldsymbol{Q}$ *for all* n *prime to* N.

By virtue of the result of Hecke [30] quoted in Remark 3.60, (7.5.8) is satisfied if ψ is a primitive character modulo (N). Let $\tau = \begin{bmatrix} 0 & -1 \\ N & 0 \end{bmatrix}$, and let β be the element of $\mathrm{End}(A_S)$ associated with $X_{SS}(\tau)$. Further let ρ denote the complex conjugation. By Prop. 3.55, we have

(7.5.10) $a_n^\sigma = \psi(n)^\sigma a_n^{\sigma\rho}$ *for every* $\sigma \in \mathfrak{Z}$ *and for every* n *prime to* N.

Therefore we see that $K^\rho = K$ and $a_n^{\sigma\rho} = a_n^{\rho\sigma}$ for every $\sigma \in \mathfrak{Z}$, so that K is totally real if ρ is the identity map on K. If it is not, K must be a CM-field in the sense of § 5.5, on account of Prop. 5.11. By Prop. 3.57, we have $f_\sigma | [\tau]_2 T'(n)_2 = a_n^{\sigma\rho} f_\sigma | [\tau]_2$ for all n prime to N. Therefore, by (7.5.8) and Cor. 3.44, we see that

(7.5.11) *For every* $\sigma \in \mathfrak{Z}$, $f_\sigma | [\tau]_2$ *is a constant multiple of* $f_{\sigma\rho}$.

It follows that $[\tau]_2$ sends W onto itself, and hence A is stable under β. Therefore, by means of the same argument as in the proof of Th. 7.11, we obtain the first part of the following

THEOREM 7.16. *The notation being as in Th.* 7.14, *suppose that the condition* (7.5.8) *is satisfied. Then the zeta-function of* A *over* $\boldsymbol{Q}$ *coincides, up to a finite number of Euler factors, with* $\prod_{\sigma \in \mathfrak{Z}} L(s, f_\sigma)$. *Moreover, if* ψ *is the trivial character,* K *is totally real. If* ψ *is not trivial, then* K *is a totally imaginary quadratic extension of a totally real algebraic number field* K', *and there exists an abelian variety* A' *such that* A *is isogenous to* $A' \times A'$, *and* $\mathrm{End}_{\boldsymbol{Q}}(A')$ *contains an isomorphic image of* K'.

PROOF. If ψ is trivial, K must be totally real on account of (7.5.10). Suppose that ψ is not trivial. For every q prime to N, let σ_q be as in (7.3.8), and η_q the element of $\mathrm{End}(A_S)$ associated with $J_{SS}(\sigma_q)$. By Prop. 7.8, we have

(7.5.12) $$\beta = \beta^\sigma \eta_q \text{ if } \zeta^\sigma = \zeta^q \text{ for } \zeta = e^{2\pi i/N}.$$

Since ψ is non-trivial, we see that $\eta_q \neq \mathrm{id.}$ on A. Let μ be the restriction of β to A. Then $\mu^\sigma \neq \mu$ for some $\sigma \in \mathrm{Gal}(\boldsymbol{Q}(\zeta)/\boldsymbol{Q})$, so that μ is different from ± 1 on A. Put $A' = (1+\mu)A$. Then $A' \neq 0$, and since $\mu^2 = 1$, $\dim(A') < \dim(A) = [K : \boldsymbol{Q}]$. From (7.5.11), we obtain

(7.5.13) $\mu\theta(a)=\theta(a^{\rho})\mu$ *for every* $a\in K$.

Assume that K is totally real. Then $\theta(a)$ gives an endomorphism of A' for every $a\in K$. Therefore K is embeddable into $\mathrm{End}_{\boldsymbol{Q}}(A')$. But this is impossible on account of the following

LEMMA 7.17. *Let K be a totally real algebraic number field, and A' an abelian variety defined over (a subfield of) $\boldsymbol{C}$. If there exists an isomorphism of K into* $\mathrm{End}_{\boldsymbol{Q}}(A')$ *which maps the identity element of K to the identity element of* $\mathrm{End}(A)$, *then* $[K:\boldsymbol{Q}]$ *divides* $\dim(A')$.

PROOF. Let R and R^0 denote respectively the representations of $\mathrm{End}_{\boldsymbol{Q}}(A)$ on $\mathfrak{D}(A)$ and on the first cohomology group of A. Then R^0 is equivalent to the direct sum of R and its complex conjugate $\bar{R}$. Restrict R and R^0 to the image of K in $\mathrm{End}_{\boldsymbol{Q}}(A)$, which we identify with K. Then R is equivalent to a direct sum of several isomorphisms of K into $\boldsymbol{C}$. Since K is totally real, we see that R is equivalent to $\bar{R}$. On the other hand, since R^0 is a rational representation, we see that $\mathrm{tr}(R(a))=\mathrm{tr}(R^0(a))/2\in\boldsymbol{Q}$ for every $a\in K$. Therefore the degree of R must be a multiple of $[K:\boldsymbol{Q}]$. Since $\dim(A)$ is exactly the degree of R, we obtain our assertion.

Coming back to the proof of Th. 7.16, we can therefore conclude that K is a CM-field in the sense of § 5.5. Take an element b of K so that $0\neq b=-\bar{b}$ and $\theta(b)\in\mathrm{End}(A)$. By (7.5.13), we have $\theta(b)A'=(1-\mu)\theta(b)A=(1-\mu)A$. Therefore $A=(1+\mu)A+(1-\mu)A=A'+\theta(b)A'$, and hence A is isogenous to $A'\times A'$. Further, if $a\in K$ and $a^{\rho}=a$, we have $\theta(a)A'\subset A'$ by (7.5.13), so that $\mathrm{End}_{\boldsymbol{Q}}(A')$ contains the isomorphic image of $\{a\in K\mid a^{\rho}=a\}$. This completes the proof of Th. 7.16.

Let k be the subfield of $\boldsymbol{Q}(e^{2\pi i/N})$ defined in Prop. 7.8. Then β is defined over k by that proposition, and hence A' is defined over k. Moreover, let $K'=\{a\in K\mid a^{\rho}=a\}$, and let $\theta'(a)$ be the restriction of $\theta(a)$ to A' for every $a\in K'$. Then θ' is an isomorphism of K' into $\mathrm{End}_{\boldsymbol{Q}}(A')$, and $\theta'(a)$ is defined over k for every $a\in K'$.

Now it is natural to consider the zeta-function of A' over k. Since a general discussion of this question has been made in T. Miyake [50][15], we shall only treat here the simplest case, in a somewhat different formulation.

Besides (7.5.8), let us make the following assumptions:

(7.5.14) ψ *is a character of* $(\boldsymbol{Z}/N\boldsymbol{Z})^{\times}$ *of order 2 such that* $\psi(-1)=1$.

(7.5.15) $\mathfrak{h}^*$ *corresponds to the kernel of* ψ, *so that* $[\mathfrak{g}^{\times}:\mathfrak{h}^*]=2$, *and hence*

$$\Gamma'=\left\{\begin{bmatrix} a & b \\ c & d\end{bmatrix}\in SL_2(\boldsymbol{Z})\,\middle|\,\psi(a)=1,\ c\equiv 0 \bmod (N)\right\}.$$

15) This paper treats also V_S and A_S for a more general type of S than (7.3.5).

THEOREM 7.18. *The notation being as in Th.* 7.14 *and Th.* 7.16, *suppose that the conditions* (7.5.8), (7.5.14), *and* (7.5.15) *are satisfied. Let k be the quadratic extension of $\boldsymbol{Q}$ corresponding to ψ. Then A' is defined over k, and A'^{ε} is isogenous to A' over k for the generator ε of* Gal $(k/\boldsymbol{Q})$. *Moreover, the zeta-function of A' over k coincides, up to a finite number of Euler factors, with the product* $\prod_{\sigma\in\mathfrak{J}} L(s, f_\sigma)$.

Note that k is *real*, since $\psi(-1)=1$.

PROOF. As is mentioned above, β (and hence its restriction μ to A) is defined over k, so that $A'=(1+\mu)A$ is defined over k. Let q be a positive integer such that $\psi(q)=-1$. Let η_q be as in the proof of Th. 7.16. Then $\eta_q=-1$ on A, since the cusp forms f_σ, for all $\sigma\in\mathfrak{J}$, are contained in $S_2(\Gamma_0', \psi)$. Therefore, from (7.5.12) we obtain, if ε is the generator of Gal $(k/\boldsymbol{Q})$,

$$\mu^{\varepsilon}=-\mu. \tag{7.5.16}$$

It follows that $A'^{\varepsilon}=[(1+\mu)A]^{\varepsilon}=(1-\mu)A=\theta(b)A'$, where b is an element of K considered in the proof of Th. 7.16. Therefore A'^{ε} is isogenous to A' over k. For every prime ideal $\mathfrak{p}$ in k, let $\varphi_{\mathfrak{p}}$ denote the Frobenius endomorphism of $\tilde{A}=\mathfrak{p}(A)$ of degree $N(\mathfrak{p})$, and R_l' the l-adic representation of End $(\tilde{A})$. From (7.5.1) we obtain $\tilde{\theta}(a_p)=\tilde{\xi}_p=\pi_p+\psi(p)\pi_p^*$ on $\tilde{A}$ for every rational prime p not contained in $\mathfrak{B}_S$. Therefore, if $N(\mathfrak{p})=p^2$, we have $\varphi_{\mathfrak{p}}=\pi_p^2$, so that

$$\begin{aligned}(*)\qquad \det[1-u^2R'(\varphi_{\mathfrak{p}})]&=\det[1-u\cdot R_l'(\pi_p)]\cdot\det[1+u\cdot R_l'(\pi_p)]\\ &=\det[1-u\cdot R_l'(\pi_p)]\cdot\det[1-\psi(p)u\cdot R_l'(\pi_p^*)]\\ &=\det[1-u\cdot R_l'(\tilde{\xi}_p)+\psi(p)pu^2].\end{aligned}$$

Let $T^W(p)_2$ be the restriction of $T'(p)_2$ to W. By the same reasoning as in the proof of Th. 7.11, we see that the right hand side of (*) coincides with

$$(**)\qquad \det[1-T^W(p)_2u+\psi(p)pu^2]^2.$$

On the other hand, if $(p)=\mathfrak{p}\mathfrak{p}'$ in k, both $\varphi_{\mathfrak{p}}$ and $\varphi_{\mathfrak{p}'}$ can be identified with π_p, so that

$$\begin{aligned}\det[1-u\cdot R_l'(\varphi_{\mathfrak{p}})]&\cdot\det[1-u\cdot R_l'(\varphi_{\mathfrak{p}'})]\\ &=\det[1-u\cdot R_l'(\pi_p)]^2\\ &=\det[1-u\cdot R_l'(\pi_p)]\cdot\det[1-\psi(p)u\cdot R_l'(\pi_p^*)]\\ &=\det[1-u\cdot R_l'(\tilde{\xi}_p)+\psi(p)pu^2],\end{aligned}$$

which coincides with (**), for the same reason as above. It follows that $\zeta(s; A/k)$ is, up to a finite number of Euler factors, given by the product

$\prod_{\sigma\in\mathfrak{F}} L(s, f_\sigma)^2$. Now A is isogenous to $A'\times A'$ over k. Therefore, if $\varphi_\mathfrak{p}^0$ denotes the restriction of $\varphi_\mathfrak{p}$ to $\mathfrak{p}(A')$, and R_l^0 the l-adic representation of End$(\mathfrak{p}(A'))$, we have

$$\det[1-u\cdot R_l^0(\varphi_\mathfrak{p}^0)]^2=\det[1-u\cdot R_l'(\varphi_\mathfrak{p})]\,,$$

so that

$$\det[1-T^W(p)_2u+\psi(p)pu^2]$$
$$=\begin{cases}\det[1-u^2\cdot R_l^0(\varphi_\mathfrak{p}^0)] & \text{if } N(\mathfrak{p})=p^2\,,\\ \det[1-u\cdot R_l^0(\varphi_\mathfrak{p}^0)]\cdot\det[1-u\cdot R_l^0(\varphi_{\mathfrak{p}'}^0)] & \text{if } (p)=\mathfrak{p}\mathfrak{p}'\,.\end{cases}$$

Therefore we obtain our assertion for $\zeta(s\,;A'/k)$.

We also notice that

$$\det[1-u\cdot R_l^0(\varphi_\mathfrak{p}^0)]=\det[1-u\cdot R_l^0(\varphi_{\mathfrak{p}'}^0)] \quad \text{if} \quad (p)=\mathfrak{p}\mathfrak{p}'\,, \tag{7.5.17}$$

since $\det[1-u\cdot R_l'(\pi_p)]$ equals both sides squared, or since A' is isogenous to A'^ε over k.

Now we have $[\tau]_2^2=1$, so that, by (7.5.11), we obtain

$$f_\sigma\,|\,[\tau]_2=\gamma f_{\sigma\rho}\,,\qquad f_{\sigma\rho}\,|\,[\tau]_2=\gamma^{-1}f_\sigma \tag{7.5.18}$$

with a constant γ. (Prop. 3.40 implies that $|\gamma|=1$, though we do not need this fact.) Therefore, if we put $L(s, A')=\prod_{\sigma\in\mathfrak{F}} L(s, f_\sigma)$, $m=[K:\boldsymbol{Q}]$, and $R(s, A')=\Gamma(s)^m(2\pi)^{-ms}N^{ms/2}L(s, A')$, then

$$R(s, A')=R(2-s, A')\,. \tag{7.5.19}$$

As an example, let us consider the case where

$$\dim(S_2(\Gamma_0', \psi))=2\,. \tag{7.5.20}$$

If Γ' is as in (7.5.15), we have

$$S_2(\Gamma')=S_2(\Gamma_0')+S_2(\Gamma_0', \psi)\,.$$

Therefore, by Th. 3.51, the $T'(n)_{2,\psi}$ form an algebra $\mathfrak{A}$ of rank 2 over $\boldsymbol{Q}$. Under the assumption (7.5.8), $\mathfrak{A}$ must be semi-simple, so that $\mathfrak{A}$ is isomorphic to a quadratic field, or to $\boldsymbol{Q}\oplus\boldsymbol{Q}$. Th. 7.16 implies that the latter case is impossible, since $\boldsymbol{Q}$ is not totally imaginary. Thus $\mathfrak{A}$ must be isomorphic to a quadratic extension K of $\boldsymbol{Q}$. Then, by Th. 7.14, we find an abelian subvariety A of A_S of dimension 2, and an isomorphism θ of K into $\mathrm{End}_{\boldsymbol{Q}}(A)$. By Th. 7.16 and Th. 7.18, K must be imaginary, and A is isogenous to $E\times E$ with an elliptic curve E defined over a real quadratic field k. If A_0 denotes the jacobian variety of $\Gamma_0'\backslash\mathfrak{H}^*$, then the jacobian A_S of $\Gamma'\backslash\mathfrak{H}^*$ is isogenous to $A\times A_0$. An elliptic curve E of this type has very interesting properties, which we shall discuss in § 7.7 along with some more examples of A and A'.

In the above, we have started our discussion from a common eigen-function f of the Hecke operators on $S_2(\Gamma')$ and obtained an abelian variety A. Instead, we can start from an abelian subvariety of A_S as follows.

PROPOSITION 7.19. *Let S and Γ' be as in* (7.3.5-6) *with* $\mathfrak{h}^* = \mathfrak{g}^\times$, $t \geqq 1$, *and A_S the jacobian variety of V_S as before. Further let ξ_n be the endomorphism of A_S corresponding to the Hecke operator $T'(n)_2$ on $S_2(\Gamma')$. If A is an abelian subvariety of A_S rational over $\boldsymbol{Q}$, then A is stable under ξ_n for all n prime to N. Moreover, if X denotes the subspace of $S_2(\Gamma')$ corresponding to A, and $T^X(n)$ the restriction of $T'(n)_2$ to X, then $\zeta(s; A/\boldsymbol{Q})$ coincides, up to a finite number of Euler factors, with* $\det(\sum_{(n,N)=1} T^X(n)n^{-s})$.

PROOF. Let p be a rational prime not dividing N. To prove the first assertion, it is sufficient to show that $\xi_p(A) \subset A$. Suppose that $\xi_p(A) \not\subset A$, and put $A^* = \xi_p(A) + A$. Then A^* is an abelian subvariety of A_S, and $\dim(A) < \dim(A^*)$. Let tilde denote reduction modulo p. If π_p and π_p^* are as in the proof of Th. 7.11, we have $\pi_p(\tilde{A}) = \pi_p^*(\tilde{A}) = \tilde{A}$, since A is rational over $\boldsymbol{Q}$, so that $\tilde{\xi}_p(\tilde{A}) \subset \tilde{A}$ by (7.5.1). (Note that $\eta_p = \mathrm{id.}$ on account of the assumption $\mathfrak{h}^* = \mathfrak{g}^\times$.) But $\tilde{A}^* = \tilde{\xi}_p(\tilde{A}) + \tilde{A}$ has the same dimension as A^* by the general theory of reduction modulo p (see [69]), which is a contradiction. Therefore $\xi_p(A) \subset A$. Now consider A_S as a complex torus $S_2(\Gamma')/L$ as in the proof of Th. 7.14. Then A corresponds to a vector subspace X of $S_2(\Gamma')$ stable under the $T'(n)_2$ for all n prime to N. Then we obtain the last assertion by means of the same argument as in the proof of Th. 7.11.

Since the $T^X(n)$ for all n prime to N form a commutative semi-simple algebra, we can find a basis $\{f_1, \cdots, f_r\}$ of X over $\boldsymbol{C}$ formed by common eigen-functions of all such $T^X(n)$. Put $f_\nu \mid T^X(n) = a_{\nu n} f_\nu$ with $a_{\nu n} \in \boldsymbol{C}$. Then, for each fixed ν,

$$\left\{ g \in S_2(\Gamma') \,\middle|\, g \mid T'(n)_2 = a_{\nu n} g \text{ for all } n \text{ prime to } N \right\}$$

is stable under the $T'(n)_2$ for all n (not necessarily prime to N). Therefore we can find a common eigen-function g_ν of all $T'(n)_2$ such that $g_\nu \mid T'(n)_2 = b_{\nu n} g_\nu$ with $b_{\nu n} = a_{\nu n}$ if $(n, N) = 1$. By Th. 3.43, we may assume that $g_\nu(z) = \sum_{n=1}^\infty b_{\nu n} e^{2\pi i n z/t}$. This shows that $\zeta(s; A/\boldsymbol{Q})$ for the above A coincides, up to a finite number of Euler factors, with the product $\prod_{\nu=1}^r L(s, g_\nu)$. The functions g_ν may not be contained in X. It should also be noted that $\sum_{(n,N)=1} b_{\nu n} e^{2\pi i n z}$ belongs to $S_2(\Gamma_0(N^2))$ (cf. Hecke [30, Satz 19]).

7.6. $\mathfrak{l}$-adic representations

We shall first generalize the notion of l-adic coordinate system of an abelian variety A, by considering everything relative to an algebraic number field embedded in $\mathrm{End}_{\boldsymbol{Q}}(A)$. Let A be an abelian variety defined over any field, F an algebraic number field of finite degree, and θ an isomorphism of F into $\mathrm{End}_{\boldsymbol{Q}}(A)$ which maps the identity element of F to the identity element of $\mathrm{End}(A)$. Put $g=\dim(A)$, and $h=[F:\boldsymbol{Q}]$. By [81, § 5.1, Prop. 2], $2g$ is a multiple of d; put $2g=dh$. The same proposition asserts:

(7.6.1) *The characteristic polynomial of $\theta(x)$, for every $x \in F$, is the d-th power of the principal polynomial of x over $\boldsymbol{Q}$; especially* $\deg(\theta(x))=N_{F/\boldsymbol{Q}}(x)^d$.

(See Appendix N° 10 for the notation deg( ).)

Let $\mathfrak{o}$ denote the maximal order in F. Suppose that

(7.6.2) $$\theta(\mathfrak{o}) \subset \mathrm{End}(A).$$

For an integral ideal (or an integer) $\mathfrak{a}$ in F, put

(7.6.3) $$A[\mathfrak{a}]=\{t \in A \mid \theta(\mathfrak{a})t=0\}, \qquad A[\mathfrak{a}^\infty]=\bigcup_{n=1}^{\infty} A[\mathfrak{a}^n].$$

It can easily be seen that $A[\mathfrak{a}\mathfrak{b}]=A[\mathfrak{a}]+A[\mathfrak{b}]$ if $\mathfrak{a}$ is prime to $\mathfrak{b}$ (cf. [81, p. 61, Prop. 18]).

PROPOSITION 7.20. *If A is defined over a field whose characteristic is either* 0 *or prime to $\mathfrak{a}$, then $A[\mathfrak{a}]$ is isomorphic to the direct sum of d copies of $\mathfrak{o}/\mathfrak{a}$.*

PROOF. It is sufficient to prove the assertion in the case where $\mathfrak{a}$ is a power of a prime ideal $\mathfrak{l}$. By elementary divisor theory, $A[\mathfrak{l}^n]$, for a positive integer n, is isomorphic to

(*) $$\mathfrak{o}/\mathfrak{l}^{m_1} \oplus \cdots \oplus \mathfrak{o}/\mathfrak{l}^{m_s} \qquad (0<m_1 \leqq \cdots \leqq m_s \leqq n).$$

In [81, p. 56, Prop. 10], we have proved:

(7.6.4) *$A[\mathfrak{a}]$ is of order $N(\mathfrak{a})^d$ under the assumption on the characteristic of the field of definition for A.*

Therefore we have $m_1+\cdots+m_s=nd$. On the other hand, (*) implies that $A[\mathfrak{l}]$ is isomorphic to $(\mathfrak{o}/\mathfrak{l})^s$, hence by (7.6.4), we have $s=d$. Since $m_i \leqq n$, we obtain $m_1=\cdots=m_s=n$, q. e. d.

Now, for every prime ideal $\mathfrak{l}$ in F, let $F_{\mathfrak{l}}$ (resp. $\mathfrak{o}_{\mathfrak{l}}$) denote the $\mathfrak{l}$-adic completion of F (resp. $\mathfrak{o}$). We fix a vector space W over F of dimension d, and an $\mathfrak{o}$-lattice D in W. (An *$\mathfrak{o}$-lattice in W* is a finitely generated $\mathfrak{o}$-submodule of W which spans W over F.) We put $W_{\mathfrak{l}}=W \otimes_F F_{\mathfrak{l}}$, and $D_{\mathfrak{l}}=D \otimes_{\mathfrak{o}} \mathfrak{o}_{\mathfrak{l}}$. From the above proposition, we obtain easily

(7.6.5) *If A is defined over a field whose characteristic is either* 0, *or prime to $\mathfrak{l}$, then there exists an exact sequence*

$$0 \longrightarrow D_{\mathfrak{l}} \longrightarrow W_{\mathfrak{l}} \xrightarrow{\mathfrak{t}} A[\mathfrak{l}^{\infty}] \longrightarrow 0\,.$$

(In brief, $A[\mathfrak{l}^{\infty}]$ is isomorphic to $(F_{\mathfrak{l}}/\mathfrak{o}_{\mathfrak{l}})^{d}$.)

We call such an exact sequence, or the map $\mathfrak{t}$, an $\mathfrak{l}$-adic coordinate system of A.

Let Y be the subring of $\mathrm{End}_{\boldsymbol{Q}}(A)$ consisting of all the elements which commute with the elements of $\theta(F)$. Every element ξ of $Y \cap \mathrm{End}(A)$ induces an endomorphism of $A[\mathfrak{l}^{\infty}]$, which is obtained from a unique element $R_{\mathfrak{l}}(\xi)$ of $\mathrm{End}(W_{\mathfrak{l}}, F_{\mathfrak{l}})$ stable on $D_{\mathfrak{l}}$. In this way we obtain an F-linear homomorphism

$$R_{\mathfrak{l}}:\ Y \longrightarrow \mathrm{End}(W_{\mathfrak{l}}, F_{\mathfrak{l}}) \qquad (\cong M_d(F_{\mathfrak{l}}))\,.$$

If $K=\boldsymbol{Q}$ and $\mathfrak{l}=l\boldsymbol{Z}$ with a rational prime l, this is the l-adic representation of Weil [92, N° 31].

PROPOSITION 7.21. *For every $\xi \in Y$, the characteristic polynomial f_{ξ} of $R_{\mathfrak{l}}(\xi)$ has coefficients in F, and is independent of $\mathfrak{l}$. Moreover, $N_{F/\boldsymbol{Q}}(f_{\xi})$ (understood in an obvious sense) is exactly the characteristic polynomial of ξ in the sense of* [92, N° 67].

PROOF is given in [78, 11.9].

PROPOSITION 7.22. *The restriction of $R_{\mathfrak{l}}$ to any simple subalgebra Z of Y, containing $\theta(F)$, is faithful, and equivalent to the direct sum of a multiple of the reduced representation of Z over F and (possibly) a zero-representation. Moreover, the restriction of $R_{\mathfrak{l}}$ to Z can be extended to an $F_{\mathfrak{l}}$-linear representation*

$$Z \otimes_F F_{\mathfrak{l}} \longrightarrow M_d(F_{\mathfrak{l}})$$

which is equivalent to a multiple of the reduced representation of $Z \otimes_F F_{\mathfrak{l}}$ over $F_{\mathfrak{l}}$ modulo a zero-representation.

This follows from Prop. 7.21, by means of the same reasoning as in [81, § 5.1, Lemma 1].

Suppose now that A and the elements of $\theta(F) \cap \mathrm{End}(A)$ are defined over an algebraic number field k of finite degree. Then $\mathrm{Gal}(\bar{\boldsymbol{Q}}/k)$ acts on $A[\mathfrak{l}^{\infty}]$. Therefore we obtain a representation

$$\mathfrak{R}_{\mathfrak{l}}:\ \mathrm{Gal}(\bar{\boldsymbol{Q}}/k) \longrightarrow \mathrm{End}(D_{\mathfrak{l}}, \mathfrak{o}_{\mathfrak{l}})^{\times} \qquad (\cong GL_d(\mathfrak{o}_{\mathfrak{l}}))\,.$$

Let B be the set of all the prime ideals in k for which A has defect. Take a prime ideal $\mathfrak{p}$ in k which does not belong to B and which is prime to $N(\mathfrak{l})$. Let $\mathfrak{P}$ be a prime divisor of $\bar{\boldsymbol{Q}}$ which divides $\mathfrak{p}$, and σ a Frobenius element of $\mathrm{Gal}(\bar{\boldsymbol{Q}}/k)$ with respect to $\mathfrak{P}$. Further let $\tilde{A}$ denote the abelian variety obtained from A by reduction modulo $\mathfrak{p}$. Then we can define an isomorphism $\tilde{\theta}: F \rightarrow \mathrm{End}_{\boldsymbol{Q}}(\tilde{A})$ by $\tilde{\theta}(a)=\mathfrak{p}(\theta(a))$ for every $a \in K$ such that $\theta(a) \in \mathrm{End}(A)$. Therefore an $\mathfrak{l}$-adic representation $R_{\mathfrak{l}}'$ of the commutor of $\tilde{\theta}(F)$ in $\mathrm{End}_{\boldsymbol{Q}}(\tilde{A})$ can be defined as above. Let $\varphi_{\mathfrak{p}}$ denote the Frobenius endomorphism of $\tilde{A}$ of degree $N(\mathfrak{p})$. By [81, § 11.1, Prop. 14], we have a commutative diagram:

$$\text{(7.6.6)} \qquad \begin{array}{ccc} W_{\mathfrak{l}}/D_{\mathfrak{l}} & \longrightarrow & A[\mathfrak{l}^{\infty}] \\ \text{id.}\downarrow & & \downarrow \text{reduction modulo } \mathfrak{P} \\ W_{\mathfrak{l}}/D_{\mathfrak{l}} & \longrightarrow & \tilde{A}[\mathfrak{l}^{\infty}] \end{array}$$

If $t \in A[\mathfrak{l}^{\infty}]$, we have $\mathfrak{P}(t^{\sigma})=\varphi_{\mathfrak{p}}(\tilde{t})$. Therefore, if we define $\mathfrak{R}_{\mathfrak{l}}$ and $R_{\mathfrak{l}}'$ with respect to the horizontal arrows of (7.6.6), we obtain

$$\text{(7.6.7)} \qquad \mathfrak{R}_{\mathfrak{l}}(\sigma)=R_{\mathfrak{l}}'(\varphi_{\mathfrak{p}}).$$

(Note that $\varphi_{\mathfrak{p}}$ belongs to the commutor of $\tilde{\theta}(F)$.) This shows that $\mathfrak{R}_{\mathfrak{l}}(\sigma)$ is uniquely determined by $\mathfrak{P}$, hence we obtain the first part of the following

PROPOSITION 7.23. *Let $\mathfrak{K}(\mathfrak{l})$ denote the subfield of $\bar{\boldsymbol{Q}}$ corresponding to the kernel of $\mathfrak{R}_{\mathfrak{l}}$. Then a prime ideal $\mathfrak{p}$ in k is unramified in $\mathfrak{K}(\mathfrak{l})$ if $\mathfrak{p}$ does not belong to B and is prime to $N(\mathfrak{l})$. Moreover, for such a prime ideal $\mathfrak{p}$, let σ be a Frobenius element of* $\mathrm{Gal}(\bar{\boldsymbol{Q}}/k)$ *with respect to any prime divisor $\mathfrak{P}$ of $\mathfrak{p}$ in $\bar{\boldsymbol{Q}}$. Then the characteristic polynomial of $\mathfrak{R}_{\mathfrak{l}}(\sigma)$ has coefficients in $\mathfrak{o}$, and depends only on $\mathfrak{p}$ (i.e., it is independent of the choice of $\mathfrak{l}$ and $\mathfrak{P}$).*

This is a generalization of [81, § 18.5, Prop. 18]. The assertion concerning $\mathfrak{R}_{\mathfrak{l}}(\sigma)$ follows from (7.6.7) and Prop. 7.21.

The notation being as above, let $H_{\mathfrak{p}}(u)$ denote the characteristic polynomial of $\mathfrak{R}_{\mathfrak{l}}(\sigma)$. Then we can define *the zeta-function of A over k* relative to $\theta: F \rightarrow \mathrm{End}_{\boldsymbol{Q}}(A)$ by

$$\zeta(s\,;A/k, F)=\prod_{\mathfrak{p} \notin B} N(\mathfrak{p})^{ds} \cdot H_{\mathfrak{p}}(N(\mathfrak{p})^{s})^{-1}.$$

If $F=\boldsymbol{Q}$, this is exactly $\zeta(s\,; A/k)$ defined in § 7.5. It is of course natural to extend the Hasse-Weil conjecture to $\zeta(s\,; A/k, F)$.

Observe that $\zeta(s\,; A/k, F)$ depends only on the isogeny class of A over k. Therefore the assumption (7.6.2) is inessential, since for a given (A, θ), we can always find another (A', θ') satisfying (7.6.2) by an isogeny rational over k (see [81, § 7.1, Prop. 7]).

Now, since $H_{\mathfrak{p}}$ is the characteristic polynomial of $\mathfrak{R}_{\mathfrak{l}}(\sigma)$, we see that $\zeta(s\,; A/k, F)$ is analogous to the Artin L-functions of finite normal extensions of algebraic number fields. Therefore the determination of $\zeta(s\,; A/k, F)$

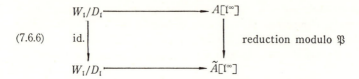

provides a certain reciprocity-law for the extensions $\mathfrak{K}(\mathfrak{l})$ of k, which are not necessarily abelian, as we already emphasized in [81, § 18.5] and in [73, § 6.3]. For further discussion of this topic, we refer the reader to Taniyama [87], the author [76], [78], [79], [80], and Serre [65].

Coming back to (A, θ) which is not necessarily defined over k, but over any field, suppose that A has a polarization $\mathcal{C}$ with the following property:

(7.6.8) *If* $*$ *denotes the involution of* $\mathrm{End}_{\boldsymbol{Q}}(A)$ *defined by* $\mathcal{C}$ (see Appendix N° 13), *then* $\theta(a)^* = \theta(a)$ *for every* $a \in F$.

Since $*$ is a positive involution of $\mathrm{End}_{\boldsymbol{Q}}(A)$, F must be totally real. For the rational prime l divisible by $\mathfrak{l}$, put $W_l = W \otimes_{\boldsymbol{Q}} \boldsymbol{Q}_l$, and $D_l = D \otimes_{\boldsymbol{Z}} \boldsymbol{Z}_l$. Then we obtain an l-adic coordinate system

$$0 \longrightarrow D_l \longrightarrow W_l \longrightarrow A[l^\infty] \longrightarrow 0 \qquad \text{(exact)}.$$

Take a divisor X in $\mathcal{C}$. By Weil [92, N° 76], we can associate with X a non-degenerate alternating form

$$E_l: \ W_l \times W_l \rightarrow \boldsymbol{Q}_l$$

such that $E_l(x, y) \in \boldsymbol{Z}_l$ for every $(x, y) \in D_l \times D_l$, and

$$E_l(R_l(\lambda)x, y) = E_l(x, R_l(\lambda^*)y) \tag{7.6.9}$$

for every $\lambda \in \mathrm{End}_{\boldsymbol{Q}}(A)$, where R_l is the l-adic representation of $\mathrm{End}_{\boldsymbol{Q}}(A)$. Now $W_{\mathfrak{l}}$ can be identified with a subspace of W_l in a natural way. Restrict E_l to $W_{\mathfrak{l}} \times W_{\mathfrak{l}}$. By [75, I, Lemma 1.2], we can find a non-degenerate alternating form

$$S_{\mathfrak{l}}: \ W_{\mathfrak{l}} \times W_{\mathfrak{l}} \rightarrow F_{\mathfrak{l}}$$

such that

$$E_l(x, y) = \mathrm{Tr}_{F_{\mathfrak{l}}/\boldsymbol{Q}_l}(S_{\mathfrak{l}}(x, y)) \qquad ((x, y) \in W_{\mathfrak{l}} \times W_{\mathfrak{l}}).$$

Suppose now that A, X, and the elements of $\theta(F) \cap \mathrm{End}(A)$ are all rational over a finite field with q elements. Let φ be the Frobenius endomorphism of A of degree q. Then

$$S_{\mathfrak{l}}(R_{\mathfrak{l}}(\varphi)x, y) = S_{\mathfrak{l}}(x, R_{\mathfrak{l}}(\varphi^*)y).$$

It follows that $R_{\mathfrak{l}}(\varphi)$ and $R_{\mathfrak{l}}(\varphi^*)$ have the same characteristic polynomial.

We shall now consider A_S, f, A, K, and θ as in Th. 7.14. Let $\mathcal{C}_S$ be the canonical polarization of A_S, and $*$ the involution of $\mathrm{End}_{\boldsymbol{Q}}(A_S)$ defined by $\mathcal{C}_S$. Consider A_S as a complex torus $\boldsymbol{C}^g/L$, and take a Riemann form E_S on $\boldsymbol{C}^g$ corresponding to a divisor in $\mathcal{C}_S$. $\boldsymbol{C}^g$ can be identified with $S_2(\Gamma')$. By (6) and (9) of Prop. 7.2, we have ${}^t X_{SS}(\alpha) = X_{SS}(\alpha^{-1}) = X_{SS}(\alpha^\iota)$ for every $\alpha \in \Delta'$, so that

$$E_S([\Gamma'\alpha\Gamma']_2 x, y) = E_S(x, [\Gamma'\alpha^t\Gamma']_2 y) \qquad ((x, y) \in \boldsymbol{C}^g \times \boldsymbol{C}^g).$$

In view of (3.3.13), this implies especially

(7.6.10) $\xi_q = \eta_q \xi_q^*$ *if* q *is prime to* N.

If we denote by E the restriction of E_S to the subspace of $\boldsymbol{C}^g$ corresponding to A, then we have, by (7.5.10),

(7.6.11) $E(\theta(a_q)x, y) = E(x, \theta(a_q^\rho)y)$.

Let Y be a divisor on A corresponding to E, and φ_Y the isogeny of A to its Picard variety associated to Y. Then (7.6.11) implies that ${}^t\theta(a_q)\varphi_Y = \varphi_Y\theta(a_q^\rho)$. Changing Y by algebraic equivalence, we may assume that Y is rational over $\bar{\boldsymbol{Q}}$. Then ${}^t\theta(a_q)\varphi_{Y^\sigma} = \varphi_{Y^\sigma} \cdot \theta(a_q^\rho)$ for every $\sigma \in \mathrm{Gal}(\bar{\boldsymbol{Q}}/\boldsymbol{Q})$. Let X be the sum of all distinct Y^σ with $\sigma \in \mathrm{Gal}(\bar{\boldsymbol{Q}}/\boldsymbol{Q})$. Then X determines a polarization $\mathcal{C}$ of A rational over $\boldsymbol{Q}$, and ${}^t\theta(a_q)\varphi_X = \varphi_X\theta(a_q^\rho)$. If we denote by * the involution of $\mathrm{End}_{\boldsymbol{Q}}(A)$ defined by $\mathcal{C}$, then we have

(7.6.12) $\theta(a_q^\rho) = \theta(a_q)^*$ *if* q *is prime to* N.

The corresponding relation holds on $(\tilde{A}, \tilde{\theta})$, by reduction modulo p. Now we can re-formulate Theorems 7.15 and 7.18 as follows.

THEOREM 7.24. *Let f, A, K, and θ be as in Th.* 7.14. *Suppose that* $\Gamma' = \Gamma_0(N)$, *and* (7.5.9) *is satisfied. Then, up to a finite number of Euler factors,* $\zeta(s; A/\boldsymbol{Q}, K)$ *coincides with* $L(s, f)$.

PROOF. We have $[K:\boldsymbol{Q}] = \dim(A)$, so that $d = 2$. By our assumption, K is totally real, and $\theta(a)^* = \theta(a)$ for every $a \in K$. Let $R_\mathfrak{l}$ and $R'_\mathfrak{l}$ denote the $\mathfrak{l}$-adic representations of $\mathrm{End}(A)$ and $\mathrm{End}(\tilde{A})$, respectively. From (7.5.1) we obtain

$$R'_\mathfrak{l}(\pi_p + \pi_p^*) = R_\mathfrak{l}(\theta(a_p)) = a_p 1_2,$$

so that

$$\det[1 - u \cdot R'_\mathfrak{l}(\pi_p)] \cdot \det[1 - u \cdot R'_\mathfrak{l}(\pi_p^*)] = (1 - a_p u + p u^2)^2.$$

Since $R'_\mathfrak{l}(\pi_p)$ and $R'_\mathfrak{l}(\pi_p^*)$ have the same characteristic polynomial as remarked above, we obtain

$$\det[1 - u \cdot R'_\mathfrak{l}(\pi_p)] = 1 - a_p u + p u^2,$$

which proves our theorem.

THEOREM 7.25. *Under the assumptions* (7.5.8), (7.5.14), *and* (7.5.15), *let A', K', and k be as in Th.* 7.16 *and Th.* 7.18. *Then, up to a finite number of Euler factors,* $\zeta(s; A'/k, K')$ *coincides with* $L(s, f)L(s, f_\rho)$, *where ρ is the complex conjugation, and f_ρ is as in Th.* 7.14.

PROOF. Let $\varphi_{\mathfrak{p}}$ and $\varphi_{\mathfrak{p}}^0$ be as in the proof of Th. 7.18, $\mathfrak{l}$ a prime ideal in K', and $R_{\mathfrak{l}}$ (resp. $R'_{\mathfrak{l}}$, $R_{\mathfrak{l}}^0$) the $\mathfrak{l}$-adic representation of $\mathrm{End}(A)$ (resp. $\mathrm{End}(\tilde{A})$, $\mathrm{End}(\tilde{A}')$). Then, as in the proof of Th. 7.18, we have

(7.6.13) $$\det[1_2 - u\cdot R_{\mathfrak{l}}^0(\varphi_{\mathfrak{p}}^0)]^2 = \det[1_4 - u\cdot R'_{\mathfrak{l}}(\varphi_{\mathfrak{p}})]\,,$$

(7.6.14) $$\det[1_4 - u\cdot R_{\mathfrak{l}}(\theta(a_p)) + p\cdot\psi(p)u^2 1_4] = \begin{cases} \det[1_4 - u^2 R'_{\mathfrak{l}}(\varphi_{\mathfrak{p}})] & \text{if } N(\mathfrak{p}) = p^2\,, \\ \det[1_4 - u\cdot R'_{\mathfrak{l}}(\varphi_{\mathfrak{p}})]\cdot\det[1_4 - u\cdot R'_{\mathfrak{l}}(\varphi_{\mathfrak{p}'})] & \text{if } (p) = \mathfrak{p}\mathfrak{p}'\,. \end{cases}$$

By Prop. 7.21 and Prop. 7.22, $R_{\mathfrak{l}}(\theta(a_p))$ has a_p and a_p^ρ as characteristic roots, both with multiplicity 2. Therefore the left hand side of (7.6.14) equals

$$(1 - a_p u + p\cdot\psi(p)u^2)^2(1 - a_p^\rho u + p\cdot\psi(p)u^2)^2\,.$$

Further, for the same reason as in (7.5.17), $R_{\mathfrak{l}}^0(\varphi_{\mathfrak{p}}^0)$ and $R_{\mathfrak{l}}^0(\varphi_{\mathfrak{p}'}^0)$ have the same characteristic polynomial if $(p) = \mathfrak{p}\mathfrak{p}'$. Therefore we obtain

(7.6.15) $$\det[1_2 - u\cdot R_{\mathfrak{l}}^0(\varphi_{\mathfrak{p}}^0)] = \det[1_2 - u\cdot R_{\mathfrak{l}}^0(\varphi_{\mathfrak{p}'}^0)] = 1 - a_p u + p u^2 \quad \text{if } (p) = \mathfrak{p}\mathfrak{p}'\,,$$
$$\det[1_2 - u^2 R_{\mathfrak{l}}^0(\varphi_{\mathfrak{p}}^0)] = (1 - a_p u + p u^2)(1 - a_p^\rho u - p u^2) \quad \text{if } N(\mathfrak{p}) = p^2\,,$$

hence our assertion.

EXAMPLE 7.26. Consider $\Gamma_0(N)$ for $N = 23, 29, 31$. Then the genus of V_S $(=\Gamma_0(N)\backslash\mathfrak{H}^*)$ is 2. It has been shown by Doi [14] that A_S is simple, and $\mathrm{End}_{\boldsymbol{Q}}(A_S)$ is isomorphic to $\boldsymbol{Q}(\sqrt{5})$, $\boldsymbol{Q}(\sqrt{2})$, $\boldsymbol{Q}(\sqrt{5})$, respectively; moreover, $\mathrm{End}(A_S)$ is isomorphic to the maximal order in these fields. Therefore we can put $A = A_S$ with these fields as K. There exists an element $f(z) = \sum_{n=1}^\infty a_n e^{2\pi i n z}$ of $S_2(\Gamma_0(N))$ such that $f\,|\,T'(n)_2 = a_n f$ for all n, with a_n in K. Then Th. 7.23 implies that $\zeta(s;\, A_S/\boldsymbol{Q}, K)$ is essentially $L(s, f) = \sum_{n=1}^\infty a_n n^{-s}$ (if we use this f to define $\theta: K \to \mathrm{End}_{\boldsymbol{Q}}(A)$); $\zeta(s;\, A_S/\boldsymbol{Q})$ is essentially $L(s, f)L(s, f_\sigma)$, where $f_\sigma(z) = \sum_{n=1}^\infty a_n^\sigma e^{2\pi i n z}$, and σ is the generator of $\mathrm{Gal}(K/\boldsymbol{Q})$.

REMARK 7.27. (A) In the setting of Th. 7.25, we have

$$L(s, f)L(s, f_\rho) = \prod_{p\mid N}(1 - a_p p^{-s})^{-1}(1 - a_p^\rho p^{-s})^{-1}$$
$$\times\prod_{p\nmid N}(1 - a_p p^{-s} + \psi(p)p^{1-2s})^{-1}(1 - a_p^\rho p^{-s} + \psi(p)p^{1-2s})^{-1}\,.$$

Put $R(s) = N^s(2\pi)^{-2s}\Gamma(s)^2 L(s, f)L(s, f_\rho)$. By Th. 3.66, we obtain $R(s) = R(2-s)$. If N is a prime, we have $k = \boldsymbol{Q}(\sqrt{N})$, and $N \equiv 1 \bmod (4)$. Moreover, $a_N a_N^\rho = N$ by virtue of a result of Hecke [30, Satz 61].

It may be conjectured that, if N is a prime, the abelian variety A' has good reduction for the prime ideal $(\sqrt{N})$, hence for all prime ideals in $\boldsymbol{Q}(\sqrt{N})$, and that the factor $(1-a_N N^{-s})^{-1}(1-a_N^\rho N^{-s})^{-1}$ is the Euler factor $\zeta(s; A'/k, K')$ for $(\sqrt{N})$; this means that one has exactly $\zeta(s; A'/k, K')=L(s,f)L(s,f_\rho)$ without worrying about bad primes. We shall come back again to this question at the end of the next section.

REMARK 7.27. (B) The jacobian variety A_S of V_S has often some points of finite order, rational over $\boldsymbol{Q}$, which are obtained in the following way. For simplicity, let us assume that

$$S=\boldsymbol{Q}^\times\cdot U', \qquad U'=\left\{\begin{bmatrix} a & b \\ c & d\end{bmatrix}\in U \,\middle|\, c\equiv 0 \bmod (N)\right\},$$

so that $\Gamma_S=\boldsymbol{Q}^\times\cdot\Gamma_0(N)$. Let ψ be a character of $(\boldsymbol{Z}/N\boldsymbol{Z})^\times$ of order r such that $\psi(-1)=1$, and $\mathfrak{k}$ the subgroup of $\mathfrak{g}^\times$ corresponding to the kernel of ψ in an obvious sense. Put

$$T=\boldsymbol{Q}^\times\cdot\left\{\begin{bmatrix} a & b \\ c & d\end{bmatrix}\in U' \,\middle|\, a\in\mathfrak{k}\right\}.$$

Then we have

$$\Gamma_T=\boldsymbol{Q}^\times\cdot\left\{\begin{bmatrix} a & b \\ c & d\end{bmatrix}\in \Gamma_0(N) \,\middle|\, \psi(a)=1\right\},$$

and V_T is a cyclic covering of V_S of order r; every element of $\mathrm{Gal}(V_T/V_S)$, as a birational automorphism of V_T, is defined over $\boldsymbol{Q}$. Suppose that

(7.6.16) *V_T is unramified over V_S.*

Let P be the projection map of V_T to V_S. Put $F=\boldsymbol{Q}(e^{2\pi i/r})$. Then the function field $F(V_T)$ is generated over $F(V_S)\circ P$ by an element h such that $h^r\in F(V_S)\circ P$, and $h\circ\lambda=e^{2\pi i/r}h$, where λ is a generator of $\mathrm{Gal}(V_T/V_S)$. (For the notation $F(V_S)$, see Appendix N° 4.) Let div_T (resp. div_S) denote the divisor of a function on V_T (resp. V_S). Then $P(\mathrm{div}_T(h))=r\mathfrak{a}$ with a divisor $\mathfrak{a}$ of V_S rational over F, and $r\mathfrak{a}$ is linearly equivalent to 0. For every $\sigma\in\mathrm{Gal}(F/\boldsymbol{Q})$, $(h^\sigma)^r\in F(V_S)\circ P$, so that $h^\sigma=fh$ with an element f of $F(V_S)\circ P$. Then we see that $\mathfrak{a}^\sigma$ is linearly equivalent to $\mathfrak{a}$. Therefore, if t denotes the point of A_S corresponding to the divisor class of $\mathfrak{a}$, then t is rational over $\boldsymbol{Q}$, and $rt=0$. If $ct=0$ for a positive integer $c<r$, then $c\mathfrak{a}=\mathrm{div}_S(g)$ with $g\in F(V_S)$, hence $P(\mathrm{div}_T(h^c))=r\cdot\mathrm{div}_S(g)$. It follows that $\mathrm{div}_T(h^c)=\mathrm{div}_T(g\circ P)$, so that $h^c\in F(V_S)\circ P$, a contradiction. Therefore t must be of order r. Thus we have shown

(7.6.17) *The jacobian A_S of V_S has a point of order r rational over $\boldsymbol{Q}$ under the assumption* (7.6.16).

The verification of (7.6.16) in each case can easily be done by checking which parabolic and elliptic elements of $\Gamma_0(N)$ are contained in Γ_T. For instance, V_T is unramified over V_S at the cusps, if N is square-free.

Let, for example, $N=23, 29, 31$. By the above principle, we can find a point t of A_S of order $r=11, 7, 5$, respectively, rational over $\boldsymbol{Q}$. Let $K=\boldsymbol{Q}(\sqrt{5}), \boldsymbol{Q}(\sqrt{2}), \boldsymbol{Q}(\sqrt{5})$, as considered in Ex. 7.26, and let $f(z)=\sum_{n=1}^{\infty} a_n e^{2\pi i n z}$, with $a_1=1$, be a common eigen-function of all $T'(n)_2$ on $S_2(\Gamma_0(N))$. Define an isomorphism θ of K onto $\mathrm{End}_{\boldsymbol{Q}}(A_S)$ with respect to f, as in Th. 7.14. Consider reduction modulo p of A and t for a prime p not dividing Nr. If π_p and π_p^* are as above, we have $\pi_p\tilde{t}=\tilde{t}$ and $\pi_p^*\tilde{t}=p\tilde{t}$, so that $[1+p-\tilde{\theta}(a_p)]\tilde{t}=0$ by (7.5.1). Let $\mathfrak{a}$ be the integral ideal in K generated by r and $1+p-a_p$ for all such p. Then $\tilde{\theta}(\mathfrak{a})\tilde{t}=0$. Since $\tilde{t}$ is of order r, $\mathfrak{a}$ can not be the identity ideal. Therefore we obtain

(7.6.18) $1-a_p+p\equiv 0 \bmod (\mathfrak{l})$ *for every prime* p *not dividing* Nr, *where* $\mathfrak{l}$ *is a prime ideal in* K *dividing* r.

(Actually the congruence is true also for $p=r$.)

Similarly, let $N=11, 14, 15, 17, 19, 21$. In these cases A_S is an elliptic curve. The above principle ensures the existence of a point of A_S, rational over $\boldsymbol{Q}$, of order $r=5, 3, 4, 4, 3, 2$, respectively. Then we obtain, by a similar and simpler argument,

(7.6.19) *If* $f(z)=\sum_{n=1}^{\infty} a_n e^{2\pi i n z}$, *with* $a_1=1$, *is a generator of* $S_2(\Gamma_0(N))$ *for these values of* N, *then* $1-a_p+p\equiv 0 \bmod (r)$ *for every prime* p *not dividing* Nr.

7.7. Construction of class fields over real quadratic fields

We shall now show that certain points of finite order on the abelian variety A' of Th. 7.16 and Th. 7.18 generate non-cyclotomic abelian extensions of real quadratic fields. Let us first recall the properties of A' and other symbols.

N: a positive integer.
ψ: a character of $(\boldsymbol{Z}/N\boldsymbol{Z})^\times$ of order 2 such that $\psi(-1)=1$.
f: an element of $S_2(\Gamma_0(N), \psi)$, i. e., a cusp form of level N satisfying

$$f((az+b)(cz+d)^{-1})(cz+d)^{-2}=\psi(d)f(z) \qquad \text{for all } \begin{bmatrix} a & b \\ c & d \end{bmatrix}\in\Gamma_0(N).$$

We assume that f is a common eigen-function of all the Hecke operators $T'(n)_{2,\psi}$ on $S_2(\Gamma_0(N), \psi)$, and

$$f(z)=\sum_{n=1}^{\infty} a_n e^{2\pi i n z}, \qquad f\,|\,T'(n)_{2,\psi}=a_n f \qquad (\text{cf. Th. } 3.43)\,.$$

Further we assume that the algebra of all $T'(n)_{2,\psi}$ can be generated only by the subset $\{T'(n)_{2,\psi}\,|\, n \text{ prime to } N\}$. (See Remark 3.60, especially the results of Hecke mentioned there.)

k: the real quadratic field corresponding to the kernel of ψ.
ε: the generator of Gal$(k/\boldsymbol{Q})$.
K: the field generated by the numbers a_n over $\boldsymbol{Q}$ for all n.
ρ: the complex conjugation.
$K' = \{x \in K \mid x^\rho = x\}$.

We have shown that K' is totally real, and K is totally imaginary. Further we have

(7.7.1) $$a_n^\rho = \psi(n)a_n \text{ if } n \text{ is prime to } N.$$

In Th. 7.14, we have obtained an abelian variety A and an isomorphism θ of K into $\mathrm{End}_{\boldsymbol{Q}}(A)$; A and $\theta(a)$ for all $a \in K$ are rational over $\boldsymbol{Q}$; the couple (A, θ) is characterized by (1) and (2) of Th. 7.14. Further A has an automorphism μ, rational over k, such that

(7.7.2) $$\mu^2 = 1, \qquad \mu \cdot \theta(a) = \theta(a^\rho)\mu \qquad (a \in K),$$

(7.7.3) $$\mu^\varepsilon = -\mu.$$

We put

(7.7.4) $$A' = (1+\mu)A.$$

Then we have seen that A' is an abelian subvariety of A rational over k, and

(7.7.5) $$A = A' + A'^\varepsilon, \qquad A'^\varepsilon = (1-\mu)A.$$

Denote by $\theta'(a)$ the restriction of $\theta(a)$ to A' for every $a \in K'$. Then θ' is an isomorphism of K' into $\mathrm{End}_{\boldsymbol{Q}}(A')$. We can also define an isomorphism θ'^ε of K' into $\mathrm{End}_{\boldsymbol{Q}}(A'^\varepsilon)$ by $\theta'^\varepsilon(a) = \theta'(a)^\varepsilon$. Obviously $\theta'^\varepsilon(a)$ is the restriction of $\theta(a)$ to A'^ε.

Let p be a rational prime not dividing N, and $\mathfrak{p}$ a prime ideal in k, dividing p. As is noted in § 7.5,

(7.7.6) *A and A' have good reduction modulo $\mathfrak{p}$.*

Let the tilde denote reduction modulo $\mathfrak{p}$, and π_p the Frobenius endomorphism of $\tilde{A}$ of degree p. Further let π_p^* be the element of $\mathrm{End}(\tilde{A})$ such that $\pi_p^*\pi_p = p$. From (7.5.1) we obtain

(7.7.7) $$\pi_p + \psi(p)\pi_p^* = \tilde{\theta}(a_p).$$

Let $\mathfrak{o}$ and $\mathfrak{o}'$ be the maximal orders in K and K', respectively. In general $\theta(\mathfrak{o})$ may not be contained in $\mathrm{End}(A)$. However, by changing A by a suitable isogeny over $\boldsymbol{Q}$, we may assume that condition. In fact, let m be a positive integer such that $\theta(m\mathfrak{o}) \subset \mathrm{End}(A)$. By [81, § 7.1, Prop. 7, Prop. 8], we can find an abelian variety A_1, an isomorphism θ_1 of K into $\mathrm{End}_{\boldsymbol{Q}}(A_1)$, and an isogeny λ of A to A_1 such that: (i) $\lambda \cdot \theta(a) = \theta_1(a)\lambda$ for all $a \in K$; (ii) $\theta_1(\mathfrak{o}) \subset \mathrm{End}(A_1)$; (iii) $\mathrm{Ker}(\lambda) = \{t \in A \mid \theta(m\mathfrak{o})t = 0\}$; (iv) A_1, λ, and $\theta_1(a)$ for every $a \in \mathfrak{o}$ are rational over $\boldsymbol{Q}$. Observe that μ maps $\mathrm{Ker}(\lambda)$ onto itself. Therefore we can define an automorphism μ_1 of A_1, rational over k, by $\mu_1\lambda = \lambda\mu$. Change A, θ, μ for A_1, θ_1, μ_1, and write them again A, θ, μ. This change does not disturb (7.7.2, 3). Defining again A' by (7.7.4), we have still (7.7.5~7). Of course the new A may no longer be a subvariety of the jacobian A_S of the curve V_S, which we do not need in the following discussion. All what we need is (A, θ), (A', θ'), an automorphism μ of A, and a cusp form $f(z) = \sum_{n=1}^{\infty} a_n e^{2\pi inz}$, which satisfy the above (7.7.2~7) and

(7.7.8) $$\theta(\mathfrak{o}) \subset \mathrm{End}(A), \qquad \theta'(\mathfrak{o}') \subset \mathrm{End}(A').$$

In other words, these conditions and (7.7.9) below are the "axioms" of our theory, for which we have (not yet, so far as (7.7.9) is concerned) shown the existence of the objects satisfying them.[16)]

Let $\mathfrak{d}$ denote the different of K relative to K'. Put

$$\mathfrak{b}_0 = \{x \in \mathfrak{o} \mid x^\rho = -x\}.$$

Let $\mathfrak{b}$ be the ideal of $\mathfrak{o}$ generated by $\mathfrak{b}_0$. Then $\mathfrak{d} \subset \mathfrak{b}$. Observe that $a_n \in \mathfrak{b}_0$ if $\psi(n) = -1$.

PROPOSITION 7.28. *For every* $x \in \mathfrak{b}_0$, *one has* $-N_{K/\boldsymbol{Q}}(x) \in N_{k/\boldsymbol{Q}}(k^\times)$. *Moreover,* $-N(\mathfrak{b}) \in N_{k/\boldsymbol{Q}}(k^\times)$.

PROOF. Let $x \in \mathfrak{b}_0$. Then $x^{-1}\mathfrak{b}_0 \subset K'$. Let $\mathfrak{e}$ be the fractional ideal in K' generated by $x^{-1}\mathfrak{b}_0$ over $\mathfrak{o}'$. Then we see that $\mathfrak{b} = x\mathfrak{o}\mathfrak{e}$, hence $N(\mathfrak{b}) = N_{K/\boldsymbol{Q}}(x)N(\mathfrak{e})^2$. Therefore, to prove our proposition, it is sufficient to prove $-N_{K/\boldsymbol{Q}}(x) \in N_{k/\boldsymbol{Q}}(k^\times)$. Take a basis $\{\omega_1, \cdots, \omega_n\}$ of $\mathfrak{D}(A')$ rational over k. Then $\{\omega_1^\varepsilon, \cdots, \omega_n^\varepsilon\}$ is a basis of $\mathfrak{D}(A'^\varepsilon)$. By (7.7.2) and (7.7.5), we see that $\theta(x)$ maps A' to A'^ε, and A'^ε to A'. We can put $\omega_h^\varepsilon \circ \theta(x) = \sum_{i=1}^n y_{hi}\omega_i$ $(h = 1, \cdots, n)$ with y_{hi} in k. Applying ε to this relation, we obtain $\omega_h \circ \theta(x) = \sum_i y_{hi}^\varepsilon \omega_i^\varepsilon$, hence $\omega_h \circ \theta(x^2) = \sum_{i,j} y_{hi}^\varepsilon y_{ij}\omega_j$. Now the representation of K' on $\mathfrak{D}(A')$ is equivalent to the regular representation of K' over $\boldsymbol{Q}$. Therefore

$$-N_{K/\boldsymbol{Q}}(x) = N_{K'/\boldsymbol{Q}}(x^2) = \det(y_{hi}) \cdot \det(y_{hi})^\varepsilon \in N_{k'/\boldsymbol{Q}}(k^\times),$$

q. e. d.

16) Probably it is not always best to assume (7.7.8). The reader should consider the condition and the change of A, θ, μ rather tentative, or made just for the sake of simplicity. The same remark applies to (7.7.9, 15) below.

REMARK 7.28′. Hecke [30, Satz 61] proved that $a_N a_N^\rho = N$ if N is a prime. Therefore we have $N \in N_{K/K'}(K^\times)$. It is noteworthy that this is reciprocal to Prop. 7.28.

Let us now make the following assumption:

(7.7.9) $\mathfrak{b}$ *is prime to* 2.

The fields K and K' are determined by f, so that (7.7.9) can be viewed as a condition on f. If $\mathfrak{q}$ is a prime ideal in K prime to 2, then $\mathfrak{d}$ is not divisible by $\mathfrak{q}^2$, by a well-known property of the different. Therefore, under the assumption (7.7.9), $\mathfrak{b}$ is a square-free ideal in K, and $\mathfrak{b}^\rho = \mathfrak{b}$. Put $\mathfrak{c} = N_{K/K'}(\mathfrak{b})$. Then $\mathfrak{c}$ is a square-free integral ideal in K', and $\mathfrak{c}\mathfrak{o} = \mathfrak{b}^2$. Put

$$\mathfrak{x} = \{t \in A \mid \theta(\mathfrak{b})t = 0\} = A[\mathfrak{b}],$$

$$\mathfrak{y} = \mathfrak{x} \cap A', \qquad \mathfrak{z} = \mathfrak{x} \cap A'^{\varepsilon}.$$

We see easily that $\mathfrak{y}$ and $\mathfrak{z}$ are $\mathfrak{o}'$-modules, and $\mathfrak{x}$ is an $\mathfrak{o}$-module.

PROPOSITION 7.29. *The $\mathfrak{o}'$-modules $\mathfrak{y}$ and $\mathfrak{z}$ are isomorphic to $\mathfrak{o}'/\mathfrak{c}$, and* $\mathfrak{x} = \mathfrak{y} \oplus \mathfrak{z}$.

PROOF. By Prop. 7.20, we see that $\mathfrak{x}$ is isomorphic to $(\mathfrak{o}/\mathfrak{b})^2$. Observe that $\mathfrak{o}/\mathfrak{b}$ is isomorphic to $\mathfrak{o}'/\mathfrak{c}$. From our definition of $\mathfrak{b}$, we see that $\mathfrak{x}$ is stable under μ. Therefore $(1+\mu)\mathfrak{x} \subset \mathfrak{y}$ and $(1-\mu)\mathfrak{x} \subset \mathfrak{z}$. Since $\mathfrak{b}$ is prime to 2, every element t of $\mathfrak{x}$ can be written as $t = 2s$ with $s \in \mathfrak{x}$, so that $t = (1+\mu)s + (1-\mu)s \in \mathfrak{y} + \mathfrak{z}$. If $u \in \mathfrak{y} \cap \mathfrak{z}$, we have $2u = 0$, so that $u = 0$. This proves that $\mathfrak{x} = \mathfrak{y} \oplus \mathfrak{z}$. Extend ε to an automorphism δ of $\bar{\boldsymbol{Q}}$. Then δ gives an $\mathfrak{o}'$-isomorphism of $\mathfrak{y}$ to $\mathfrak{z}$. Therefore both $\mathfrak{y}$ and $\mathfrak{z}$ must be isomorphic to $\mathfrak{o}'/\mathfrak{c}$.

Let F denote the field generated over k by the coordinates of the points of $\mathfrak{x}$. By the above proposition, we can find an element s of $\mathfrak{y}$ and an element t of $\mathfrak{z}$ so that

(7.7.10) $$\mathfrak{y} = \theta(\mathfrak{o}')s, \qquad \mathfrak{z} = \theta(\mathfrak{o}')t.$$

Then we have $F = k(s, t)$. It is this extension F of k whose class-field-theoretical structure we are going to investigate.

THEOREM 7.30. *The field F is an abelian extension of k, which is unramified at every finite prime of k not dividing $N(\mathfrak{c})N$. Moreover, if m is a positive rational integer prime to $N(\mathfrak{c})N$, and $\sigma = \left(\frac{F/k}{(m)}\right)$, then $x^\sigma = mx$ for every $x \in \mathfrak{x}$.*

PROOF. By Prop. 7.23, every prime ideal in k, prime to $N(\mathfrak{c})N$, is unramified in F. Observe that the map $a \mapsto \theta(a)s$ defines an isomorphism of

$\mathfrak{o}'/\mathfrak{c}$ onto $\mathfrak{y}$. Let $\tau \in \mathrm{Gal}(\bar{Q}/k)$. By our definition of $\mathfrak{y}$ and $\mathfrak{z}$, τ maps $\mathfrak{y}$ and $\mathfrak{z}$ onto themselves. Therefore F is a Galois extension of k, and one has

$$s^{\tau} = \theta(g_{\tau})s\,, \qquad t^{\tau} = \theta(h_{\tau})t \tag{7.7.11}$$

with elements g_{τ} and h_{τ} of $\mathfrak{o}'$ prime to $\mathfrak{c}$. We see easily that $\tau \mapsto (g_{\tau}, h_{\tau})$ defines an isomorphism of $\mathrm{Gal}(F/k)$ to a subgroup of $(\mathfrak{o}'/\mathfrak{c})^{\times 2}$, and hence F is abelian over k. Let p be a rational prime not dividing $N(\mathfrak{c})N$, $\mathfrak{p}$ a prime ideal in k dividing p, and $\mathfrak{P}$ a prime divisor of $\bar{Q}$ which extends $\mathfrak{p}$. Let $\tilde{X}$ or $\mathfrak{P}(X)$ denote the object obtained from X by reduction modulo $\mathfrak{P}$. Further let π_p and π_p^* be as in (7.7.7), and σ a Frobenius element of $\mathrm{Gal}(\bar{Q}/k)$ for $\mathfrak{P}$. Then $\sigma = \left(\frac{F/k}{\mathfrak{p}}\right)$ on F.

(I) First consider the case $\psi(p) = -1$, so that $N(\mathfrak{p}) = p^2$. From (7.7.7) we obtain $\pi_p^2 - p = \pi_p(\pi_p - \pi_p^*) = \pi_p \cdot \tilde{\theta}(a_p)$. By (7.7.1), we have $a_p \in \mathfrak{b}_0$, so that $(\pi_p^2 - p)\tilde{x} = 0$ for all $x \in \mathfrak{x}$. Since $\pi_p^2 x = \mathfrak{P}(x^{\sigma})$, we have $\mathfrak{P}(x^{\sigma} - px) = 0$ for all $x \in \mathfrak{x}$. By [81, § 11.1, Prop. 13], reduction modulo $\mathfrak{P}$ defines an isomorphism of $\mathfrak{x}$ onto $\mathfrak{P}(\mathfrak{x})$. Therefore we have $x^{\sigma} = px$ for all $x \in \mathfrak{x}$.

(II) Next suppose that $\psi(p) = 1$. Then $(p) = \mathfrak{p}\mathfrak{p}^{\varepsilon}$ in k. Let δ be an element of $\mathrm{Gal}(\bar{Q}/Q)$ which coincides with ε on k. Then $F^{\delta} = F$, and $\delta^{-1}\sigma\delta = \left(\frac{F/k}{\mathfrak{p}^{\varepsilon}}\right)$ on F. Now, for every prime ideal $\mathfrak{l}$ in K' dividing $\mathfrak{c}$, consider $\mathfrak{l}$-adic coordinate systems on A' and on A'^{ε}. We have $\mathfrak{l}\mathfrak{o} = \mathfrak{L}^2$ with a prime ideal $\mathfrak{L}$ in K. By our definition of $\mathfrak{b}$, we can find an element λ of $\mathfrak{b}_0$ such that λ is divisible by $\mathfrak{L}$ but not by $\mathfrak{L}^2$. By (7.7.2), we have $\theta(\lambda)\mu = -\mu\theta(\lambda)$. Then we see that $\theta(\lambda)$ maps $A'[\mathfrak{l}]$ into $A'^{\varepsilon}[\mathfrak{l}]$, and $A'^{\varepsilon}[\mathfrak{l}]$ into $A'[\mathfrak{l}]$. (For the notation $A'[\mathfrak{l}]$, see (7.6.3).) Moreover, we have

$$\mathrm{Ker}(\theta(\lambda)) \cap A[\mathfrak{l}] = A[\mathfrak{L}] = \mathfrak{x} \cap A[\mathfrak{l}]\,,$$

so that

$$\mathrm{Ker}(\theta(\lambda)) \cap A'[\mathfrak{l}] = \mathfrak{y} \cap A'[\mathfrak{l}] \cong \mathfrak{o}'/\mathfrak{l}\,,$$

$$\mathrm{Ker}(\theta(\lambda)) \cap A'^{\varepsilon}[\mathfrak{l}] = \mathfrak{z} \cap A'^{\varepsilon}[\mathfrak{l}] \cong \mathfrak{o}'/\mathfrak{l}\,.$$

Put $\mathfrak{y}_{\mathfrak{l}} = \mathfrak{y} \cap A'[\mathfrak{l}]$ and $\mathfrak{z}_{\mathfrak{l}} = \mathfrak{z} \cap A'^{\varepsilon}[\mathfrak{l}]$. Since $\lambda\mathfrak{b} \subset \mathfrak{c}\mathfrak{o}$, we see that

$$\theta(\lambda)(A'[\mathfrak{l}]) \subset \mathfrak{z}_{\mathfrak{l}}\,, \qquad \theta(\lambda)(A'^{\varepsilon}[\mathfrak{l}]) \subset \mathfrak{y}_{\mathfrak{l}}\,.$$

Comparing the order of the modules, we obtain exact sequences

$$0 \longrightarrow \mathfrak{y}_{\mathfrak{l}} \longrightarrow A'[\mathfrak{l}] \xrightarrow{\theta(\lambda)} \mathfrak{z}_{\mathfrak{l}} \longrightarrow 0\,,$$

$$0 \longrightarrow \mathfrak{z}_{\mathfrak{l}} \longrightarrow A'^{\varepsilon}[\mathfrak{l}] \xrightarrow{\theta(\lambda)} \mathfrak{y}_{\mathfrak{l}} \longrightarrow 0\,.$$

Take elements u and v so that

$$\mathfrak{y}_{\mathfrak{l}} = \theta(\mathfrak{o}')u, \qquad A'[\mathfrak{l}] = \mathfrak{y}_{\mathfrak{l}} + \theta(\mathfrak{o}')v.$$

Applying δ, we obtain

$$\mathfrak{z}_{\mathfrak{l}} = \theta(\mathfrak{o}')u^{\delta}, \qquad A'^{\varepsilon}[\mathfrak{l}] = \mathfrak{z}_{\mathfrak{l}} + \theta(\mathfrak{o}')v^{\delta}.$$

The symbols $\sigma = \left(\frac{F/k}{\mathfrak{p}}\right)$ and g_σ, h_σ being as above, we have $u^{\sigma} = \theta(g_\sigma)u$, $u^{\delta\sigma} = \theta(h_\sigma)u^{\delta}$. Put $v^{\sigma} = \theta(c)u + \theta(d)v$ with c and d in $\mathfrak{o}'$. Then $(\theta(\lambda)v)^{\sigma} = \theta(\lambda)v^{\sigma} = \theta(d)\theta(\lambda)v$. On the other hand, since $\theta(\lambda)v \in \mathfrak{z}$, we have $(\theta(\lambda)v)^{\sigma} = \theta(h_\sigma)\theta(\lambda)v$. Since $\theta(\lambda)v$ generates $\mathfrak{z}_{\mathfrak{l}}$ over $\theta(\mathfrak{o}')$, we see that $d \equiv h_\sigma \bmod \mathfrak{l}$. Thus we obtain $v^{\sigma} = \theta(c)u + \theta(h_\sigma)v$ with $c \in \mathfrak{o}'$. Similarly $v^{\delta\sigma} = \theta(e)u^{\delta} + \theta(g_\sigma)v^{\delta}$ with $e \in \mathfrak{o}'$. In other words, if we define $\mathfrak{l}$-adic representations $\mathfrak{R}_{\mathfrak{l}}^1$ and $\mathfrak{R}_{\mathfrak{l}}^2$ of $\mathrm{Gal}(\bar{\boldsymbol{Q}}/k)$ on A' and on A'^{ε} as in § 7.6, then

$$\mathfrak{R}_{\mathfrak{l}}^1(\sigma) \equiv \begin{bmatrix} g_\sigma & c \\ 0 & h_\sigma \end{bmatrix}, \qquad \mathfrak{R}_{\mathfrak{l}}^2(\sigma) \equiv \begin{bmatrix} h_\sigma & e \\ 0 & g_\sigma \end{bmatrix} \mod \mathfrak{l}.$$

By (7.6.7) and (7.6.15), we have

$$a_p \equiv g_\sigma + h_\sigma, \qquad p \equiv g_\sigma h_\sigma \qquad \mod \mathfrak{l}.$$

This holds for all prime factors $\mathfrak{l}$ of $\mathfrak{c}$. Therefore

$$a_p \equiv g_\sigma + h_\sigma, \qquad p \equiv g_\sigma h_\sigma \qquad \mod \mathfrak{c}. \tag{7.7.12}$$

Now we have

$$u^{\delta\sigma\delta^{-1}} = \theta(h_\sigma)u, \qquad v^{\delta\sigma\delta^{-1}} = \theta(e)u + \theta(g_\sigma)v,$$

$$(u^{\delta})^{\delta^{-1}\sigma\delta} = \theta(g_\sigma)u^{\delta}, \qquad (v^{\delta})^{\delta^{-1}\sigma\delta} = \theta(c)u^{\delta} + \theta(h_\sigma)v^{\delta},$$

so that

$$\mathfrak{R}_{\mathfrak{l}}^1(\delta\sigma\delta^{-1}) \equiv \mathfrak{R}_{\mathfrak{l}}^2(\sigma), \qquad \mathfrak{R}_{\mathfrak{l}}^2(\delta^{-1}\sigma\delta) \equiv \mathfrak{R}_{\mathfrak{l}}^1(\sigma) \qquad \mod \mathfrak{l}. \tag{7.7.13}$$

Therefore

$$\mathfrak{R}_{\mathfrak{l}}^1(\sigma \cdot \delta\sigma\delta^{-1}) \equiv \begin{bmatrix} p & * \\ 0 & p \end{bmatrix} \qquad \mod \mathfrak{l},$$

$$\mathfrak{R}_{\mathfrak{l}}^2(\sigma \cdot \delta^{-1}\sigma\delta) \equiv \begin{bmatrix} p & * \\ 0 & p \end{bmatrix} \qquad \mod \mathfrak{l}.$$

Since $\delta\sigma\delta^{-1} = \delta^{-1}\sigma\delta = \left(\frac{F/k}{\mathfrak{p}^{\varepsilon}}\right)$ on F, we have, if $\tau = \left(\frac{F/k}{(p)}\right)$, then $\tau = \sigma \cdot \delta\sigma\delta^{-1} = \sigma \cdot \delta^{-1}\sigma\delta$, so that $x^{\tau} = px$ for every $x \in \mathfrak{y} + \mathfrak{z} = \mathfrak{x}$.

(III) Let m be a positive integer prime to $N(\mathfrak{c})N$, and $\sigma = \left(\frac{F/k}{(m)}\right)$. We can find a rational prime p, prime to $N(\mathfrak{c})N$, such that $p \equiv m \bmod N(\mathfrak{c})N_{k/\boldsymbol{Q}}(\mathfrak{f})$, where $\mathfrak{f}$ is the finite part of the conductor of F over k. (Note that $N_{k/\boldsymbol{Q}}(\mathfrak{f})$ is divisible only by the prime factors of $N(\mathfrak{c})N$, since every prime ideal in k prime to $N(\mathfrak{c})N$ is unramified in F.) Then $\sigma = \left(\frac{F/k}{(p)}\right)$, and hence, by the results of (I) and (II), $x^\sigma = px = mx$ for every $x \in \mathfrak{x}$. This completes the proof.

COROLLARY 7.31. *Let $\mathfrak{c} \cap \boldsymbol{Z} = q\boldsymbol{Z}$ with a positive integer q, and let ζ be a primitive q-th root of unity. Then $\zeta \in F$, and $F \neq k(\zeta)$.*

PROOF. Let $\mathfrak{p}$ be a prime ideal in k not dividing qN. If $\sigma = \left(\frac{F/k}{\mathfrak{p}}\right) = \mathrm{id.}$, we have $g_\sigma \equiv h_\sigma \equiv 1 \bmod \mathfrak{c}$, so that $N(\mathfrak{p}) \equiv 1 \bmod q\boldsymbol{Z}$. By class field theory, this shows that $k(\zeta) \subset F$. Take a rational prime p not dividing N such that $p \equiv -1 \bmod q\boldsymbol{Z}$, and put $\tau = \left(\frac{F/k}{(p)}\right)$. Then $\tau = \mathrm{id.}$ on $k(\zeta)$, but $s^\tau = -s$ by Th. 7.30, so that $\tau \neq \mathrm{id.}$ on F. Therefore $F \neq k(\zeta)$.

Let $\mathfrak{r}$ denote the ring of all algebraic integers in k, and $\mathfrak{r}_\mathfrak{p}$, for a prime ideal $\mathfrak{p}$ in k, the $\mathfrak{p}$-completion of $\mathfrak{r}$. For every integral ideal $\mathfrak{a}$ in k, define a subgroup $\mathfrak{u}(\mathfrak{a})$ of the idele group $k_A^\times$ of k by

$$\mathfrak{u}(\mathfrak{a}) = \{(x_\mathfrak{p}) \in \textstyle\prod_\mathfrak{p} \mathfrak{r}_\mathfrak{p}^\times \mid x_\mathfrak{p} - 1 \in \mathfrak{r}_\mathfrak{p}\mathfrak{a} \text{ for all } \mathfrak{p}\}\,. \tag{7.7.14}$$

LEMMA 7.32. *Let F be an abelian extension of k, and $\mathfrak{w}$ the subgroup of $k_A^\times$ corresponding to F. Suppose that $\mathfrak{u}(\mathfrak{a}\mathfrak{l}^n) \subset \mathfrak{w}$ with an integral ideal $\mathfrak{a}$ in k, a prime ideal $\mathfrak{l}$ in k, and an integer $n > 1$. Further suppose that $[F:k]$ is prime to $N(\mathfrak{l})$. Then $\mathfrak{u}(\mathfrak{a}\mathfrak{l}) \subset \mathfrak{w}$.*

PROOF. Our assertion follows immediately from the equalities

$$[F:k] = [k_A^\times : \mathfrak{w}] \quad \text{and} \quad [\mathfrak{u}(\mathfrak{a}\mathfrak{l}) : \mathfrak{u}(\mathfrak{a}\mathfrak{l}^n)] = N(\mathfrak{l})^{n-1}\,.$$

Put $q = N(\mathfrak{c})$. We shall obtain further information on the conductor of F under the following set of assumptions:

(7.7.15) (i) *q is a prime;* (ii) *N is square-free;* (iii) *N is prime to $q(q-1)$;* (iv) $\psi(a) = \left(\frac{a}{N}\right)$.

Then $k = \boldsymbol{Q}(\sqrt{N})$. Since $\psi(-1) = 1$, we have $N \equiv 1 \bmod (4)$. By Prop. 7.28, we have $q\mathfrak{r} = \mathfrak{q}\mathfrak{q}^\varepsilon$ with distinct (principal) prime ideals $\mathfrak{q}$ and $\mathfrak{q}^\varepsilon$ in k.

Observe that $\mathfrak{o}'/\mathfrak{c}$ is canonically isomorphic to $\boldsymbol{Z}/q\boldsymbol{Z}$. Therefore the numbers g_σ, h_σ of (7.7.11) can be taken from $\boldsymbol{Z}$, modulo $q\boldsymbol{Z}$. Thus we have an injective homomorphism

$$\mathrm{Gal}(F/k) \ni \tau \longrightarrow (g_\tau, h_\tau) \in (\boldsymbol{Z}/q\boldsymbol{Z})^{\times 2}\,. \tag{7.7.16}$$

It follows that $[F:k]$ divides $(q-1)^2$.

THEOREM 7.33. *Let* $\mathfrak{i}$ *denote the product of the two archimedean primes of* k. *Then the conductor of* F *over* k *is exactly* $\mathfrak{q}\mathfrak{q}^{\varepsilon}\mathfrak{i}$.

PROOF. Let $\mathfrak{f}$ denote the conductor of F over k. On account of Cor. 7.31, $\mathfrak{f}$ must be of the form

$$\mathfrak{f}=\mathfrak{i}\cdot\mathfrak{q}^{a}(\mathfrak{q}^{\varepsilon})^{b}\cdot\textstyle\prod_{\mathfrak{m}|N}\mathfrak{m}^{c}$$

with integers $a>0$, $b>0$, $c\geqq 0$ (c depending on $\mathfrak{m}$), where $\mathfrak{m}$ runs over all prime ideals in k dividing N. Since $[F:k]$ divides $(q-1)^2$, we see, in view of (7.7.15) and Lemma 7.32, that a, b, c are $\leqq 1$, i.e., $\mathfrak{f}$ is of the form $\mathfrak{i}\mathfrak{q}\mathfrak{q}^{\varepsilon}\mathfrak{n}$ with a square-free ideal $\mathfrak{n}$ dividing N. To show that $\mathfrak{n}=\mathfrak{r}$, take any prime ideal $\mathfrak{m}$ dividing N, and let $\mathfrak{m}'$ be the ideal such that $\mathfrak{m}\mathfrak{m}'=\sqrt{N}\cdot\mathfrak{r}$. Let $x\in\mathfrak{r}_{\mathfrak{m}}^{\times}$. Since $\mathfrak{r}/\mathfrak{m}\mathfrak{m}'$ is isomorphic to $\boldsymbol{Z}/N\boldsymbol{Z}$, we can find a positive rational integer y such that $y\equiv x \bmod \mathfrak{m}$, and $y\equiv 1 \bmod q\mathfrak{m}'$. By Th. 7.30, we have $\left(\frac{F/k}{(y)}\right)=1$. Let x' denote the image of x by the natural injection $\mathfrak{r}_{\mathfrak{m}}^{\times}\to k_A^{\times}$. Then

$$[x', k]=[y^{-1}x', k]=\left(\frac{F/k}{il(y^{-1}x')}\right)=\left(\frac{F/k}{(y^{-1})}\right)=1 \qquad (\text{see } \S\, 5.2)\,.$$

This shows that $\mathfrak{r}_{\mathfrak{m}}^{\times}$ is contained in the subgroup of $k_A^{\times}$ corresponding to F, so that $\mathfrak{m}$ is unramified in F. This completes the proof.

Our next task is to investigate whether F is actually the *maximal* ray-class-field of conductor $\mathfrak{i}\mathfrak{q}\mathfrak{q}^{\varepsilon}$. Let u_0 denote the fundamental unit of k, and ν_n the smallest positive integer such that $u_0^{\nu_n}$ is totally positive and $u_0^{\nu_n}\equiv 1 \bmod (\mathfrak{q}\mathfrak{q}^{\varepsilon})^n$. Further let F_n denote the maximal ray class field modulo $\mathfrak{i}\cdot(\mathfrak{q}\mathfrak{q}^{\varepsilon})^n$ over k, i.e., the subfield of k_{ab} corresponding to the subgroup $k^{\times}k_{\infty+}^{\times}\cdot\mathfrak{u}((\mathfrak{q}\mathfrak{q}^{\varepsilon})^n)$ of $k_A^{\times}$, where $\mathfrak{u}(\ )$ is as in (7.7.14). If c_k denotes the class number of k, one has

$$[F_n:k]=2c_k(q-1)^2q^{2n-2}/\nu_n\,. \tag{7.7.17}$$

In the numerical examples for small N (see below), we notice that u_0-1 or u_0^2-1 is divisible by q, according as u_0 is totally positive or not. Observe that $u_0^2-1=u_0\cdot\mathrm{Tr}_{k/\boldsymbol{Q}}(u_0)$, if $N_{k/\boldsymbol{Q}}(u_0)=-1$.

PROPOSITION 7.34. *Suppose that* $N_{k/\boldsymbol{Q}}(u_0)=-1$, *and* u_0^2-1 *is divisible by* q. *Let* q^m *be the highest power of* q *which divides* $\mathrm{Tr}_{k/\boldsymbol{Q}}(u_0)$, *and* $\mathfrak{F}$ *the union of the fields* F_n *for all* n. *(In other words,* $\mathfrak{F}$ *is the largest abelian extension of* k *in which only* $\mathfrak{q}$, $\mathfrak{q}^{\varepsilon}$, *and the archimedean primes are ramified.) Then* $\mathfrak{F}$ *is generated over* F_m *by the* q^n*-th roots of unity for all* n.

PROOF. We have $u_0^2=1+q^m v$ with an algebraic integer v prime to q. Therefore it can easily be shown, by induction on n, that $\nu_{n+m}=2q^n$, and hence $[F_{n+m}:k]=c_k(q-1)^2q^{n+2m-2}$ by (7.7.17). Put $\zeta_n=\exp[2\pi i/q^n]$. Then $k(\zeta_{n+m})\cap F_m=k(\zeta_m)$. In fact, if $k(\zeta_{n+m})\cap F_m$ is larger than $k(\zeta_m)$, then F_m must contain $k(\zeta_{m+1})$. Take a rational prime p so that $p\equiv 1+q^m \bmod (q^{m+1})$ and $\psi(p)=-1$. Then the prime ideal $p\mathfrak{r}$ in k decomposes completely in F_m but not in $k(\zeta_{m+1})$, since $p^2\not\equiv 1 \bmod (q^{m+1})$. Thus $k(\zeta_{n+m})$ and F_m are linearly disjoint over $k(\zeta_m)$, so that

$$[F_m(\zeta_{n+m}):k]=q^n\cdot[F_m:k]=[F_{n+m}:k].$$

Since $F_m(\zeta_{n+m})\subset F_{n+m}$, we obtain $F_m(\zeta_{n+m})=F_{n+m}$, which proves our proposition.

Observe that every element of $(\mathfrak{r}/q\mathfrak{r})^\times$ is represented by a totally positive element of $\mathfrak{r}$. For every totally positive element α of $\mathfrak{r}$ prime to q, consider $\sigma=\left(\frac{F/k}{\alpha\mathfrak{r}}\right)$, and then the element (g_σ, h_σ) of $(\mathfrak{o}'/\mathfrak{c})^{\times 2}$ as in (7.7.11), or of $(\boldsymbol{Z}/q\boldsymbol{Z})^{\times 2}$ as in (7.7.16). Observe also that $\mathfrak{r}/q\mathfrak{r}$ is isomorphic to $(\boldsymbol{Z}/q\boldsymbol{Z})^2$. Thus we obtain a sequence of homomorphisms

(7.7.18) $$(\boldsymbol{Z}/q\boldsymbol{Z})^{\times 2}\longrightarrow(\mathfrak{r}/q\mathfrak{r})^\times\longrightarrow \mathrm{Gal}(F/k)\longrightarrow(\mathfrak{o}'/\mathfrak{c})^{\times 2}\longrightarrow(\boldsymbol{Z}/q\boldsymbol{Z})^{\times 2},$$

$$\alpha\longmapsto\sigma=\left(\frac{F/k}{\alpha\mathfrak{r}}\right)\longmapsto(g_\sigma, h_\sigma).$$

The first and last arrows are isomorphisms, determined up to the change of factors. Recall also that the map $\sigma\longmapsto(g_\sigma, h_\sigma)$ does not depend on the choice of the points s and t. For every $(x, y)\in(\boldsymbol{Z}/q\boldsymbol{Z})^{\times 2}$ with x and y in $(\boldsymbol{Z}/q\boldsymbol{Z})^\times$, denote by $(g(x, y), h(x, y))$ the element of $(\boldsymbol{Z}/q\boldsymbol{Z})^{\times 2}$ corresponding to (x, y) through the composed map of the homomorphisms of (7.7.18). In this way, we obtain a homomorphism

(7.7.19) $$(x, y)\longmapsto(g(x, y), h(x, y))$$

of $(\boldsymbol{Z}/q\boldsymbol{Z})^{\times 2}$ into itself. We can naturally ask the following question (if k is of class number one):

(7.7.20) *Is the map* (7.7.19) *the identity map, up to the change of x and y?*

We shall later show that this is the case at least for $N=29, 53, 61, 73, 89, 97$. First we prove a few simple propositions.

PROPOSITION 7.35. *If $[F:k]=(q-1)^2$, $N_{k/\boldsymbol{Q}}(u_0)=-1$, and the class number of k is one, then $F=F_1$, and u_0^2-1 is divisible by q.*

PROOF. Since $F\subset F_1$ and $\nu_1\geqq 2$, we have

$$[F:k]\leqq[F_1:k]=2(q-1)^2/\nu_1\leqq(q-1)^2.$$

Therefore, if $[F:k]=(q-1)^2$, we have $\nu_1=2$ and $F=F_1$.

PROPOSITION 7.36. *The map* (7.7.19) *has the following properties:*

(a) $g(x, x) = h(x, x) = x$;

(b) $g(x, y)h(x, y) = xy$;

(c) $g(x, y) = h(y, x)$.

PROOF. The first relation follows from Th. 7.30. In (7.7.18), we can take α so that $N_{k/\boldsymbol{Q}}(\alpha)$ is a rational prime p not dividing qN. If $x \equiv \alpha \bmod \mathfrak{q}$ and $y \equiv \alpha \bmod \mathfrak{q}^{\varepsilon}$, then $y \equiv \alpha^{\varepsilon} \bmod \mathfrak{q}$, so that $xy \equiv p \bmod q\boldsymbol{Z}$. From (7.7.12) we obtain $p \equiv g(x, y)h(x, y) \bmod q\boldsymbol{Z}$, hence (b). The proof of Th. 7.30 shows that if $\sigma = \left(\frac{F/k}{\mathfrak{p}}\right)$ and $\tau = \left(\frac{F/k}{\mathfrak{p}^{\varepsilon}}\right)$, then $g_{\sigma} \equiv h_{\tau}$, $h_{\sigma} \equiv g_{\tau} \bmod \mathfrak{c}$. This proves (c).

PROPOSITION 7.37. *There exists a rational integer b such that*

$$g(x, y) = x^{1-b}y^{b}, \qquad h(x, y) = x^{b}y^{1-b}.$$

PROOF. Since $(\boldsymbol{Z}/q\boldsymbol{Z})^{\times}$ is cyclic, we have $g(x, y) = x^{a}y^{b}$, $h(x, y) = x^{b}y^{a}$ with integers a, b, on account of (c) of Prop. 7.36. From (a) of Prop. 7.36, we obtain $a + b \equiv 1 \bmod (q-1)$, hence our assertion.

PROPOSITION 7.38. *The answer to the question* (7.7.20) *is affirmative if and only if*

$$a_p \equiv \mathrm{Tr}_{k/\boldsymbol{Q}}(\alpha) \mod \mathfrak{c} \tag{7.7.21}$$

for every rational prime p not dividing qN, and for every totally positive element α of $\mathfrak{r}$ such that $\psi(p) = 1$ and $N_{k/\boldsymbol{Q}}(\alpha) = p$. Moreover, (7.7.21) *is satisfied by all such p and α, if it is satisfied by at least one α such that $\alpha/\alpha^{\varepsilon}$ generates $(\mathfrak{r}/\mathfrak{q})^{\times}$.*

(By virtue of the generalized Dirichlet theorem, one can always find an element α of $\mathfrak{r}$ such that $N_{k/\boldsymbol{Q}}(\alpha)$ is a rational prime not dividing qN, and $\alpha/\alpha^{\varepsilon}$ generates $(\mathfrak{r}/\mathfrak{q})^{\times}$.)

PROOF. Let $N_{k/\boldsymbol{Q}}(\alpha) = p$ and $\psi(p) = 1$ with a totally positive element α of $\mathfrak{r}$ and a rational prime p. Further let $\sigma = \left(\frac{F/k}{\alpha\mathfrak{r}}\right)$, and $\alpha \equiv x_0$, $\alpha^{\varepsilon} \equiv y_0 \bmod \mathfrak{q}$ with rational integers x_0 and y_0. Then $\mathrm{Tr}_{k/\boldsymbol{Q}}(\alpha) \equiv x_0 + y_0 \bmod q\boldsymbol{Z}$, and

$$g(x_0, y_0) + h(x_0, y_0) \equiv g_{\sigma} + h_{\sigma} \equiv a_p \mod \mathfrak{c}$$

by (7.7.12). Therefore, if (7.7.19) is the identity map, we obtain (7.7.21). Conversely, suppose that (7.7.21) is true for α, and $\alpha/\alpha^{\varepsilon}$ is of order $q-1$ in $(\mathfrak{r}/\mathfrak{q})^{\times}$. Then

$$x_0 + y_0 \equiv g(x_0, y_0) + h(x_0, y_0) \qquad \bmod q\boldsymbol{Z}.$$

By (b) of Prop. 7.36, we may assume, changing x and y if necessary, that

N	$[K:\boldsymbol{Q}]$	K'	$N(\mathfrak{c})$	u_0	p	$\left(\frac{N}{p}\right)$	a_p
29	2	$\boldsymbol{Q}$	5	$\frac{5+\sqrt{29}}{2}$	2	$-$	$\sqrt{-5}$
					3	$-$	$-\sqrt{-5}$
					5	$+$	-3
					7	$+$	2
					11	$-$	$\sqrt{-5}$
					13	$+$	-1
					17	$-$	$-2\sqrt{-5}$
37	2	$\boldsymbol{Q}$	1	$6+\sqrt{37}$	2	$-$	$2i$
					3	$+$	-1
					5	$-$	$-2i$
					7	$+$	3
					11	$+$	-3
					13	$-$	$-6i$
					17	$-$	$2i$
41	2	$\boldsymbol{Q}$	2	$32+5\sqrt{41}$	2	$+$	-1
					3	$-$	$2\sqrt{-2}$
					5	$+$	2
					7	$-$	$-2\sqrt{-2}$
					11	$-$	$2\sqrt{-2}$
					13	$-$	$-4\sqrt{-2}$
53	4	$\boldsymbol{Q}(\sqrt{2})$	7	$\frac{7+\sqrt{53}}{2}$	2	$-$	$\sqrt{-3+\sqrt{2}}$
					7	$+$	$-2-\sqrt{2}$
					11	$+$	$3\sqrt{2}$
					13	$+$	$1-2\sqrt{2}$
					17	$+$	-3
					29	$+$	$-3+3\sqrt{2}$
61	4	$\boldsymbol{Q}(\sqrt{3})$	13	$\frac{39+5\sqrt{61}}{2}$	2	$-$	$\sqrt{-4-\sqrt{3}}$
					3	$+$	$-1-\sqrt{3}$
					5	$+$	$\sqrt{3}$
73	4	$\boldsymbol{Q}(\sqrt{5})$	89	$1068+125\sqrt{73}$	2	$+$	$(-1+\sqrt{5})/2$
					3	$+$	$(1+\sqrt{5})/2$
					5	$-$	$\sqrt{(-19+\sqrt{5})/2}$
89	6	$\boldsymbol{Q}(a_2)$	5	$500+53\sqrt{89}$	2	$+$	$a^3+a^2-3a-1=0$
					3	$-$	$a^6+17a^4+83a^2+125=0$
97	6	$\boldsymbol{Q}(a_2)$	467	$5604+569\sqrt{97}$	2	$+$	$a^3-3a-1=0$
					3	$+$	$a^3-3a-1=0$
					5	$-$	$a^6+27a^4+204a^2+467=0$

$$x_0 \equiv g(x_0, y_0), \qquad y_0 \equiv h(x_0, y_0) \qquad \bmod q\boldsymbol{Z}.$$

If b is as in Prop. 7.37, we have $x_0 \equiv x_0^{1-b} y_0^b \bmod q\boldsymbol{Z}$, so that $(x_0/y_0)^b \equiv 1 \bmod q\boldsymbol{Z}$, hence $b \equiv 0 \bmod (q-1)$. Therefore we have $g(x, y) = x$, $h(x, y) = y$ for all x and y in $(\boldsymbol{Z}/q\boldsymbol{Z})^\times$. This completes the proof.

The table in p. 207 gives the Fourier coefficients a_p for every prime level $N \leqq 97$ with $\psi(x) = \left(\frac{x}{N}\right)$. Note that $S_2(\Gamma_0(N), \psi) = \{0\}$ for all primes $N < 29$. These values a_p have been computed by Doi and Naganuma (by hand) and by Trotter (by computer) by means of the trace-formula of Eichler and Selberg. For each N in the table, the Hecke operators $T'(n)_{2,\psi}$ generate, over $\boldsymbol{Q}$, a field whose degree equals the dimension of $S_2(\Gamma_0(N), \psi)$ over $\boldsymbol{C}$. (This is not necessarily true for larger N.) Therefore we find unique (A, θ), K, f belonging to N and ψ. The uniqueness should be understood as follows: K is unique up to the conjugacy over $\boldsymbol{Q}$; $f = \sum_{n=1}^{\infty} a_n e^{2\pi i n z}$ is unique up to the conjugacy of the a_n over $\boldsymbol{Q}$. We have

$$[K : \boldsymbol{Q}] = 2 \cdot [K' : \boldsymbol{Q}] = \dim (S_2(\Gamma_0(N), \psi)).$$

In the table, the values of a_p, or the irreducible equations for them, are given with respect to a fixed conjugate of K, and a fixed isomorphism θ.

We observe that $u_0^2 - 1 = u_0 \cdot \mathrm{Tr}_{k/\boldsymbol{Q}}(u_0)$ is divisible by $N(\mathfrak{c})$ for all these N. For example, if $N = 61$, we have $a_2 = \sqrt{-4-\sqrt{3}}$, so that $\mathfrak{c} = (-4-\sqrt{3})$ and $N(\mathfrak{c}) = 13$. On the other hand, $u_0 = (39 + 5\sqrt{61})/2$, hence $N_{k/\boldsymbol{Q}}(u_0) = 39$. Now we see that (7.7.9) and (7.7.15) are satisfied for $N = 29, 53, 61, 73, 89, 97$. For these N, the class number of $k = \boldsymbol{Q}(\sqrt{N})$ is one.

THEOREM 7.39. *The following assertions hold (at least) for* $N = 29, 53, 61, 73, 89, 97$.

(1) *The map* (7.7.19) *is the identity map, up to the change of* x *and* y.

(2) $F = F_1$, *i.e.,* F *is the maximal class field over* $k = \boldsymbol{Q}(\sqrt{N})$ *of conductor* $\mathfrak{q}\mathfrak{q}^{\varepsilon}\mathfrak{i}$.

(3) *There is no abelian variety defined over* $\boldsymbol{Q}$ *which is isogenous to* A' *over* $\overline{\boldsymbol{Q}}$.

(4) A' *is simple, and* $\mathrm{End}_{\boldsymbol{Q}}(A') = \theta'(K')$. *(This assertion holds also for* $N = 37, 41$.)

PROOF. We shall discuss here only the case $N = 97$; the other cases can be treated in a similar and simpler way. If $N = 97$, the table shows that $K' = \boldsymbol{Q}(r)$, $r = -(\omega + \omega^{-1})$, $\omega = e^{2\pi i/9}$, $q = 467$. The number r satisfies the equation

$$(*) \qquad X^3 - 3X - 1 = 0$$

which has r, $2 - r^2 = -(\omega^2 + \omega^{-2})$, $r^2 - r - 2 = -(\omega^4 + \omega^{-4})$ as its roots. Since

$N_{K'/Q}(20-r)=17\cdot 467$, there is a unique prime ideal which is a common divisor of $20-r$ and 467. Choosing θ suitably, we may take this prime ideal as $\mathfrak{c}$. The table shows that a_3 is a root of (*). Now one has

$$r\equiv 20\,,\qquad 2-r^2\equiv 69\,,\qquad r^2-r-2\equiv 378 \mod \mathfrak{c}\,.$$

Our theory tells that $X^2-a_pX+p\equiv 0 \bmod \mathfrak{c}$ has roots in $\mathfrak{o}'/\mathfrak{c}$ if $\psi(p)=1$. Therefore a_3 must be r, since the congruence has no solutions if a_3 is $2-r^2$ or r^2-r-2. Solving the congruence $X^2-20X+3\equiv 0 \bmod (467)$, we obtain $(x, y)=(97, 390)$ as solutions. One has $3=\alpha\alpha^{\varepsilon}$ with $\alpha=10+\sqrt{97}$, which is totally positive, and $\mathrm{Tr}_{k/Q}(\alpha)=20\equiv a_3 \bmod \mathfrak{c}$. But 97/390 is of order 233 modulo 467. Therefore the argument of the proof of Prop. 7.38 shows that the exponent b of Prop. 7.37 must be divisible by 233, i.e.,

$$(**)\qquad g(x, y)=\pm x\,,\qquad h(x, y)=\pm y\,.$$

Similarly, checking the solvability of $X^2-a_2X+2\equiv 0 \bmod \mathfrak{c}$, we find that a_2 is either $2-r^2$ or r^2-r-2. If $a_2=r^2-r-2$, the congruence has solutions $(x, y)=(197, 339)$. On the other hand, $2=\beta\beta^{\varepsilon}$ with $\beta=(69+7\sqrt{97})/2$ which is totally positive. Then we observe that (197, 339) does not fit the relation (**). Therefore we must have $a_2=2-r^2$, so that $\mathrm{Tr}_{k/Q}(\beta)=69\equiv a_2 \bmod \mathfrak{c}$. The congruence $X^2-69X+2\equiv 0 \bmod (467)$ has 412 and 433 as solutions. Since 412/433 is a primitive root modulo 467, we obtain the first assertion by virtue of Prop. 7.38. Then, by Prop. 7.35, we have $F=F_1$.

To study the structure of $\mathrm{End}_Q(A')$, we consider the p-th power Frobenius endomorphism φ_p of A' modulo $\mathfrak{p}$, where $\mathfrak{p}$ is a prime ideal in $k=Q(\sqrt{97})$ such that $N(\mathfrak{p})=p$, $\psi(p)=1$. On account of (7.7.7), φ_p satisfies the equation $X^2-a_pX+p=0$, where we identify a_p with $\tilde{\theta}'(a_p)$. If $p=2$, we have $a_2=2-r^2$. Then we see easily that $p=2$ remains prime in K' and decomposes into two prime ideals $\mathfrak{P}$ and $\bar{\mathfrak{P}}$ in the field $K'(\varphi_2)$. It can easily be seen that $K'(\varphi_2)$ is not Galois over Q, and K' is the only non-trivial subfield of $K'(\varphi_2)$. Since φ_2 is divisible by $\mathfrak{P}$ or $\bar{\mathfrak{P}}$, but not by $\mathfrak{P}\bar{\mathfrak{P}}=(2)$, we see that $K'(\varphi_2)=Q(\varphi_2^n)$ for every positive integer n. Now every element of $\mathrm{End}_Q(A' \bmod \mathfrak{p})$ commutes with φ_p^n for sufficiently large n. Therefore, by [81, § 5.1, Prop. 1], we obtain

$$(***)\qquad \mathrm{End}_Q(A' \bmod \mathfrak{p})=Q(\varphi_p^n)=K'(\varphi_p)$$

for $p=2$. If $p=3$, we have a prime ideal $\mathfrak{t}=((1-\omega)(1-\omega^{-1}))$ in K' such that $\mathfrak{t}^3=(3)$, and $\mathfrak{t}$ decomposes into two prime ideals in $K'(\varphi_3)$. Then, by the same reasoning as above we obtain (***) for $p=3$. But $a_2=-(\omega^2+\omega^{-2})\equiv 1 \bmod \mathfrak{t}$, so that $X^2-a_2X+2\equiv 0 \bmod \mathfrak{t}$ is irreducible. It follows that $\mathfrak{t}$ remains prime in $K'(\varphi_2)$. Therefore $K'(\varphi_2)$ is not isomorphic to $K'(\varphi_3)$. This shows that $\mathrm{End}_Q(A')=K'$, and hence A' is simple.

To prove (3), let B be an abelian variety defined over $\boldsymbol{Q}$, and λ an isogeny of A' to B, rational over $\bar{\boldsymbol{Q}}$. Then we can find an isogeny λ' of B to A' such that $\lambda'\lambda=\deg(\lambda)\cdot \mathrm{id}_{A'}$, where $\mathrm{id}_{A'}$ denotes the identity map of A'. Let ξ denote the restriction of $\theta(a_5)$ to A'. Since $a_5 \in \mathfrak{b}_0$, and $N_{K/\boldsymbol{Q}}(a_5)=467$, ξ is an isogeny of A' to A'^{ε} of degree 467. Extend ε to an automorphism δ of $\bar{\boldsymbol{Q}}$. Then $\lambda'\lambda^{\delta}\xi$ is an endomorphism of A', so that $\lambda'\lambda^{\delta}\xi=\theta'(e)$ with an element e of $\mathfrak{o}'$. By (7.6.1), we have

$$N_{K'/\boldsymbol{Q}}(e)^2=\deg(\theta'(e))=\deg(\lambda'\lambda^{\delta}\xi)=467\cdot\deg(\lambda)^6\,,$$

which is a contradiction, since 467 is a prime. This completes the proof.

In the case $N=89$, there is a possibility that one should take 5^3 instead of $q=5$, and consider the congruence

$$a_p\equiv \mathrm{Tr}_{k/\boldsymbol{Q}}(\alpha) \mod \mathfrak{c}^3$$

instead of (7.7.21). Then the coordinates of some points of order 5^3 on A would generate F_3 over k.

Recently W. Casselman [4] has proved that the abelian varieties A' for $N=29, 53, 61, 73, 89, 97$ have good reduction for all prime ideals in $k=\boldsymbol{Q}(\sqrt{N})$. As a consequence of this result, we shall now show that $\zeta(s\,;A'/k)$ is exactly $L(s,f)L(s,f_\rho)=(\sum_{n=1}^{\infty}a_n n^{-s})(\sum_{n=1}^{\infty}a_n^\rho n^{-s})$ if $N=29$. In Th. 7.25, we have seen the coincidence of the Euler factors of $\zeta(s\,;A'/k,K')$ and $L(s,f)L(s,f_\rho)$ for all the prime ideals in k other than $\mathfrak{n}=\sqrt{N}\cdot\mathfrak{r}$. Now we see that A' and A'^{ε} have the same reduction modulo $\mathfrak{n}$. If $a\in\mathfrak{o}$ and $a^\rho=-a$, then $\theta(a)$ defines an isogeny of A' to A'^{ε}. Taking this modulo $\mathfrak{n}$, we obtain an element of $\mathrm{End}(\mathfrak{n}(A'))$, which, together with $\tilde{\theta}'(K')$, generates a subfield $\mathfrak{K}$ of $\mathrm{End}_{\boldsymbol{Q}}(\mathfrak{n}(A'))$ isomorphic to K. Let φ be the Frobenius endomorphism of $\mathfrak{n}(A')$ of degree N. Since the elements of $\mathfrak{K}\cap\mathrm{End}(\mathfrak{n}(A'))$ are defined over the prime field, φ commutes with those elements, so that φ is contained in $\mathfrak{K}$ by virtue of [81, p. 39, Prop. 1]. (If $N=29$, one has $\dim(A')=1$, so that the assertion follows from (5.1.5).) Therefore φ has an element φ_0 of K as an eigen-value. By the Weil theorem, we have $|\varphi_0^\tau|^2=N$ for every isomorphism τ of K into $\boldsymbol{C}$. If $N=29$, we have $K=\boldsymbol{Q}(\sqrt{-5})$, so that the condition $|\varphi_0|^2=29$ implies $\varphi_0=\pm3\pm2\sqrt{-5}$. Put $\alpha=(29+5\sqrt{29})/2=\sqrt{29}\cdot u_0$, and $\sigma=\left(\frac{F/k}{\mathfrak{n}}\right)$. Then α is totally positive, and $\alpha\equiv 2 \mod 5\mathfrak{r}$, so that $(g_\sigma,h_\sigma)\equiv(2,2)\mod(5)$. By virtue of (7.6.7) with (5) as $\mathfrak{l}$, we obtain $\mathrm{Tr}_{K/\boldsymbol{Q}}(\varphi_0)\equiv g_\sigma+h_\sigma\equiv 4 \mod (5)$. It follows that $\varphi_0=-3\pm2\sqrt{-5}$. This agrees with the Fourier coefficient a_N of our cusp form f, given by Hecke [30, pp. 904-905]. Thus we have the exact equality $\zeta(s\,;A'/k)=L(s,f)L(s,f_\rho)$ for the elliptic curve A' in the case $N=29$.

We can also verify that $\mathrm{End}_{\boldsymbol{Q}}(\mathfrak{n}(A'))$ is actually isomorphic to K for the above six values of N.

It need hardly be said that, in this section, we have merely begun the investigation of the mysterious connection between real quadratic fields and the cusp forms of Hecke's "Nebentypus". Also, while the above discussion has been restricted to the case of weight 2, there is some numerical evidence connecting cusp forms of weight > 2 with real quadratic fields in a similar way. The author hopes to treat this question on some other occasion.

7.8. The zeta-function of an abelian variety of *CM*-type

Let A be an abelian variety of dimension n, defined over an algebraic number field k, such that $\mathrm{End}_{\boldsymbol{Q}}(A)$ is isomorphic to a CM-field K of degree $2n$. We shall now determine the zeta-function of A over k.[17] We fix a polarization $\mathcal{C}$ of A and an isomorphism θ of K into $\mathrm{End}_{\boldsymbol{Q}}(A)$, and define couples (K, Φ), (K^*, Φ^*) as in § 5.5. Further we assume (5.5.10) and the following two conditions:

(7.8.1) *All the elements of* $\theta(K) \cap \mathrm{End}(A)$ *and* $\mathcal{C}$ *are rational over* k;

(7.8.2) $K^* \subset k$.

(Actually (7.8.2) follows from (7.8.1), and if A is simple, the converse is true, see [81, § 8.5, Prop. 30]; cf. (5.1.3) when A is an elliptic curve.) Take a $\boldsymbol{Z}$-lattice $\mathfrak{a}$ in K and an isomorphism ξ of $\boldsymbol{C}^n/u(\mathfrak{a})$ onto A as in (5.5.9), where u is defined by (5.5.8). Put

$$\eta(y) = \det(\Phi^*(y)) \qquad (y \in K_A^{*\times}),$$

$$\mu(x) = \eta(N_{k/K^*}(x)) \qquad (x \in k_A^\times).$$

Recall that $\eta(y) \in K_A^\times$ for every $y \in K_A^{*\times}$. Since (K^*, Φ^*) is a CM-type, if we denote by ρ the complex conjugation in $K^\times$ and its obvious extension to $K_A^\times$, we have

$$(7.8.3) \qquad \mu(x)\mu(x)^\rho = N_{k/\boldsymbol{Q}}(x) \qquad (x \in k_A^\times).$$

PROPOSITION 7.40. (1) *Every point of finite order on* A *is rational over* k_{ab}.
(2) *For* $x \in k_A^\times$, *there exists a unique element* α *of* $K^\times$ *such that* $\alpha \cdot \mu(x)^{-1}\mathfrak{a} = \mathfrak{a}$, $\alpha\alpha^\rho = N(il(x))$, *and* $\xi(u(v))^{[x,k]} = \xi(u(\alpha \cdot \mu(x)^{-1}v))$ *for all* $v \in K/\mathfrak{a}$.

The map $x \mapsto \alpha$ defines obviously a homomorphism of $k_A^\times$ into $K^\times$.

---

17) If the reader is interested only in the one-dimensional case, he can simplify the whole discussion by assuming that A is an elliptic curve, K is an imaginary quadratic field, θ is normalized in the sense of § 5.1, $u(a) = a$, $K^* = K$, $\Phi = \Phi^* = \mathrm{id.}$, and $\mu(x) = N_{k/K}(x)$. The polarization can be disregarded; Th. 5.4 can be used instead of Th. 5.15.

PROOF. Let k' be the field generated over k by the coordinates of all the points of finite order on A, i.e., the points $\xi(u(v))$ for all $v \in K/\mathfrak{a}$. For $x \in k_A^\times$, let τ be an element of $\mathrm{Gal}(k'/k)$ such that $\tau = [x, k]$ on $k' \cap k_{ab}$. Put $y = N_{k/K^*}(x)$. Then $\tau = [y, K^*]$ on $k' \cap K^*_{ab}$. By Th. 5.15, there exists an isomorphism ξ' of $C^n/u(\mu(x)^{-1}\mathfrak{a})$ onto A^τ such that $(A^\tau, C^\tau, \theta^\tau)$ is of type $(K, \Phi; \mu(x)^{-1}\mathfrak{a}, N(il(y))\zeta)$ with respect to ξ', and $\xi(u(v))^\tau = \xi'(u(\mu(x)^{-1}v))$ for all $v \in K/\mathfrak{a}$, where ζ is as in Th. 5.15. Since $\tau = \mathrm{id.}$ on k, we have $(A^\tau, C^\tau, \theta^\tau) = (A, C, \theta)$. Therefore we can find a linear transformation T in C^n such that: (i) $T(u(\mu(x)^{-1}\mathfrak{a})) = u(\mathfrak{a})$; (ii) $\xi' = \xi \circ T$; (iii) T commutes with the elements of $\Phi(K)$; (iv) T sends the Riemann form E of (5.5.15) to $N(il(y)) \cdot E$. Then $T = \Phi(\alpha)$ with an element α of $K^\times$, so that $\alpha \cdot \mu(x)^{-1}\mathfrak{a} = \mathfrak{a}$. On account of (iv), we have $\alpha\alpha^\rho = N(il(y)) = N(il(x))$. From (ii), we obtain

$$\xi(u(v))^\tau = \xi'(u(\mu(x)^{-1}v)) = \xi(u(\alpha \cdot \mu(x)^{-1}v)) \qquad (v \in K/\mathfrak{a}).$$

Put, for every positive integer n,

$$W_n = \{w \in K_A^\times \mid w\mathfrak{a} = \mathfrak{a},\ wv = v \text{ for all } v \in n^{-1}\mathfrak{a}/\mathfrak{a}\}. \tag{7.8.4}$$

Let W_1' be the projection of W_1 to the non-archimedean part of $K_A^\times$, and let z be the non-archimedean part of $\alpha \cdot \mu(x)^{-1}$. Then $z \in W_1'$ and $\xi(u(v))^\tau = \xi(u(zv))$ for all $v \in K/\mathfrak{a}$. Obviously the element z of W_1' is uniquely determined by the last equality. Moreover, we see easily that the map $\tau \mapsto z$ defines a homomorphism of $\mathrm{Gal}(k'/k)$ into W_1'. This is injective, since k' is generated by the coordinates of $\xi(u(v))$, and hence τ is completely determined by $\xi(u(v))^\tau$. It follows that $\mathrm{Gal}(k'/k)$ is abelian, which proves (1). Then the above α has the property of (2). The uniqueness of such an α is obvious.

PROPOSITION 7.41. *For $\lambda = 1, \cdots, n$, define a $C^\times$-valued function ψ_λ on $k_A^\times$ by*

$$\psi_\lambda(x) = (\alpha/\mu(x))_\lambda \qquad (x \in k_A^\times)$$

with the element α of (2) of Prop. 7.40, which is unique for x, where $(\ )_\lambda$ denotes the component of an idele at the λ-th archimedean prime of K (with any ordering). Then ψ_λ is a continuous homomorphism of $k_A^\times$ into $C^\times$ trivial on $k^\times$ (i.e., ψ_λ is a Grössen-character of k).

PROOF. It is obvious that ψ_λ is a homomorphism. If $x \in k^\times$, we have $[x, k] = \mathrm{id.}$ in (2) of Prop. 7.40, so that we can put $\alpha = \mu(x)$, hence $\psi_\lambda(x) = 1$. If $x \in k_\infty^\times$, we have again $[x, k] = \mathrm{id.}$, and $\alpha = 1$, so that $\psi_\lambda(x) = (\mu(x)^{-1})_\lambda$. Now take a positive integer $n > 2$, and let $k^{(n)}$ be the field generated over k by the coordinates of $\xi(u(v))$ for all $v \in n^{-1}\mathfrak{a}/\mathfrak{a}$. Since $k^{(n)} \subset k_{ab}$, $k^{(n)}$ corresponds, by class field theory, to an open subgroup Y of $k_A^\times$ containing $k^\times k_\infty^\times$. Let x be an element of Y such that $\mu(x) \in W_n$ and $x_\infty = 1$. Let $\sigma = [x, k]$, and let α be as in (2) of Prop. 7.40. Then $\mathfrak{a} = \alpha\mathfrak{a}$, $\alpha\alpha^\rho = 1$, and if $v \in n^{-1}\mathfrak{a}/\mathfrak{a}$, we have

$$\xi(u(v))=\xi(u(v))^{\sigma}=\xi(u(\alpha\cdot\mu(x)^{-1}v))=\xi(u(\alpha v)),$$

so that $(\alpha-1)\mathfrak{a}\subset n\mathfrak{a}$. Observe that α is a unit of K, and $|\alpha^{\tau}|=1$ for every isomorphism τ of K into $\boldsymbol{C}$, on account of (2) of Prop. 5.11. Therefore α must be a root of unity. Since $n>2$, we have $\alpha=1$, so that $\psi_{\lambda}(x)=1$. This proves the continuity of ψ_{λ} (and also that the kernel of the map $x\mapsto\alpha$ is open), and completes the proof.

We can now attach to ψ_{λ} an L-function of k as follows. (For detailed discussions about such L-functions, the reader is referred to [6] and [99].) For every prime ideal $\mathfrak{p}$ in k, let $k_{\mathfrak{p}}$ denote the $\mathfrak{p}$-completion of k, and $\mathfrak{o}_{\mathfrak{p}}$ the maximal compact subring of $k_{\mathfrak{p}}$. Consider $k_{\mathfrak{p}}$ a subgroup of $k_{A}^{\times}$ in a natural way. Then we say that ψ_{λ} is *unramified at* $\mathfrak{p}$ if $\psi_{\lambda}(\mathfrak{o}_{\mathfrak{p}}^{\times})=1$. This is so for all except a finite number of $\mathfrak{p}$. Then we define the L-function $L(s,\psi_{\lambda})$ by

$$L(s,\psi_{\lambda})=\prod_{\mathfrak{p}}[1-\psi_{\lambda}(c_{\mathfrak{p}})N(\mathfrak{p})^{-s}]^{-1},$$

where the product is taken over all $\mathfrak{p}$ where ψ_{λ} is unramified, and $c_{\mathfrak{p}}$ is a prime element of $k_{\mathfrak{p}}$. Observe that $\psi_{\lambda}(c_{\mathfrak{p}})$ does not depend on the choice of $c_{\mathfrak{p}}$. It is a classical fact, first proved by Hecke, that $L(s,\psi_{\lambda})$ can be holomorphically continued to the whole s-plane and satisfies a functional equation.

THEOREM 7.42. *The notation being as above, ψ_{λ} is unramified at $\mathfrak{p}$ if and only if A has good reduction modulo $\mathfrak{p}$. Further the zeta-function of A over k coincides exactly with the product*

$$\prod_{\lambda=1}^{n}L(s,\psi_{\lambda})L(s,\bar{\psi}_{\lambda}).$$

PROOF. Let $\mathfrak{p}$ and $c_{\mathfrak{p}}$ be as above, and $\sigma=[c_{\mathfrak{p}},k]$. Suppose that A has good reduction modulo $\mathfrak{p}$. Define $\varphi_{\mathfrak{p}}$, R_{l}', $\mathfrak{R}_{l}$, and $\mathfrak{K}(l)$ for every rational prime l as in § 7.6 (with $\boldsymbol{Q}$ and l as F and $\mathfrak{l}$). Suppose that $\mathfrak{p}$ is prime to l. By Prop. 7.23, $\mathfrak{p}$ is unramified in $\mathfrak{K}(l)$. Since $\mathfrak{K}(l)\subset k_{ab}$, we see that σ induces a Frobenius element of $\mathrm{Gal}(\mathfrak{K}(l)/k)$ for $\mathfrak{p}$. Therefore we have $\mathfrak{R}_{l}(\sigma)=R_{l}'(\varphi_{\mathfrak{p}})$ by (7.6.7). If α is defined for $c_{\mathfrak{p}}$ by (2) of Prop. 7.40, we have $\xi(u(v))^{\sigma}=\xi(u(\alpha\cdot\mu(c_{\mathfrak{p}})^{-1}v))$ for all $v\in K/\mathfrak{a}$. Since the l-component of $c_{\mathfrak{p}}$ is 1, we have $\xi(u(v))^{\sigma}=\theta(\alpha)\cdot\xi(u(v))$ for all $v\in l^{-n}\mathfrak{a}/\mathfrak{a}$, $n=1,2,\cdots$. It follows that $\varphi_{\mathfrak{p}}=\tilde{\theta}(\alpha)$. Therefore, if X is an indeterminate,

$$\begin{aligned}\det[1-R_{l}'(\varphi_{\mathfrak{p}})X]&=\prod_{\lambda=1}^{n}(1-(\alpha)_{\lambda}X)(1-(\bar{\alpha})_{\lambda}X)\\&=\prod_{\lambda=1}^{n}(1-\psi_{\lambda}(c_{\mathfrak{p}})X)(1-\bar{\psi}_{\lambda}(c_{\mathfrak{p}})X),\end{aligned}$$

hence our proof is completed if we prove the first assertion. For that purpose, we use the result due to Serre and Tate [66]:

(7.8.5) $\mathfrak{p}$ *is unramified in* $\mathfrak{K}(l)$ *if and only if A has good reduction modulo* $\mathfrak{p}$.

Let $y \in \mathfrak{o}^\times$, and let α be an element of $K^\times$ determined for this y as in Prop. 7.40. Further let $H_l = \bigcup_{m=0}^{\infty} l^{-m}\mathfrak{a}$. Suppose that A has good reduction modulo $\mathfrak{p}$. By (7.8.5), we have $[y, k] = \text{id.}$ on $\mathfrak{R}(l)$, so that $\xi(u(v)) = \xi(u(v))^{[y,k]} = \xi(u(\alpha \cdot \mu(y)^{-1}v))$ for all $v \in H_l/\mathfrak{a}$. Then the l-component of $\alpha \cdot \mu(y)^{-1}$ is equal to 1, so that $\alpha = 1$, hence $\psi_\lambda(y) = 1$. Conversely, if ψ_λ is unramified at $\mathfrak{p}$, then $\psi_\lambda(y) = 1$, so that $\alpha = \mu(y)_\lambda = 1$. Therefore $\xi(u(v))^{[y,k]} = \xi(u(v))$ for all $v \in H_l/\mathfrak{a}$, i.e., $[y, k] = \text{id.}$ on $\mathfrak{R}(l)$ for all $y \in \mathfrak{o}_\mathfrak{p}^\times$. By (7.8.5), A has good reduction modulo $\mathfrak{p}$. This completes the proof.

For further discussion about the conductor of ψ_λ, the reader is referred to Deuring [12] (in the one-dimensional case), and Serre and Tate [66] (in the general case).

THEOREM 7.43. *The notation being as above, let F be the maximal real subfield of K, and $\mathfrak{o}_F$ the maximal order of F. Suppose that $\theta(\mathfrak{o}_F) \subset \mathrm{End}(A)$, and the natural injection $K \to \boldsymbol{C}$ coincides with the map $\alpha \mapsto (\alpha)_1$ of Prop. 7.41. Then*

$$\zeta(s\,;\ A/k, F) = L(s, \psi_1)L(s, \bar{\psi}_1)\,.$$

For the definition of $\zeta(s\,;\ A/k, F)$, see § 7.6. The assumption about $\theta(\mathfrak{o}_F)$ is not essential, since we can always find a model satisfying this condition by changing A by an isogeny over k.

PROOF. With the same notation as in the above proof, take a prime ideal $\mathfrak{l}$ in F dividing l, and define $R'_\mathfrak{l}$ as in § 7.6. Since $\varphi_\mathfrak{p} = \tilde{\theta}(\alpha)$, we have, on account of Prop. 7.21,

$$\begin{aligned}(7.8.6)\qquad \det[1 - R'_\mathfrak{l}(\varphi_\mathfrak{p})X] &= (1-\alpha X)(1-\bar{\alpha}X) \\ &= [1-\psi_1(c_\mathfrak{p})X][1-\bar{\psi}_1(c_\mathfrak{p})X]\,,\end{aligned}$$

hence our theorem.

Let k' be a subfield of k containing K^*. If (A, C, θ) is rational over k', we can define characters ψ'_λ of $k'^\times_A$ in the same manner as above. Then it is easy to verify

$$(7.8.7)\qquad \psi_\lambda = \psi'_\lambda \circ N_{k/k'}\,.$$

We can actually prove a stronger result as follows:

THEOREM 7.44. *The notation being as above, let M be a subfield of k containing K^*. Then the following two conditions are equivalent:*

(1) *There exists a continuous homomorphism φ of $M_A^\times$ into $\boldsymbol{C}^\times$ trivial on $M^\times$ (i.e., a Größen-character of M) such that $\psi_\lambda = \varphi \circ N_{k/M}$.*

(2) *All the points of A of finite order are rational over $M_{ab} \cdot k$.*

Moreover, if these conditions are satisfied, the number of characters φ as in (1), *for a fixed λ, is exactly* $[M_{ab} \cap k : M]$.

We note that the case $M = K^*$ is most interesting, see the discussion after the proof.

PROOF. Assume the existence of φ as in (1). Let $\sigma \in \mathrm{Aut}(\boldsymbol{C}/M_{ab} \cdot k)$. Take $z \in k_A^\times$ so that $\sigma = [z, k]$ on k_{ab}, and put $s = N_{k/M}(z)$. Since $\sigma = \mathrm{id}.$ on M_{ab}, s is contained in the closure of $M^\times M_\infty^\times$. We can find an open subgroup T of the finite part of $M_A^\times$ so that $\varphi(T) = 1$. Then $s \in M^\times M_\infty^\times T$, so that $s = \beta r t$ with $\beta \in M^\times$, $r \in M_\infty^\times$, and $t \in T$. Since $N_{k/M}(k_\infty^\times) = M_\infty^\times$, we have $r = N_{k/M}(y)$ for some $y \in k_\infty^\times$. Put $x = zy^{-1}$, and define α as in Prop. 7.40 for this x. Then $(\alpha/\mu(x))_\lambda = \psi_\lambda(x) = \varphi(\beta t) = 1$, since $\varphi(M^\times T) = 1$. Put

$$\mu'(a) = \eta(N_{M/K^*}(a)) \qquad \text{for} \quad a \in M_A^\times .$$

Then $\mu(x) = \mu'(N_{k/M}(x)) = \mu'(\beta t)$. Since $\mu'(t)_\lambda = 1$ and $\mu'(\beta) \in K^\times$, we obtain $\mu'(\beta) = \alpha$. Therefore $\alpha/\mu(x) = \mu'(t)^{-1}$. Since $\sigma = [x, k]$ on k_{ab}, we have $\xi(u(v))^\sigma = \xi(u(\mu'(t)^{-1}v))$ for all $v \in K/\mathfrak{a}$. Now we can replace T by any of its open subgroups, especially by

$$T_v = \{w \in T \mid \mu'(w)v = v\}$$

for any fixed $v \in K/\mathfrak{a}$. Then we see that $\xi(u(v))$ is invariant under σ for every $v \in K/\mathfrak{a}$, which implies (2).

Conversely, suppose that (2) is satisfied, and put $S = M^\times \cdot N_{k/M}(k_A^\times)$. Then S is the subgroup of $M_A^\times$ corresponding to $M_{ab} \cap k$ by class field theory. Let $\sigma = [s, M]$ with any $s \in S$. Then $\sigma = \mathrm{id}.$ on $M_{ab} \cap k$, so that σ can be extended uniquely to an element τ of $\mathrm{Gal}(M_{ab} \cdot k/k)$. By the assumption (2), $\xi(u(v))^\tau$ is meaningful for every $v \in K/\mathfrak{a}$. We can therefore repeat the proof of Prop. 7.40, with $\mu'(s)$ and $N(il(s))$ in place of $\mu(x)$ and $N(il(x))$. Then we obtain an element α of $K^\times$ such that

$$\alpha\alpha^\rho = N(il(s)), \qquad \alpha \cdot \mu'(s)^{-1}\mathfrak{a} = \mathfrak{a},$$

$$\xi(u(v))^\tau = \xi(u(\alpha \cdot \mu'(s)^{-1}v)) \qquad (v \in K/\mathfrak{a}).$$

Obviously α is unique for s. Define $\varphi_\lambda : S \to \boldsymbol{C}^\times$ by

$$\varphi_\lambda(s) = (\alpha/\mu'(s))_\lambda .$$

By the same reasoning as in the proof of Prop. 7.41, we can show that φ_λ is trivial on $M^\times$, and $\varphi_\lambda(s) = (\mu'(s)^{-1})_\lambda$ for $s \in M_\infty^\times$. Now define $k^{(n)}$ as in the proof of Prop. 7.41. By our assumption, we have $k^{(n)} \subset M_{ab} \cdot k$. Let U be the open subgroup of $M_A^\times$ corresponding to $k^{(n)} \cap M_{ab}$. Then $U \subset S$. Let s be an element of U such that $s_\infty = 1$ and $\mu'(s) \in W_n$, where W_n is as in (7.8.4). Then, by the same argument as in the proof of Prop. 7.41, we can show that $\varphi_\lambda(s) = 1$, which proves the continuity of φ_λ. By our definition of φ_λ, we have $\psi_\lambda = \varphi_\lambda \circ N_{k/M}$. Since $S = M^\times \cdot N_{k/M}(k_A^\times)$, the homomorphism $\varphi_\lambda : S \to \boldsymbol{C}^\times$ is completely determined by ψ_λ. Therefore our problem is reduced to the possibility of extending φ_λ to $M_A^\times$. This can be settled by the following lemma, on account of the equality $[M_A^\times : S] = [M_{ab} \cap k : M]$.

LEMMA 7.45. *Let G be a commutative topological group, H an open subgroup of G of finite index, and φ a continuous homomorphism of H into $\boldsymbol{C}^\times$. Then there are exactly $[G : H]$ continuous homomorphisms of G into $\boldsymbol{C}^\times$ which coincide with φ on H.*

PROOF. Decompose G/H into the product of finite cyclic groups $P_1, \cdots, P_r$ of order $m_1, \cdots, m_r$, respectively, and take, for each i, an element a_i of G which generates P_i modulo H. Let c_i be any m_i-th root of $\varphi(a_i^{m_i})$, and put

$$\varphi'(ha_1^{e_1} \cdots a_r^{e_r}) = \varphi(h)c_1^{e_1} \cdots c_r^{e_r} \qquad (h \in H,\ e_i \in \boldsymbol{Z}).$$

It is now easy to verify that φ' is a well-defined continuous homomorphism of G into $\boldsymbol{C}^\times$, and $\varphi' = \varphi$ on H. It is also clear that the number of such extensions is $[G : H]$, and any extension of φ to G can be obtained in such a manner.

Let us now show that, for any given A, there is a model which is isomorphic to A over $\bar{\boldsymbol{Q}}$, and satisfies the conditions of Th. 7.44 with K^* as M. For a given (A, C, θ), we can always find points $t_1, \cdots, t_r$ of A of finite order so that the structure

$$Q = (A, C, \theta\, ;\ t_1, \cdots, t_r)$$

has no automorphism other than the identity map. With any such t_i, let k' be the field of moduli of Q (see p. 130). Then there exists a structure

$$Q' = (A', C', \theta'\, ;\ t_1', \cdots, t_r')$$

which is isomorphic to Q and defined over k'; moreover, such a Q' is unique up to isomorphisms rational over k' (see [75, II, 1.5]). By Cor. 5.16, $k' \subset K_{ab}^*$. Moreover, we have

(7.8.8) *All the points of A' of finite order are rational over K_{ab}^*.*

To see this, take an isomorphism $\xi' : \boldsymbol{C}^n/u(\mathfrak{a}) \to A'$ so that (A', C', θ') is of type $(K, \Phi\, ;\ \mathfrak{a}, \zeta)$ with respect to ξ'. Let $\sigma \in \mathrm{Aut}(\boldsymbol{C}/K_{ab}^*)$. Apply Th. 5.15 to Q' with $s = 1$. Then we find an isomorphism ξ'' of $\boldsymbol{C}^n/u(\mathfrak{a})$ to A' such that (A', C', θ') is of type $(K, \Phi\, ;\ \mathfrak{a}, \zeta)$ with respect to ξ'', and $\xi'(u(v))^\sigma = \xi''(u(v))$ for all $v \in K/\mathfrak{a}$. Then we obtain an automorphism γ of A' such that $\xi'' = \gamma \circ \xi'$. It can easily be seen that γ is an automorphism of Q', so that $\gamma = 1$. It follows that $\xi' = \xi''$, hence $\xi'(u(v))$ is invariant under σ for every $v \in K/\mathfrak{a}$, which proves (7.8.8).

Thus A' and k' satisfy the condition (2) of Th. 7.44 with K^* as M. (We know even that $k' \subset K^*_{ab}$.)

Under certain circumstances, we can take k' to be the field of moduli of $(A, \mathcal{C}, \theta)$. For example, assume the following set of conditions:

(7.8.9) (i) $\mathrm{End}(A) \cap \theta(K) = \theta(\mathfrak{o}_K)$ *with the maximal order* $\mathfrak{o}_K$ *in* K; (ii) $\mathfrak{o}_K$ *has no roots of unity other than* ± 1; (iii) $\mathfrak{o}_K$ *has a prime ideal* $\mathfrak{h}$ *such that* $N(\mathfrak{h}) = 3$.

Take an element b of K that generates $\mathfrak{h}^{-1}\mathfrak{a}/\mathfrak{a}$, and put $t = \xi(u(b))$. Let γ be an automorphism of $(A, \mathcal{C}, \theta; t)$. Then $\gamma = \theta(\varepsilon)$ with a root of unity ε contained in $\mathfrak{o}_K$, and $t = \gamma t$, so that $\varepsilon b \equiv b \bmod \mathfrak{a}$. Since $\varepsilon = \pm 1$ and b is of order 3, we have $\varepsilon = 1$. Therefore $(A, \mathcal{C}, \theta; t)$ has no automorphism other than the identity map. On the other hand, we have:

(7.8.10) $(A, \mathcal{C}, \theta; t)$ *and* $(A, \mathcal{C}, \theta)$ *have the same field of moduli*.

To see this, let σ be an element of $\mathrm{Aut}(\boldsymbol{C})$ which is the identity map on the field of moduli of $(A, \mathcal{C}, \theta)$. Then there is an isomorphism δ of $(A, \mathcal{C}, \theta)$ to $(A^\sigma, \mathcal{C}^\sigma, \theta^\sigma)$. We see that the point t has the property

$$\mathfrak{h} = \{\alpha \in \mathfrak{o}_K \mid \theta(\alpha)t = 0\},$$

and $\pm t$ are the only points of A satisfying this condition. Therefore $\delta^{-1}t^\sigma = \pm t$. It follows that either δ or $-\delta$ gives an isomorphism of $(A, \mathcal{C}, \theta; t)$ to $(A^\sigma, \mathcal{C}^\sigma, \theta^\sigma; t^\sigma)$, hence (7.8.10).

Now, by the principle explained above, we obtain a structure $(A', \mathcal{C}', \theta'; t')$ which is isomorphic to $(A, \mathcal{C}, \theta; t)$ and defined over the field of moduli of $(A, \mathcal{C}, \theta)$, and which satisfies the condition (2) of Th. 7.44 with K^* as M.

In particular, if A is an elliptic curve and j its invariant, then the field of moduli of $(A, \mathcal{C}, \theta)$ is $K(j)$. In this case, the number of characters of $K_A^\times$ as in (1) of Th. 7.44 is exactly $[K(j):K]$, the class number of K. For $K = \boldsymbol{Q}(\sqrt{-d})$ with d square-free, the condition (iii) of (7.8.9) is satisfied if and only if $d \not\equiv 1 \bmod (3)$.

Next let us give an example for which (2) of Th. 7.44 is not satisfied. We have just shown the existence of an elliptic curve E, defined over $K(j_E)$, whose points of finite order are all rational over K_{ab}. Take an elliptic curve E' defined over $K(j_E)$ and isomorphic to E over $\bar{\boldsymbol{Q}}$. Suppose that E' also satisfies (2) of Th. 7.44 with K as M, i.e., all the points of E' of finite order are rational over K_{ab}. Then we see easily that any isomorphism λ of E to E' is rational over K_{ab}. But this is not always the case, since the smallest field of definition for λ containing $K(j_E)$ is not necessarily contained in K_{ab}. (For example take any μ such that $\mu^2 \in K(j_E)$ and $\mu \notin K_{ab}$, and define an isomorphism λ as in Prop. 4.1.) Thus E' cannot satisfy (2) of Th. 7.44 with K as M, for such a choice of λ.

For an elliptic curve E with complex multiplications, Deuring [12, IV] determined the zeta-function of E over a field which does not contain the imaginary quadratic field in question. We shall now generalize this result in the following form:

THEOREM 7.46. *The notation being as in Th. 7.43, let k_0 be an algebraic number field of finite degree, over which $(A, \mathcal{C})$ is rational. Suppose that A is simple, $\theta(\mathfrak{o}_F) \subset \mathrm{End}(A)$, every element of $\theta(\mathfrak{o}_F)$ is rational over k_0, and $k_0 \cap K^*$ is the maximal real subfield of K^*. Define the characters ψ_λ of $(k_0K^*)_A^\times$ as above with k_0K^* as k. Then $\zeta(s; A/k_0, F)$ coincides, up to finitely many Euler factors, with $L(s, \psi_1)$. More precisely, for almost all primes $\mathfrak{q}$ of k_0, the Euler $\mathfrak{q}$-factor of $\zeta(s; A/k_0, F)$ is the product of the Euler $\mathfrak{p}$-factors of $L(s, \psi_1)$ for the prime factors $\mathfrak{p}$ of $\mathfrak{q}$ in k_0K^*.*

Note that every element of $\mathrm{End}(A)$ is rational over k_0K^*, see [81, § 8.5, Prop. 30]. A typical example is the case where A is an elliptic curve, and $k_0 = \boldsymbol{Q}(j)$ (see (ii) of Th. 5.7). One has $K^* = K$ and $k_0K^* = K(j)$ in this case. The "bad Euler factors" will be discussed after the proof.

PROOF. Put $k = k_0K^*$. Then $[k : k_0] = 2$. Let ρ denote the complex conjugation, and τ an element of $\mathrm{Gal}(k_{ab}/k_0)$ which is non-trivial on k. Since $\theta(K) = \mathrm{End}_{\boldsymbol{Q}}(A)$, we can define an automorphism ε of K by $\theta(a)^\tau = \theta(a^\varepsilon)$. We have $\tau = \rho$ on K^*, so that

$$\mathrm{tr}\, \Phi(a^\varepsilon) = \mathrm{tr}\, \Phi(a)^\tau = \mathrm{tr}\, \Phi(a)^\rho = \mathrm{tr}\, \Phi(a^\rho) \qquad (a \in K).$$

Since A is simple, this implies that $\varepsilon = \rho$ on K, by virtue of [81, § 8.2, Prop. 26]. Therefore $\theta(a)^\tau = \theta(a^\rho)$. Put $\xi_0 = \xi \circ u$. Then τ^{-1} induces an automorphism of the module $\xi_0(K/\mathfrak{a})$, which is semi-linear with respect to the action of $\theta(a)$. Therefore, $w \mapsto \xi_0^{-1}(\xi_0(w^\rho)^{\tau^{-1}})$ is an isomorphism of $K/\mathfrak{a}^\rho$ to $K/\mathfrak{a}$, which is linear with respect to the action of the elements of the order of $\mathfrak{a}$. (Note that $\mathfrak{a}$ and $\mathfrak{a}^\rho$ have the same order, on account of the assumption $\theta(\mathfrak{o}_F) \subset \mathrm{End}(A)$.) Thus we obtain an element z of $K_A^\times$ such that $z\mathfrak{a}^\rho = \mathfrak{a}$, and $\xi_0(zw)^\tau = \xi_0(w^\rho)$ for all $w \in K/\mathfrak{a}^\rho$, i. e., $\xi_0(v) = \xi_0(zv^\rho)^\tau$ for all $v \in K/\mathfrak{a}$. Now, for every $x \in k_A^\times$, x^τ is a meaningful element of $k_A^\times$, and $\tau[x^\tau, k] = [x, k]\tau$. Note also that $\mu(x^\tau) = \mu(x)^\rho$. Define α as in Prop. 7.40. Then

$$\begin{aligned}\xi_0(v)^{[x^\tau, k]} &= \xi_0(zv^\rho)^{\tau[x^\tau, k]} = \xi_0(zv^\rho)^{[x, k]\tau} \\ &= \xi_0(\alpha \cdot \mu(x)^{-1}zv^\rho)^\tau = \xi_0(\alpha^\rho \cdot \mu(x^\tau)^{-1}v) \qquad (v \in K/\mathfrak{a}).\end{aligned}$$

Further $\alpha^\rho\alpha = N(il(x^\tau))$, and $\alpha^\rho \cdot \mu(x^\tau)^{-1}\mathfrak{a} = \mathfrak{a}$. Therefore α^ρ is the element of $K^\times$ corresponding to x^τ, hence

$$\psi_\lambda(x^\tau) = \psi_\lambda(x)^\rho \,. \tag{7.8.11}$$

Let $\mathfrak{p}$, $c_\mathfrak{p}$, $\tilde{A}$, $\varphi_\mathfrak{p}$, and R_l' be as in the proof of Th. 7.42 and Th. 7.43, assuming that A has good reduction modulo $\mathfrak{p}$, hence ψ_λ is unramified at $\mathfrak{p}$. Let $\mathfrak{q}$ be the restriction of $\mathfrak{p}$ to k_0, and $\varphi_\mathfrak{q}$ the Frobenius endomorphism of $\tilde{A}$ of degree $N(\mathfrak{q})$. Suppose that $\mathfrak{p} \neq \mathfrak{p}^\tau$. We can take $c_\mathfrak{p}^\tau$ as $c_{\mathfrak{p}^\tau}$. Since $\varphi_\mathfrak{p} = \varphi_\mathfrak{q}$ in this case, we have, by (7.8.6) and (7.8.11),

$$(*) \qquad \det[1 - R_l'(\varphi_\mathfrak{q})X] = [1 - \psi_1(c_\mathfrak{p})X][1 - \bar{\psi}_1(c_\mathfrak{p})X]$$
$$= [1 - \psi_1(c_\mathfrak{p})X][1 - \psi_1(c_\mathfrak{p}^\tau)X] \,.$$

Next suppose that $\mathfrak{p} = \mathfrak{p}^\tau$ and $N(\mathfrak{p}) = N(\mathfrak{q})^2$. Put $\alpha = \psi(c_\mathfrak{p})$. Then $\alpha = \psi(c_\mathfrak{p}) = \psi(c_\mathfrak{p}^\tau) = \alpha^\rho$, so that $\alpha \in F$. Now we have $\varphi_\mathfrak{q}^2 = \varphi_\mathfrak{p}$, so that, by (7.8.6),

$$\det[1 - R_l'(\varphi_\mathfrak{q}^2)X] = (1 - \alpha X)^2 \,.$$

Let $\sigma = \left(\frac{k/k_0}{\mathfrak{q}}\right)$. Then we have $\theta(a)^\sigma = \theta(a^\rho)$, so that $\varphi_\mathfrak{q}$ does not commute with $\tilde{\theta}(a)$ for $a \in K$, $\notin F$. It follows that $R_l'(\varphi_\mathfrak{q})$ is not a scalar matrix, hence

$$(**) \qquad \det[1 - R_l'(\varphi_\mathfrak{q})X] = 1 - \alpha X^2 = 1 - \psi_1(c_\mathfrak{p})X^2 \,.$$

Taking the product of $(*)$ and $(**)$ with $X = N(\mathfrak{q})^{-s}$ for all "good" $\mathfrak{q}$, we obtain our assertion.

It remains to discuss the "bad Euler factors", for which the last statement of our theorem does not hold. In view of Th. 7.42, it is sufficient to consider the primes $\mathfrak{q}$ of k_0 such that ψ_1 is unramified at the prime factors of $\mathfrak{q}$ in k. The above discussion shows that the bad factors may occur for the primes of k_0 ramified in k. Other primes are actually "good". In fact, one has

PROPOSITION 7.47. *The notation and the assumptions being as in Th.* 7.46, *let* $\mathfrak{q}$ *be a prime ideal in* k_0. *Then* A *has good reduction modulo* $\mathfrak{q}$ *if and only if* $\mathfrak{q}$ *is unramified in* k, *and* A *has good reduction modulo the prime factors of* $\mathfrak{q}$ *in* k. *The last statement of Th.* 7.46 *holds for such a prime* $\mathfrak{q}$.

PROOF. Let $\mathfrak{K}_0(l)$ (resp. $\mathfrak{K}(l)$) be the field generated over k_0 (resp. k) by the coordinates of the points of A of order l^m for all positive integers m. We see easily that every element of End(A) is defined over $\mathfrak{K}_0(l)$. By [81, § 8.5, Prop. 30], $K^* \subset \mathfrak{K}_0(l)$, hence $\mathfrak{K}_0(l) = \mathfrak{K}(l)$. Our assertions follow directly from this fact and the result of Serre and Tate [66] (see (7.8.5)).

7.9. Supplementary remarks

A. Change of model and the field of definition

In § 7.5, we have determined the zeta-function of a special model V_S of $\Gamma'\backslash\mathfrak{H}^*$ over $\boldsymbol{Q}$. Actually there exist curves V defined over algebraic number fields k of finite degree birationally equivalent to V_S over $\bar{\boldsymbol{Q}}$, but not necessarily over k. Therefore one can naturally ask about the determination of the zeta-function of any such V over k. The same question may be asked for the abelian varieties A_S, or its factors A, A' considered in §§ 7.5–7.6. The complete solution of this question seems rather difficult. We shall discuss here only special cases.

(I) Let S, V_S, and A_S be as in §§ 7.3–7.5. Let k be a finite abelian extension of $\boldsymbol{Q}$ of conductor (r), and let $m=[k:\boldsymbol{Q}]$. Then there are m characters $\chi_1, \cdots, \chi_m$ of $(\boldsymbol{Z}/N\boldsymbol{Z})^\times$ such that

$$(1-u^f)^{m/f}=\prod_{i=1}^m (1-\chi_i(p)u)$$

for every rational prime p, not dividing r, which decomposes into m/f prime ideals in k, where u is an indeterminate. If $\mathfrak{p}$ denotes such a prime ideal in k, $\varphi_\mathfrak{p}$ (resp. π_p) denotes the Frobenius endomorphism of $\tilde{A}_S$ of degree $N(\mathfrak{p})$ (resp. p), and R'_l denotes the l-adic representation of $\mathrm{End}(\tilde{A}_S)$, then one has

$$\prod_{\mathfrak{p}|p} \det[1-u^f R'_l(\varphi_\mathfrak{p})]=\prod_{i=1}^m \det[1-u\cdot\chi_i(p)R'_l(\pi_p)]\,.$$

Therefore, if we put, for $f(z)=\sum_{n=1}^\infty a_n e^{2\pi i n z/t}\in S_k(\Gamma'')$,

$$L(s, f, \chi)=\sum_{n=1}^\infty a_n\cdot\chi(n)n^{-s}$$

as in § 3.6, and if $\{h_1, \cdots, h_\kappa\}$ is as in (7.5.4′), then the zeta-function of V_S (or A_S) over k coincides, up to a finite number of Euler factors, with the product

$$\prod_{i=1}^m \prod_{\nu=1}^\kappa L(s, h_\nu, \chi_i)\,,$$

which is holomorphic on the whole s-plane, and satisfies a functional equation, on account of Remark 3.58, Prop. 3.64, and Th. 3.66.

(II) In the next place, consider an arbitrary quadratic extension k of $\boldsymbol{Q}$, of conductor (r). By virtue of a result of Weil [94], one can construct an abelian variety B_S defined over $\boldsymbol{Q}$ and an isomorphism λ of A_S onto B_S defined over k such that $\lambda^\sigma=-\lambda$ for the generator σ of $\mathrm{Gal}(k/\boldsymbol{Q})$. The couple (B_S, λ) is unique up to isomorphisms over $\boldsymbol{Q}$. If ψ_p is the Frobenius endomorphism of $\tilde{B}_S$ of degree p, we have, for almost all p, $\psi_p\tilde{\lambda}=\chi(p)\tilde{\lambda}\pi_p$, where χ is the character of $(\boldsymbol{Z}/r\boldsymbol{Z})^\times$ corresponding to k. Therefore the zeta-function of B_S over $\boldsymbol{Q}$ coincides, up to a finite number of Euler factors, with the product

$$\prod_{\nu=1}^{\kappa} L(s, h_\nu, \chi),$$

which is holomorphic on the whole s-plane, and satisfies a functional equation. We can of course make a similar consideration by taking a factor A of A_S, considered in Th. 7.14, in place of A_S.

B. Rational points of an elliptic curve

The group of rational points of an elliptic curve defined over an algebraic number field, a function field, or a local field, has been a subject of extensive study. An excellent survey of this topic is given by Cassels [5], in which the reader can find references up to 1966. Here we content ourselves with mentioning only

THE CONJECTURE OF BIRCH AND SWINNERTON-DYER [1]: *If the zeta-function $\zeta(s; E/\boldsymbol{Q})$ of an elliptic curve E defined over $\boldsymbol{Q}$ has a zero of order $h \geqq 0$ at $s=1$, then the group of rational points of E over $\boldsymbol{Q}$ has rank h.*

They verified the conjecture for many curves, especially for curves of type $y^2 = x^3 - Dx$.

If E is a curve V_S of genus 1 isomorphic to $\mathfrak{H}^*/\Gamma_0(N)$ for N belonging to the values of (7.5.6), then $\zeta(s; E/\boldsymbol{Q})$ is, possibly up to bad factors, given by

$$\sum_{n=1}^{\infty} a_n n^{-s} = \Gamma(s)^{-1}(2\pi)^s \int_0^{\infty} f(iy) y^{s-1} dy$$

with an element $f(z) = \sum_{n=1}^{\infty} a_n e^{2\pi i n z}$ of $S_2(\Gamma_0(N))$. The last integral is convergent for all s (see Proof of Th. 3.66). Since V_S is of genus 1, we have $\mathrm{div}\,(f(z)dz) = 0$, so that $\mathrm{div}\,(f)$ can be obtained from the formula of Prop. 2.16. Checking the elliptic points of $\Gamma_0(N)$, we see easily that f has no zero on the imaginary axis, except at ∞. Since $f(iy) = \sum_{n=1}^{\infty} a_n e^{-2\pi n y}$ takes real values, it follows that $\zeta(s; E/\boldsymbol{Q})$ does not vanish at $s=1$. Birch has verified that this fact is in agreement with the above conjecture.

C. The Euler factors for the primes where the variety has bad reduction

To define the zeta-function of a curve or an abelian variety, we have considered only the primes for which the variety has good reduction. It is of course natural to seek the Euler factors also for the "bad" primes. Néron [53] has shown that an abelian variety over a local (or global) field has a model which has the "best behavior" for the reduction process modulo the prime in question. By means of this result, one can define the "conductor" of an abelian variety over a number field, and its Euler factors for bad primes (at least for elliptic curves). For details, the reader is referred to Ogg [54], Serre and Tate [66], and Weil [98]. With such factors and the notion of Tamagawa number, the above conjecture of Birch and Swinnerton-Dyer can be formulated in a more precise form (see [1] and the article by Swinnerton-Dyer in [6]).

CHAPTER 8
THE COHOMOLOGY GROUP ASSOCIATED WITH CUSP FORMS

8.1. Cohomology groups of Fuchsian groups

We shall now construct a certain cohomology group isomorphic to $S_k(\Gamma)$, which was first found by Eichler. Here k is any (odd or even) integer $\geqq 2$. To define it, we start with the usual definition of the cohomology group $H^i(G, X)$ with an arbitrary group G and a left G-module X. We fix an associative ring R with an identity element, and denote by $R[G]$ the group ring of G over R. In later applications, R will be $\boldsymbol{Z}$ or a field. We assume that X is an $R[G]$-module, and denote by $C^i(G, X)$, for an integer $i \geqq 0$, the R-module of all maps of $G^i = G \times \cdots \times G$ (the product of i copies) into X; we understand that $C^0(G, X) = X$. For $u \in C^i(G, X)$, define an element ∂u of $C^{i+1}(G, X)$ by

$$\partial u(\alpha) = (\alpha - 1)u \qquad \text{if } i = 0,$$

$$\begin{aligned}\partial u(\alpha_1, \alpha_2, \cdots, \alpha_{i+1}) = \alpha_1 \cdot u(\alpha_2, \cdots, \alpha_{i+1}) &\\ + \textstyle\sum_{j=1}^{i} (-1)^j u(\alpha_1, \cdots, \alpha_{j-1}, \alpha_j\alpha_{j+1}, \cdots, \alpha_{i+1}) &\\ + (-1)^{i+1} u(\alpha_1, \cdots, \alpha_i) & \qquad \text{if } i > 0.\end{aligned}$$

It can easily be verified that $\partial\partial = 0$. Put

$$Z^i(G, X) = \{u \in C^i(G, X) \mid \partial u = 0\},$$

$$B^i(G, X) = \begin{cases} 0 & \text{if } i = 0, \\ \partial(C^{i-1}(G, X)) & \text{if } i > 0, \end{cases}$$

$$H^i(G, X) = Z^i(G, X)/B^i(G, X),$$

$$X^G = \{x \in X \mid \alpha x = x \text{ for all } \alpha \in G\}.$$

We call $H^i(G, X)$ *the i-th cohomology group of G with coefficients in X.* Clearly $H^0(G, X)$ and $Z^0(G, X)$ can be identified with X^G. We observe that $Z^1(G, X)$ consists of all maps u of G into X such that

(8.1.1) $$u(\alpha\beta) = u(\alpha) + \alpha u(\beta) \qquad (\alpha, \beta \in G),$$

and $B^1(G, X)$ consists of all maps v of G into X such that

(8.1.2) $$v(\alpha) = (\alpha - 1)x_v \qquad (\alpha \in G)$$

with an element x_v of X independent of α. From (8.1.1), we obtain $u(1)=0$, and

$$u(\alpha^{-1})=-\alpha^{-1}u(\alpha) \qquad (\alpha\in G). \tag{8.1.3}$$

Now fix a subset Q of G, which may or may not be empty, and denote by $C_Q^1(G, X)$ the R-submodule of $C^1(G, X)$ consisting of the elements u with the following property:

$$u(\pi)\in(\pi-1)X \textit{ for every } \pi\in Q. \tag{8.1.4}$$

Then we put

$$Z_Q^1(G, X)=Z^1(G, X)\cap C_Q^1(G, X),$$

$$B_Q^2(G, X)=\partial(C_Q^1(G, X)),$$

$$H_Q^0(G, X)=H^0(G, X),$$

$$H_Q^1(G, X)=Z_Q^1(G, X)/B^1(G, X),$$

$$H_Q^2(G, X)=Z^2(G, X)/B_Q^2(G, X).$$

Note that $B^1(G, X)\subset Z_Q^1(G, X)$.

Now we shall consider the case where G is a Fuchsian group of the first kind. Here we understand that G is a subgroup of $SL_2(\boldsymbol{R})/\{\pm1\}$, and not of $SL_2(\boldsymbol{R})$. We denote by P the set of all parabolic elements of G. We are going to establish an "isogeny" of $H_Q^i(G, X)$, with a certain subset Q of P, to a certain cohomology group defined with respect to a simplicial complex on $\mathfrak{H}$. If $\mathfrak{H}/G$ is compact, and G has no elliptic elements, such an isogeny is actually an isomorphism and a special case of a well-known isomorphism due to Hopf, Eilenberg, MacLane, and Eckmann. It is therefore our task to modify the standard argument so that the difficulty arising from parabolic and elliptic elements of G can be eliminated.

Take a set $\{\varepsilon_1, \cdots, \varepsilon_r\}$ of representatives of elliptic elements of G, i.e., a minimal set such that every elliptic element of G is conjugate in G to a power of some ε_j. Let e_j be the order of ε_j, and E the least common multiple of $e_1, \cdots, e_r$. We put $E=1$ if $\{\varepsilon_j\}$ is empty. Let $\mathfrak{H}^*$ be the union of $\mathfrak{H}$ and the cusps of G. Let $c_1, \cdots, c_m$ be the points of $\mathfrak{H}^*/G$ corresponding to the cusps of G. Take a small open disc D_k on $\mathfrak{H}^*/G$ containing c_k so that the closures of $D_1, \cdots, D_m$ are disjoint from each other. For example, if ∞ is a cusp of G corresponding to c_k, we can take D_k to be the image of $\{z\in\mathfrak{H}^* \mid \operatorname{Im}(z)>y\}$ for a suitably large y, as described in § 1.3. Let $\mathfrak{H}_0$ be the inverse image of $(\mathfrak{H}^*/G)-(\bigcup_{k=1}^m D_k)$ by the map $\mathfrak{H}^*\to\mathfrak{H}^*/G$. We make a simplicial complex K with the underlying space $\mathfrak{H}_0$ so that the following conditions (8.1.5-8) are satisfied:

(8.1.5) *Every element of G induces a simplicial map of K onto itself.*

(8.1.6) *The fixed point of ε_j on $\mathfrak{H}$ is a 0-simplex of K; we denote it by d_j.*

(8.1.7) *There exists a 1-chain t_k of K which is mapped onto the boundary of D_k.*

(8.1.8) *There exists a fundamental domain for $\mathfrak{H}_0/G$ whose closure consists of a finite number of simplexes of K.*

One can construct such a K, for example, by taking a fundamental domain for $\mathfrak{H}^*/G$ as considered in the proof of Th. 2.20, and removing the parts corresponding to the D_k.

Let $(A_i, \partial, \boldsymbol{a})$ be the chain complex, with coefficients in R, obtained from K, with the usual boundary operator ∂ and the (unit) augmentation $\boldsymbol{a}$ defined by $\boldsymbol{a}(\sum_j s_j p_j)=\sum_j s_j$ for $s_j \in R$ and 0-simplexes p_j. Since $\mathfrak{H}_0$ is homeomorphic to a Euclidean plane, we have an exact sequence

(8.1.9) $$0 \longrightarrow A_2 \xrightarrow{\ \partial\ } A_1 \xrightarrow{\ \partial\ } A_0 \xrightarrow{\ \boldsymbol{a}\ } R \longrightarrow 0 .$$

In view of (8.1.5), A_i becomes an $R[G]$-module, and ∂ commutes with the action of $R[G]$. By (8.1.7), we have

(8.1.10) $$\partial t_k = \pi_k(q_k) - q_k \qquad (k=1, \cdots, m)$$

with a 0-simplex q_k and an element π_k of P. Then every parabolic element of G is conjugate to a power of some π_k. Put $Q=\{\pi_1, \cdots, \pi_m\}$.

Let $A^i(X)$ denote the module of all $R[G]$-linear maps of A_i into X, and let $\partial: A^i(X) \to A^{i+1}(X)$ be defined by $\partial u = u\partial$ for $u \in A^i(X)$. Further let $A^1_Q(X)$ be the submodule consisting of all u in $A^1(X)$ such that $u(t_k) \in (\pi_k - 1)X$ for every k. Then we put

$$Z^i(K, X) = \{u \in A^i(X) \mid \partial u = 0\} ,$$

$$B^i(K, X) = \partial A^{i-1}(X) ,$$

$$Z^1_Q(K, X) = Z^1(K, X) \cap A^1_Q(X) ,$$

$$B^2_Q(K, X) = \partial A^1_Q(X) ,$$

$$H^0_Q(K, X) = Z^0(K, X) ,$$

$$H^1_Q(K, X) = Z^1_Q(K, X)/B^1(K, X) ,$$

$$H^2_Q(K, X) = Z^2(K, X)/B^2_Q(K, X) .$$

Note that $B^1(K, X) \subset Z^1_Q(K, X)$ in view of (8.1.10).

PROPOSITION 8.1. *There exists, for* $i=0, 1, 2$, *an* R-*homomorphism* g^i *of* $H_Q^i(K, X)$ *into* $H_Q^i(G, X)$, *and an* R-*homomorphism* f^i *of* $H_Q^i(G, X)$ *into* $H_Q^i(K, X)$ *such that*

$$g^i \circ f^i = E\cdot(\text{identity map of } H_Q^i(G, X)),$$

$$f^i \circ g^i = E\cdot(\text{identity map of } H_Q^i(K, X)).$$

Especially, if R *is a field whose characteristic is* 0 *or prime to* E, $H_Q^i(G, X)$ *is isomorphic to* $H_Q^i(K, X)$.

PROOF. First consider a well-known chain complex $(M_i, \partial, \boldsymbol{a})$ consisting of the following:

(8.1.11) M_i, *for an integer* $i \geqq 0$, *is the free* R-*module generated by all the ordered sets* $[\alpha_0, \alpha_1, \cdots, \alpha_i]$ *of* $i+1$ *elements of* G.

(8.1.12) $\partial: M_i \to M_{i-1}$ *is defined by*

$$\partial[\alpha_0, \cdots, \alpha_i] = \textstyle\sum_{\nu=0}^{i} (-1)^\nu [\alpha_0, \cdots, \alpha_{\nu-1}, \alpha_{\nu+1}, \cdots, \alpha_i].$$

(8.1.13) $\boldsymbol{a}(\sum_\nu b_\nu[\alpha_\nu]) = \sum_\nu b_\nu$ *for* $\sum_\nu b_\nu[\alpha_\nu] \in M_0$ *with* $b_\nu \in R$.

(8.1.14) G *acts on* M_i *by the rule* $\beta[\alpha_0, \cdots, \alpha_i] = [\beta\alpha_0, \cdots, \beta\alpha_i]$.

It is well-known that

$$\text{(8.1.15)} \qquad \cdots \xrightarrow{\partial} M_2 \xrightarrow{\partial} M_1 \xrightarrow{\partial} M_0 \xrightarrow{\boldsymbol{a}} R \longrightarrow 0 \qquad \textit{is exact}.$$

Denote by $M^i(X)$ the module of all $R[G]$-linear maps of M_i into X, and define $\partial: M^i(X) \to M^{i+1}(X)$ by $\partial u = u\partial$. For every $u \in C^i(G, X)$, put

$$\bar{u}([\alpha_0, \cdots, \alpha_i]) = \alpha_0 \cdot u(\alpha_0^{-1}\alpha_1, \alpha_1^{-1}\alpha_2, \cdots, \alpha_{i-1}^{-1}\alpha_i).$$

Then we see that $u \mapsto \bar{u}$ gives an R-isomorphism of $C^i(G, X)$ onto $M^i(X)$, and $\partial\bar{u} = \overline{\partial u}$.

We are going to define an R-linear map $f: A_i \to M_i$ such that:

$$\text{(8.1.16)} \qquad \boldsymbol{a}f = E\boldsymbol{a}, \quad f\partial = \partial f, \quad f\alpha = \alpha f \qquad (\alpha \in G),$$

$$\text{(8.1.17)} \qquad f(d_j) = (E/e_j)\cdot \textstyle\sum_{\nu=0}^{e_j-1} [\varepsilon_j^\nu] \qquad (j = 1, \cdots, r),$$

$$\text{(8.1.18)} \qquad f(t_k) = E\cdot[1, \pi_k] \qquad (k = 1, \cdots, m).$$

Such an f can be obtained by the standard argument by induction on i, with a little care about d_j and t_k. In fact, first define $f(d_j)$ by (8.1.17), and put $f(\alpha(d_j)) = \alpha f(d_j)$ for all $\alpha \in G$. Then take a finite set S_0 of 0-simplexes so that every 0-simplex, other than the elliptic points of G, can be written as $\alpha(p)$ with a unique $p \in S_0$ and a unique $\alpha \in G$. We include the points q_k of (8.1.10) in S_0. Then we put $f(\alpha(p)) = E\cdot[\alpha]$ for all $\alpha \in G$ and all $p \in S_0$.

Similarly we fix a finite set S_i of i-simplexes, for $i=1, 2$, so that the $\alpha(s)$ for all $\alpha \in G$ and all $s \in S_i$ form a free R-basis of A_i. We include the t_k in S_1. Obviously $\boldsymbol{a}f=E\boldsymbol{a}$. Therefore $\boldsymbol{a}f(\partial s)=0$ for every $s \in S_1$. By virtue of (8.1.15), we can define $f(s)$ so that $\partial f(s)=f(\partial s)$. In particular we can put $f(t_k)=E\cdot[1, \pi_k]$ without contradiction. Then we put $f(\alpha(s))=\alpha f(s)$ for every $\alpha \in G$. Next, for $s \in S_2$, we have $\partial f(\partial s)=0$, hence we can define $f(s)$ so that $\partial f(s)=f(\partial s)$, in view of (8.1.15). Then we put $f(\alpha(s))=\alpha f(s)$.

To an element $u \in C^i(G, X)$, we assign an element w of $A^i(X)$ by $w=\bar{u}\circ f$. If $u \in C^1_P(G, X)$, $w(t_k)=E\cdot u(\pi_k) \in (\pi_k-1)X$, hence $u \in A^1_Q(X)$. Moreover it can easily be seen that the correspondence $u \mapsto w$ commutes with ∂, hence defines a homomorphism f^i of $H^i_Q(G, X)$ into $H^i_Q(K, X)$.

By a similar argument, we can define an R-linear map $g: M_i \to A_i$ satisfying the following two conditions:

(8.1.19) $$\boldsymbol{a}g=\boldsymbol{a}, \qquad g\partial=\partial g, \qquad g\alpha=\alpha g \qquad (\alpha \in G),$$

(8.1.20) $g([1, \pi_k])=t_k+(\pi_k-1)b_k$ *with a* 1-*chain* b_k *such that* $\partial b_k=p_0-q_k$.

Here p_0 is a fixed 0-simplex in S_0. To define such a g, first put $g([\alpha])=\alpha(p_0)$ for all $\alpha \in G$. Then define $g([1, \alpha])$ so that $\partial g([1, \alpha])=\alpha(p_0)-p_0$, and put $g([\alpha, \beta])=g([1, \alpha^{-1}\beta])$. In particular we can define $g([1, \pi_k])$ as in (8.1.20). Since $\partial g(\partial[1, \alpha, \beta])=0$, we can define $g([1, \alpha, \beta])$ so that $\partial g([1, \alpha, \beta])=g(\partial[1, \alpha, \beta])$ in view of the exactness of (8.1.9). Then we put $g([\alpha, \beta, \gamma])=\alpha g([1, \alpha^{-1}\beta, \alpha^{-1}\gamma])$.

To an element $x \in A^i(X)$, we assign an element y of $C^i(G, X)$ so that $\bar{y}=x\circ g$. If $x \in A^1_Q(X)$, $y(\pi_k)=\bar{y}([1, \pi_k])=x(t_k)+(\pi_k-1)x(b_k) \in (\pi_k-1)X$, hence $y \in C^1_Q(G, X)$. Moreover, we can easily verify that the correspondence $x \mapsto y$ commutes with ∂, hence defines a homomorphism g^i of $H^i_Q(K, X)$ into $H^i_Q(G, X)$.

Let us now construct an R-linear map $U: M_i \to M_{i+1}$ with the following properties:

(8.1.21) $$U\alpha=\alpha U \qquad (\alpha \in G),$$

(8.1.22) $$f\circ g-E\cdot(\text{identity map})=\partial U+U\partial,$$

(8.1.23) $$U([1, \pi_k]) \in (\pi_k-1)M_2.$$

We first observe that $f(g(x))=Ex$ for $x \in M_0$. Defining $U=0$ on M_0, we see that (8.1.22) is satisfied on M_0. Let α be an element of G other than the π_k. Since

$$\partial\{f(g([1, \alpha]))-E[1, \alpha]\}=f(g(\partial[1, \alpha]))-E\partial[1, \alpha]=0,$$

we can define $U([1, \alpha])$ so that

$$\partial U([1, \alpha])=f(g([1, \alpha]))-E[1, \alpha],$$

in view of (8.1.15). If $\alpha=\pi_k$, we have to choose $U([1, \alpha])$ more specifically. Since $\partial f(b_k)=0$, we can find an element n_k of M_2 so that $\partial n_k=f(b_k)$. Put $U([1, \pi_k])=(\pi_k-1)n_k$. In view of (8.1.18) and (8.1.20), we have

$$f(g([1, \pi_k]))-E[1, \pi_k]=(\pi_k-1)f(b_k)=\partial U([1, \pi_k]).$$

Now we put $U([\alpha, \beta])=\alpha U([1, \alpha^{-1}\beta])$. Then (8.1.22) is true on M_1. Further we have to define $U([1, \alpha, \beta])$ so that

$$\partial U([1, \alpha, \beta])=f(g([1, \alpha, \beta]))-E[1, \alpha, \beta]-U(\partial[1, \alpha, \beta]).$$

This can actually be done, since the boundary of the right hand side is 0. Putting $U([\alpha, \beta, \gamma])=\alpha U([1, \alpha^{-1}\beta, \alpha^{-1}\gamma])$, we obtain the desired U.

Let $x\in Z^i(G, X)$. Then there exists an element y of $C^{i-1}(G, X)$ such that $\bar{y}=\bar{x}\circ U$. By (8.1.22), we obtain $\bar{x}\circ f\circ g-E\bar{x}=\partial\bar{y}$. (If $i\leqq 1$, $y=0$.) If $i=2$, one has

$$y(\pi_k)=\bar{x}(U([1, \pi_k]))=(\pi_k-1)\bar{x}(n_k)\in(\pi_k-1)X,$$

hence $y\in C_Q^1(G, X)$. This shows that $g^i\circ f^i=E\cdot$(identity map) for $i=0, 1, 2$.

Similarly we obtain an R-linear map $V: A_i\to A_{i+1}$ with the following properties:

(8.1.24) $$V\alpha=\alpha V \qquad (\alpha\in G),$$

(8.1.25) $$g\circ f-E\cdot(\text{identity map})=\partial V+V\partial,$$

(8.1.26) $$V(t_k)=0.$$

Since $\boldsymbol{a}\circ g\circ f=E\cdot\boldsymbol{a}$ on A_0, we can define $V(s)$ for $s\in S_0$ so that $\partial V(s)=g(f(s))-Es$. In particular we can put $V(q_k)=Eb_k$. As for d_j, we take a 1-chain h_j in A_1 so that $\partial h_j=p_0-d_j$, and put $V(d_j)=(E/e_j)\cdot\sum_{\nu=0}^{e_j-1}\varepsilon_j^\nu(h_j)$. Then we can put $V(\alpha(p))=\alpha V(p)$ for $\alpha\in G$ and for an arbitrary 0-simplex p without contradiction. By the procedure similar to the construction of U, we define V on S_1 and S_2 so that (8.1.25) is satisfied, and put $V(\alpha(s))=\alpha V(s)$ for $\alpha\in G$, $s\in S_i$. The choice (8.1.26) is possible in view of (8.1.18) and (8.1.20). Note that $V=0$ on A_2.

Let $u\in Z^i(K, X)$. Then $u\circ g\circ f-Eu=\partial(u\circ V)$. If $i=2$, we have $u(V(t_k))=0$, hence $u\circ V\in A_Q^1(X)$. This proves that $f^i\circ g^i=E\cdot$(identity map), and completes the proof of Prop. 8.1.

Actually the isomorphism of $H_Q^0(G, X)$ and $H_Q^0(K, X)$ can be seen immediately. In fact, if $w\in Z^0(K, X)$, then $w(p)$ is independent of p. Therefore $\gamma w(p)=w(\gamma(p))=w(p)$ for all $\gamma\in G$, hence $w(p)\in X^G=H_Q^0(G, X)$. Conversely, any element of X^G corresponds to an element of $H_Q^0(K, X)$. Thus *$H_Q^0(K, X)$ is always isomorphic to $X^G=H_Q^0(G, X)$.*

PROPOSITION 8.2. *Let Y be the R-submodule of X generated by $(\alpha-1)X$ for all $\alpha \in G$. Then $H_Q^2(K, X)$ is isomorphic to X/Y.*

PROOF. Take a fundamental domain F for $\mathfrak{H}_0$ modulo G as described in (8.1.8). We may assume:

(8.1.27) *F is simply connected;*

(8.1.28) *If $a_1, \cdots, a_\mu$ are the 2-simplexes contained in F, one has*

$$\partial(\textstyle\sum_{i=1}^{\mu} a_i) = \sum_{k=1}^{m} \alpha_k(t_k) + \sum_{l=1}^{\lambda} (\beta_l - 1)s_l$$

with some α_k, $\beta_l \in G$ and some $s_l \in A_1$.

Then G is generated by the β_l and $\alpha_k \pi_k \alpha_k^{-1}$, and A_2 is generated by the $\gamma(a_i)$ for all i and all $\gamma \in G$. Therefore an element u of $Z^2(K, X)$ is determined by the values $u(a_i)$. Let us put $u(F) = \sum_{i=1}^{\mu} u(a_i)$. Suppose $u(F) \in Y$. Then there exist elements y_k and z_l of X such that

$$u(F) = \textstyle\sum_{k=1}^{m} (\alpha_k \pi_k \alpha_k^{-1} - 1) y_k + \sum_{l=1}^{\lambda} (\beta_l - 1) z_l\,. \tag{8.1.29}$$

We can find an element w of $A_Q^1(X)$ so that $u = \partial w$, $w(t_k) = (\pi_k - 1)\alpha_k^{-1} y_k$, and $w(s_l) = z_l$. In fact, we first define the values of w at t_k and s_l as specified. Then we set the values of w at the 1-simplexes lying inside F, one by one, so that $u(a_j) = w(\partial a_j)$. This is possible in view of (8.1.29). Then extend w to the whole A_1 by the property $w\gamma = \gamma w$ for all $\gamma \in G$. Thus $u \in B_Q^2(K, X)$, if $u(F) \in Y$. Conversely, if $u = \partial w$ with $w \in A_Q^1(X)$, we have

$$u(F) = w(\partial F) = \textstyle\sum_{k=1}^{m} w(\alpha_k(t_k)) + \sum_{l=1}^{\lambda} (\beta_l - 1) w(s_l) \in Y\,.$$

This completes the proof.

PROPOSITION 8.3. *Suppose that R is a field, and X is a finite dimensional vector space over R. Let g be the genus of $G\backslash\mathfrak{H}^*$, Y be as in Prop. 8.2, and let*

$$\zeta = \dim(X^G)\,, \qquad \zeta' = \dim(X/Y)\,,$$

$$\xi_j = \dim(\{x \in X \mid \varepsilon_j x = x\}) \qquad (j = 1, \cdots, r)\,,$$

$$\eta_k = \dim((\pi_k - 1)X) \qquad (k = 1, \cdots, m)\,,$$

where $\dim(\ )$ *denotes the dimension over R, and the ε_j (resp. π_k) are representatives for the elliptic (resp. parabolic) elements of G as in the above discussion. Then*

$$\dim(H_Q^1(K, X)) = (2g-2)\dim(X) + \zeta + \zeta' + \textstyle\sum_{k=1}^{m} \eta_k + \sum_{j=1}^{r} (\dim(X) - \xi_j)\,.$$

PROOF. Let K be as above, and N_i the number of G-inequivalent i-simplexes in K. Then we see easily that

$$N_0-N_1+N_2+m=2-2g,$$
$$\dim(A^0(X))=N_0\cdot\dim(X)-\textstyle\sum_{j=1}^{r}(\dim(X)-\xi_j),$$
$$\dim(A_Q^1(X))=N_1\cdot\dim(X)-\textstyle\sum_{k=1}^{m}(\dim(X)-\eta_k),$$
$$\dim(A^2(X))=N_2\cdot\dim(X).$$

Further we have

$$\textstyle\sum_{i=0}^{2}(-1)^i\dim(H_Q^i(K,X))=\dim(A^0(X))-\dim(A_Q^1(X))+\dim(A^2(X)).$$

Our assertion now follows immediately from these relations, Prop. 8.2, and the isomorphism of $H_Q^0(K, X)$ with X^G.

Let P be the set of all parabolic elements of G. Then we have

(8.1.30) $$H_P^1(G, X)=H_Q^1(G, X).$$

To prove this, it is sufficient to show that $Z_P^1(G, X)=Z_Q^1(G, X)$. Obviously $Z_P^1(G, X)\subset Z_Q^1(G, X)$. Let $u\in Z_Q^1(G, X)$, and $\pi\in Q$. Then $u(\pi)=(\pi-1)x$ with $x\in X$, so that, by (8.1.1), $u(\pi^m)=(1+\pi+\cdots+\pi^{m-1})u(\pi)=(\pi^m-1)x$ for any positive integer m, and by (8.1.3), $u(\pi^{-m})=-\pi^{-m}u(\pi^m)=(\pi^{-m}-1)x$. Therefore, for every $\alpha\in G$ and every $\mu\in \boldsymbol{Z}$, we have $u(\alpha\pi^\mu\alpha^{-1})=(\alpha\pi^\mu\alpha^{-1}-1)(\alpha x-u(\alpha))$. Now every element of P is of the form $\alpha\pi^\mu\alpha^{-1}$ with $\pi\in Q$, $\alpha\in G$, and $\mu\in\boldsymbol{Z}$. Therefore $u\in Z_P^1(G, X)$, so that $Z_P^1(G, X)=Z_Q^1(G, X)$, q. e. d.

8.2. The correspondence between cusp forms and cohomology classes

For $\begin{bmatrix} u \\ v \end{bmatrix}\in\boldsymbol{C}^2$ and for every integer $n\geqq 0$, let us define an $(n+1)$-dimensional column vector $\begin{bmatrix} u \\ v \end{bmatrix}^n$ by

$$\begin{bmatrix} u \\ v \end{bmatrix}^n={}^t(u^n, u^{n-1}v, \cdots, u^{n-k}v^k, \cdots, uv^{n-1}, v^n).$$

Then we can define a representation $\rho_n: GL_2(\boldsymbol{C})\to GL_{n+1}(\boldsymbol{C})$ by

(8.2.1) $$\rho_n(\alpha)\begin{bmatrix} u \\ v \end{bmatrix}^n=\left(\alpha\begin{bmatrix} u \\ v \end{bmatrix}\right)^n.$$

If $n=0$, we understand that $\begin{bmatrix} u \\ v \end{bmatrix}^0=1$, and $\rho_0(\alpha)=1$ for every $\alpha\in G$. There exists a unique non-degenerate bilinear form on $\boldsymbol{C}^{n+1}$, represented by a real matrix Θ_n such that

(8.2.2) $${}^t\begin{bmatrix} u \\ v \end{bmatrix}^n\cdot\Theta_n\cdot\begin{bmatrix} x \\ y \end{bmatrix}^n=\det\begin{bmatrix} u & x \\ v & y \end{bmatrix}^n.$$

We see easily that $\Theta_0=1$, and

(8.2.3) $${}^t\Theta_n=(-1)^n\Theta_n\,,$$

(8.2.4) $${}^t\rho_n(\alpha)\Theta_n\rho_n(\alpha)=\det(\alpha)^n\Theta_n\,,$$

(8.2.5) $$\begin{bmatrix}\alpha(z)\\1\end{bmatrix}^n=j(\alpha,z)^{-n}\rho_n(\alpha)\begin{bmatrix}z\\1\end{bmatrix}^n\qquad(\alpha\in GL_2(\boldsymbol{R}),\ z\in\mathfrak{H})\,,$$

$$GL_2(\boldsymbol{R})\cap\operatorname{Ker}(\rho_n)=\begin{cases}\{1_2\} & \text{if } n \text{ is odd}\,,\\ \{\pm 1_2\} & \text{if } n \text{ is even}\,.\end{cases}$$

Let Γ be a discrete subgroup of $SL_2(\boldsymbol{R})$ which is a Fuchsian group of the first kind, and $\bar{\Gamma}=\Gamma/(\Gamma\cap\{\pm 1\})$. Let P (resp. $\bar{P}$) denote the set of all parabolic elements of Γ (resp. $\bar{\Gamma}$). Let X be a $\bar{\Gamma}$-module, which we consider a Γ-module in a natural way. (This means that if $-1\in\Gamma$, -1 acts as the identity map of X.) Consider the following condition on X:

(8.2.6) *If* $x\in X$ *and* $2x=0$, *then* $x=0$.

Under this assumption, if $u\in Z^1(\Gamma,X)$ and $-1\in\Gamma$, then $0=u((-1)^2)=u(-1)+u(-1)$, so that $u(-1)=0$. Then u can be considered as an element of $Z^1(\bar{\Gamma},X)$ in a natural way. Therefore $Z_P^1(\Gamma,X)$ (resp. $B^1(\Gamma,X)$) can be identified with $Z_{\bar{P}}^1(\bar{\Gamma},X)$ (resp. $B^1(\bar{\Gamma},X)$) in a natural way, so that $H_P^1(\Gamma,X)$ can be identified with $H_{\bar{P}}^1(\bar{\Gamma},X)$.

Now we consider a representation Ψ of Γ into $GL_r(\boldsymbol{R})$, with any $r>0$, satisfying the following two conditions:

(8.2.7) Ψ *maps* Γ *into a compact subgroup of* $GL_r(\boldsymbol{R})$;

(8.2.8) *The kernel of* Ψ *is of finite index in* Γ *if* Γ *has cusps.*

Then we denote by $S_k(\Gamma,\Psi)$ the vector space of all holomorphic maps f of $\mathfrak{H}$ into $\boldsymbol{C}^r$ satisfying the following two conditions:

(8.2.9) $f(\alpha(z))j(\alpha,z)^{-k}=\Psi(\alpha)f(z)$ *for all* $\alpha\in\Gamma$;

(8.2.10) *The components of* f *belong to* $S_k(\operatorname{Ker}(\Psi))$, *if* Γ *has cusps. (This is meaningful in view of* (8.2.8).)

The vector space $S_k(\Gamma_0',\psi)$ of § 3.5 is an example of $S_k(\Gamma,\Psi)$. In § 9.2, we shall give an example of Ψ such that $\operatorname{Ker}(\Psi)$ is not of finite index in Γ and Γ has no cusp.

If Ψ is absolutely irreducible and $-1\in\Gamma$, then $S_k(\Gamma,\Psi)\neq\{0\}$ only when $\Psi(-1)=(-1)^k$. Further, if Ψ is the direct sum of two representations Ψ_1 and Ψ_2, then $S_k(\Gamma,\Psi)$ can be identified with the direct sum of $S_k(\Gamma,\Psi_1)$ and $S_k(\Gamma,\Psi_2)$. Therefore, without losing much generality, we shall hereafter assume

$$\Psi(-1)=(-1)^k \text{ if } -1\in\Gamma. \tag{8.2.11}$$

By virtue of the assumption (8.2.7), we find a positive definite real symmetric matrix P such that ${}^t\Psi(\alpha)P\Psi(\alpha)=P$ for all $\alpha\in\Gamma$. Then we define a positive definite hermitian inner product on $S_k(\Gamma,\Psi)$ (depending on P), by

$$(f,g)=\int_{\Gamma\backslash\mathfrak{H}} {}^tfP\bar{g}\cdot y^{k-2}dxdy \qquad (f,\ g\in S_k(\Gamma,\Psi)\,;\ z=x+iy)\,.$$

This is a generalization of the Petersson inner product of § 3.4; the convergence of the integral can be shown in a similar way. Hereafter we fix Γ and Ψ, and consider $S_{n+2}(\Gamma,\Psi)$ with a non-negative integer n. Our principal aim of this section is to find an isomorphism of $S_{n+2}(\Gamma,\Psi)$ to the cohomology group $H_P^1(\Gamma,X)$ with a suitable Γ-module X. First we define, for every $f\in S_{n+2}(\Gamma,\Psi)$, a holomorphic vector differential form $\mathfrak{d}(f)$ with values in $\boldsymbol{C}^r\otimes\boldsymbol{C}^{n+1}$ by

$$\mathfrak{d}(f)=f\otimes\begin{bmatrix} z\\ 1\end{bmatrix}^n dz\,. \tag{8.2.12}$$

If $n=0$, we understand that $\mathfrak{d}(f)=f(z)dz$. Put

$$W=P\otimes\Theta_n\,,\qquad \chi(\alpha)=\Psi(\alpha)\otimes\rho_n(\alpha)\qquad(\alpha\in\Gamma)\,. \tag{8.2.13}$$

In view of (8.2.4), (8.2.5), and (8.2.9), we obtain

$${}^t\chi(\alpha)W\chi(\alpha)=W\qquad(\alpha\in\Gamma)\,, \tag{8.2.14}$$

$$\mathfrak{d}(f)\circ\alpha=\chi(\alpha)\mathfrak{d}(f)\qquad(\alpha\in\Gamma)\,, \tag{8.2.15}$$

where $\circ\alpha$ means the transform of a differential form by α. Since $\chi(\alpha)$ is real, we have also

$$\operatorname{Re}(\mathfrak{d}(f))\circ\alpha=\chi(\alpha)\cdot\operatorname{Re}(\mathfrak{d}(f))\qquad(\alpha\in\Gamma)\,, \tag{8.2.16}$$

where Re( ) stands for the real part. Therefore we can define an $\boldsymbol{R}$-valued $\boldsymbol{R}$-bilinear form $A(f,g)$ on $S_{n+2}(\Gamma,\Psi)$ by

$$A(f,g)=\int_{\Gamma\backslash\mathfrak{H}} {}^t\operatorname{Re}(\mathfrak{d}(f))\wedge W\cdot\operatorname{Re}(\mathfrak{d}(g))\,. \tag{8.2.17}$$

In view of (8.2.2), we have ${}^t\mathfrak{d}(f)\wedge W\overline{\mathfrak{d}(g)}=-(2i)^{n+1}\cdot{}^tfP\bar{g}\cdot y^n dx\wedge dy$, so that

$$A(f,g)=(2i)^{n-1}[(f,g)+(-1)^{n+1}(g,f)]\,, \tag{8.2.18$_a$}$$

$$A(f,g)=(-1)^{n+1}A(g,f)\,, \tag{8.2.18$_b$}$$

$$A(f,i^{n-1}g)=2^n\cdot\operatorname{Re}((f,g))\,. \tag{8.2.18$_c$}$$

Therefore, $A(f,g)$ is non-degenerate.

Now we consider $\boldsymbol{R}^r \otimes \boldsymbol{R}^{n+1}$ (resp. $\boldsymbol{C}^r \otimes \boldsymbol{C}^{n+1}$) as an $\boldsymbol{R}[\Gamma]$-module (resp. $\boldsymbol{C}[\Gamma]$-module) through the representation χ, and also as an $\boldsymbol{R}[\bar{\Gamma}]$-module (resp. $\boldsymbol{C}[\bar{\Gamma}]$-module), on account of our assumption $\Psi(-1)=(-1)^n$ if $-1 \in \Gamma$. Hereafter we denote this $\boldsymbol{R}[\Gamma]$-module (resp. $\boldsymbol{C}[\Gamma]$-module) by X (resp. X_C). If it is necessary to specify n and Ψ, we write $X = X_n^\Psi$.

Fix any point z_0 of $\mathfrak{H}$. For $f \in S_{n+2}(\Gamma, \Psi)$, put

$$F(z) = \int_{z_0}^{z} \mathfrak{d}(f) + v$$

with any fixed vector v of X_C. Since $\mathfrak{d}(f)$ is holomorphic, $F(z)$ is independent of the choice of the path of integral. For every $\alpha \in \Gamma$, we have, by (8.2.15),

$$F(\alpha(z)) = \int_{\alpha(z_0)}^{\alpha(z)} \mathfrak{d}(f) + \int_{z_0}^{\alpha(z_0)} \mathfrak{d}(f) + v = \chi(\alpha) F(z) + t(\alpha),$$

where $t(\alpha) = \int_{z_0}^{\alpha(z_0)} \mathfrak{d}(f) + [1 - \chi(\alpha)]v$. Therefore we see that

$$t(\alpha\beta) = t(\alpha) + \chi(\alpha) t(\beta),$$

so that $t \in Z^1(\Gamma, X_C)$. We observe also that the change of v (and hence the change of z_0) affects t only by an addition of an element of $B^1(\Gamma, X_C)$. Suppose that Γ has a cusp s. Take $\rho \in SL_2(\boldsymbol{R})$ so that $\rho(s) = \infty$, and $\rho^{-1}\begin{bmatrix} 1 & h \\ 0 & 1 \end{bmatrix}\rho$, with $h > 0$, generates $\{\gamma \in \operatorname{Ker}(\Psi) \mid \gamma(s) = s\}$ (see § 2.1). We can put $j(\rho^{-1}, z)^{-n-2} f(\rho^{-1}(z)) = \Phi(q)$ with a holomorphic $\boldsymbol{C}^r$-valued function $\Phi(q)$ in $q = e^{\pi i z/h}$, on account of (8.2.10). Then putting $p(w) = j(\rho^{-1}, w)^n \cdot \rho^{-1}(w)^n$, we have

$$\begin{aligned} \int_{z_0}^{z} f(w) w^n dw &= \int_{\rho(z_0)}^{\rho(z)} f(\rho^{-1}(w)) j(\rho^{-1}, z)^{-n-2} p(w) dw \\ &= \int_{\rho(z_0)}^{\rho(z)} \Phi(e^{\pi i w/h}) p(w) dw. \end{aligned}$$

Since $p(w)$ is a polynomial in w, and $\Phi(0) = 0$, the integral has a limit when $\rho(z)$ tends to ∞, i.e., z tends to s (with respect to the topology of $\mathfrak{H}^*$). Therefore we can meaningfully put $F(s) = \lim_{z \to s} F(z)$. Then

$$F(s) = F(\pi(s)) = \chi(\pi) F(s) + t(\pi).$$

This proves that $t \in Z_P^1(\Gamma, X_C)$.

Taking $\operatorname{Re}(\mathfrak{d}(f))$ in place of $\mathfrak{d}(f)$, put

$$\mathfrak{f}(z) = \int_{z_0}^{z} \operatorname{Re}(\mathfrak{d}(f)) + a \tag{8.2.19}$$

with any fixed $a \in X$. Then

$$\mathfrak{f}(\alpha(z)) = \chi(\alpha)\mathfrak{f}(z) + u(\alpha) \qquad (\alpha \in \Gamma) \tag{8.2.20}$$

with an element u of $Z_P^1(\Gamma, X)$. As is shown above, the cohomology class of u is uniquely determined by f, and independent of the choice of z_0. Therefore we can define an $\boldsymbol{R}$-linear map φ of $S_{n+2}(\Gamma, \Psi)$ into $H_P^1(\Gamma, X)$ by

$$\varphi(f) = \text{the cohomology class of } u.$$

THEOREM 8.4. *For every (even or odd) $n \geqq 0$ and every representation Ψ of Γ satisfying* (8.2.7, 8, 11), *the map φ is an $\boldsymbol{R}$-linear isomorphism of $S_{n+2}(\Gamma, \Psi)$ onto $H_P^1(\Gamma, X_n^\Psi)$.*

A result of this type, in a somewhat different form, was first given by Eichler [18] in the case where n is even and Ψ is trivial. The theorem in the present form, under some restrictive conditions, was proved in the previous papers:

I. [71, Th. 1] when n is even and Ψ is trivial.

II. [74, Th. 2] when n is even and Γ has no cusps. This method is applicable to the case of odd n.

III. [48, Prop. 4.4] when Γ has no cusps. This includes also the case of the product of several copies of $\mathfrak{H}$. A further generalization was given by Matsushima and Murakami [47] for discontinuous groups acting on a bounded symmetric domain with compact quotient.

Here we shall prove the above theorem only in the case where $\mathrm{Ker}(\Psi)$ is of finite index in Γ. This together with the previously known results will give a complete proof.

Let f and g be elements of $S_{n+2}(\Gamma, \Psi)$. Define $\mathfrak{f}$ and u as in (8.2.19) and (8.2.20). Similarly, put

$$\mathfrak{g}(z) = \int_{z_0}^{z} \mathrm{Re}[\mathfrak{d}(f)] + b$$

with any fixed $b \in X$. Then

$$\mathfrak{g}(\alpha(z)) = \chi(\alpha)\mathfrak{g}(z) + v(\alpha) \qquad (\alpha \in \Gamma) \tag{8.2.21}$$

with an element v of $Z_P^1(\Gamma, X)$. Since $d\mathfrak{f} = \mathrm{Re}[\mathfrak{d}(f)]$ and $d\mathfrak{g} = \mathrm{Re}[\mathfrak{d}(g)]$, we have

$$A(f, g) = \int_{\Gamma \backslash \mathfrak{H}} {}^t d\mathfrak{f} \wedge W d\mathfrak{g}.$$

Take a fundamental domain Π for $\Gamma \backslash \mathfrak{H}$ constructed in the proof of Th. 2.20. Here we do not take small circles around cusps and elliptic points as considered there. Since $d({}^t\mathfrak{f} W d\mathfrak{g}) = {}^t d\mathfrak{f} \wedge W d\mathfrak{g}$, we have

$$A(f, g) = \int_{\partial \Pi} {}^t\mathfrak{f} W d\mathfrak{g},$$

where $\partial \Pi$ is the boundary of Π. As is observed in the proof of Th. 2.20, we have $\partial\Pi=\sum_\lambda[S_\lambda-\sigma_\lambda(S_\lambda)]$, with 1-simplexes S_λ and elements σ_λ of Γ, so that

$$A(f, g)=\sum_\lambda\int_{S_\lambda}{}^t\mathfrak{f}Wd\mathfrak{g}-\sum_\lambda\int_{\sigma_\lambda(S_\lambda)}{}^t\mathfrak{f}Wd\mathfrak{g}\,.$$

By virtue of (8.2.20), (8.2.21), and (8.2.14),

$$\int_{\sigma_\lambda(S_\lambda)}{}^t\mathfrak{f}Wd\mathfrak{g}=\int_{S_\lambda}{}^t(\mathfrak{f}\circ\sigma_\lambda)W\cdot d(\mathfrak{g}\circ\sigma_\lambda)$$

$$=\int_{S_\lambda}{}^t\mathfrak{f}Wd\mathfrak{g}+\int_{S_\lambda}{}^tu(\sigma_\lambda)W\chi(\sigma_\lambda)d\mathfrak{g}\,,$$

hence, by (8.1.3) and (8.2.14),

$$A(f, g)=\sum_\lambda{}^tu(\sigma_\lambda^{-1})W\int_{S_\lambda}d\mathfrak{g}\,. \tag{8.2.22}$$

Now suppose that $\varphi(f)=0$. Then, choosing the constant vector a of (8.2.19) suitably, we may put $u=0$. Then (8.2.22) implies $A(f, g)=0$ for every $g\in S_{n+2}(\Gamma, \Psi)$. Since $A(f, g)$ is non-degenerate, f must be 0. This proves that the map φ is injective.

In the next place, we compute the dimension of $H_P^1(\Gamma, X)$, assuming that Ψ is trivial. In this case $X=\boldsymbol{R}^{n+1}$, and $\chi=\rho_n$. (The condition "$\Psi(-1)=(-1)^n$ if $-1\in\Gamma$" then implies that $-1\notin\Gamma$ if n is odd.) We are going to show

(8.2.23) *The dimension of $H_P^1(\Gamma, X)$ over $\boldsymbol{R}$ is twice the dimension of $S_{n+2}(\Gamma)$ over $\boldsymbol{C}$.*

Let $\varepsilon_1, \cdots, \varepsilon_r$, and $\pi_1, \cdots, \pi_m$ be defined for Γ as in § 8.1. Let ξ_j, η_k, ζ, and ζ' be as in Prop. 8.3. Since $H_P^1(\Gamma, X)=H_{\bar{P}}^1(\Gamma, X)$, we have, by (8.1.30), Prop. 8.2, and Prop. 8.3,

$$\dim(H_P^1(\Gamma, X))=(2g-2)(n+1)+\zeta+\zeta'+\textstyle\sum_{k=1}^m\eta_k+\sum_{j=1}^r(n+1-\xi_j)\,. \tag{8.2.24}$$

Suppose first that $n=0$. Then $\eta_k=0$, and $\xi_j=\zeta=\zeta'=1$, hence $\dim(H_P^1(\Gamma, X))=2g$, so that (8.2.23) is true. Next suppose that $n>0$. The Jordan canonical form of (the matrix representing) π_k is $\begin{bmatrix}1 & 1\\ 0 & 1\end{bmatrix}$ or $\begin{bmatrix}-1 & 1\\ 0 & -1\end{bmatrix}$ according as the corresponding cusp is regular or irregular (see § 2.1). Therefore, looking at the form of $\rho_n\left(\begin{bmatrix}\pm1 & 1\\ 0 & \pm1\end{bmatrix}\right)$, we see easily that

$$\eta_k=\begin{cases} n+1 & \text{if } n \text{ is odd and the cusp is irregular,}\\ n & \text{otherwise.}\end{cases}$$

To determine ξ_j, let e_j be the order of ε_j. Then $\rho_n(\varepsilon_j)$ has $n+1$ characteristic roots $\omega^n, \omega^{n-2}, \cdots, \omega^{2-n}, \omega^{-n}$ with a root of unity ω, whose order is e_j or $2e_j$

according as e_j is odd or even. Therefore $n+1-\xi_j$ is the number of these roots different from 1. We can show that

$$n+1-\xi_j=2\cdot[(n+2)(e_i-1)/2e_i],$$

where $[x]$ means the largest integer $\leqq x$. We omit the details of the verification of this formula, since it is quite elementary and rather tedious. Finally we have $\zeta=\zeta'=0$. To show this, let $x\in X^{\Gamma}$, i.e., $\rho_n(\alpha)x=x$ for all $\alpha\in\Gamma$. Put $p(z)={}^t x\Theta_n\begin{bmatrix} z \\ 1 \end{bmatrix}^n$. By (8.2.5), $p(\alpha(z))j(\alpha, z)^n=p(z)$ for all $\alpha\in\Gamma$. Therefore $p\in S_{-n}(\Gamma)$ if Γ has no cusps. If s is a cusp of Γ, take an element ρ of $SL_2(\mathbf{R})$ so that $\rho(s)=\infty$, and $\rho^{-1}\begin{bmatrix} 1 & h \\ 0 & 1 \end{bmatrix}\rho$ generates $\{\gamma\in\Gamma \mid \gamma(s)=s\}$. Put $t(z)=p(\rho^{-1}(z))j(\rho^{-1}, z)^n$. Then t is a polynomial in z, and $t(z+2h)=t(z)$. Therefore t must be a constant. It follows that $p\in G_{-n}(\Gamma)$. Since $G_{-n}(\Gamma)=\{0\}$ for $n>0$ by Th. 2.23 and Th. 2.25, we have $p=0$, so that $x=0$. This proves that $X^{\Gamma}=\{0\}$, hence $\zeta=0$. To show that $\zeta'=0$, let Y be as in Prop. 8.2. Let x be an element of X such that ${}^t y\Theta_n x=0$ for all $y\in Y$. Then for every $w\in X$ and every $\alpha\in\Gamma$, we have

$$0={}^t[(\rho_n(\alpha^{-1})-1)w]\Theta_n x={}^t w\Theta_n(\rho_n(\alpha)-1)x,$$

so that $(\rho_n(\alpha)-1)x=0$, hence $x\in X^{\Gamma}$. Since $X^{\Gamma}=\{0\}$, this proves that $Y=X$, hence $\zeta'=0$.

Thus we have determined ξ_j, η_k, ζ and ζ'. Putting these numbers into (8.2.24), and comparing the result with $\dim(S_{n+2}(\Gamma))$ given in Th. 2.24 and Th. 2.25, we obtain (8.2.23). Since we have already seen that φ is injective, this completes the proof of Th. 8.4 for the trivial Ψ. The case of non-trivial Ψ will be proved in the next section.

8.3. Action of double cosets on the cohomology group

Let Γ_1 and Γ_2 be commensurable Fuchsian groups of the first kind, given as subgroups of $SL_2(\mathbf{R})$, and Δ a semi-group contained in $GL_2^+(\mathbf{R})$, and containing Γ_1 and Γ_2, such that $\alpha\Gamma_1\alpha^{-1}$ is commensurable with Γ_1 for every $\alpha\in\Delta$. We assume that Δ is stable under the main involution ι (see p. 72) of $M_2(\mathbf{R})$. Let R be an arbitrary associative ring with identity, $R[\Delta]$ the semi-group-ring (monoid ring) of Δ over R, and X an $R[\Delta]$-module. We are going to define an R-linear map

(8.3.1) $$(\Gamma_1\alpha\Gamma_2)_X:\ H^1_{P_1}(\Gamma_1, X)\longrightarrow H^1_{P_2}(\Gamma_2, X),$$

where P_i is the set of all parabolic elements of Γ_i. Let $\Gamma_1\alpha\Gamma_2=\bigcup_{i=1}^d \Gamma_1\alpha_i$ be a disjoint union. For every $u\in Z^1_{P_1}(\Gamma_1, X)$, define a map v of Γ_2 into X as follows. Given $\gamma\in\Gamma_2$, let $\alpha_i\gamma=\gamma_i\alpha_j$ with some j and some $\gamma_i\in\Gamma_1$.

Obviously $\alpha_i \mapsto \alpha_j$ is a permutation of $\{\alpha_1, \cdots, \alpha_d\}$. Put

$$v(\gamma) = \sum_{i=1}^{d} \alpha_i^{\iota} u(\gamma_i). \tag{8.3.2}$$

It can be verified in a straightforward way that $v \in Z^1(\Gamma_2, X)$; moreover, $v \in B^1(\Gamma_2, X)$ if $u \in B^1(\Gamma_1, X)$. Further, the cohomology class of v does not depend on the choice of α_i. To see this, let $\beta_i = \delta_i \alpha_i$ with $\delta_i \in \Gamma_1$. Then $\beta_i \gamma = \delta_i \gamma_i \alpha_j = \delta_i \gamma_i \delta_j^{-1} \beta_j$, and

$$\begin{aligned} \sum_i \beta_i^{\iota} u(\delta_i \gamma_i \delta_j^{-1}) &= \sum_i \alpha_i^{\iota} \delta_i^{-1} u(\delta_i \gamma_i \delta_j^{-1}) \\ &= v(\gamma) + (\gamma - 1) \sum_i \alpha_i^{\iota} u(\delta_i^{-1}). \end{aligned} \tag{8.3.3}$$

Thus we obtain the same cohomology class as before. We shall now show that $v \in Z_{P_2}^1(\Gamma_2, X)$. Let $\pi \in P_2$ and $\alpha_i \pi = \xi_i \alpha_j$ with $\xi_i \in \Gamma_1$. In view of (8.3.3), it is sufficient to show that $\sum_i \alpha_i^{\iota} u(\xi_i) \in (\pi - 1)X$ with a *special choice* of α_i. (Note that we may choose the α_i depending even on π.) Therefore we take a subgroup Λ generated by π, and consider disjoint coset decompositions

$$\Gamma_1 \alpha \Gamma_2 = \bigcup \Gamma_1 \zeta \Lambda, \qquad \Gamma_1 \zeta \Lambda = \bigcup_{\nu=0}^{m-1} \Gamma_1 \zeta \pi^{\nu},$$

where m is the smallest positive integer such that $\pi^m \in \zeta^{-1} \Gamma_1 \zeta$; therefore m may depend on ζ. Then we take $\{\zeta \pi^{\nu}\}$ to be $\{\alpha_i\}$. Since $\zeta \pi^{\nu} \pi = \zeta \pi^{\nu+1}$ for $\nu < m-1$, and $\zeta \pi^{m-1} \pi = (\zeta \pi^m \zeta^{-1}) \zeta$, we have $\sum_i \alpha_i^{\iota} u(\xi_i) = \sum_{\zeta} \zeta^{\iota} u(\zeta \pi^m \zeta^{-1})$. We have $\zeta \pi^m \zeta^{-1} \in P_1$, so that $u(\zeta \pi^m \zeta^{-1}) = (\zeta \pi^m \zeta^{-1} - 1) y_{\zeta}$ with an element y_{ζ} of X. Since $\zeta^{\iota} \zeta \pi^m = \pi^m \zeta^{\iota} \zeta$, we have

$$\sum_i \alpha_i^{\iota} u(\xi_i) = \sum_{\zeta} (\pi^m - 1) \zeta^{\iota} y_{\zeta} \in (\pi - 1) X, \qquad \text{q. e. d.}$$

Thus we have shown that v determines an element of $H_{P_2}^1(\Gamma_2, X)$ independent of the choice of $\{\alpha_i\}$. Therefore we define $(\Gamma_1 \alpha \Gamma_2)_X$ to be the map which assigns the cohomology class of v to the cohomology class of u.

The notation being as above, let Ψ be a multiplicative map of Δ into $GL_r(\boldsymbol{R})$ which maps Γ_1 and Γ_2 into compact subgroups of $GL_r(\boldsymbol{R})$. Define $\chi(\alpha)$ for $\alpha \in \Delta$ by (8.2.13), and put $k = n+2$. Suppose that $\Psi(-1) = (-1)^n$ if $-1 \in \Delta$. We can now define a $\boldsymbol{C}$-linear map $[\Gamma_1 \alpha \Gamma_2]_{k,\Psi}$ of $S_k(\Gamma_1, \Psi)$ to $S_k(\Gamma_2, \Psi)$ by

$$f \mid [\Gamma_1 \alpha \Gamma_2]_{k,\Psi} = \det(\alpha)^{k-1} \sum_{i=1}^{d} \Psi(\alpha_i^{\iota}) f(\alpha_i(z)) j(\alpha_i, z)^{-k} \qquad (f \in S_k(\Gamma_1, \Psi)). \tag{8.3.4}$$

It can be verified in a straightforward way that the right hand side belongs to $S_k(\Gamma_2, \Psi)$, and is independent of the choice of $\{\alpha_i\}$; moreover we have

$$\mathfrak{d}(f \mid [\Gamma_1 \alpha \Gamma_2]_{k,\Psi}) = \sum_{i=1}^{d} \chi(\alpha_i^{\iota}) \mathfrak{d}(f) \circ \alpha_i. \tag{8.3.5}$$

PROPOSITION 8.5. *The diagram*

$$\begin{array}{ccc} S_k(\Gamma_1, \Psi) & \xrightarrow{[\Gamma_1\alpha\Gamma_2]_{k,\Psi}} & S_k(\Gamma_2, \Psi) \\ \downarrow \varphi_1 & & \downarrow \varphi_2 \\ H^1_{P_1}(\Gamma_1, X) & \xrightarrow{(\Gamma_1\alpha\Gamma_2)_X} & H^1_{P_2}(\Gamma_2, X) \end{array}$$

is commutative, where φ_1 *and* φ_2 *are the maps defined in* § 8.2, *and* $X = X_n^{\Psi}$.

PROOF. Let $f \in S_k(\Gamma_1, \Psi)$, and $g = f \mid [\Gamma_1\alpha\Gamma_2]_{k,\Psi}$. Define $\mathfrak{f}$ and u by (8.2.19) and (8.2.20). Let $\gamma \in \Gamma_2$, and $\alpha_i\gamma = \gamma_i\alpha_j$ with $\gamma_i \in \Gamma_1$ as above. Then $\varphi_2(g)$ is represented by a cocycle w which is given by

$$w(\gamma) = \int_{z_0}^{\gamma(z_0)} \mathrm{Re}\,(\mathfrak{d}(g))\,.$$

By (8.3.5) and (8.2.20), we have

$$\begin{aligned} w(\gamma) &= \textstyle\sum_{i=1}^d \chi(\alpha_i^\iota) \displaystyle\int_{z_0}^{\gamma(z_0)} \mathrm{Re}\,(\mathfrak{d}(f)) \circ \alpha_i \\ &= \textstyle\sum_{i=1}^d \chi(\alpha_i^\iota)[\mathfrak{f}(\alpha_i\gamma(z_0)) - \mathfrak{f}(\alpha_i(z_0))] \\ &= \textstyle\sum_{i=1}^d \chi(\alpha_i^\iota)[\mathfrak{f}(\gamma_i\alpha_j(z_0)) - \mathfrak{f}(\alpha_i(z_0))] \\ &= \textstyle\sum_{i=1}^d \chi(\alpha_i^\iota)[u(\gamma_i) + \chi(\gamma_i)\mathfrak{f}(\alpha_j(z_0)) - \mathfrak{f}(\alpha_i(z_0))] \\ &= \textstyle\sum_{i=1}^d \chi(\alpha_i^\iota)u(\gamma_i) + [\chi(\gamma) - 1]x\,, \end{aligned}$$

where $x = \sum_{i=1}^d \chi(\alpha_i^\iota)\mathfrak{f}(\alpha_i(z_0))$. Thus w belongs to the same cohomology class as the cocycle v determined by (8.3.2). This proves our proposition.

Let us now complete the proof of Th. 8.4 for non-trivial Ψ. Let $\Gamma_0 = \mathrm{Ker}\,(\Psi)$, and let P_0 be the set of all parabolic elements of Γ_0. Consider $\Gamma_0\alpha\Gamma$ by taking α to be the identity element. We see that $S_k(\Gamma_0, \Psi)$ is the direct sum of r copies of $S_k(\Gamma_0)$, hence the map

$$\varphi_0 : \ S_k(\Gamma_0, \Psi) \longrightarrow H^1_{P_0}(\Gamma_0, X)$$

is surjective by what we have already proved. By virtue of this fact and Prop. 8.5, it is sufficient to show that $(\Gamma_0 \cdot 1 \cdot \Gamma)_X$ is surjective. Therefore, let $\Gamma = \bigcup_{i=1}^d \Gamma_0\alpha_i$, and $t \in Z_P^1(\Gamma, X)$. Let u be the restriction of t to Γ_0. Define v by (8.3.2) with $\gamma \in \Gamma$ and $\gamma_i \in \Gamma_0$. Then

$$v(\gamma) = \textstyle\sum_{i=1}^d \chi(\alpha_i^\iota)t(\alpha_i\gamma\alpha_j^{-1}) = d \cdot t(\gamma) + (\chi(\gamma) - 1) \sum_{i=1}^d t(\alpha_i^{-1})\,.$$

This implies that v belongs to the same cohomology class as $d \cdot t$, hence $(\Gamma_0 \cdot 1 \cdot \Gamma)_X$ is surjective. This completes the proof of Th. 8.4.

8.4. The complex torus associated with the space of cusp forms

Let Γ be a discrete subgroup of $SL_2(\boldsymbol{R})$, which is a Fuchsian group of the first kind, and P the set of all parabolic elements of Γ. We consider a Γ-module D, which is a free $\boldsymbol{Z}$-module of finite rank. Put $D_{\boldsymbol{R}} = D \otimes_{\boldsymbol{Q}} \boldsymbol{R}$. Then the natural injection of $Z_P^1(\Gamma, D)$ into $Z_P^1(\Gamma, D_{\boldsymbol{R}})$ defines a $\boldsymbol{Z}$-linear map

(8.4.1) $$j: \ H_P^1(\Gamma, D) \longrightarrow H_P^1(\Gamma, D_{\boldsymbol{R}}).$$

Regard $H_P^1(\Gamma, D_{\boldsymbol{R}})$ as a vector space over $\boldsymbol{R}$.

PROPOSITION 8.6. *The image of $H_P^1(\Gamma, D)$ by j is a lattice (i.e., a discrete subgroup of maximal rank) of $H_P^1(\Gamma, D_{\boldsymbol{R}})$, and* Ker$(j)$ *is finite.*

PROOF. The group Γ has a finite set of generators, say $\{\sigma_1, \cdots, \sigma_m\}$. (For example, the elements $\{\gamma_\lambda\}$ of the formula (1) in the proof of Th. 2.20 form a set of generators of Γ, cf. Ex. 1.35.) Then every element u of $Z_P^1(\Gamma, X)$, with any Γ-module X, is completely determined by $u(\sigma_1), \cdots, u(\sigma_m)$. This shows that $Z_P^1(\Gamma, D)$ (resp. $Z_P^1(\Gamma, D_{\boldsymbol{R}})$) is finitely generated over $\boldsymbol{Z}$ (resp. $\boldsymbol{R}$). Further we obtain an $\boldsymbol{R}$-linear injective map

$$u \longmapsto (u(\sigma_1), \cdots, u(\sigma_m))$$

of $Z_P^1(\Gamma, D_{\boldsymbol{R}})$ into $D_{\boldsymbol{R}}^m$. The conditions (8.1.1) and (8.1.4) can be written in the form

(1) $$\textstyle\sum_{i=1}^m E_{hi} u(\sigma_i) = 0 \qquad (h = 1, 2, \cdots)$$

with $\boldsymbol{R}$-linear endomorphisms E_{hi} of $D_{\boldsymbol{R}}$ which are stable on D. Similarly, $B^1(\Gamma, D_{\boldsymbol{R}})$ is characterized by the equations

(2) $$\textstyle\sum_{i=1}^m F_{hi} u(\sigma_i) = 0 \qquad (h = 1, 2, \cdots)$$

with maps F_{hi} of the same type. Put

$$Z' = \{u \in Z_P^1(\Gamma, D_{\boldsymbol{R}}) \mid u(\gamma) \in D \text{ for all } \gamma \in \Gamma\},$$

$$B' = Z' \cap B^1(\Gamma, D_{\boldsymbol{R}}).$$

From (1) and (2), we see that Z' (resp. B') is a lattice of $Z_P^1(\Gamma, D_{\boldsymbol{R}})$ (resp. $B^1(\Gamma, D_{\boldsymbol{R}})$), and hence Z'/B' can be identified with a lattice of $H_P^1(\Gamma, D_{\boldsymbol{R}}) = Z_P^1(\Gamma, D_{\boldsymbol{R}})/B^1(\Gamma, D_{\boldsymbol{R}})$. Let Q be a finite subset of P considered in § 8.1. We can find a positive integer t such that

$$t \cdot [D \cap (\pi - 1) D_{\boldsymbol{R}}] \subset (\pi - 1) D$$

for every $\pi \in Q$. By the same argument as in the end of § 8.1, we can show that $t \cdot Z' \subset Z_P^1(\Gamma, D)$. Therefore the image of $H_P^1(\Gamma, D)$ contains $t \cdot (Z'/B')$, and is contained in Z'/B', hence our first assertion. To prove the finiteness of Ker(j), define a map $\lambda: D_{\mathbf{R}} \to D_{\mathbf{R}}^m$ by

$$\lambda(x) = ((\sigma_1 - 1)x, \cdots, (\sigma_m - 1)x) \qquad (x \in D_{\mathbf{R}}).$$

Then we can find a positive integer r such that

$$r \cdot [\lambda(D_{\mathbf{R}}) \cap D^m] \subset \lambda(D).$$

Let $u \in B' \cap Z_P^1(\Gamma, D)$. Then $(u(\sigma_1), \cdots, u(\sigma_m)) = \lambda(x)$ for some $x \in D_{\mathbf{R}}$. Since $\lambda(x) \in \lambda(D_{\mathbf{R}}) \cap D^m$, we can find an element y of D so that $r \cdot \lambda(x) = \lambda(y)$. Then $r \cdot u \in B^1(\Gamma, D)$. Since $Z_P^1(\Gamma, D)$ is finitely generated over $\mathbf{Z}$, this implies that Ker(j) is finite, which completes the proof.

From the above proposition, we obtain especially

$$H_P^1(\Gamma, D_{\mathbf{R}}) = H_P^1(\Gamma, D) \otimes_{\mathbf{Z}} \mathbf{R}. \tag{8.4.2}$$

Now suppose that the above D satisfies the following condition

(8.4.3) *The $\mathbf{R}[\Gamma]$-module $D_{\mathbf{R}}$ is isomorphic to the direct sum of a finite number of modules $X_{n_1}^{\Psi_1}, \cdots, X_{n_s}^{\Psi_s}$ of the type discussed in* § 8.2.

Then, by Th. 8.4, there exists an $\mathbf{R}$-linear isomorphism μ of $H_P^1(\Gamma, D_{\mathbf{R}})$ onto the direct sum

$$\mathfrak{S} = S_{n_1+2}(\Gamma, \Psi_1) \oplus \cdots \oplus S_{n_s+2}(\Gamma, \Psi_s).$$

Put $L = \mu(j(H_P^1(\Gamma, D)))$. Then L is a lattice in $\mathfrak{S}$, so that we obtain a complex torus $\mathfrak{S}/L$.

Let α be an element of $GL_2^+(\mathbf{R})$ such that $\alpha^{-1}\Gamma\alpha$ is commensurable with Γ. Then $\Gamma\alpha\Gamma$ acts both on $\mathfrak{S}$ and on $H_P^1(\Gamma, D_{\mathbf{R}})$. The action commutes with μ by Prop. 8.5. Moreover, it is stable on $H_P^1(\Gamma, D)$, if $\alpha' D \subset D$. Therefore the action of $\Gamma\alpha\Gamma$ defines an endomorphism of $\mathfrak{S}/L$.

For example, let Γ be a subgroup of $SL_2(\mathbf{Z})$ of finite index, and let $D = \mathbf{Z}^{n+1}$ with $n \geq 0$. Through the representation ρ_n, we can regard D as a Γ-module. Then the $\mathbf{R}[\Gamma]$-module $D_{\mathbf{R}}$ is nothing but X_n^{Ψ} with the trivial representation as Ψ, so that $\mathfrak{S} = S_{n+2}(\Gamma)$. Therefore we obtain a lattice L of $S_{n+2}(\Gamma)$ which is stable under $(\Gamma\alpha\Gamma)_{n+2}$ for every $\alpha \in M_2(\mathbf{Z}) \cap GL_2^+(\mathbf{R})$. This proves the statement (3.5.20), which we needed for the proof of Th. 3.48.

It was also shown in [71] that $S_{n+2}(\Gamma)/L$ has a structure of abelian variety if n is even. In [74], this result was generalized to the case of $\mathfrak{S}/L$ of a more general type. For further discussion of the cohomology of this type, the reader may be referred to the papers mentioned in p. 234, Verdier [88], Kuga [41], and Deligne [9]. One should also note the investigations in the higher dimensional case by Matsushima, Murakami, Raghunathan, and Garland.

CHAPTER 9
ARITHMETIC FUCHSIAN GROUPS

9.1. Unit groups of simple algebras

So far, our number-theoretical investigation has been restricted to the Fuchsian groups of congruence type contained in $GL_2(\boldsymbol{Q})$. We shall now show, without detailed proofs, that most of our results can be generalized to arithmetic Fuchsian groups obtained from quaternion algebras. In this section, we shall discuss the group of units of an order in an arbitrary simple algebra over an algebraic number field.

Let B be a simple algebra over $\boldsymbol{Q}$. Then we can define the adele ring B_A and the idele group $B_A^\times$ of B as follows (cf. [96], [99]). Put

$$B_\infty = B_{\boldsymbol{R}} = B \otimes_{\boldsymbol{Q}} \boldsymbol{R}\,,$$

$$B_p = B \otimes_{\boldsymbol{Q}} \boldsymbol{Q}_p \qquad (p: \text{rational prime}).$$

Take any $\boldsymbol{Z}$-lattice $\mathfrak{x}$ in B, and put $\mathfrak{x}_p = \mathfrak{x} \otimes_{\boldsymbol{Z}} \boldsymbol{Z}_p$, and

$$\mathfrak{o}_p(\mathfrak{x}) = \{a \in B_p \mid \mathfrak{x}_p a \subset \mathfrak{x}_p\}\,.$$

Then B_A is the subring of $B_\infty \times \prod_p B_p$ consisting of all the elements $(a_\infty, \cdots, a_p, \cdots)$ such that $a_p \in \mathfrak{o}_p(\mathfrak{x})$ for all except a finite number of p. B_A contains a subring

$$\mathfrak{o}(\mathfrak{x}) = B_\infty \times \prod_p \mathfrak{o}_p(\mathfrak{x})\,,$$

which is a locally compact ring with respect to the usual product topology. We introduce a topology into B_A by taking $\mathfrak{o}(\mathfrak{x})$ to be an open subring of B_A. One can also define B_A to be simply $B \otimes_{\boldsymbol{Q}} \boldsymbol{A}$. Now $B_A^\times$, as an abstract group, is just the group of all invertible elements of B_A. In other words, $B_A^\times$ consists of all the elements $(a_\infty, \cdots, a_p, \cdots)$ such that $a_p \in \mathfrak{o}_p(\mathfrak{x})^\times$ for all except a finite number of p. $B_A^\times$ has a subgroup

$$\mathfrak{o}(\mathfrak{x})^\times = B_\infty^\times \times \prod_p \mathfrak{o}_p(\mathfrak{x})^\times\,,$$

which is a locally compact group with respect to the usual product topology. We introduce a topology into $B_A^\times$ by taking $\mathfrak{o}(\mathfrak{x})^\times$ to be an open subgroup of $B_A^\times$. The definition of the topological ring B_A and the topological group $B_A^\times$ does not depend on the choice of $\mathfrak{x}$. It should also be noted that the topology of $B_A^\times$ is not induced from that of B_A.

Hereafter we write $G_{\boldsymbol{Q}}$ for $B^{\times}$, and put

$$G_{\infty}=B_{\infty}^{\times}, \qquad G_p=B_p^{\times}, \qquad G_{\boldsymbol{A}}=B_{\boldsymbol{A}}^{\times}.$$

One can regard $G_{\boldsymbol{Q}}$ as the group of $\boldsymbol{Q}$-rational points of an algebraic group G defined over $\boldsymbol{Q}$, and $G_{\boldsymbol{A}}$ as the adelization of G. If the reader is not familiar with the general theory of algebraic groups and their adelization, he may consider $G_{\boldsymbol{Q}}$ and $G_{\boldsymbol{A}}$ just new symbols for $B^{\times}$ and $B_{\boldsymbol{A}}^{\times}$. Denote by G_0 the non-archimedean part of $G_{\boldsymbol{A}}$, and by $G_{\infty+}$ the identity component of G_{∞}. These symbols are in agreement with those of Ch. 6, if $B=M_2(\boldsymbol{Q})$.

We identify B (resp. $G_{\boldsymbol{Q}}$) with a subset of $B_{\boldsymbol{A}}$ (resp. $G_{\boldsymbol{A}}$) by means of the diagonal injection $x\mapsto(x, x, x, \cdots)$.

Let F denote the center of B, and ν the reduced norm of B to F. The map ν can be naturally extended to a map of $B_{\boldsymbol{A}}$ to $F_{\boldsymbol{A}}$, which we denote again by ν. (Note that $\nu=\det$, if $B=M_2(\boldsymbol{Q})$.) Put

$$G_{\boldsymbol{A}}^u=\{x\in G_{\boldsymbol{A}} \mid \nu(x)=1\},$$
$$G_{\boldsymbol{Q}}^u=\{x\in G_{\boldsymbol{Q}} \mid \nu(x)=1\}.$$

Then the following theorem is fundamental and well-known.

THEOREM 9.1. (1) *$G_{\boldsymbol{Q}}$ is a discrete subgroup of $G_{\boldsymbol{A}}$.*

(2) *$G_{\boldsymbol{Q}}^u\backslash G_{\boldsymbol{A}}^u$ is compact if B is a division algebra.*

(3) *For any open subgroup S of $G_{\boldsymbol{A}}$ containing G_{∞}, the orbit space $G_{\boldsymbol{Q}}\backslash G_{\boldsymbol{A}}/S$ is finite.*

(4) *For any open subgroup T of $G_{\boldsymbol{A}}^u$ containing G_{∞}^u, the orbit space $G_{\boldsymbol{Q}}^u\backslash G_{\boldsymbol{A}}^u/T$ is finite.*

For the proof, see Weil [96], [99]. These facts can be generalized to reductive algebraic groups, see Borel [2], Borel and Harish-Chandra [3], Mostow and Tamagawa [52], and Godement [24].

Let g be the number of archimedean primes of F, and let $F_{\infty}=F\otimes_{\boldsymbol{Q}}\boldsymbol{R}$. Then we can put

$$B_{\infty}=B_{\infty 1}\oplus\cdots\oplus B_{\infty g}, \tag{9.1.1}$$

$$F_{\infty}=F_{\infty 1}\oplus\cdots\oplus F_{\infty g}, \tag{9.1.2}$$

where $F_{\infty i}$ is the center of $B_{\infty i}$; $F_{\infty i}$ is either $\boldsymbol{R}$ or $\boldsymbol{C}$; $B_{\infty i}$ belongs to the algebras of the following three types: $M_n(\boldsymbol{R})$, $M_n(\boldsymbol{C})$, $M_n(\boldsymbol{H})$, where $\boldsymbol{H}$ denotes the division ring of Hamilton quaternions. Put $G_{\infty i}=B_{\infty i}^{\times}$. Then $G_{\infty}=G_{\infty 1}\times\cdots\times G_{\infty g}$, and $G_{\infty i}$ belongs to the following three types: $GL_n(\boldsymbol{R})$, $GL_n(\boldsymbol{C})$, $GL_n(\boldsymbol{H})$. Put $G_0^u=G_{\boldsymbol{A}}^u\cap G_0$, $G_{\infty}^u=G_{\infty}\cap G_{\boldsymbol{A}}^u$, $G_{\infty i}^u=G_{\infty i}\cap G_{\boldsymbol{A}}^u$. Then $G_{\infty}^u=G_{\infty 1}^u\times\cdots\times G_{\infty g}^u$. Now fix any open compact subgroup T_0 of G_0^u, and put $T=T_0G_{\infty}^u$, $\Gamma_T=T\cap G_{\boldsymbol{Q}}^u$.

PROPOSITION 9.2. *Let Γ denote the projection of Γ_T to G_∞. Then Γ is a discrete subgroup of G_∞^u. Moreover, $\Gamma\backslash G_\infty^u$ is compact if B is a division algebra.*

PROOF. By (1) of Th. 9.1, G_Q^u is discrete in G_A^u, so that Γ_T is a discrete subgroup of $T_0 \times G_\infty^u$. Since T_0 is compact, we see, by (3) of Prop. 1.10, that the projection Γ of Γ_T to G_∞^u is discrete in G_∞^u. On account of (4) of Th. 9.1, $G_Q^u T$ is an open closed subset of G_A^u. Suppose that B is a division algebra. By (2) of Th. 9.1 and Prop. 1.3, one has $G_Q^u T = G_Q^u K$ with a compact subset K of T. Since $T = T_0 G_\infty^u$, we can take K in the form $K = T_0 H$ with a compact subset H of G_∞^u. Write every element of G_A^u in the form (x, y) with $x \in G_0^u$ and $y \in G_\infty^u$. Since $G_\infty \subset G_Q^u T_0 H$, every element $(1, y)$ with $y \in G_\infty^u$ can be written as $(1, y) = (\alpha, \alpha)(t, h)$ with $\alpha \in G_Q^u$, $t \in T_0$, and $h \in H$. Then $\alpha \in G_Q^u \cap T = \Gamma_T$. Since $y = \alpha h$, this shows that $G_\infty^u = \Gamma H$. By Prop. 1.3, $\Gamma\backslash G_\infty^u$ is compact.

9.2. Fuchsian groups obtained from quaternion algebras

By a *quaternion algebra* over a field k, we understand a central simple algebra over k of rank 4. Let $\bar{k}$ denote the algebraic closure of k. Then, an algebra R over k is a quaternion algebra if and only if $R \otimes_k \bar{k}$ is isomorphic to $M_2(\bar{k})$ over $\bar{k}$. A quaternion algebra R over k is either isomorphic to $M_2(k)$, or a division algebra. Let tr and ν denote the reduced trace and the reduced norm of R to k. Then we can define an involution ι of R over k (i.e., a k-linear one-to-one map of R to itself such that $(xy)^\iota = y^\iota x^\iota$, $x = (x^\iota)^\iota$) by

$$x + x^\iota = \mathrm{tr}(x) \qquad (x \in R).$$

In fact, if f is any $\bar{k}$-linear isomorphism of $R \otimes_k \bar{k}$ to $M_2(\bar{k})$, and $f(x) = \begin{bmatrix} a & b \\ c & d \end{bmatrix}$, then we see that $\mathrm{tr}(x) = \mathrm{tr}(f(x))$, and hence

$$f(x^\iota) = \mathrm{tr}(f(x)) - f(x) = \begin{bmatrix} d & -b \\ -c & a \end{bmatrix} = j \cdot {}^t f(x) j^{-1},$$

where $j = \begin{bmatrix} 0 & -1 \\ 1 & 0 \end{bmatrix}$. It follows that ι defines an involution of R over k, which we call *the main involution of* R. We can easily verify that $\nu(x) = xx^\iota$, $(y^{-1}xy)^\iota = y^{-1}x^\iota y$ for all $x \in R$, $y \in R^\times$.

Coming back to the simple algebra B and its center F of § 9.1, we now make the following assumptions:

(9.2.1) *F is totally real;*

(9.2.2) *B is a quaternion algebra over F.*

Then, all the components $F_{\infty i}$ of (9.1.2) must be $\boldsymbol{R}$, so that the components $B_{\infty i}$ of (9.1.1) is either $M_2(\boldsymbol{R})$ or $\boldsymbol{H}$, since $M_2(\boldsymbol{R})$ and $\boldsymbol{H}$ are the only quaternion algebras over $\boldsymbol{R}$. Changing the order of $B_{\infty i}$, if necessary, we may assume

$$B_{\infty i}=\begin{cases} M_2(\boldsymbol{R}) & (1\leqq i\leqq r), \\ \boldsymbol{H} & (r<i\leqq g), \end{cases}$$

where $g=[F:\boldsymbol{Q}]$, and r is the number of the archimedean primes of F which are unramified in B (see the explanation below about P_B). In the following discussion, we always assume

$$r>0, \tag{9.2.3}$$

and fix, *once for all*, the identification of $B_{\infty i}$ with $M_2(\boldsymbol{R})$ or $\boldsymbol{H}$.

The groups G_∞, $G_{\infty+}$, and G_∞^u in the present case can be written as

$$G_\infty=GL_2(\boldsymbol{R})^r\times(\boldsymbol{H}^\times)^{g-r},$$

$$G_{\infty+}=GL_2^+(\boldsymbol{R})^r\times(\boldsymbol{H}^\times)^{g-r},$$

$$G_\infty^u=SL_2(\boldsymbol{R})^r\times(\boldsymbol{H}^u)^{g-r},$$

where $\boldsymbol{H}^u=\{x\in\boldsymbol{H}\mid\nu(x)=1\}$.

PROPOSITION 9.3. *The notation being as in Prop.* 9.2, *let* Γ' *be the projection of* Γ *to the factor* $SL_2(\boldsymbol{R})^r$ *of* $G_{\infty+}$, *under the assumption* (9.2.1, 2, 3). *Then* Γ' *is a discrete subgroup of* $SL_2(\boldsymbol{R})^r$. *Moreover, if* B *is a division algebra,* $\Gamma'\backslash SL_2(\boldsymbol{R})^r$ *is compact.*

This follows immediately from Prop. 9.2, and Prop. 1.10, since $\boldsymbol{H}^u$ is compact.

Now let $GL_2^+(\boldsymbol{R})$ act on $\mathfrak{H}$ as before. Then $GL_2^+(\boldsymbol{R})^r$ acts on $\mathfrak{H}^r$ componentwise. Put

$$G_{A+}=G_0G_{\infty+}, \qquad G_{Q+}=G_Q\cap G_{A+}.$$

We define the action of an element α of G_{Q+} on $\mathfrak{H}^r$ to be the action of the projection of α to the factor $GL_2^+(\boldsymbol{R})^r$ of $G_{\infty+}$. Observe that $F^\times$ is contained in G_{Q+}, and coincides with the set of all elements of G_{Q+} which act trivially on $\mathfrak{H}^r$. We denote by τ_i the injection of F into $\boldsymbol{R}$ obtained by identifying $F_{\infty i}$ with $\boldsymbol{R}$. Then

$$G_{Q+}=\{\alpha\in B\mid\nu(\alpha)^{\tau_i}>0\ (1\leqq i\leqq r)\}.$$

PROPOSITION 9.4. *Let* K *be a totally imaginary quadratic extension of* F, *and* q *an* F*-linear isomorphism of* K *into* B. *Then* $q(K^\times)$ *is contained in* G_{Q+}, *and every element of* $q(K^\times)$, *not contained in* F, *has a unique fixed point* w *on* $\mathfrak{H}^r$, *which is common to all such elements of* $q(K^\times)$. *Moreover,* $q(K^\times)=\{\gamma\in G_{Q+}\mid\gamma(w)=w\}$. *Conversely, if an element* α *of* G_{Q+}, *not contained in* F, *has a fixed point on* $\mathfrak{H}^r$, *then* $F(\alpha)$ *is isomorphic to a totally imaginary* *quadratic extension of F.*

We call w *the fixed point of* $q(K^\times)$ *on* $\mathfrak{H}^r$.

PROOF. Let $a \in K^\times$, $\notin F$, $\alpha = q(a)$, and let $\alpha^{(1)}, \cdots, \alpha^{(r)}$ be the projections of α to $B_{\infty 1}, \cdots, B_{\infty r}$. Let σ_i be an isomorphism of K into $\boldsymbol{C}$ which coincides with τ_i on F. Since K is totally imaginary, we see that the eigen-values of $\alpha^{(i)}$ are a^{σ_i} and $a^{\sigma_i \rho}$, where ρ denotes the complex conjugation. Therefore, $\alpha^{(i)}$ gives an elliptic transformation on $\mathfrak{H}$ (see § 1.2), and hence has a unique fixed point w_i on $\mathfrak{H}$. Put $w = (w_1, \cdots, w_r)$. Let $\beta \in q(K^\times)$. Then $\beta(w) = \beta\alpha(w) = \alpha(\beta(w))$, so that $\beta(w)$ is a fixed point of α on $\mathfrak{H}^r$. Since w is the only fixed point of α, we have $\beta(w) = w$. Suppose $\gamma(w) = w$ with $\gamma \in G_{\boldsymbol{Q}+}$. Observe that the isotropy subgroup

$$\{\xi \in GL_2^+(\boldsymbol{R}) \mid \xi(w_i) = w_i\}$$

is isomorphic to $\boldsymbol{R}^\times \cdot SO(2)$ (see § 1.2), and hence commutative. It follows that γ commutes with every element of $q(K^\times)$. Since $q(K)$ is its commutor in B, γ must belong to $q(K)$. Now, conversely let $\alpha \in G_{\boldsymbol{Q}+}$, $\alpha \notin F$, $\alpha(z) = z$ with $z \in \mathfrak{H}^r$. Let $\alpha^{(i)}$ denote the projection of α to $B_{\infty i}$. Observe that $\alpha^{(i)}$ does not belong to the center $\boldsymbol{R}$ of $B_{\infty i}$. Therefore $\alpha^{(1)}, \cdots, \alpha^{(r)}$ are elliptic, so that none of the eigen-values of $\alpha^{(1)}, \cdots, \alpha^{(r)}$ can be real, hence the first r archimedean primes of F corresponding to $\tau_1, \cdots, \tau_r$ are ramified in $F(\alpha)$. The remaining $g-r$ archimedean primes of F are ramified in every quadratic subfield of B, since they correspond to the factors $\boldsymbol{H}$ of B_∞. Therefore we obtain the last assertion.

One can naturally ask the following question: (i) How many quaternion algebras B, with a given r, over F can be obtained? (ii) What type of quadratic extension K of F is embeddable in B?

To answer these questions, let F_v denote the completion of F with respect to an archimedean or a non-archimedean prime v of F. Put $B_v = B \otimes_F F_v$. Let P_B denote the set of all v such that B_v is a division algebra. A prime v contained (resp. not contained) in P_B is said to be *ramified* (resp. *unramified*) in B. Then the following assertions hold:

(9.2.4) *P_B is a finite set consisting of an even number of primes.*

(9.2.5) *For any finite set P with an even number of archimedean or non-archimedean primes of F, there exists a quaternion algebra B over F, unique up to F-linear isomorphisms, such that $P = P_B$.*

(9.2.6) *A quadratic extension K of F is F-linearly embeddable in B if and only if $K \otimes_F F_v$ is a field for every prime $v \in P_B$.*

These results are special cases of Hasse's theorems on simple algebras over algebraic number fields (see for example [99]). Observe that P_B contains exactly $g-r$ archimedean primes which correspond to the factors $B_{\infty r+1}=\cdots=B_{\infty g}=\boldsymbol{H}$. The set P_B can be empty; we have then $B=M_2(F)$. Therefore B is a division algebra if P_B is not empty, especially if $g>r$.

Let us now consider the case $r=1$. Then we see that B is either a division algebra, or isomorphic to $M_2(\boldsymbol{Q})$. Therefore the group Γ' of Prop. 9.3 is always a Fuchsian group of the first kind; $\Gamma'\backslash\mathfrak{H}$ is compact unless B is isomorphic to $M_2(\boldsymbol{Q})$. We consider F as a subfield of $\boldsymbol{R}$, and assume that the projection of F to the first factor $GL_2^+(\boldsymbol{R})$ of $G_{\infty+}$ (i.e., τ_1 in the above notation) is the identity map of F. This assumption is not absolutely necessary, but simplifies our discussion.

Let K, q, and w be as in Prop. 9.4. In view of the assumption just made, we obtain

$$q(\mu)\begin{bmatrix} w \\ 1 \end{bmatrix}=\mu\begin{bmatrix} w \\ 1 \end{bmatrix}, \qquad \text{or} \quad q(\mu)\begin{bmatrix} w \\ 1 \end{bmatrix}=\bar{\mu}\begin{bmatrix} w \\ 1 \end{bmatrix}$$

for all $\mu\in K$. (We are considering K as an *algebraic number field* in the sense of 0.4, so that K is a subfield of $\boldsymbol{C}$.) We call q *normalized* if $q(\mu)\begin{bmatrix} w \\ 1 \end{bmatrix}=\mu\begin{bmatrix} w \\ 1 \end{bmatrix}$ for all $\mu\in K$. If q is not normalized, its "complex conjugate" q' defined by $q'(\mu)=q'(\bar{\mu})$ is normalized. Thus the non-trivial fixed points of $G_{\boldsymbol{Q}+}$ on $\mathfrak{H}$ are in one-to-one correspondence with the normalized embeddings of totally imaginary quadratic extensions of F into B. Our present discussion generalizes that of § 4.4, except that we have nothing here corresponding to the elliptic curves considered there. Anyway we shall be interested in the values of automorphic functions at these points.

Before going further, let us insert an example of the representation Ψ of § 8.2. Let p_i denote the projection map of $G_{\boldsymbol{Q}}$ into the i-th factor of G_∞. Observe that p_i is injective. Assume $g>1$, and let Γ and Γ' be as in Prop. 9.3. Then p_1 is an isomorphism of Γ to Γ'. If $i>1$, p_i maps Γ into $\boldsymbol{H}^u$. It is well-known that there is a homomorphism f of $\boldsymbol{H}^u$ onto

$$SO(3)=\{X\in GL_3(\boldsymbol{R})\mid {}^tXX=1_3\},$$

such that $\mathrm{Ker}(f)=\{\pm1\}$. Therefore $f\circ p_i\circ p_1^{-1}$, for $i>1$, maps Γ' into a compact subgroup of $GL_3(\boldsymbol{R})$. This gives an example of Ψ considered in § 8.2. One can further obtain some interesting examples of Γ'-modules D satisfying (8.4.3), which are composed of these $f\circ p_i\circ p_1^{-1}$ by the operation of direct sum and tensor product. But we shall not go into details of such modules in this book.

Let us identify $F_A^\times$ with a subgroup of G_A, and denote by F^c the closure of $F^\times F_{\infty+}^\times$ in $F_A^\times$. It can easily be verified that $F^c G_{\infty+}$ is the closure of $F^\times G_{\infty+}$ in G_A. Now denote by $\mathfrak{Z}$ the set of all open subgroups S of G_{A+} containing $F^c G_{\infty+}$ and such that $S/F^c G_{\infty+}$ is compact. Put, for each $S \in \mathfrak{Z}$,

$$\Gamma_S = S \cap G_{Q+}. \tag{9.2.7}$$

Observe that $F^\times \subset \Gamma_S$.

PROPOSITION 9.5. *For any $S \in \mathfrak{Z}$, the group $\Gamma_S/F^\times$, as a transformation group on $\mathfrak{H}$, is a Fuchsian group of the first kind.*

PROOF. Since Γ_S and $\Gamma_{S'}$ are commensurable for any two members S and S' of $\mathfrak{Z}$, it is sufficient to prove our assertion for one S. Take any maximal order $\mathfrak{o}$ in B, and put $R = G_{\infty+} \times \prod_p \mathfrak{o}_p^\times$, where $\mathfrak{o}_p = \mathfrak{o} \otimes_Z Z_p$, and $T = R \cap G_A^u$, $S = F^\times R$, $\Gamma_R = R \cap G_{Q+}$, $\Gamma_T = T \cap G_Q^u$. Then $S \in \mathfrak{Z}$, and $\Gamma_S = F^\times \Gamma_R$. Let E denote the group of all units in F. Then $\nu(x) \in E$ if $x \in \Gamma_R$. Put $E^{(2)} = \{e^2 \mid e \in E\}$, and $\Gamma' = \{\gamma \in \Gamma_R \mid \nu(\gamma) \in E^{(2)}\}$. Then $[\Gamma_R : \Gamma']$ is finite, since $[E : E^{(2)}]$ is finite. If $\gamma \in \Gamma'$, then $\nu(\gamma) = e^2$ with $e \in E$, so that $\nu(e^{-1}\gamma) = 1$. Since $e \in R$, $e^{-1}\gamma$ is contained in Γ_T. This proves that $\Gamma' \subset E\Gamma_T$. From our definition of Γ' and Γ_T, we obtain $E\Gamma_T \subset \Gamma'$, so that $\Gamma' = E\Gamma_T$. Therefore

$$[\Gamma_S : F^\times \Gamma_T] = [F^\times \Gamma_R : F^\times \Gamma'] \leqq [\Gamma_R : \Gamma'] < \infty.$$

As is seen above, by virtue of Prop. 9.3, Γ_T is a Fuchsian group of the first kind. This proves our proposition.

Define a homomorphism σ of G_A to $\mathrm{Gal}(F_{ab}/F)$ by

$$\sigma(x) = [\nu(x)^{-1}, F] \qquad (x \in G_A).$$

(For the notation $[s, F]$ with $s \in F_A^\times$, see § 5.2.) We see that $F^\times \cdot \nu(S)$ is an open subgroup of $F_A^\times$ of finite index for every $S \in \mathfrak{Z}$. By class field theory, it corresponds to a subfield of F_{ab} of finite degree over F, which we denote by k_S. Then Lemmas 6.16 and 6.17 are true in the present case. It should also be noted that Lemma 6.15 is true with G_A^u in place of $SL_2(A)$, by virtue of the approximation theorem due to Eichler [15] and Kneser [39].

We are ready to state the first main theorem of this section, which is a generalization of the discussion of § 6.7 and Th. 6.31.

THEOREM 9.6. *There exists a system*

$$\{V_S, \varphi_S, J_{TS}(x), (S, T \in \mathfrak{Z};\ x \in G_{A+})\},$$

formed by the objects satisfying the following conditions.

(1) *For each $S \in \mathfrak{Z}$, (V_S, φ_S) is a model of $\mathfrak{H}^*/\Gamma_S$ in the sense of* § 6.7, *where $\mathfrak{H}^*$ is $\mathfrak{H}$ or $\mathfrak{H} \cup Q \cup \{\infty\}$, according as B is a division algebra or not.*

(2) *V_S is defined over k_S.*

(3) *$J_{TS}(x)$, defined if and only if $xSx^{-1}\subset T$, is a morphism of V_S onto $V_T^{\sigma(x)}$, rational over k_S, and has the following properties:*

(3 a) *$J_{TS}(x)$ is the identity map if $x\in S$;*

(3 b) *$J_{TS}(x)^{\sigma(y)}\circ J_{SR}(y)=J_{TR}(xy)$;*

(3 c) *$J_{TS}(\alpha)[\varphi_S(z)]=\varphi_T(\alpha(z))$ for every $\alpha\in G_{Q+}$ and every $z\in\mathfrak{H}$ (if $\alpha S\alpha^{-1}\subset T$).*

(4) *Let K be a totally imaginary quadratic extension of F, q a normalized F-linear isomorphism of K into B, and w the fixed point of $q(K^\times)$ on $\mathfrak{H}$ (see Prop. 9.4). Then, for every $S\in\mathfrak{Z}$, $\varphi_S(w)$ is rational over K_{ab}. Moreover, for every $u\in K_A^\times$, one has*

$$\varphi_T(w)^{[u,K]}=J_{TS}(q(u)^{-1})[\varphi_S(w)],$$

where $T=q(u)^{-1}Sq(u)$.

The system is unique in the following sense.

THEOREM 9.7. *If two systems $\{V_S, \varphi_S, J_{TS}(x)\}$ and $\{V_S', \varphi_S', J_{TS}'(x)\}$ satisfy all the conditions of the above theorem, then there exists, for each $S\in\mathfrak{Z}$, a biregular isomorphism P_S of V_S to V_S', rational over k_S, such that*

$$\varphi_S'=P_S\circ\varphi_S, \qquad J_{TS}'(x)\circ P_S=P_T^{\sigma(x)}\circ J_{TS}(x)$$

for all S, T of $\mathfrak{Z}$ and all $x\in G_{A+}$ satisfying $xSx^{-1}\subset T$.

It is easy to give a generalization of Prop. 6.33, which may be left to the reader as an exercise.

In the next place, to generalize Th. 6.23, put

$$\mathfrak{F}_S=\{f\circ\varphi_S\mid f\in k_S(V_S)\}, \qquad \mathfrak{F}=\bigcup_{S\in\mathfrak{Z}}\mathfrak{F}_S.$$

Then (1) of Th. 9.6 implies that $\boldsymbol{C}\mathfrak{F}_S$ is the field of all automorphic functions with respect to Γ_S. Also we have $k_S=F_{ab}\cap\mathfrak{F}_S$, and $F_{ab}=\boldsymbol{C}\cap\mathfrak{F}$.

For every $x\in G_{A+}$, we can define an automorphism $\tau(x)$ of $\mathfrak{F}$ over F by

$$(f\circ\varphi_T)^{\tau(x)}=f^{\sigma(x)}\circ J_{TS}(x)\circ\varphi_S \qquad (f\in k_T(V_T),\ S=x^{-1}Tx). \tag{9.2.8}$$

PROPOSITION 9.8. *The symbol $\tau(x)$ has the following properties.*

(i) *$\tau(xy)=\tau(x)\tau(y)$, i.e., τ defines a homomorphism of G_{A+} into* Aut$(\mathfrak{F}/F)$.

(ii) *$\tau(x)=\sigma(x)$ on F_{ab}.*

(iii) *$h^{\tau(\alpha)}(z)=h(\alpha(z))$ for every $h\in\mathfrak{F}$, $\alpha\in G_{Q+}$, and $z\in\mathfrak{H}$.*

PROOF. The equality (ii) follows directly from the definition (9.2.8); (i) from (3 b) of Th. 9.6; (iii) from (3 c) of Th. 9.6.

Now Th. 6.23 and Th. 6.31 can be generalized as follows.

THEOREM 9.9. *The sequence*

$$1 \longrightarrow F^c G_{\infty+} \longrightarrow G_{A+} \xrightarrow{\tau} \mathrm{Aut}(\mathfrak{F}/F) \longrightarrow 1$$

is exact. The map τ *is continuous, and induces a topological isomorphism of* $G_{A+}/F^c G_{\infty+}$ *onto* $\mathrm{Aut}(\mathfrak{F}/F)$. *Moreover, for every* $S \in \mathfrak{Z}$, *one has*

(1) $S = \{x \in G_{A+} \mid h^{\tau(x)} = h$ *for all* $h \in \mathfrak{F}_S\}$, *i. e.*, $\tau(S) = \mathrm{Gal}(\mathfrak{F}/\mathfrak{F}_S)$.

(2) $\mathfrak{F}_S = \{h \in \mathfrak{F} \mid h^{\tau(x)} = h$ *for all* $x \in S\}$.

THEOREM 9.10. *Let* K, q, *and* w *be as in* (4) *of Th.* 9.6. *Then, for every* $h \in \mathfrak{F}$, *defined and finite at* w, *the value* $h(w)$ *belongs to* K_{ab}, *and*

$$h(w)^{[u,K]} = h^{\tau(q(u)^{-1})}(w)$$

for every $u \in K_A^\times$.

Th. 9.10 follows immediately from (4) of Th. 9.6.

PROOF of Th. 9.9. It is straightforward to see that $h^{\tau(x)} = h$ for $h \in \mathfrak{F}_S$ and $x \in S$. Conversely, suppose that $\tau(x) = \mathrm{id.}$ on $\mathfrak{F}_S$. By (ii) of Prop. 9.8, $\tau(x) = \mathrm{id.}$ on k_S. By the generalization of Lemma 6.17 mentioned above, we have $x = s\alpha$ for some $s \in S$ and $\alpha \in G_{Q+}$. Then, for every $f \in k_S(V_S)$, we have

$$f \circ \varphi_S = (f \circ \varphi_S)^{\tau(s\alpha)} = f \circ J_{ST}(s\alpha) \circ \varphi_T = f \circ J_{ST}(\alpha) \circ \varphi_T\,,$$

where $T = \alpha^{-1} S \alpha$, so that $\varphi_S = J_{ST}(\alpha) \circ \varphi_T$, and hence $\varphi_S(z) = \varphi_S(\alpha(z))$ for all $z \in \mathfrak{H}$. Therefore $\alpha \in \Gamma_S$, so that $x \in S$. This proves the first equality of (1) of Th. 9.9. It follows from this result that

$$\mathrm{Ker}(\tau) = \textstyle\bigcap_{S \in \mathfrak{Z}} S = F^c G_{\infty+}\,.$$

Now we can repeat the proof of Th. 6.23, and obtain the surjectivity and the continuity of τ. If $B \neq M_2(\boldsymbol{Q})$, we can dispense with the discussion about cusps. The equality (2) of Th. 9.9 follows from (1) and Prop. 6.11.

PROPOSITION 9.11. (i) *Let* G^c *denote the closure of* $G_{Q+} G_{\infty+}$ *in* G_A. *Then*

$$G^c = F^c G_{Q+} G_A^u = \{x \in G_{A+} \mid \nu(x) \in F^c\}\,.$$

(ii) *For every* $S \in \mathfrak{Z}$, $G^c \cap S$ *is the closure of* $\Gamma_S G_{\infty+}$ *in* G_A.

(iii) $\tau(G^c \cap S) = \{\sigma \in \mathrm{Aut}(\mathfrak{F}/F) \mid \sigma = \mathrm{id.}$ on $F_{ab} \cdot \mathfrak{F}_S\} = \mathrm{Gal}(\mathfrak{F}/F_{ab} \cdot \mathfrak{F}_S)$.

To prove this we need

LEMMA 9.12. *Let* E_+ *be the group of all totally positive units of* F, E_0 *the projection of* E_+ *to the non-archimedean part of* $F_A^\times$, *and* $\bar{E}_0$ *the closure of* E_0 *in* $F_A^\times$. *For a positive integer* n, *put*

$$\bar{E}_0^{(n)} = \{x^n \mid x \in \bar{E}_0\}\,, \qquad F^{c(n)} = \{x^n \mid x \in F^c\}\,.$$

Then $F^c = \bar{E}_0 F^\times F_{\infty+}^\times$, $\bar{E}_0 = E_0 \bar{E}_0^{(n)}$, *and* $F^c = F^\times F^{c(n)}$ *for every positive integer* n.

PROOF. Let $\{U_m\}_{m=1}^{\infty}$ be a family of compact groups which form a basis of neighborhoods of the identity in the non-archimedean part of $F_A^\times$. Let $x \in F^c$. Then, for every m, there exists an element y_m of $F^\times$ such that $y_m^{-1}x \in U_m F_{\infty+}^\times$. Put $e_m = y_1^{-1}y_m$. Then $e_m \in E_+$ and $e_m^{-1}y_1^{-1}x \in U_m F_{\infty+}^\times$. Therefore the non-archimedean part of $y_1^{-1}x$ belongs to $\bar{E}_0$. This shows that $F^c \subset \bar{E}_0 F^\times F_{\infty+}^\times$. Since the opposite inclusion is obvious, we obtain the first assertion. Next, since $\{x^n \mid x \in E_0\}$ is of finite index in E_0, we have $[E_0\bar{E}_0^{(n)} : \bar{E}_0^{(n)}] < \infty$. We see also that $\bar{E}_0^{(n)}$ is closed, since it is the image of the compact set $\bar{E}_0$ under the continuous map $x \mapsto x^n$. Therefore $E_0\bar{E}_0^{(n)}$ is closed, hence the second assertion. The last assertion follows easily from the first and second ones.

PROOF of Prop. 9.11. Since $F^\times \subset G_{Q+}$, we have $F^c \subset G^c$. The strong approximation theorem, mentioned above, (of which Lemma 6.15 is a special case) implies that $G_A^u \subset G_{Q+}U$ for any open subgroup U of G_A, so that $G_A^u \subset G^c$. Therefore we obtain

$$F^c G_{Q+} G_A^u \subset G^c \subset \{x \in G_{A+} \mid \nu(x) \in F^c\}.$$

Let $x \in G_{A+}$ and $\nu(x) \in F^c$. By Lemma 9.12, $\nu(x) = ab^2$ with $a \in F^\times$ and $b \in F^c$. We see that a is totally positive. By virtue of the norm theorem of simple algebras (see, for example, [99, p. 206, Prop. 3]), we have $a = \nu(\alpha)$ for some $\alpha \in B^\times = G_Q$. Then $\nu(b^{-1}\alpha^{-1}x) = 1$, so that $x = b\alpha \cdot (\alpha^{-1}b^{-1}x) \in F^c G_{Q+} G_A^u$, which proves (i). Next, let $S \in \mathcal{Z}$. For every open subgroup U of G_{A+}, we have $G^c \subset G_{Q+}U$. Therefore, if $U \subset S$, we have $G^c \cap S \subset (G_{Q+} \cap S) \cdot U = \Gamma_S U$, so that $G^c \cap S$ is contained in the closure of $\Gamma_S G_{\infty+}$. Since the opposite inclusion is obvious, we obtain (ii). By (i), we have $G^c = \{x \in G_{A+} \mid \sigma(x) = 1\}$. This together with (1) of Th. 9.9 proves (iii).

EXAMPLE 9.13. Let $m = 7$, 9, or 11, and let $F = F_m = Q(\zeta + \zeta^{-1})$ with $\zeta = e^{2\pi i/m}$. Then $[F : Q] = 3, 3, 5$, respectively. Since $[F : Q]$ is odd, there exists, by virtue of (9.2.5), a unique quaternion algebra B over F which is

unramified at { all non-archimedean primes of F,
the archimedean prime of F corresponding to the identity map of F,

ramified at all the remaining archimedean primes of F.

Take a maximal order $\mathfrak{o}$ in B, by which we mean a maximal subring of B that is a free Z-module of rank $[B : Q]$. Put $\mathfrak{o}_p = \mathfrak{o} \otimes_Z Z_p$ for every rational prime p, and $U = G_{\infty+} \times \prod_p \mathfrak{o}_p^\times$, $S = F^\times U$. Since U is open in G_A, we see that $F^c G_{\infty+} \subset S$. Moreover, $U/G_{\infty+}$ is compact, so that $S \in \mathcal{Z}$. It can be shown that $G_A = G_Q U$ and $F_A^\times = F^\times \cdot \nu(U)$, so that $k_S = F$. (This follows from the fact that the class number of F in the narrow sense is one.) We see that $\Gamma_S = F^\times \Gamma(\mathfrak{o})$, where

$$\Gamma(\mathfrak{o}) = \{\gamma \in \mathfrak{o} \mid \gamma\mathfrak{o} = \mathfrak{o},\ \nu(\gamma) > 0\}.$$

Now one can prove that $\Gamma_S/F^\times$ is a "triangle group" generated by three elliptic elements $\gamma_2, \gamma_3, \gamma_m$ of order $2, 3, m$ such that $\gamma_2\gamma_3\gamma_m = 1$; $\mathfrak{H}/\Gamma_S$ is of genus 0; every elliptic element of $\Gamma_S/F^\times$ is conjugate in $\Gamma_S/F^\times$ to a power of γ_2, γ_3, or γ_m. Let z_2, z_3, and z_m be the fixed points of γ_2, γ_3, and γ_m on $\mathfrak{H}$ respectively. Since $\mathfrak{H}/\Gamma_S$ is of genus 0, there is a Γ_S-automorphic function on $\mathfrak{H}$ which gives a biregular isomorphism of $\mathfrak{H}/\Gamma_S$ onto the complex projective line V. One can normalize such a function φ by the condition

(9.2.9) $$\varphi(z_2) = 1, \qquad \varphi(z_3) = 0, \qquad \varphi(z_m) = \infty.$$

Then (V, φ) can be taken as the member (V_S, φ_S) of the system of Th. 9.6 for the present B.

On account of (9.2.6), for every totally imaginary quadratic extension K of F, there exists a normalized F-linear isomorphism q of K into B. Moreover, one can take q so that $q(\mathfrak{o}_K) \subset \mathfrak{o}$, where $\mathfrak{o}_K$ denotes the maximal order in K. If q and q' are such F-linear isomorphisms of K into B, there is an element α of G_{Q+} such that $\alpha^{-1}q(\mu)\alpha = q'(\mu)$ for all $\mu \in K$. Then there exists a fractional ideal $\mathfrak{a}$ in K such that $q(\mathfrak{a})\mathfrak{o} = \alpha\mathfrak{o}$. The ideal $\mathfrak{a}$ is principal if and only if $\gamma^{-1}q(\mu)\gamma = q'(\mu)$ for all $\mu \in K$, with an element γ of Γ_S. In this way we can show that, if h is the class number of K, there are exactly h points $w_1, \cdots, w_h$, modulo Γ_S-equivalence, which represent the fixed points of $q(K^\times)$ for all such q satisfying $q(\mathfrak{o}_K) \subset \mathfrak{o}$. From (4) of Th. 9.6, we obtain

(9.2.10) *The values $\varphi(w_1), \cdots, \varphi(w_h)$ form a complete set of conjugates of $\varphi(w_1)$ over K, and $K(\varphi(w_1))$ is the maximal unramified abelian extension of K.*

For q, q', α, and $\mathfrak{a}$ as above, let w be the fixed point of $q(K^\times)$, and $\sigma = \left(\frac{K(\varphi(w))/K}{\mathfrak{a}}\right)$. Then Th. 9.10, or (4) of Th. 9.6, implies

(9.2.11) $$\varphi(w)^\sigma = \varphi(\alpha^{-1}(z)).$$

We observe that (9.2.10, 11) are similar to Th. 5.5 and (5.4.2). Actually it can also be shown that $\{\varphi(w_1), \cdots, \varphi(w_h)\}$ is a complete set of conjugates of $\varphi(w_1)$ not only over K, but also over F. Thus φ is an analogue of the modular function j. The condition (9.2.9) corresponds to $j(i) = 1$, $j(e^{2\pi i/3}) = 0$, $j(\infty) = \infty$. Finally we note that in the case $m = 7$, $\Gamma_S/F^\times$ is the Fuchsian group with the least measure of the fundamental domain, which was mentioned at the end of § 2.5.

Unfortunately, the proof of Th. 9.6 is too long and intricate to include in this book. It needs a detailed analysis of certain families of abelian varieties parametrized by the variable z on $\mathfrak{H}$. These abelian varieties play, to some extent, the role of elliptic curves in Chapter 6. The proof of Th. 9.7 is comparatively easy; it may be a good exercise to give a proof at least in the simplest case $B=M_2(\boldsymbol{Q})$. Actually we can generalize our theory to the case of algebraic groups whose arithmetic subgroups act on a product of Siegel upper half spaces. For details, the reader is referred to [77], [78], [80]. As for Ex. 9.11, see [77, 3.18].

We can of course propose a further generalization to the whole family of semi-simple or reductive algebraic groups whose arithmetic subgroups act on bounded symmetric domains. The case of unitary groups over algebras with involutions of the second kind has been treated by K. Miyake [49]. Therefore, roughly speaking, the theory has been established for one half of the family of all bounded symmetric domains of classical type. It seems quite likely that it can be extended to the remaining half. It is not clear, however, whether the semi-simple groups of exceptional type can be included in this framework.

The theory of Hecke operators can be developed also for the groups Γ_S of the above type. We can then construct Dirichlet series, similar to those of Ch. 3, which have Euler product and functional equation (see [74]). Further, these Dirichlet series, for cusp forms of weight 2, provide, the zeta-functions of the curves V_S of Th. 9.6, exactly in the same manner as in §§ 7.4, 7.5. For details, the reader is referred to [77], [80, 2.23], and [50]. Finally we mention that the curves V_S, or rather the above theorems, are in close connection with Ihara's recent investigation [34].

APPENDIX

The purpose of this part is to recall a few elementary facts on algebraic varieties, especially on algebraic curves and abelian varieties. We do not mean to present an introduction to algebraic geometry for the reader who is totally unfamiliar with the subject. Our intention is merely to remind a more experienced reader of some fundamental definitions, after Weil's Foundations [90], and to make sure what terminology we are using, and what results are referred to in the text.

1. We fix a *universal domain* Ω, which is an algebraically closed field of infinite transcendence degree over the prime field. If the characteristic is 0, we often take the complex number field $\boldsymbol{C}$ as Ω. By *a field*, we always mean, except when the contrary is stated, a subfield of Ω over which Ω is of infinite transcendence degree. If k is a field and $x=(x_1, \cdots, x_n)$ is a set of elements x_i of Ω, we denote by $k(x)=k(x_1, \cdots, x_n)$ the field generated by $x_1, \cdots, x_n$ over k, which is again a field in that sense. We say that $k(x)$ is a *regular extension of* k, if k is algebraically closed in $k(x)$, and $k(x)$ is a separable algebraic extension of a purely transcendental extension of k, or equivalently, if $k(x)$ is linearly disjoint from the algebraic closure of k, over k.

Consider an affine space $\mathfrak{A}_n$ and a projective space $\mathfrak{P}_n$ over Ω, of dimension n, with a fixed coordinate system. Let $a=(a_1, \cdots, a_n)$ and $b=(b_1, \cdots, b_n)$ be points of $\mathfrak{A}_n$. We say that b is a *specialization of a over* a field k, if $F(b_1, \cdots, b_n)=0$ for every polynomial $F(X_1, \cdots, X_n)$ with coefficients in k such that $F(a_1, \cdots, a_n)=0$. Then we denote by $[a\rightarrow b\,;\, k]$ the ring of all elements of the form $P(a)/Q(a)$ with polynomials P and Q with coefficients in k such that $Q(b)\neq 0$. For a point $x=(x_0, x_1, \cdots, x_n)$ of $\mathfrak{P}_n$, let $a_\lambda(x)$ denote the point $(x_0/x_\lambda, x_1/x_\lambda, \cdots, x_n/x_\lambda)$ of $\mathfrak{A}_{n+1}$, whenever $x_\lambda\neq 0$. For $x\in\mathfrak{P}_n$ and $y\in\mathfrak{P}_n$, we say that y is a *specialization of x over k*, if there is an index λ such that $x_\lambda\neq 0$, $y_\lambda\neq 0$, and $a_\lambda(y)$ is a specialization of $a_\lambda(x)$ over k. More generally, put

$$\mathfrak{X}=\mathfrak{P}_{n_1}\times\cdots\times\mathfrak{P}_{n_r}\times\mathfrak{A}_{m_1}\times\cdots\times\mathfrak{A}_{m_s},$$

and let $x=(x^{(1)}, \cdots, x^{(r+s)})$ and $y=(y^{(1)}, \cdots, y^{(r+s)})$ be points of $\mathfrak{X}$, where $x^{(i)}$ and $y^{(i)}$ are points of $\mathfrak{P}_{n_i}$ or $\mathfrak{A}_{m_{i-r}}$ according as $i\leqq r$ or $i>r$. Then we say that y is a *specialization of x over k*, if $y'=(a_{\lambda_1}(y^{(1)}), \cdots, a_{\lambda_r}(y^{(r)}), y^{(r+1)}, \cdots, y^{(r+s)})$ is a specialization of $x'=(a_{\lambda_1}(x^{(1)}), \cdots, a_{\lambda_r}(x^{(r)}), x^{(r+1)}, \cdots, x^{(r+s)})$ over k, for some $\lambda_1, \cdots, \lambda_r$. We then put

$$[x\rightarrow y\,;\, k]=[x'\rightarrow y'\,;\, k],$$

since this ring does not depend on the choice of $\lambda_1, \cdots, \lambda_r$. We denote by $k(x)$ the field $k(a_{\lambda_1}(x^{(1)}), \cdots, a_{\lambda_r}(x^{(r)}), x^{(r+1)}, \cdots, x^{(r+s)})$.

2. A set V of points of $\mathfrak{X}$ is called a *variety* (or an *algebraic variety*) if there exists a field k and a point x of $\mathfrak{X}$ such that

(i) *V is the set of all specializations of x over k;*

(ii) *$k(x)$ is a regular extension of k.*

(The condition (ii) implies that V is absolutely irreducible in the usual terminology.) If V, x, and k are in this situation, we say that: V is *defined* (or *rational*) *over* k; k is a *field of definition* (or *of rationality*) for V; x is a *generic point of V over k*; V is *the locus of x over k*. The transcendence degree of $k(x)$ over k is uniquely determined by V, and called *the dimension of V*. A variety contained in V is called a *subvariety* of V. A point of V is a zero-dimensional subvariety of V, and vice versa. A subvariety of $\mathfrak{A}_n$ (resp. $\mathfrak{P}_n$) is called an *affine* (resp. a *projective*) *variety*. We say that a projective variety V is defined by equations $F_i(X_0, \cdots, X_n)=0$ $(i=1, \cdots, t)$ if these polynomials generate, over $\Omega[X_0, \cdots, X_n]$, the ideal of all the polynomials vanishing on V.

3. If two varieties V and W are given, we can find a common field k of rationality for V and W; further we can find a generic point x of V over k and a generic point y of W over k such that $k(x)$ is linearly disjoint with $k(y)$ over k. Then the set-theoretical product $V\times W$ is the locus of (x, y) over k, so that it is a variety. A subvariety T of $V\times W$ is called a *rational map* of V to W, defined over k, if, for a generic point (u, v) of T over k, one has $k(u, v)=k(u)$, and u is a generic point of V over k. We say that T is *defined at* a point a of V, if there exists a point b of W such that $(a, b)\in T$, and

$$[v\to b;\ k]\subset[u\to a;\ k].$$

The point b is uniquely determined by a under that condition, so that we put $b=T(a)$. Especially we always have $T(u)=v$. If S is a rational map of W to a variety X defined over k, and if S is defined at v, then we denote by $S\circ T$ the locus of $(u, S(v))$ over k, which is a rational map of V to X. We call T a *morphism* if T is everywhere defined on V. T is called *birational* if $k(u)=k(v)$, and v is generic on W over k. If that is so, we denote by T^{-1} the locus of (v, u) over k, which is a rational map of W to V. We say that V is *birationally equivalent to W over k*, if there is a birational map of V to W defined over k. T is called a (*biregular*) *isomorphism* if it is birational, and both T and T^{-1} are morphisms.

4. A rational map of V to the affine 1-space $\mathfrak{A}_1$ is called a *function* (or rather a *meromorphic function*) on V. All the functions on V form a field, not contained in Ω unless $\dim(V)=0$, which is denoted by $\Omega(V)$. All the elements of $\Omega(V)$ rational over a field k of definition for V form a subfield of $\Omega(V)$, denoted by $k(V)$. Then $k(V)$ is linearly disjoint from Ω over k, and $\Omega(V)=\Omega\cdot k(V)$. For a generic point x of V over k, the map $f\mapsto f(x)$ gives an isomorphism of $k(V)$ onto $k(x)$.

5. If V is the locus of x over k, and $a\in V$, we say that a is a *simple point* of V, or a is simple on V, if there exists a birational map T of V to a subvariety W of A_n satisfying the following conditions:

(i) T is defined at a, and T^{-1} is defined at $T(a)$;

(ii) If $b=T(a)$, $y=T(x)$, and $r=\dim(V)$, then there are $n-r$ polynomials $F_i(X_1,\cdots,X_n)$ $(i=1,\cdots,n-r)$, with coefficients in k, such that $F_i(y)=0$ $(i=1,\cdots,n-r)$, and

$$\operatorname{rank}\left[\frac{\partial F_i}{\partial X_j}(b)\right]_{i,j}=n-r.$$

This definition does not depend on the choice of W and T. V is called *non-singular* if every point of V is simple.

If the universal domain is $\boldsymbol{C}$, $\mathfrak{A}_n$ and $\mathfrak{P}_n$ are viewed as complex manifolds. Then every non-singular variety of dimension r has a natural structure of complex manifold of complex dimension r. Every projective variety is compact.

6. Let V be a variety defined over k, and σ an isomorphism of k into Ω. Take a generic point x of V over k. Then we can extend σ to an isomorphism τ of $k(x)$ into Ω. Put $x'=x^\tau$. Then the locus V' of x' over k^σ is meaningful, and depends only on V and σ, i.e., it does not depend on the choice of x and τ. We put $V'=V^\sigma$, and call it the transform of V under σ. If T is a rational map of V to W rational over k, we can define T^σ and W^σ, and observe that T^σ is a rational map of V^σ to W^σ. Especially if $f\in k(V)$, then f^σ is a function on V^σ. If T is defined at a point a of V rational over k, then T^σ is defined at a^σ, and $T(a)^\sigma=T^\sigma(a^\sigma)$.

The symbols, V, x, and k being as above, let W be another variety with a generic point y over k. Suppose that there is an isomorphism ξ of $k(W)$ to $k(V)$, which induces an automorphism ρ of k. Then there exists a birational map J_ξ of V to W^ρ which is characterized by the property

(6.1) $f^\xi=f^\rho\circ J_\xi$ *for every* $f\in k(W)$.

To show this, define an isomorphism τ of $k(y)$ to $k(x)$ by $f(y)^\tau=f^\xi(x)$ for $f\in k(W)$, so that the diagram

$$\begin{array}{ccccc} f & \in k(W) & \xrightarrow{\quad\xi\quad} & k(V)\ni & g \\ \downarrow & \downarrow & & \downarrow & \downarrow \\ f(y) & \in k(y) & \xrightarrow{\quad\tau\quad} & k(x)\ni & g(x) \end{array}$$

is commutative. Since y^τ is generic on W^ρ over k, and $k(y^\tau)=k(x)$, we obtain a birational map J_ξ of V to W^ρ, defined over k, such that $J_\xi(x)=y^\tau$. Then we have $f^\xi(x)=f(y)^\tau=f^\rho(y^\tau)=f^\rho(J_\xi(x))$, hence (6.1).

If η is an isomorphism of $k(X)$ to $k(W)$ with another variety X defined over k, which induces an automorphism σ of k, then $J_\eta: W\to X^\sigma$ and $J_{\eta\xi}: V\to X^{\sigma\rho}$ is meaningful, and

(6.2) $$J_{\eta\xi}=J_\eta^\rho\circ J_\xi\,.$$

7. Suppose that the characteristic of Ω is $p>0$, and let $q=p^e$ with an integer e. Then $a\mapsto a^q$ is an automorphism of Ω. We denote by V^q the transform of a variety V under this automorphism. (In the usual circumstances, V^q will not be confused with the product of q copies of V.) If $e>0$, and V is the locus of x over k, we can define a morphism F of V to V^q rational over k by $F(x)=x^q$, which is called the q-th *power morphism* (or *the Frobenius morphism of degree* q) of V to V^q.

Let T be a rational map of V to W, and F' the q-th power morphism of W to W^q. Then we have

(7.1) $$F'\circ T=T^q\circ F,$$

(where T^q is of course the transform of T under the q-th power automorphism of Ω). In other words, the following diagram is commutative:

$$\begin{array}{ccc} V & \xrightarrow{\quad T\quad} & W \\ {\scriptstyle F}\downarrow & & \downarrow{\scriptstyle F'} \\ V^q & \xrightarrow{\quad T^q\quad} & W^q \end{array}$$

8. If W is a variety of dimension n, there exist an n-dimensional vector space $\mathrm{Dif}(W)$ over $\Omega(W)$ and an Ω-linear map $d: \Omega(W)\to\mathrm{Dif}(W)$ with the following properties:

(8.1) $d(fg)=f\cdot dg+g\cdot df \qquad (f, g\in\Omega(W))$,

(8.2) *$\{df_1, \cdots, df_n\}$ is a basis of* $\mathrm{Dif}(W)$ *over $\Omega(W)$ if and only if $\Omega(V)$ is separably algebraic over $\Omega(f_1, \cdots, f_n)$.*

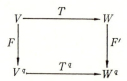

The couple (Dif (W), d) is uniquely determined by W, up to isomorphisms. An element ω of Dif (W), called a *differential form on W of degree one*, can be written as $\omega=\sum_i g_i df_i$ with g_i and f_i in $\Omega(W)$. Let W be defined over a field k. The form ω is called *rational over k*, if g_i and f_i are chosen so as to be contained in $k(W)$. Let Dif $(W; k)$ denote the elements of Dif (W) rational over k. Then Dif $(W)=$ Dif $(W; k)\otimes_{k(W)}\Omega(W)$. An isomorphism σ of k into Ω induces an isomorphism of Dif $(W; k)$ to Dif $(W^\sigma; k^\sigma)$ by $\omega^\sigma=\sum_i g_i^\sigma\cdot df_i^\sigma$. We say that ω is *finite at* a point a of W if $\omega=\sum_i g_i\cdot df_i$ with functions g_i and f_i which are defined at a. Let T be a rational map of a variety V into W. If there is a point c of V such that T is defined at c and ω is finite at $T(c)$, then we can define an element $\omega\circ T$ of Dif (V) by $\omega\circ T=\sum_i (g_i\circ T)\cdot d(f_i\circ T)$. We denote $\omega\circ T$ also by $\delta T(\omega)$. If V, W, ω, and T are rational over k, and σ is an isomorphism of k into Ω, then $(\omega\circ T)^\sigma=\omega^\sigma\circ T^\sigma$. We call a differential form ω on a projective variety W *holomorphic*, or *of the first kind*, if ω is everywhere finite on W. We denote by $\mathfrak{D}(W)$ the set of all holomorphic elements of Dif (W), and put $\mathfrak{D}(W; k)=\mathfrak{D}(W)\cap$ Dif $(W; k)$ for any field k of rationality for W. Then $\mathfrak{D}(W)=\mathfrak{D}(W; k)\otimes_k\Omega$.

9. A variety V is called an *algebraic curve*, or simply a *curve*, if V is of dimension one. If a field k of definition for V is perfect, we can find a non-singular projective curve which is birationally equivalent to V over k.

Let V be a projective non-singular curve defined over k. Then all the notions and results of § 2.3 can be generalized to the present situation. In fact, we only have to replace W, K, and $\boldsymbol{C}$ by V, $\Omega(V)$, and Ω. The divisors on V and the symbols div (f), $L(A)$, $l(A)$, etc., can be defined in the same manner, without any modification, except for the following point: The relation (2.3.1) should be

(9.1) $df=0$ *if and only if* $\Omega(V)$ *is inseparable over* $\Omega(f)$.

This is of course a special case of (8.2). The *genus* of V is defined, for example, by Prop. 2.13, or (2.3.2). Then Prop. 2.11, Th. 2.12, and Prop. 2.14 are true. A divisor on V is also called a *0-cycle* on V.

10. A projective variety A is called an *abelian variety* if there exist morphisms $f: A\times A\rightarrow A$ and $g: A\rightarrow A$ which define a group structure on A by $f(x, y)=x+y$, $g(x)=-x$. Additive notation is used, since any such group structure on a projective variety can be shown to be commutative. The neutral element is accordingly denoted by 0. If the variety A, and the morphisms f and g are defined over a field k, then we say that the abelian variety *A is defined over k*.

Let A and B be two abelian varieties. By a *homomorphism* of A into B, or an *endomorphism* when $A=B$, we understand a morphism λ of A into B satisfying $\lambda(x+y)=\lambda(x)+\lambda(y)$. If λ is birational, we call it an *isomorphism*, or an *automorphism* when $A=B$. Suppose A and B have the same dimension. Then a homomorphism λ of A into B is surjective if and only if $\mathrm{Ker}(\lambda)$ is finite. Such a λ is called an *isogeny* of A to B. If A, B, and λ are rational over a field k, and x is a generic point of A over k, then we put

$$\deg(\lambda)=[k(x):\ k(\lambda(x))] \qquad (=[k(A):\ k(B)\circ\lambda]).$$

The integer $\deg(\lambda)$ does not depend on the choice of k and x. If $\deg(\lambda)$ is prime to the characteristic of k, then $\mathrm{Ker}(\lambda)$ is of order $\deg(\lambda)$. If there exists an isogeny of A to B, A and B are said to be *isogenous*.

We denote by $\mathrm{End}(A)$ the ring of all endomorphisms of A, and put

$$\mathrm{End}_{\boldsymbol{Q}}(A)=\mathrm{End}(A)\otimes_{\boldsymbol{Z}}\boldsymbol{Q}.$$

11. Let A be an abelian variety of dimension n with $\boldsymbol{C}$ as the universal domain. Then A, as a complex manifold, is isomorphic to a complex torus $\boldsymbol{C}^n/L$, with a lattice L in $\boldsymbol{C}^n$. Here, by a lattice in $\boldsymbol{C}^n$, we understand a discrete subgroup of $\boldsymbol{C}^n$ which is a free $\boldsymbol{Z}$-module of rank $2n$. Let $\boldsymbol{Q}L$ denote the $\boldsymbol{Q}$-linear span of L. Then $\mathrm{End}(A)$ (resp. $\mathrm{End}_{\boldsymbol{Q}}(A)$) can be identified with the ring of all $\boldsymbol{C}$-linear transformations in $\boldsymbol{C}^n$ which send L into L (resp. $\boldsymbol{Q}L$ into $\boldsymbol{Q}L$). Therefore we obtain two faithful representations of $\mathrm{End}_{\boldsymbol{Q}}(A)$:

$$R:\ \mathrm{End}_{\boldsymbol{Q}}(A)\longrightarrow \mathrm{End}(\boldsymbol{C}^n,\boldsymbol{C}) \qquad (\cong M_n(\boldsymbol{C})),$$

$$R^0:\ \mathrm{End}_{\boldsymbol{Q}}(A)\longrightarrow \mathrm{End}(\boldsymbol{Q}L,\boldsymbol{Q}) \qquad (\cong M_{2n}(\boldsymbol{Q})).$$

We call R (resp. R^0) *the complex* (resp. *rational*) *representation* of $\mathrm{End}_{\boldsymbol{Q}}(A)$. It can easily be seen that R (resp. R^0) is equivalent to the representation of $\mathrm{End}_{\boldsymbol{Q}}(A)$ on $\mathfrak{D}(A)$ (resp. on the first cohomology group of A). From Lemma 3.49, it follows that R^0 is equivalent to the direct sum of R and its complex conjugate.

An arbitrary complex torus $\boldsymbol{C}^n/L$ has a structure of an abelian variety if and only if there exists an $\boldsymbol{R}$-valued $\boldsymbol{R}$-bilinear form $E(x,y)$ on $\boldsymbol{C}^n$ satisfying the following three conditions:

(11.1) $E(x,y)=-E(y,x)$.

(11.2) *The value $E(x,y)$ is an integer for every $(x,y)\in L\times L$.*

(11.3) *The $\boldsymbol{R}$-bilinear form $E(x,\sqrt{-1}\,y)$ in (x,y) is symmetric and positive definite.*

We call such a form E a *Riemann form* on $\boldsymbol{C}^n/L$.

12. A *divisor* of an algebraic variety V is an element of the free $\boldsymbol{Z}$-module formally generated by all the subvarieties of V of codimension one. Let A be an abelian variety defined over a subfield of $\boldsymbol{C}$, isomorphic to a complex torus $\boldsymbol{C}^n/L$. Take a basis $\{g_1, \cdots, g_n\}$ of the vector space $\boldsymbol{C}^n$ over $\boldsymbol{R}$, and define real coordinate functions $x_1, \cdots, x_{2n}$ on $\boldsymbol{C}^n$ by $u = \sum_{i=1}^{2n} x_i(u) g_i$ for $u \in \boldsymbol{C}^n$. Then, for a Riemann form E on $\boldsymbol{C}^n/L$, there exists a divisor X of A whose cohomology class is represented by the differential 2-form $\sum_{i<j} E(g_i, g_j) dx_i \wedge dx_j$. (Here we identify A with $\boldsymbol{C}^n/L$ for simplicity.) Since E is unique for X, we say that X *determines* E (with respect to the fixed isomorphism of A onto $\boldsymbol{C}^n/L$), if X and E are in this situation. Let two divisors X and X' on A determine Riemann forms E and E'. Then the following three conditions are equivalent:

(i) *X is algebraically equivalent to X';*

(ii) *X is homologous to X';*

(iii) $E = E'$.

13. Let A be an abelian variety defined over a field of any characteristic. A *polarization* of A is a set $\mathcal{C}$ of divisors of A satisfying the following three conditions:

(13.1) *$\mathcal{C}$ contains an ample divisor* (in the sense of Weil [90, p. 286]).

(13.2) *If X and X' belong to $\mathcal{C}$, there exist positive integers m and m' such that mX is algebraically equivalent to $m'X'$.*

(13.3) *$\mathcal{C}$ is maximal under the conditions* (13.1, 13.2).

A *polarized abelian variety* is a structure $(A, \mathcal{C})$ formed by an abelian variety A and its polarization $\mathcal{C}$. If $\mathcal{C}$ is a polarization of A, there always exists a divisor X_0 in $\mathcal{C}$ such that every X in $\mathcal{C}$ is algebraically equivalent to mX_0 with a positive integer m. Such an X_0 is called a *basic polar divisor of $\mathcal{C}$.*

If the universal domain is $\boldsymbol{C}$, and if A is identified with a complex torus $\boldsymbol{C}^n/L$, the condition (13.1) is equivalent to

(13.1′) *Every X in $\mathcal{C}$ determines a Riemann form.*

Let E be the Riemann form determined by a divisor in $\mathcal{C}$. Then we can define an involution (i.e., an anti-automorphism of order one or two) ρ of $\mathrm{End}_{\boldsymbol{Q}}(A)$ by $E(\lambda x, y) = E(x, \lambda^\rho y)$ for $\lambda \in \mathrm{End}_{\boldsymbol{Q}}(A)$. Here we identify $\mathrm{End}_{\boldsymbol{Q}}(A)$ with a subalgebra of $\mathrm{End}(\boldsymbol{C}^n, \boldsymbol{C})$ as in N° 11. We call *ρ the involution of* $\mathrm{End}_{\boldsymbol{Q}}(A)$ *determined by $\mathcal{C}$,* since it is independent of the choice of X and $\boldsymbol{C}^n/L$. One can actually define such an involution also in the case of positive characteristics. For the detailed discussion of this and other topics concerning abelian varieties, the reader is referred to Weil [92], [95], and Lang [43].

REFERENCES

[1] B. J. Birch and H. P. F. Swinnerton-Dyer, Notes on elliptic curves (I), (II), J. Reine Angew. Math., **212** (1963), 7-25, **218** (1965), 79-108.

[2] A. Borel, Some finiteness properties of adele groups over number fields, Publ. Math. I. H. E. S. no. 16 (1963), 101-126.

[3] A. Borel and Harish-Chandra, Arithmetic subgroups of algebraic groups, Ann. of Math., **75** (1962), 485-535.

[4] W. Casselman, Some new abelian varieties with good reduction, to appear.

[5] J. W. S. Cassels, Diophantine equations with special reference to elliptic curves, J. London Math. Soc., **41** (1966), 193-291.

[6] J. W. S. Cassels and A. Fröhlich (ed.), Algebraic number theory, Proc. of an instructional conference organized by London Math. Soc., 1967.

[7] C. Chevalley, Introduction to the theory of algebraic functions of one variable, Amer. Math. Soc. Surveys, No. 6, 1951.

[8] R. Dedekind, Erläuterungen zu den Fragmenten XXVIII, in B. Riemann, Ges. Math. Werke, 2 Aufl. Leipzig 1892, 466-478 (=R. Dedekind, Ges. Math. Werke, I, Vieweg 1930, 159-172).

[9] P. Deligne, Formes modulaires et représentations l-adiques, Sém. Bourbaki, exp. 355, fév. 1969.

[10] M. Deuring, Die Typen der Multiplikatorenringe elliptischer Funktionenkörper, Abh. Math. Sem. Univ. Hamburg, **14** (1941), 197-272.

[11] M. Deuring, Die Struktur der elliptischen Funktionenkörper und Klassenkörper der imaginären quadratischen Zahlkörper, Math. Ann., **124** (1952), 393-426.

[12] M. Deuring, Die Zetafunktion einer algebraischen Kurve vom Geschlechte Eins, I, II, III, IV, Nachr. Akad. Wiss. Göttingen, (1953) 85-94, (1955) 13-42, (1956) 37-76, (1957) 55-80.

[13] M. Deuring, Die Klassenkörper der komplexen Multiplikation, Enzyclopädie Math. Wiss. Neue Aufl. Band I-2, Heft 10-II, Stuttgart, 1958.

[14] K. Doi, On the jacobian varieties of the fields of elliptic modular functions, Osaka Math. J., **15** (1963), 249-256.

[15] M. Eichler, Allgemeine Kongruenzklasseneinteilungen der Ideale einfacher Algebren über algebraischen Zahlkörpern und ihre L-Reihen, J. Reine Angew. Math., **179** (1938), 227-251.

[16] M. Eichler, Quaternäre quadratische Formen und die Riemannsche Vermutung für die Kongruenzzetafunktion, Arch. Math., **5** (1954), 355-366.

[17] M. Eichler, Über die Darstellbarkeit von Modulformen durch Thetareihen, J. Reine Angew. Math., **195** (1956), 156-171.

[18] M. Eichler, Eine Verallgemeinerung der Abelschen Integrale, Math. Zeitschr., **67** (1957), 267-298.

[19] M. Eichler, Quadratische Formen und Modulfunktionen, Acta Arithm., **4** (1958), 217-239.

[20] M. Eichler, Einführung in die Theorie der algebraischen Zahlen und Funktionen, Basel and Stuttgart, 1963.

[21] R. Fricke, Die elliptischen Funktionen und ihre Anwendungen II, Leipzig and Berlin, 1922.

[22] R. Fricke and F. Klein, Vorlesungen über die Theorie der automorphen Funktionen I, II, Leipzig, 1897-1912.

[23] R. Godement, Les fonctions ζ des algèbres simples, II, Sém. Bourbaki exp. 176, fév. 1959.

[24] R. Godement, Domaines fondamentaux des groupes arithmétiques, Sém. Bourbaki, exp. 257, mai 1963.

[25] H. Hasse, Neue Begründung der komplexen Multiplikation I, II, J. Reine Angew. Math., **157** (1927), 115-139, **165** (1931), 64-88.

[26] E. Hecke, Zur Theorie der elliptischen Modulfunktionen, Math. Ann., **97** (1926), 210-242 (=Math. Werke, 428-460).

[27] E. Hecke, Theorie der Eisensteinschen Reihen höherer Stufe und ihre Anwendung auf Funktionentheorie und Arithmetik, Abh. Math. Sem. Hamburg, **5** (1927), 199-224 (=Math. Werke, 461-486).

[28] E. Hecke, Über die Bestimmung Dirichletscher Reihen durch ihre Funktionalgleichung, Math. Ann., **112** (1936), 664-699 (=Math. Werke, 591-626).

[29] E. Hecke, Über Modulfunktionen und die Dirichletschen Reihen mit Eulerscher Produktentwicklung I, II, Math. Ann., **114** (1937), 1-28, 316-351 (=Math. Werke, 644-707).

[30] E. Hecke, Analytische Arithmetik der positiven quadratischen Formen, Danske Vidensk. Selsk. Mathem.-fys. Meddel. XVII, 12, Copenhagen, 1940 (=Math. Werke, 789-918).

[31] C. Hermite, Sur quelques formules relatives à la transformation des fonctions elliptiques, J. math. pures appl., 2 Ser., **3** (1858), 26-36 (=Oeuvre I, 487-496).

[32] A. Hurwitz, Grundlagen einer independenten Theorie der Elliptischen Modulfunktionen und Theorie der Multiplikator-Gleichungen erster Stufe, Math. Ann., **18** (1881), 528-592 (=Math. Werke I, 1-66).

[33] Y. Ihara, Hecke polynomials as congruence ζ functions in elliptic modular case, Ann. of Math., **85** (1967), 267-295.

[34] Y. Ihara, On congruence monodromy problems I, II, lecture notes, Univ. of Tokyo, 1968-69.

[35] K. Iwasawa, Daisu-Kansu-Ron (The theory of algebraic functions, in Japanese), Tokyo, 1952.

[36] J. Igusa, Kroneckerian model of fields of elliptic modular functions, Amer. J. Math., **81** (1959), 561-577.

[37] H. Jacquet and R.P. Langlands, Automorphic forms on $GL(2)$, lecture notes in mathematics, Springer, Berlin-Heidelberg-New York, 1970.

[38] F. Klein and R. Fricke, Vorlesungen über die Theorie der Modulfunktionen I, II, Leipzig, 1890-92.

[39] M. Kneser, Starke Approximation in algebraischen Gruppen I, J. Reine Angew. Math., **218** (1965), 190-203.

[40] S. Koizumi and G. Shimura, On specializations of abelian varieties, Sci. Papers Coll. of Gen. Ed. Univ. of Tokyo, **9** (1959), 187-211.

[41] M. Kuga, Fibre varieties over a symmetric space whose fibres are abelian varieties, lecture notes, Univ. of Chicago, 1964-65.

[42] M. Kuga and G. Shimura, On the zeta function of a fibre variety whose fibres are abelian varieties, Ann. of Math., **82** (1965), 478-539.

[43] S. Lang, Abelian varieties, New York, 1959.

[44] H. Maass, Über eine neue Art von nichtanalytischen automorphen Funktionen und die Bestimmung Dirichletscher Reihen durch Funktionalgleichungen, Math.

Ann., **121** (1949), 141-183.

[45] H. Maass, Automorphe Funktionen von mehreren Veränderlichen und Dirichletsche Reihen, Abh. Math. Sem. Hamburg, **16** (1949), 72-100.

[46] H. Maass, Die Differentialgleichungen in der Theorie der elliptischen Modulfunktionen, Math. Ann., **125** (1953), 235-263.

[47] Y. Matsushima and S. Murakami, On vector bundle valued harmonic forms and automorphic forms on symmetric riemannian manifolds, Ann. of Math., **78** (1963), 365-416.

[48] Y. Matsushima and G. Shimura, On the cohomology groups attached to certain vector valued differential forms on the product of the upper half planes, Ann. of Math., **78** (1963), 417-449.

[49] K. Miyake, On models of certain automorphic function fields, to appear in Acta Math.

[50] T. Miyake, Decomposition of Jacobian varieties and Dirichlet series of Hecke type, to appear in Amer. J.

[51] L. J. Mordell, On Ramanujan's empirical expansions of modular functions, Proc. Cambridge Phil. Soc., **19** (1920), 117-124.

[52] G. D. Mostow and Tamagawa, On the compactness of arithmetically defined homogeneous spaces, Ann. of Math., **76** (1962), 446-463.

[53] A. Néron, Modeles minimaux des variétés abéliennes sur les corps locaux et globaux, Publ. Math. I. H. E. S. no. 21 (1964), 5-128.

[54] A. P. Ogg, Elliptic curves and wild ramification, Amer. J. Math., **89** (1967), 1-21.

[55] H. Petersson, Zur analytischen Theorie der Grenzkreisgruppen I, II, III, IV, V, Math. Ann., **115** (1938), 23-67, 175-204, 518-572 ,670-709, Math. Zeitschr., **44** (1939), 127-155.

[56] H. Petersson, Konstruktion der sämtlichen Lösungen einer Riemannschen Funktionalgleichung durch Dirichlet-Reihen mit Eulerscher Produktentwicklung I, II, III, Math. Ann., **116** (1939), 401-412, Math. Ann., **117** (1940/41), 39-64, 277-300.

[57] H. Poincaré, Oeuvres II, 1916.

[58] Pjateckii-Shapiro and Shafarevic, Galois theory of transcendental extensions and uniformization, (in Russian), Izv. Akad. Nauk SSSR Ser. Mat., **30** (1966), 671-704. (Amer. Math. Soc. Translations, **69** (1968), 111-145.)

[59] K. Ramachandra, Some applications of Kronecker's limit formula, Ann. of Math., **80** (1964), 104-148.

[60] S. Ramanujan, On certain arithmetical functions, Trans. Cambridge Phil. Soc., 22 (1916), 159-184 (=Collected Papers, 136-162).

[61] R. A. Rankin, Contributions to the theory of Ramanujan's function $\tau(n)$ and similar arithmetical functions I, II, III, Proc. Cambridge Phil. Soc., **35** (1939), 351-356, 357-372, **36** (1940), 150-151.

[62] A. Schoeneberg, Das Verhalten von mehrfachen Thetareihen bei Modulsubstitutionen, Math. Ann., **116** (1939), 511-523.

[63] A. Selberg, Harmonic analysis and discontinuous groups in weakly symmetric riemannian spaces with applications to Dirichlet series, J. Indian Math. Soc., **20** (1956), 47-87.

[64] A. Selberg, On the estimation of Fourier coefficients of modular forms, Proc. Symp. Pure Math. VIII, Theory of numbers, Amer. Math. Soc., 1965, 1-15.

[65] J.-P. Serre, Abelian l-adic representations and elliptic curves, lecture notes, New York, 1968.

[66] J.-P. Serre and J. Tate, Good reduction of abelian varieties, Ann. of Math., **88**

(1968), 492–517.

[67] J. A. Shalika and S. Tanaka, On an explicit construction of a certain class of automorphic forms, Amer. J. Math., **91** (1969), 1049–1076.

[68] H. Shimizu, On zeta functions of quaternion algebras, Ann. of Math., **81** (1965), 166–193.

[69] G. Shimura, Reduction of algebraic varieties with respect to a discrete valuation of the basic field, Amer. J. Math., **77** (1955), 134–176.

[70] G. Shimura, Correspondances modulaires et les fonctions ζ de courbes algébriques, J. Math. Soc. Japan, **10** (1958), 1–28.

[71] G. Shimura, Sur les intégrales attachées aux formes automorphes, J. Math. Soc. Japan, **11** (1959), 291–311.

[72] G. Shimura, On the theory of automorphic functions, Ann. of Math., **70** (1959), 101–144.

[73] G. Shimura, On the zeta-functions of the algebraic curves uniformized by certain automorphic functions, J. Math. Soc. Japan, **13** (1961), 275–331.

[74] G. Shimura, On Dirichlet series and abelian varieties attached to automorphic forms, Ann. of Math., **76** (1962), 237–294.

[75] G. Shimura, On the field of definition for a field of automorphic functions, I, II, III, Ann. of Math., **80** (1964), 160–189, **81** (1965), 124–165, **83** (1966), 377–385.

[76] G. Shimura, A reciprocity law in non-solvable extensions, J. Reine Angew. Math., **221** (1966), 209–220.

[77] G. Shimura, Construction of class fields and zeta functions of algebraic curves, Ann. of Math., **85** (1967), 58–159.

[78] G. Shimura, Algebraic number fields and symplectic discontinuous groups, Ann. of Math., **86** (1967), 503–592.

[79] G. Shimura, Local representations of Galois groups, Ann. of Math., **89** (1969), 99–124.

[80] G. Shimura, On canonical models of arithmetic quotients of bounded symmetric domains, Ann. of Math., **91** (1970), 144–222.

[81] G. Shimura and Y. Taniyama, Complex multiplication of abelian varieties and its applications to number theory, Publ. Math. Soc. Japan, no. 6, 1961.

[82] C. L. Siegel, Discontinuous groups, Ann. of Math., **44** (1943), 674–689 (=Ges. Abh. II, 390–405).

[83] C. L. Siegel, Some remarks on discontinuous groups, Ann. of Math., **46** (1945), 708–718 (=Ges. Abh. III, 67–77).

[84] C. L. Siegel, A simple proof of $\eta(-1/\tau)=\eta(\tau)\sqrt{\tau/i}$, Mathematica, **1** (1954), p. 4 (=Ges. Abh. III, p. 188).

[85] G. Springer, Introduction to Riemann surfaces, 1957.

[86] T. Tamagawa, On the ζ-functions of a division algebra, Ann. of Math., **77** (1963), 387–405.

[87] Y. Taniyama, L-functions of number fields and zeta functions of abelian varieties, J. Math. Soc. Japan, **9** (1957), 330–366.

[88] J.-L. Verdier, Sur les intégrales attachées aux formes automorphes, Sém. Bourbaki, exp. 216, fév. 1961.

[89] H. Weber, Lehrbuch der Algebra III, 2nd ed., 1908.

[90] A. Weil, Foundations of algebraic geometry, Amer. Math. Soc. Coll. Publ. no. 29, 2nd ed., Providence, 1962.

[91] A. Weil, Sur les courbes algébriques et les variétés qui s'en déduisent, Paris, 1948.

[92] A. Weil, Variétés abéliennes et courbes algébriques, Paris, 1948.
[93] A. Weil, Jacobi sums as "Grössencharaktere", Trans. Amer. Math. Soc., **73** (1952), 487-495.
[94] A. Weil, The field of definition of a variety, Amer. J. Math., **78** (1956), 509-524.
[95] A. Weil, Introduction à l'étude des variétés kählériennes, Paris, 1958.
[96] A. Weil, Adeles and algebraic groups, lecture notes, Institute for Advanced Study, Princeton, 1961.
[97] A. Weil, Sur certains groupes d'opérateurs unitaires, Acta Math., **111** (1964), 143-211.
[98] A. Weil, Über die Bestimmung Dirichletscher Reihen durch Funktionalgleichungen, Math. Ann., **168** (1967), 149-156.
[99] A. Weil, Basic number theory, Grundl. Math. Wiss., 144, Berlin-Heidelberg-New York, 1967.
[100] A. Weil, Sur une formule classique, J. Math. Soc. Japan, **20** (1968), 400-402.
[101] A. Weil, Zeta-functions and Mellin transforms, Algebraic geometry (Bombay Coll., 1968), Tata Institute of Fundamental Research, Bombay, 1969, 409-426.
[102] H. Weyl, Die Idee der Riemannschen Fläche, 3rd ed., Berlin, 1955.

INDEX